Thermal and Non-Thermal Treatments to Preserve and Encourage Bioactive Compounds in Fruit and Vegetables Based Products

Thermal and Non-Thermal Treatments to Preserve and Encourage Bioactive Compounds in Fruit and Vegetables Based Products

Editors

Marina Cano-Lamadrid
Francisco Artés-Hernández

MDPI • Basel • Beijing • Wuhan • Barcelona • Belgrade • Manchester • Tokyo • Cluj • Tianjin

Editors
Marina Cano-Lamadrid
Universidad Miguel
Hernández de Elche
Spain

Francisco Artés-Hernández
Universidad Politécnica de
Cartagena
Spain

Editorial Office
MDPI
St. Alban-Anlage 66
4052 Basel, Switzerland

This is a reprint of articles from the Special Issue published online in the open access journal *Foods* (ISSN 2304-8158) (available at: https://www.mdpi.com/journal/foods/special_issues/thermal_nonthermal_treatment_fruit_vegetable).

For citation purposes, cite each article independently as indicated on the article page online and as indicated below:

LastName, A.A.; LastName, B.B.; LastName, C.C. Article Title. *Journal Name* **Year**, *Volume Number*, Page Range.

ISBN 978-3-0365-6830-0 (Hbk)
ISBN 978-3-0365-6831-7 (PDF)

Contents

About the Editors

Marina Cano-Lamadrid

Marina Cano-Lamadrid is a Juan de la Cierva-Formación postdoctoral Ph.D. researcher (FJC2020-043764-I) from the Spanish Ministry of Education. She is currently working at the Postharvest and Refrigeration Group of the Department of Agronomical Engineering and the Institute of Plant Biotechnology of the Universidad Politécnica de Cartagena (Spain). She holds a Ph.D. in Food Science and Technology from Miguel Hernández University (Spain) and a Ph.D. in Sciences from Wroclaw University of Science and Technology (Poland). Her research interests include food analysis, food processing, antioxidants, food technology, food safety, food quality, food loss, and food waste. She has published more than 60 papers in peer-reviewed journals.

Francisco Artés-Hernández

Francisco Artés-Hernández is an Agricultural Engineer and has a PhD in Food Technology. He is the full Professor on Food Technology and the Head of the Postharvest and Refrigeration Group of the Universidad Politécnica de Cartagena (Spain). His research interest is focused on the optimization of sustainable postharvest technologies to preserve and enhance the quality and safety of whole, fresh-cut, and minimally processed fruit and vegetables, including by-product revalorization and investigating abiotic stresses to enhance phytochemicals. He has participated in 28 European or National Research Programs, being principal researcher in 4 of them, and in +70 R&D Projects with companies, being principal researcher in half of them. He has published over 120 manuscripts in high impact journals listed in the JCR-SCI (H=32) and over 350 papers in conferences and other journals. He has directed 15 Doctoral Theses with high scientific production and has four invention patents.

Preface to "Thermal and Non-Thermal Treatments to Preserve and Encourage Bioactive Compounds in Fruit and Vegetables Based Products"

Fruit- and vegetable-based products (F&Vs) have been conventionally processed using thermal techniques such as pasteurization, scalding, and/or drying, ensuring microbial safety and/or enzyme deactivation. Although thermal treatments are the most cost-effective tools, they could also reduce bioactive compounds, nutrients, and even sensory attributes. Nowadays, non-thermal food-processing technologies such as UV light and high-pressure processing have been proposed to develop food products with extended shelf-life and preserved/encouraged bioactive compounds, while preserving the sensory properties. To contribute to the knowledge of novel thermal and non-thermal techniques in the production of F&Vs and their derivatives, we present this Special Issue, "Thermal and Non-Thermal Treatments to Preserve and Encourage Bioactive Compounds in Fruit- and Vegetable-Based Products".

The 13 papers (divided into three blocks: thermal and non-thermal treatments on whole fruits; thermal and non-thermal treatments on fruit and vegetable byproducts; and thermal and non-thermal treatments on fruit-based beverages and purées) published in this Special Issue are a strong representation of the research activities addressing the topic of thermal and non-thermal treatments for preserving and encouraging bioactive compounds in F&V commodities (whole, byproducts, and F&V-based products). Finally, most of the authors who have contributed to this issue concluded that further research in their topic is required if we are to optimize the conditions of "green" and sustainable technologies for maintaining the functionality and safety of fruit- and vegetable-based products. All the sources of funding of this Special Issue are described in each paper and the editors are grateful to all the contributors and their particular financial support.

Marina Cano-Lamadrid and Francisco Artés-Hernández
Editors

foods

MDPI

Editorial

Thermal and Non-Thermal Treatments to Preserve and Encourage Bioactive Compounds in Fruit- and Vegetable-Based Products

Marina Cano-Lamadrid and Francisco Artés-Hernández *

Postharvest and Refrigeration Group, Department of Agronomical Engineering, Institute of Plant Biotechnology, Universidad Politécnica de Cartagena, 30203 Cartagena, Murcia, Spain
* Correspondence: fr.artes-hdez@upct.es; Tel.: +34-968-325-509

Fruit- and vegetable-based products (F&Vs) have been conventionally processed using thermal techniques such as pasteurization, scalding, or/and drying, ensuring microbial safety and/or enzyme deactivation. Although thermal treatments are the most cost-effective tools, they could also reduce bioactive compounds, nutrients, and even sensory attributes. Nowadays, non-thermal food-processing technologies such as UV light and high-pressure processing have been proposed to develop food products with extended shelf life and preserved/encouraged bioactive compounds, while preserving the sensory properties. To contribute to the knowledge of novel thermal and non-thermal techniques in the production of F&Vs and their derivatives, we present this Special Issue, "Thermal and Non-Thermal Treatments to Preserve and Encourage Bioactive Compounds in Fruit- and Vegetable-Based Products". Among the 13 published papers, three main research topics have been covered: (i) thermal and non-thermal treatments on whole fruits; (ii) thermal and non-thermal treatments on fruit and vegetable byproducts; (iii) thermal and non-thermal treatments on fruit-based products (beverages and purées).

There are four papers within the first topic, dealing with the effect of thermal and non-thermal treatments on whole fruits [1–4]. The first study, by Bedrníček et al. [1], investigated the influence of several thermally treated fresh garlic varieties (treated during ageing process—15 days with a temperature gradient ranging from 30 to 82 °C) on the physicochemical parameters of black garlic. During this thermal process, the health-promoting properties of fresh garlic are usually enhanced, but very little information has previously been reported about how the physicochemical properties of fresh garlic varieties affect those same properties of black garlic. The authors concluded that the antioxidant activity, the total polyphenol content, and the total soluble solids increased during ageing, while the pH level, moisture content, and lightness decreased in all the garlic varieties. Texture acceptance was significantly affected among the fresh and black garlic varieties, while color, odor, taste, and the acceptance of the intensity of the garlic aroma were not affected. It was highlighted that the fresh garlic parameters did not correlate well with the properties of black garlic; therefore, they cannot be used for its quality prediction. The second study, a review by Darré et al. [2], showed that ultraviolet radiation has been found to be a germicide and a potentially damaging agent; however, beneficial properties have also been reported in harvested commodities. UV treatments on whole fruit were shown to induce phytochemicals accumulation, including ascorbic acid, carotenoids, glucosinolates, and, more frequently, phenolic compounds. Scientific evidence related to the effect of ultraviolet radiation on whole fruit and vegetables through several factors (product type, maturity, cultivar, UV spectral region, dose, intensity, and radiation exposure) was collected. Additionally, the authors posited a need to fix the optimal operational conditions to maximize UV radiation efficacy—reducing treatment times and ensuring even radiation exposure—especially under realistic processing conditions. The third study in this topic, performed by Modesti et al. [3], aimed to summarize knowledge

Citation: Cano-Lamadrid, M.; Artés-Hernández, F. Thermal and Non-Thermal Treatments to Preserve and Encourage Bioactive Compounds in Fruit- and Vegetable-Based Products. *Foods* **2022**, *11*, 3400. https://doi.org/10.3390/foods11213400

Received: 6 September 2022
Accepted: 25 October 2022
Published: 27 October 2022

about the possibility of improving bioactive compounds by using ozone; then, they applied these findings to the health-related properties of grapes and wine. The greatest focus was given to the pre- and post-harvest treatment of table and wine grapes, in addition to the explanation of the mechanisms involved in the ozone-related response and the main secondary metabolites biosynthetic pathways. The authors highlight that the effect of ozone treatment on health-related properties and secondary metabolites accumulation depends on many factors (cultivar, ozone type, doses, and application method). Scientific evidence reports an increase in bioactive compounds with antioxidant activity, improving berry and wine functional and sensory quality. The fourth study, a review by Jacobo-Velázquez et al. [4], identified technologies that emulate a wound response (increase the content of bioactive compounds and increase detrimental reactions in fruits and vegetables) in whole F&Vs while retaining their quality attributes. The application of non-thermal technologies, such as high hydrostatic pressure, ultrasound, and pulsed electric fields, are presented in this review as tools for increasing the content of health-promoting compounds in whole F&Vs by inducing a wound-like response. Additionally, the industrial implementation and economic feasibility of these approaches were discussed.

There are three manuscripts in this Special Issue addressing the second topic—different thermal and non-thermal treatments on F&V byproducts [5,6]. The first study of this topic, by Michalska-Ciechanowska et al. [5], aimed to produce chokeberry pomace extract powders and conduct experimental and chemometric assessment of the effect of different carriers (10% of maltodextrin, inulin, and trehalose, and their combinations) and drying techniques (freeze-drying and vacuum-drying at 60 °C or 90 °C) on the polyphenols, anthocyanins, flavonols, phenolic acids, and hydroxymethyl-L-furfural of chokeberry pomace powders. Drying techniques had a stronger effect on the bioactive compound profiles than the type of carrier. Hydroxymethyl-L-furfural formation was enhanced by inulin addition during high-temperature treatment. They concluded that the addition of maltodextrin and trehalose mixture for freeze-drying and vacuum-drying at 90 °C caused the highest retention of polyphenolics and the lowest formation of hydroxymethyl-L-furfural. The second study, a review by Cano-Lamadrid et al. [6], provided comprehensive information about non-thermal technologies applied in F&V byproducts to enhance their phytochemicals and to obtain pectin. The potential use of bioactive compounds for food supplementation has also been collected as sustainable strategy to increase functional properties. Applications of ultrasound, light stress, enzyme-assisted treatment, fermentation, electro-technologies, and high pressure, among others, were included in this review. Future investigations should focus on the optimization of "green" non-thermal and sustainable technologies on the F&V byproducts' key compounds for the full utilization of raw materials in the food industry. The third study, by Cano-Lamadrid et al., evaluated the state of knowledge of and scientific evidence for the use of green extraction technologies (ultrasound-, microwave-, and enzymatic-assisted technologies) of bioactive compounds from pomegranate peel byproducts, and their potential application via the supplementation/fortification of vegetal matrixes to improve their quality, functional properties, and safety. Most studies are mainly focused on ultrasound extraction, which has been widely developed compared with microwave or enzymatic extractions; these should be studied in depth, accounting for their combinations. After extraction, pomegranate peel byproducts (in the form of powders, liquid extracts, and/or encapsulated, among others) have been incorporated into several food matrixes as a good tool to preserve "clean label" foods, improving their functional properties without altering their composition [7].

Five original papers have been published in this Special Issue which address the third topic—dealing with the effects of thermal and non-thermal treatments on fruit-based products (beverages and purées) [8–13]. In the first study, by Zhao et al. [8], the aim was to determine the effect of thermo-sonication combined with nisin and thermal pasteurization treatments on the quality attributes—including microbial and enzyme inactivation and the physicochemical, nutritional, functional, and sensory qualities—of orange juice. Apart from desirable bactericidal and enzyme inactivation effects and an improvement in the physicochemical properties and sensory quality, thermo-sonication combined with nisin increased the total polyphenols content (10.03%), total carotenoids (20.10%), and antioxidant capacity (51.10% and 10.58%, by ORAC and DPPH, respectively). In addition, total flavonoids and ascorbic acid were largely retained. In the second study, by López-Gámez et al. [9], the authors proposed the application of pulsed electric fields to carrots to obtain juices, purees, and oil-added purees with increased phenolic and carotenoid bio-accessibility. These carrot-based products were obtained by treating carrots with pulsed electric fields (five pulses of 3.5 $kV \cdot cm^{-1}$; 0.6 $kJ \cdot kg^{-1}$). Among the carrot-based products obtained, carrot juices exhibited the highest carotenoid and phenolic contents. However, caffeic and coumaric acid derivatives were highly sensitive to pulsed electric fields. The phenolic bio-accessibility reached 100% in purees obtained from the pulsed electric fields-treated carrots, whereas the further thermally treated oil-added purees exhibited the greatest carotenoid bio-accessibility (7.8%). The third study, by Artés-Hernández et al. [10], was focused on analyzing the effects of different green technologies (UV, high-pressure processing, pulsed electric fields, ultrasounds, and cold plasma) in beverage processing; they found a fortification effect (improvement of their extraction/bioaccessibility and/or different biosynthetic reactions that occurred during processing) on their health-promoting compounds. Fortification can be performed by several strategies, including physical elicitors (e.g., processing technologies), plant/algae extract supplementation, and fermentation with probiotics, among others. Thermal processing technologies are conventionally used to ensure the preservation of food safety with a long shelf life, but this frequently reduces nutritional and sensory quality. Nevertheless, green non-thermal technologies are being widely investigated to reduce costs and make possible more sustainable production processes without affecting the nutritional and sensory quality of beverages. The fourth study, by Salazar-Orbea et al. [11], aimed to discuss the positive and negative effects of thermal treatment and high-pressure processing on the stability of different polyphenol families in agro-food products obtained from strawberry and apple, two of the most-used fruits in food processing. The main findings showed that the phenolic content was affected by the processing, fruit type, polyphenol family, and storage conditions (time and temperature) of the final product. To increase shelf life, manufacturers are focusing on the optimal conditions for polyphenol stability (residual enzyme activity from HPP can affect polyphenols negatively) and product shelf life. The fifth study, by Tsikrika et al. [12], investigated the polyphenol oxidase (PPO) inactivation in five whole and peeled Irish potato cultivars using high-pressure processing (400 MPa and 600 MPa for 3 min). The highest PPO inactivation was observed after high-pressure processing at 600 MPa. No changes were observed on the total phenolic content, antioxidant activity, and glycemic indices of all the high-pressure-treated potatoes. Regarding individual phenolic acids, chlorogenic acid and ferulic acid were decreased in all studied varieties with a resulting increase in caffeic and quinic acid, and a variation in rutin and 4-coumaric acid levels depending on the cultivar and the sample type, respectively. Higher content of anthocyanins (pelargonidin-O-ferulorylrutinoside-O-hexoside and pelargonidin-O-rutinoside-O-hexoside) was observed in colored whole-potato varieties treated by high-pressure processing than those untreated. The sixth study, by Salar et al. [13], studied the effect of high hydrostatic pressure (450 and 600 MPa for 180 s) compared with thermal pasteurization (85 °C for 15 s) on the microbiological and phytochemical profile of citrus–maqui beverages. Additionally, the shelf life under refrigeration (4 °C) and room temperature (20 °C) was monitored for 90 days. Microbiological stability and physicochemical parameters were not different after processing or throughout the storage period. An increase in the reddish coloration was noticed during storage for those beverages treated by high hydrostatic pressure. Phenolic compounds showed little affect from the processing technique, being more stable when high hydrostatic pressure was used during processing. The authors concluded that high hydrostatic pressure technology could be an effective alternative to thermal treatments, achieving effective microbial inactivation and extending the shelf life of the juices by contributing to a better preservation of color and bioactive compounds.

In summary, the 13 papers published in this Special Issue are a strong representation of the research activities addressing the topic thermal and non-thermal treatments for preserving and encouraging bioactive compounds in F&V commodities (whole, byproducts, and F&V-based products). Finally, most authors who have contributed to this issue concluded that further research in their topic is required if we are to optimize the conditions of "green" and sustainable technologies for maintaining the functionality and safety of fruit- and vegetable-based products.

Author Contributions: F.A.-H. and M.C.-L. conceived and wrote this editorial. All authors have read and agreed to the published version of the manuscript.

Funding: This research was funded by the Spanish Ministry of Science and Innovation, Knowledge Generation Projects 2021, Type B Oriented Research Modality, grant number PID2021-123857OB-I00, REVALFOOD PROJECT. M.C.-L. contract has been co-financed by Juan de la Cierva-Formación (FJC2020-043764-I) from the Spanish Ministry of Education.

Acknowledgments: All sources of funding of this Special Issue are described in each paper. Editors are grateful to all the contributors and their particular financial support.

Conflicts of Interest: The authors declare no conflict of interest.

References

1. Bedrníček, J.; Laknerová, I.; Lorenc, F.; de Moraes, P.P.; Jarošová, M.; Samková, E.; Tříska, J.; Vrchotová, N.; Kadlec, J.; Smetana, P. The Use of a Thermal Process to Produce Black Garlic: Differences in the Physicochemical and Sensory Characteristics Using Seven Varieties of Fresh Garlic. *Foods* **2021**, *10*, 2703. [CrossRef] [PubMed]
2. Darré, M.; Vicente, A.R.; Cisneros-Zevallos, L.; Artés-Hernández, F. Postharvest Ultraviolet Radiation in Fruit and Vegetables: Applications and Factors Modulating Its Efficacy on Bioactive Compounds and Microbial Growth. *Foods* **2022**, *11*, 653. [CrossRef] [PubMed]
3. Modesti, M.; Macaluso, M.; Taglieri, I.; Bellincontro, A.; Sanmartin, C. Ozone and Bioactive Compounds in Grapes and Wine. *Foods* **2021**, *10*, 2934. [CrossRef] [PubMed]
4. Jacobo-Velázquez, D.A.; Benavides, J. Non-Thermal Technologies as Tools to Increase the Content of Health-Promoting Compounds in Whole Fruits and Vegetables While Retaining Quality Attributes. *Foods* **2021**, *10*, 2904. [CrossRef] [PubMed]
5. Michalska-Ciechanowska, A.; Hendrysiak, A.; Brzezowska, J.; Wojdyło, A.; Gajewicz-Skretna, A. How Do the Different Types of Carrier and Drying Techniques Affect the Changes in Physico-Chemical Properties of Powders from Chokeberry Pomace Extracts? *Foods* **2021**, *10*, 1864. [CrossRef] [PubMed]
6. Cano-Lamadrid, M.; Artés-Hernández, F. By-Products Revalorization with Non-Thermal Treatments to Enhance Phytochemical Compounds of Fruit and Vegetables Derived Products: A Review. *Foods* **2022**, *11*, 59. [CrossRef] [PubMed]
7. Cano-Lamadrid, M.; Martínez-Zamora, L.; Castillejo, N.; Artés-Hernández, F. From Pomegranate Byproducts Waste to Worth: A Review of Extraction Techniques and Potential Applications for Their Revalorization. *Foods* **2022**, *11*, 2596. [CrossRef] [PubMed]
8. Zhao, Q.; Yuan, Q.; Gao, C.; Wang, X.; Zhu, B.; Wang, J.; Sun, X.; Ma, T. Thermosonication Combined with Natural Antimicrobial Nisin: A Potential Technique Ensuring Microbiological Safety and Improving the Quality Parameters of Orange Juice. *Foods* **2021**, *10*, 1851. [CrossRef] [PubMed]
9. López-Gámez, G.; Elez-Martínez, P.; Martín-Belloso, O.; Soliva-Fortuny, R. Applying Pulsed Electric Fields to Whole Carrots Enhances the Bioaccessibility of Carotenoid and Phenolic Compounds in Derived Products. *Foods* **2021**, *10*, 1321. [CrossRef] [PubMed]
10. Artés-Hernández, F.; Castillejo, N.; Martínez-Zamora, L.; Martínez-Hernández, G.B. Phytochemical Fortification in Fruit and Vegetable Beverages with Green Technologies. *Foods* **2021**, *10*, 2534. [CrossRef] [PubMed]
11. Salazar-Orbea, G.L.; García-Villalba, R.; Tomás-Barberán, F.A.; Sánchez-Siles, L.M. High–Pressure Processing vs. Thermal Treatment: Effect on the Stability of Polyphenols in Strawberry and Apple Products. *Foods* **2021**, *10*, 2919. [CrossRef] [PubMed]
12. Tsikrika, K.; Muldoon, A.; O'Brien, N.M.; Rai, D.K. High-Pressure Processing on Whole and Peeled Potatoes: Influence on Polyphenol Oxidase, Antioxidants, and Glycaemic Indices. *Foods* **2021**, *10*, 2425. [CrossRef] [PubMed]
13. Salar, F.J.; Periago, P.M.; Agulló, V.; García-Viguera, C.; Fernández, P.S. High Hydrostatic Pressure vs. Thermal Pasteurization: The Effect on the Bioactive Compound Profile of a Citrus Maqui Beverage. *Foods* **2021**, *10*, 2416. [CrossRef] [PubMed]

foods

Review

Phytochemical Fortification in Fruit and Vegetable Beverages with Green Technologies

Francisco Artés-Hernández *, Noelia Castillejo, Lorena Martínez-Zamora and Ginés Benito Martínez-Hernández

Department of Agronomical Engineering & Institute of Plant Biotechnology, Universidad Politécnica de Cartagena, 30203 Cartagena, Spain; noelia.castillejo@upct.es (N.C.); lorena.martinez@upct.es (L.M.-Z.); ginesbenito.martinez@upct.es (G.B.M.-H.)
* Correspondence: fr.artes-hdez@upct.es; Tel.: +34-968325509

Abstract: Background: Phytochemical, bioactive and nutraceutical compounds are terms usually found in the scientific literature related to natural compounds found in plants linked to health-promoting properties. Fruit and vegetable beverages (mainly juice and smoothies) are a convenient strategy to enhance the consumption of horticultural commodities, with the possibility of being fortified with plant byproducts to enhance the content of bioactive compounds. Objective: This review aims to analyse the different green technologies applied in beverage processing with a fortification effect on their health promoting compounds. Results: Fortification can be performed by several strategies, including physical elicitors (e.g., processing technologies), plant/algae extract supplementation, and fermentation with probiotics, among others. Thermal processing technologies are conventionally used to ensure the preservation of food safety with a long shelf life, but this frequently reduces nutritional and sensory quality. However, green non-thermal technologies (e.g., UV, high-pressure processing, pulsed electric fields, ultrasounds, cold plasma, etc.) are being widely investigated in order to reduce costs and make possible more sustainable production processes without affecting the nutritional and sensory quality of beverages. Conclusions: Such green processing technologies may enhance the content of phytochemical compounds through improvement of their extraction/bioaccessibility and/or different biosynthetic reactions that occurred during processing.

Keywords: smoothies; juices; elicitors; abiotic stresses; nutraceuticals; health-promoting compounds

Citation: Artés-Hernández, F.; Castillejo, N.; Martínez-Zamora, L.; Martínez-Hernández, G.B. Phytochemical Fortification in Fruit and Vegetable Beverages with Green Technologies. *Foods* **2021**, *10*, 2534. https://doi.org/10.3390/foods10112534

Academic Editor: Sabine Martina Harrison

Received: 24 September 2021
Accepted: 19 October 2021
Published: 21 October 2021

Publisher's Note: MDPI stays neutral with regard to jurisdictional claims in published maps and institutional affiliations.

1. Introduction

More than two millennia ago, the father of modern medicine related the use of appropriate foods for therapeutic benefits (Hippocrates, 460–377 BC). In this sense, the origin of the term 'nutraceutical', first coined by Stephen DeFelice in 1989, comes from the combination of nutrients and pharmaceuticals, because these compounds have been shown to provide medical or health benefits, including the prevention and the treatment of disease [1]. Nevertheless, a regulatory definition has not been reached yet. As synonyms of this terminology, EFSA has defined a 'bioactive compound' as a type of chemical found in small amounts in plants and certain foods (such as fruits, vegetables, nuts, oils, and whole grains) and which have actions in the body that may promote good health [2]. In this sense, the term 'phytochemical' also refers to nutritive or non-nutritive, biologically active compounds present in edible natural foods, including fruit, vegetables, grains, nuts, seeds, and tea, which also prevent or delay chronic diseases in humans and animals [3]. Nevertheless, an EFSA or USDA positive opinion regarding a bioactive compound implies the need to corroborate a clear relationship based on clinical studies in healthy consumers. Therefore, the use and development of these substances and their incorporation into foods and beverages depends on their safety, adverse effects, and toxicity studies [4].

Healthy habits such as diet and active lifestyle are associated with human wellbeing. Nowadays, people have become concerned about their health and dietary habits, especially since the COVID-19 pandemic. Nutraceutical compounds contained in foods have gained increasing attention because they can provide health benefits with negligible side effects compared to traditional pharmacological therapies. Therefore, although there is no evidence on their effects as a potentially useful against SARS-CoV-2 infection, consumers are focusing on proper nutrition rich in antioxidant nutrients [5]. In this context, fruit and vegetable beverages are excellent and convenient drinks to promote the consumption of bioactive compounds. Common types of fruit and vegetable beverages mostly include juices and smoothies. While a juice usually contains the liquid extracted when fruit and vegetables are pressed and the pulp is discarded, smoothies are made of whole products in a blender, usually containing some non-juiceable ingredients. Rodríguez-Verástegui et al. [6] have defined smoothies as non-alcoholic beverages prepared from fresh or frozen fruits and or vegetables, which are blended and usually mixed with crushed ice to be immediately consumed. The main difference is fibre, which makes you feel full, which smoothies contain much of while juices typically have very small amounts. While by-products are discarded in juices, smoothies usually contain the whole edible product, which includes important bioactive compounds. Hence, high health benefits can be obtained from mixing fruit and vegetables due to the synergistic relationships between their different bioactive compounds [7]. As the conversion of fruit and vegetables into processed products also enhances the economic value of fruit, the fruit and vegetable beverages industry is rapidly growing worldwide [8].

The main issue in smoothie/juice processing is their limited shelf-life; they are susceptible to spoilage [9] and quality degradation. Moreover, quality degradation due to endogenous enzyme activity is an important factor, as while in preparation enzymes can come in contact with substrates. For this reason, thermal treatments are usually performed during processing in order to increase shelf life while keeping quality and avoiding microbial spoilage [10]. However, the application of heat treatments for long times negatively affects the nutritional and sensory quality of smoothies and juices [11]. Given the high economic and environmental impact of such technologies, alternative technologies have been recently developed to reduce costs while maintaining a high content of health-promoting compounds. In this field, new emerging eco-friendly technologies are expected to be developed during the coming years.

The horticultural industry generates a large number of by-products that constitute an excellent source of valuable bioactive compounds which deserve to be revalorized. Therefore, green extraction techniques have been developed in recent years and innovative additional systems for phytochemical extraction in fruit and vegetable beverages have been performed [12].

Therefore, this work reviews the nutraceutical fortification of fruit and vegetable beverages by using green emerging technologies to improve the bioactive content and preserve sensory quality and microbial safety while reducing costs and energy resources.

2. Nutraceutical Food Supplements

Nutraceuticals can be classified based on their natural source (from plants, animals, or microbes), their mechanism of action (antioxidant, anti-inflammatory, and anticancer capacities, mainly), or their chemical constitution [4]. With respect to their main components in terms of chemical structure, the most relevant nutraceutical compounds are carotenoids, dietary fibre, omega-3 fatty acids, flavonoids, glucosinolates, indoles, isothiocyanates, phenolic acids, plant sterols, prebiotics, probiotics, saponins, phytoestrogens, tannins, thiols, and sulphides [4]. In this context, many of these bioactive compounds can protect our bodies against disease as well as prevent food spoilage by blocking oxidation processes. These reactions are performed in different ways: preventing chain inhibition by scavenging initiating radicals, acting as electron donors, breaking chain reactions, transferring hydrogen atoms to generate stable radicals, decomposing peroxides, decreasing localized oxygen concentrations, and binding chain initiating catalyst such as metal ions [13].

The adherence to the Mediterranean dietary pattern, characterized by the presence of foods rich in such bioactive compounds, has been demonstrated to have potentially beneficial health effects and can have cardioprotective, neuroprotective, antioxidant, anti-inflammatory, and anticarcinogenic properties [14].

Food industry by-products produced in the Mediterranean basin, as an important production area of fruit and vegetables consumed worldwide, have become an important source of nutraceutical supplements used as enrichers and fortifiers in food and beverages (Figure 1). As examples, olive oil production is the main source of natural hydroxytyrosol (Figure 1), one of the most important antioxidants in the "Mediterranean Diet", even more than tea or Q10 coenzyme [15]. Similarly, the production of wine is an inexhaustible source of flavonoids, catechins and epicatechins obtained from grape seed and grape byproducts (Figure 1). Additionally, the increased production of fresh-cut and ready-to-eat convenience products obtained from fruit and vegetables, where waste accounts for more than 50%, takes a further step towards the revalorization of such amount through the extraction of their antioxidant compounds and their subsequent incorporation into food matrices. Figure 1 presents some examples of these compounds: punicalagin from pomegranate, hesperidin or naringenin from citrus peels, sulforaphane from broccoli leaves and florets, allicin from garlic and onion peels, lycopene from fried tomato or ketchup processed, β-carotene from fresh-cut carrots, and cynarine from artichokes leaves.

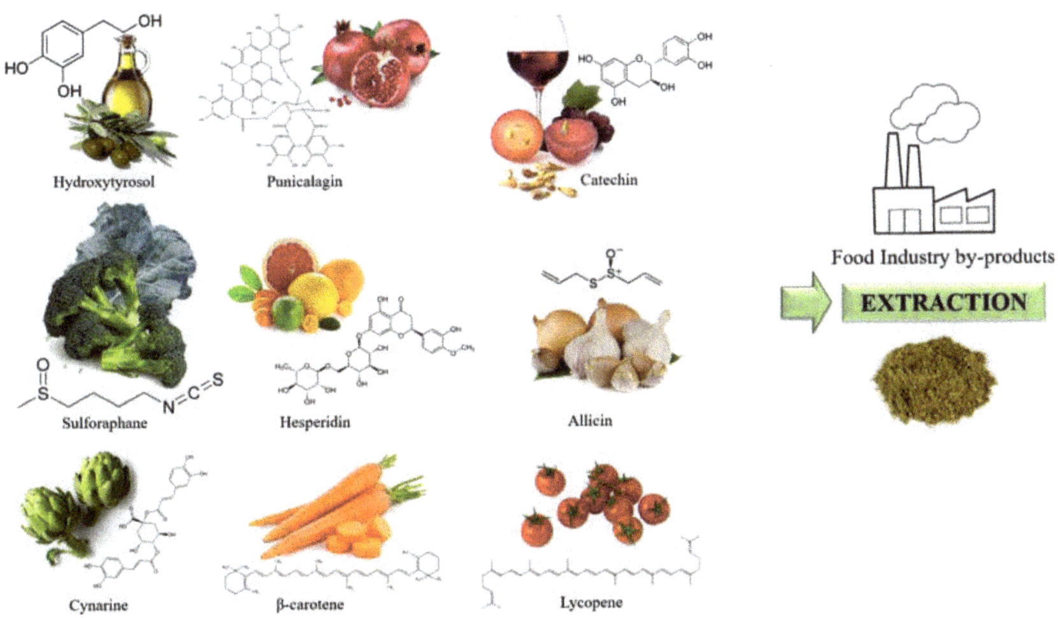

Figure 1. Some of the most common nutraceutical compounds obtained as food industry by-products used to enrich fruit and vegetable beverages.

Furthermore, it is important to emphasise that the bioavailability of these bioactive compounds is highly elevated, which also justifies their incorporation in nutraceuticals as a way to enrich processed food and beverages. In this sense, isothiocyanates have been shown to have 37–50% bioavailability [16,17], with a demonstrated anticancer potential. In addition, carotenoids have been reported to be highly available for the human body (80–90%) [18,19], while phenolics and flavonoids have shown between 50% and 80% availability of the total consumed content [16]. Indeed, hydroxytyrosol obtained from olive leaves has shown 98.5% bioavailability in vitro [20,21].

In addition, some extraction procedures improve the uptake rate of some nutraceutical molecules, as occurs with lycopene [22,23] and some vitamins [24]. Thus, as the literature is extensive, it is necessary to review how green technologies can enhance nutraceutical content and/or availability, even after their incorporation into foods and beverages.

3. Preservation of Fruit and Vegetable Beverages Using Green Technologies

Despite the high nutritional value of fruit and vegetable beverages, their shelf life is limited due to processing steps in which the blending process breaks most of the cells and leaves them unprotected against microbial and enzymatic spoilage, which contribute to impairment of sensory, nutritional, and safety parameters [25]. The main strategy to preserve quality and safety is refrigeration; however, even under low storage temperatures the deterioration rate is very high.

In this sense, heat treatments such as pasteurization and sterilization have been widely used to ensure safety and prevent spoilage [26]. The main aim of these procedures is to destroy pathogenic microorganisms (*Escherichia coli*, *Staphylococcus*, *Listeria*, *Salmonella*, *Bacillus*, and *Clostridium*, among other species) and decrease spoilage organisms that grow during their shelf life. Moreover, these thermal treatments are used to inactivate enzymes such as pectin methylesterase, polyphenoloxidase, and peroxidase, which are responsible for detrimental effects on product quality.

However, the application of high temperatures over long periods usually affects the nutritional and sensory quality of smoothies and juices [11]. In addition, the continuous use of thermal treatment in the Food Industry has a high economic cost and a high impact on carbon footprint. In the last few decades, alternative treatments, referred to as 'Green Technologies', have been developed to reduce costs and make possible more sustainable production processes. A summary of these emerging technologies, which support energy efficiency, recycling, health and safety concerns, and renewable resources, is shown in Figure 2.

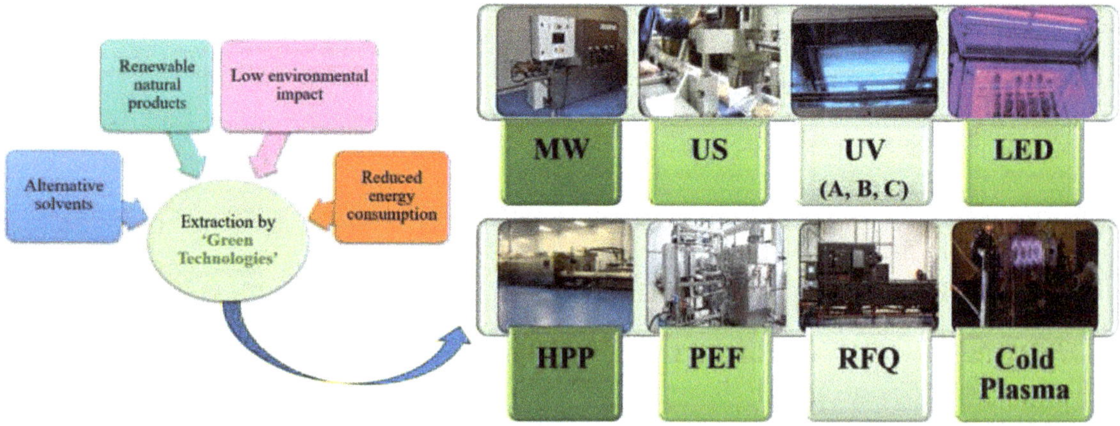

Figure 2. Green technologies to improve the healthiness of fruit and vegetable beverages. MW: microwave; US: ultrasounds; UV: ultraviolet; LED: Light Emitting Diode; HPP: High-Pressure Processing; PEF: Pulsed Electric Field; RFQ: Radiofrequency.

3.1. Microwaves

Microwave (MW) processing is a thermal treatment used for the pasteurization of liquid products that can reach high temperatures in seconds, which considerably reduces application time without altering sensory and nutritional quality [11]. For instance, continuous MW heating, combining high power and short time (3.6 kW for 93 s) reaching 90 °C, inactivated up to 96% of peroxidase activity compared to conventional heating of untreated smoothies. Furthermore, this treatment increased the viscosity of a tomato smoothie [27]. In this sense, Markowski et al. [28] related this increase in viscosity as an indicator of severe heat treatment, and the highest degradation of ascorbic acid.

3.2. Ultrasounds

Ultrasound (US) is an emerging technology characterized by its high efficiency, low and economical instrumental requirements; it has been used in juices to increase their shelf life while maintaining their nutritional and physicochemical attributes [29]. However, it decreases the viscosity of juice due to cavitation phenomena. The Ultrasound acts on cellular structures, breaking them due to extreme conditions, mainly pressure. From an industrial point of view, a less viscous liquid should behave better under this process. For instance, 147 W US treatment for 2 min affected the viscosity of a fruit smoothie and even achieved the maximum retention of anthocyanin (99%) [30]. In addition, other authors [31] have reported that ultrasound treatment was able to retain ascorbic acid (84–91%) content even more than using pulsed electric fields (PEF) (80–83%).

3.3. Ultraviolet

Ultraviolet (UV) light, especially UV-C (280–100 nm), has been widely reported on for its germicidal effects against a wide range of pathogenic and food spoilage microorganisms, although recent findings have shown that low doses of UV-C are also able to enhance the nutraceutical content in acerola fruits [32]. Furthermore, UV-B (320–280 nm) and UV-A (400–320 nm) have demonstrated properties that improve the content of the bioactive compounds in broccoli sprouts [33,34], kale sprouts [35], bell peppers [36], carrots [37], and broccoli [12], among others. UV light is an environmentally friendly technology with low costs in terms of equipment, energy and maintenance. However, due to the low penetration power and transmission of UV light, the application of this technology in beverages is limited [8].

3.4. Light Emitting Diodes (LEDs)

A Light Emitting Diode (LED) emits different light colours depending on the energy of the photons. The technological advancements in LED technology have resulted in widespread application in horticultural production systems due to its limited thermal dissipation, low energy requirements and the possibility of finely customizing the light intensity and spectral properties [38]. For instance, past research has shown that different regions of the visible spectrum are able to affect the biosynthesis of nutraceuticals, since fruit, plants, and vegetables have photoreceptors in these wavelengths that trigger the accumulation of glucosinolates, carotenoids, phenolic acids, and flavonoids, among others.

3.5. High Hydrostatic Pressure Treatment

High pressure processing (HPP) is based on Le Chatelier's principle. Low molecular weight compounds such as vitamins, minerals, and volatile compounds are rarely affected by HPP due to the low compressibility of covalent bonds. In contrast, macromolecules like proteins, lipids, or starches can similarly change their native structure under heat treatments [39]. The application of HPP in beverages ranges from 50 to 1000 MPa for several minutes and cycles, in combination with controlled temperature (<0 °C up to 100 °C). In such conditions, HPP can produce important antimicrobial effects with no consequences in the physicochemical and nutritional profile [39,40]. Furthermore, in combination with high temperatures (60–90 °C), HPP technology (300–700 MPa) can be used for food sterilisation, although this can lead to colour losses [40].

3.6. Pulsed Electric Fields

Pulsed electric field (PEF) treatment involves the application of direct current voltage pulses for very short periods (µs-ms), which results in an electric field the intensity of which depends on the gap between the electrodes. The microbial inactivation fundament is based on the electroporation of bacterial membranes producing pores in cellular tissues, leading to leakage of intracellular contents. As bacterial spores are resistant to PEF treatments, its application should be focused on pasteurization [41].

3.7. Radiofrequency

Radiofrequency (RFQ) refers to electromagnetic waves in the range of 10–300 MHz, however, the range of permitted frequencies in industrial applications is 10–50 MHz [42]. When foods with several factors affecting the electric field (food composition, salt, fat content, moisture, temperature, etc.) are exposed to an alternating electric field, dielectric heating occurs. Such heating is directly related to microbial reduction at a relatively low temperatures (60–65 °C), even able to inactivate pathogenic microorganisms such as *Listeria monocytogenes* and *E. coli* [43]. In this sense, the controlled application of this technology provides fast heating with the purpose of sterilization, pasteurization, thawing, and drying. However, lengthy exposure to high temperatures can produce loss of nutraceuticals, colour, and aroma, altering the quality of fruit and vegetable beverages.

3.8. Cold Plasma Treatment

Cold plasma (CP) technology provides a wide range of antimicrobial action while requiring low temperature changes and short times for application, which make it preferable for use with the thermolabile compounds responsible for the nutritional and sensory quality in fruit and vegetable beverages [44]. Plasma is the fourth state of matter, an ionized gas composed of ions, radicals (reactive oxygen and nitrogen species, ROS and RNS, respectively), atoms, and electrons in both excited and ground states [45]. These excited atoms and molecules emit excess energy in broad-spectrum electromagnetic radiation, including UV, when they return to a more stable state. The main parameters for plasma generation are pressure, voltage, treatment time, and type of gas [46]. Its application in juices is based on its ability to inactivate microorganisms located on food surfaces, food packaging materials, and process equipment under atmospheric pressure conditions [46,47].

4. Phytochemical Elicitors in Fruit and Vegetables Used in Beverages

Although the term 'elicitor' was originally applied to molecules able to induce the production of phytoalexins [48], nowadays it is used for compounds that stimulate any type of plant defence. Therefore, plant elicitors could be defined as external agents (biotic or abiotic) which act as indirect sources of enrichment and fortification due to plant response by synthetization of secondary metabolites because of the stress-induced stimulation.

In this sense, the first plant elicitors were described by Keen [49] as biotic stresses produced by pathogens. Nowadays, the exact molecules (polysaccharides, oligosaccharides, proteins, glycoproteins, or fatty acids) are independently added to produce such elicitation. However, the application of controlled postharvest abiotic stresses was proposed in 2003 by Cisneros-Zevallos [50] as an innovative tool to increase the biosynthesis of health-promoting compounds. In this review, we will focus on some of the abiotic stresses applied in raw plant materials used for beverages.

In this sense, abiotic stress can be classified according to its nature into three main groups: physical elicitors, chemical elicitors, and plant hormones (Figure 3).

Physical elicitation, such as wounding, temperature, gas composition, salinity, drought, high pressure, osmolarity, light conditions, and UV radiation, refers to controlled physical damage applied to the plant to trigger the activation of the plant's secondary metabolism as a defence mechanism. To confirm this behaviour, many authors have studied the correlation between these abiotic stresses and the accumulation of nutraceuticals. For instance, Jacobo-Velázquez and Cisneros-Zevallos have shown a direct relation between wounding carrot tissues and the accumulation of chlorogenic acid and its derivatives, which highly increased in combination with high temperatures (>15 °C), UV-C light, and the application of plant hormones [51,52]. Low doses of UV-A, UV-B, and UV-C in combination with wounding in fresh-cut carrot was also effective [53,54]. In addition, these authors have shown this behaviour in red prickly pears after a combination of wounding and UV-B light [55]. Furthermore, our own findings have demonstrated that the combination of UV-B and UV-C radiation separately applied can enhance glucosinolate biosynthesis in broccoli by-products [12] and phenolics in carrots [37]; furthermore, the simultaneous application of UV-B and UV-C at low doses can increase sulforaphane in broccoli and radish sprouts [56]. Moreover, hyperoxic conditions have shown to be interesting tools for improving phenolic biosynthesis in carrots [57] and tatsoi baby leaves [58], alone or combined with UV-C radiation.

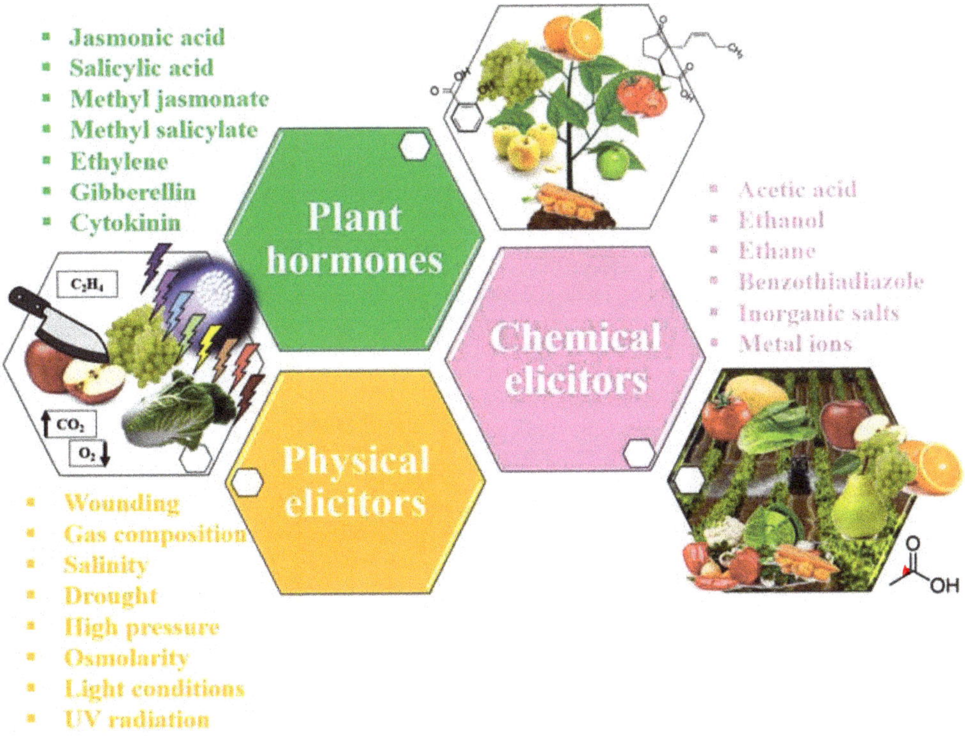

Figure 3. Summary of abiotic stresses applied as plant elicitors.

In a similar trend, acetic acid, ethanol, ethane, benzothiadiazole, inorganic salts, and metal ions have been applied to crops as chemical elicitors of major bioactive compounds to improve the quality of plants and fruits. For instance, some heavy metal salts such as $AgNO_3$ and $CdCl_2$ are able to trigger the production of phytoalexin and alkaloids; synthetic chemicals can also produce these signals [48].

Lastly, plant hormones such as jasmonic acid, salicylic acid, methyl jasmonate, methyl salicylate, ethylene, gibberellin, and cytokinin, among others, have also been studied as abiotic elicitors in plant cultivars. Thus, jasmonic acid, salicylic acid, and methyl jasmonate have been studied in combination with physical elicitors to increase the accumulation of glucosinolates in broccoli sprouts [34,59].

In this sense, most of the "Green technologies" previously detailed could act as inductors of the elicitation of phytochemicals in fruit and vegetables, which can be an enriched source of functional juices and beverages. In fact, these new technologies (HPP, US, UVC, PEF, RFQ, and CP) can extend the shelf-life of such products without the need to apply high temperatures, which helps to preserve their sensory and nutritional quality with no reduction in the phytochemical content.

5. Green Technologies as Elicitors of the Fortification of Fruit and Vegetable Beverages

5.1. Ultraviolet

The use of UV light to reduce microbial loads may not compromise the phytochemical content of beverages (Table 1). It is crucial to remember that several UV treatment parameters (UV dose, intensity, distance to the product, etc.) must be considered when evaluating the effectiveness of such treatments [60]. UV-C treatment at 1.08 kJ m^{-2} of kale juice, which reached up to 5-log reduction of inoculated *E. coli*, only led to low (<20%) total phenolic content (TPC) reduction [61]. High microbial inactivation (2.5–5.9 log reductions) was also found after UV-C treatment (10.6 kJ m^{-2}) in inoculated *E. coli*, *Pseudomonas fluorescens* and *Saccharomyces cerevisiae* in carrot-orange juice [62], while a similar UV-C dose (11.4 kJ m^{-2}) aimed to preserve the sensory quality of carrot juice during storage at 5 °C.

Higher UV-C doses in melon juice (16 kJ m^{-2}) and pineapple–mango juice (8 kJ m^{-2}) did not induce either TPC or total antioxidant capacity (TAC) changes [63,64]. UV-A treatment (1.5 J m^{-2}) did not either affect TPC and flavonoid content, while 1 log reduction of microbial spoilage was achieved [65]. Furthermore, UV treatments better preserved TPC, TAC and physicochemical quality (colour, pH, SSC, viscosity, etc.) [61,63–66]. Interestingly, a combined UV-C/UV-A treatment led to TPC and TAC enhancements of 1.8- and 4.6-fold in carrot-carob-ginger-lemon-grape juice [66]. The better preservation of phytochemical contents in these beverages may be owed to: (i) increased phytochemical extraction; (ii) impairment of some phenolic molecules; (iii) breakdown of polyphenols into smaller phenolic components; and/or (iv) antioxidant biosynthesis as a response to the free radicals produced during UV light exposure [63,66]. On the other hand, the observed mild phytochemical reductions in some studies may be explained by oxidation reactions and double bond disruption of these compounds promoted by free radicals produced during UV treatment and further photon absorption by double bonds or oxygen [61,63].

Table 1. Effects of green processing technologies on the phytochemical profiles of juice and smoothie beverages, along with other quality attributes.

Technolog	Beverage Type and Components	Treatment Conditions	Optimum Conditions	Shelf Life	Main Results Obtained	Reference
ULTRAVIOLET	Black carrot juice	254 and 365 nm; 15 W LMP lamps: $2.16 \cdot 10^{-3}$ and $1.50 \cdot 10^{-3}$ kJ/m^2 respectively; 0–60 min; 25 °C	365 nm $1.5 \cdot 10^{-3}$ kJ/m^2	-	Increased TPC No effect on colour Approx. 1 log reduction in microbial spoilage	[65]
	Carrot juice	30 W LMP lamps; 11.4 kJ/m^2	253.7 nm 11.4 kJ/m^2	12 d at 5 °C	Good sensory parameters during storage	[67]
	Carrot-carob-ginger-lemon-grape juice	280 and 365 nm; 0.6–0.4 mW LEDs; single and combined wavelength for 10–100 min; 25 °C	Combined 280/365 nm 0.77/2.2 kJ/m^2	-	Increased TPC and TAC 4.05 log reductions of inoculated *Escherichia coli* Minimum effect on physical properties	[66]
	Carrot-orange juice	30 W LMP lamps; 0–10.6 kJ/m^2; 0–15 min; 1.6 L/min; 20 °C	253.7 nm 10.6 kJ/m^2	-	Up to 2.5–5.9 log reductions in inoculated *E. coli*, *Pseudomonas fluorescens*, and *saccharomyces cerevisiae* Reduction by 20% in TPC	[62]
	Kale juice	25 W LMP lamps: 74 and 108.3 mJ/cm^2; 0.14 L/min; RT	253.7 nm 1.08 kJ/m^2	4 d at 4 °C	Up to 5-log reduction of inoculated *E. coli* No effect on viscosity, chlorophyll content, colour, TAC, PPO, and POD Increased sedimentation rate Higher PME activity	[61]
	Melon juice	15 W LMP lamps; 4 and 16 kJ/m^2; 5 and 20 min; 25 °C	254 nm 16 kJ/m^2	13 d at 5 °C	Better retention of TAC and colour No effect on TPC Moulds and yeasts did not grow	[63]
	Pineapple-mango juice	55 W LPM lamps; 0.08 kJ/m^2; 8.65 s	254 nm; 8 kJ/m^2	9 weeks at 4 °C	Minimal degradation of ascorbic acid TPC and TAC retention	[64]

Table 1. *Cont.*

Technology	Beverage Type and Components	Treatment Conditions	Optimum Conditions	Shelf Life	Main Results Obtained	Reference
HPP	Apple-carrot-zucchini-pumpkin-leek smoothie	350 MPa; 10 °C; 5 min	350 MPa 10 °C 5 min	28 d at 4 °C	High retention of vitamin C during storage Retention of antioxidants (TPC and flavonoids) Lower microbiological load Higher oxidation due to earlier clarification	[68,69]
	Apple-orange-strawberry-banana smoothie	350–600 MPa; 10 °C; 3–5 min	350 MPa 10 °C 5 min	48 h at 4 °C	Does not affect phenolics and flavonoids Preserves vitamin C and flavours Ensures microbial quality	[70]
	Apple-strawberry-banana-orange smoothie	450–600 MPa; 20 °C; 5–10 min	600 MPa 20 °C 5 min	10 h at 4 °C	Retains ascorbic acid High PPO inactivation rate (83%)	[71]
	Berries-grape-orange-strawberry-apple smoothie	100–300 MPa; −5–45 °C; 5 min	300 MPa 45 °C 5 min	15 d at 4 and 20 °C	Reduction of up to 6-log of mesophilic lactobacilli	[72]
	Carrot juice	550 MPa; <38 °C; 6 min	550 MPa <38 °C 6 min	20 d at 4 °C	Better carotenoid (α- and β-carotene), phenolic, polyacetylene and TAC retention Better preservation of nutritional compounds Higher rheological properties Better sensory attributes	[73]
	Carrot-pumpkin smoothie	300–600 MPa; 23 °C; 5 min	400 MPa 23 °C 5 min	7 d at 5 °C	Mild TPC reduction (<15%) Better TPC preservation during storage High microbial control (≈6 log lower counts) No high physicochemical changes (SSC, pH and colour)	[74]
	Grape juice	500 MPa; 45 °C; 5 min	500 MPa 45 °C 5 min	-	Reduction of 17–29% in aflatoxins	[75]

Table 1. *Cont.*

Technology	Beverage Type and Components	Treatment Conditions	Optimum Conditions	Shelf Life	Main Results Obtained	Reference
	Indian gooseberry juice	200–500 MPa; 30–60 °C; 5 min	500 MPa 30 °C 5 min	-	Increase in TPC and TAC up to 50 °C Less vitamin C degradation	[76]
	Juçara-mango juice	600 MPa; 25 °C; 5 min	600 MPa 25 °C 5 min	-	Does not affect anthocyanin content Good sensory properties	[77]
	Orange juice	0–200 MPa; 25 °C; 1 min (whole peeled orange) + 400 MPa; 40 °C; 1 min (juice)	200 MPa 25°C 1 min + 400 MPa 40 °C 1 min	-	Non-additive effect in flavonoids and vitamin C 12-fold increase in content of colourless carotenoids	[78]
	Tomato-pepper-celery-cucumber-onion-carrot-lemon beverage	100–400 MPa; <30 °C; 2–9 min	400 MPa <30 °C 2–5 min	-	Good preservation of vitamin C Slight colour change	[79]
PEF	Apple juice	20–30 kV/cm; 5–125 μs; ≤55 °C	25 kV/cm 63 μs ≤55 °C	-	>5 log reduction cycles of *E. coli*, *L. monocytogenes*, *Staphylococcus aureus*, and *Salmonella typhimurium*	[80]
	Apple-strawberry-banana smoothie	13.5–24 kV/cm; 100–290 Hz; 8.7–24.1 pulses; 3 μs; 130 L/h; <58 °C	24 kV/cm; 100 Hz 8.7 pulses <58 °C	27 d at 7 °C & 7 d at 4 °C	Highest inactivation of moulds and yeasts	[81]
	Grape juice	3 kV/cm; 238 pulses up to 500 kJ/kg; sample: 215 mL; <75 °C	3 kV/cm 238 pulses <75 °C	-	Reductions by 24–84% in aflatoxins	[75]
	Grapefruit juice	20 kV/cm; 1 kHz; 80 mL/min; <45 °C	20 kV/cm; 1 kHz <45 °C	-	Lower non-enzymatic browning and viscosity than the untreated sample	[82]
	Mango-papaya-stevia juice	20–40 kV/cm; 2.5 μs; 30 mL/min; 100–360 μs; <50 °C	21 kV/cm 360 μs <50 °C	-	Greatest content of bioactive compounds Minimal colour changes	[83]

Table 1. *Cont.*

Technology	Beverage Type and Components	Treatment Conditions	Optimum Conditions	Shelf Life	Main Results Obtained	Reference
ULTRASOUND	Apple juice	2 W/cm^2; 25 kHz; 70%; 30–60 min; 20 °C; sample: 60 mL	2 W/cm^2; 25 kHz 30 min 20 °C	-	The highest polyphenolic and sugars content Higher minerals and total carotenoids (60 min)	[84]
	Apple-carrot-stevia juice	750 W; 20 kHz; 20–80%; 15 min; sample: 100 mL	750 W; 20 kHz; 60% 15 min	-	Better phenolic profile Better radical scavenging activity	[85]
	Apple-strawberry-banana-orange smoothie	1.5 kW; 20 kHz; 40–100%; 25 °C; sample: 200 mL	1.5 kW; 20 kHz; 70% (42.7 µm) 3 min 25 °C	-	Better TPC preservation Higher flavonoid content	[86]
	Carrot juice	750 W; 20 kHz; pulses of 5 s on and 5 s off; 70%; 15 °C; sample: 250 mL	750 W; 20 kHz; 70% 15 °C	48 h at 4 °C	Enhancement of colouring pigments, sugar, chlorogenic acid and some mineral contents Decreased microbial population	[87]
	Grape-apple juice	750 W; 20 kHz; 100%; 20–40 min; sample: 100 mL	750 W; 20 kHz; 100% 20 min	-	Increased phenolic profile and TAC Higher organic acids	[88]
	Grapefruit juice	720 W; 28 kHz; 70%; 30–90 min; 20 °C	720 W; 28 kHz; 70% 90 min 20 °C	-	Improvement in sugar, carotenoid, mineral and phenolic content Decreased spoilage microbe population	[89]
	Strawberry-banana-juçara smoothie	73.5–250 W; 20 kHz; 7–19 min; <60 °C; sample: 200 mL	147 W; 20 kHz 2 min <60 °C	-	The highest anthocyanin retention	[30]
	Orange juice	33.31 W/mL; 24 kHz; 105 µm; 1–30 min; <46 °C; sample: 30 mL	33.31 W/mL; 24 kHz; 105 µm 30 min <46 °C	28 d at 5 °C	Increased phenolic and flavonoid content Higher vitamin C retention Better sensorial properties	[90]
	Nopal beverage	240 W; 42 kHz; 10–40 min; <34 °C; sample: 300 mL	240 W; 42 kHz 40 min <34 °C	28 d at 4 °C	Higher stability for bioactive compounds Ascorbic acid reduction Best acceptability Low changes in colour	[91]
	Tomato-coconut water-beetroot juice-based beverage	240 V; 37 kHz; 10 and 15 min	240 V; 37 kHz 10 min	-	High sinapic and gallic acid contents Ascorbic acid reduction 1 log reduction of yeast and mould	[92]

Table 1. *Cont.*

Technology	Beverage Type and Components	Treatment Conditions	Optimum Conditions	Shelf Life	Main Results Obtained	Reference
COLD PLASMA	Apple juice	Atmospheric jet; 65 V; 1.1 MHz; 0–0.1% oxygen-argon gas flow: 5 slm; 0–8 min	Atmospheric jet; 65 V 0.1% O_2 in Ar gas 8 min	24 h	Reduction of *C. freundii* by ~5 log cycles	[93]
	Apple juice (cashew)	Indirect plasma field under 30 kPa; 80 kHz; nitrogen gas flow: 10–50 mL/min; 5–15 min; sample: 10 mL	Indirect plasma; 30 kPa 10 mL N_2/min 5 min	-	Increased TPC and TAC Higher vitamin C retention	[94]
	Apple juice (cloudy)	Spark and glow discharge; 7.9–10.9 kV; 20–65 kHz; 1–5 min	Spark discharge 10.5 kV 5 min	28 d at 4 °C	Increased TPC and TAC PPO inactivation Lighter juice colour	[95]
	Blueberry juice	Single-electrode atmospheric jet; 11 kV; 1 kHz; 0–1% oxygen-argon gas flow: 1 L/min; 2–6 min	Single-electrode 11 kV 1% O_2 in Ar gas 6 min	-	Increased TPC and TAC Higher content of anthocyanin and vitamin C (at 0% O_2 and 2–4 min) than heat treatment 7.2 log reduction of Bacillus	[96]
	Chokeberry juice	Single-electrode atmospheric jet; 25 kHz; argon gas flow: 0.75 dm^3/min; time: 3–5 min; 24 °C	Single-electrode 5–7 cm^3 3 min	-	Polyphenolic content stability	[97]
	Coconut liquid endosperm	Atmospheric jet powered by a microwave generator; 450–650 W; air gas flow: 5 L/min; time: 0–25 min	Atmospheric jet 450 W 22–24 min	-	Reduction of initial counts of *S. enterica* and *E. coli* by 4 log cycles	[98]
	Orange juice	DBD-low-temperature plasma; 30 kV; 60 kHz; time: 3–12 s and 5–20 s; sample: 50 µL and 4 mL, respectively	DBD; 30 kV 10 s	16 d at 4 °C	Absence of *E. coli* in juice inoculated with 4.20×10^7 CFU/mL Preservation of vitamin C content	[99]
	Orange juice	DBD-atmosphere CP; 90 kV; 60 Hz; time: 30–120 s; sample: 25–50 mL; atmosphere gas: air or 65% O_2	DBD; 90 kV 2 min direct plasma 50 mL 65% O_2	24 h at 4 °C	Reduction of 4.7-log of *S. enterica* Reduced PME enzyme activity Higher vitamin C retention in air-packaging compared to the high-oxygen atmosphere.	[100]
	Orange juice with oligosaccharides	DBD-atmosphere CP (direct and indirect plasma field); 70 kV; 50 Hz; time: 15–60 s; sample: 20 mL	DBD 70 kV Direct plasma field	24 h at RT	12% oligosaccharide loss preservation of TPC, TAC and colour	[101]

Table 1. *Cont.*

Technology	Beverage Type and Components	Treatment Conditions	Optimum Conditions	Shelf Life	Main Results Obtained	Reference
COLD PLASMA	Pomegranate juice	Single-electrode atmospheric jet; 2.5 kV; 25 kHz; argon gas flow: 0.75–1.25 dm^3/min; time: 3–5 min; sample: 3–5 cm^3	Single-electrode; 2.5 kV 0.75 dm^3/min 3 min 5 cm^3	-	Greater anthocyanin stability Less colour changing with higher gas flow	[97]
	Pomegranate juice	Single-electrode atmospheric jet; 2.5 kV; 25 kHz; argon gas flow: 0.75–1.25 dm^3/min; time: 3–5 min; sample: 3–5 cm^3	Single-electrode; 2.5 kV 1 dm^3/min 5 min 3 cm^3	-	Better phenolic compound stability	[102]
	Sour cherry Marasca juice	Single-electrode atmospheric plasma jet; 2.5 kV; 25 kHz; argon gas flow: 0.75–1.25 L/min; time: 3–5 min; sample: 2–4 mL	Single-electrode; 2.5 kV 3 min 3 mL	-	Highest anthocyanin and phenol content	[103]
	Tomato juice	DBD; 10 kV; 5 min; 30 °C	DBD; 10 kV 5 min 30 °C	-	No effects on flavour and aroma Lower volatile compound release	[104]
	Tomato-coconut water-beetroot juice-based beverage	DBD; 60 kV; 50 Hz; time: 10 and 15 min; sample: 100 mL	DBD; 60 kV 10 min	-	Improvement of TPC	[92]
	White grape juice	DBD; 80 kV; 60 Hz; time: 1–4 min; 24 °C	DBD; 80 kV 1 min 24 °C	-	Higher bioactive compound levels	[105]
	Apple juice	US + PEF US: 600 W; 20 kHz; 80%; 20–44 °C; 10–30 min; sample: 95 mL PEF: 23.9–71.6 J/cm^2; 360 µs; 2–60 s; sample: 5 mL; <56 °C	US30 + PEF60	15 d at 5 °C	5.8-log reduction of *S. cerevisiae*	[106]
	Apple juice	UV + US; US + UV UV: 254 nm; 15 W lamps: 13.44 W/m^2; 5–25 min US: 120–480 W; 35 kHz; 5–25 min	US: 120 W; 5 min + UV: 254 nm; 20.2 kJ/m^2	-	5-log reduction of *A. acidoterrestris*	[107]

Table 1. *Cont.*

Technology	Beverage Type and Components	Treatment Conditions	Optimum Conditions	Shelf Life	Main Results Obtained	Reference
COMBINED TECHNOLOGIES	Apple juice (cloudy)	HPP + UV HPP: 0–300 MPa; 32 °C; sample: 13 L UV: 254 nm; 55 W lamp; 14.3–28.7 J/mL; 20 °C; sample: 70 mL	HPP: 300 MPa; 32 °C + UV: 254 nm; 28.7 J/mL	-	Increased TPC by 277.6% Reduced PME activity	[108]
	Carrot juice	CP + US CP: DBD; 70 kV; time: 3 × 4 min, US: 750 W; 20 kHz; pulses of 5 s on and 5 s off; 80%; <20 °C; 3 min; sample: 100 mL	CP: DBD; 70 kV + US: 750 W; 3 min; <20 °C	-	Better stability Higher TPC, carotenoid, lycopene, and lutein Up to 2 log reductions of mesophilic and yeast and moulds	[109]
	Cranberry juice	US + HPP US: 600–1200 W/L; 18 kHz; <25 °C; 5 min HPP: 450 MPa; 11.5 °C; 5 min	US: 1.2 kW/L; 5 min; 25 °C + HPP: 450 MPa; 5 min; 11.5 °C	-	Higher anthocyanin content Good preservation of FOS	[110]
	Mango juice	US-UV: 600 W; 20 kHz; pulses of 5 s on and 5 s off; 10 min; 3600 J/mL; sample: 100 mL UV: 254 nm; 8 W lamp;	US-UV: 600 W; 254 nm; 3.6 kJ/mL; 10 min	30 d at 4 °C	Increased the bioaccessibility of ascorbic acid, TPC, and carotenoids by 102%, 114%, and 32%, respectively Good retention up to 30 d	[111]
	Orange juice	US + PEF US: 500 W; 30 kHz; 55 °C; 10 min; sample: 800 mL PEF: 40 kV/cm; 15 Hz; 100 µs	US: 500 W; 10 min; 55 °C + PEF: 40 kV/cm	168 d at 25 °C	Lower colour differences Similar attributes to heat treatment	[112]

TPC: total phenolic content; TAC: total antioxidant capacity; PPO: polyphenol oxidase; POD: peroxidase; DBD: dielectric barrier discharge; RT: room temperature; FOS: fructooligosaccharides; PME: Pectin methylesterase.

5.2. High-Pressure Processing

HPP is a well-known green technology for its minimal impact on the phytochemical content of beverages. In addition, phytochemical increases may be expected due to plant cell disruption leading to higher extractability of these compounds. Nevertheless, special attention must be paid since within these disrupted cells degradative enzymes come in contact with their substrates, which is crucial to high enzyme inactivation rates. Thus, the polyphenol oxidase (PPO) activity of a fruit smoothie was inactivated by 83% after a 600 MPa-HPP treatment (5 min; 20 °C) [71]. Thus, it is recommended that HPP treatments be conducted at higher temperatures (usually up to 60 °C) to increase the inactivation rate of degradative enzymes like PPO, ascorbate oxidase, etc. In addition, unavoidable heat transfer among vessel and products through pressurisation and depressurisation, and the increased time to reach final pressurisation and additional dwell time, may also affect the phytochemical content and enzyme activity [71].

Vitamin C (ascorbic acid) is considered among the most thermolabile antioxidants in fruits and vegetables, being considered an indicator of the nutritional quality of beverages after processing treatments. HPP is considered an excellent green technology that does not greatly affect the vitamin C content of fruit and vegetable beverages, as heat treatments do. Hence, mild HPP treatments (350–450 MPa; 5 min; 10–20 °C) were enough to reach high vitamin C retention in different fruit–vegetables smoothies [68,70,71,79]. Furthermore, the dissolved oxygen proportion in beverages is of high interest due to the high oxidation rates of vitamin C. Thus, dissolved oxygen in a fruit smoothie was reduced approximately three-fold after HPP treatment (450–600 MPa; 5–10 min; 20 °C) with consequent better vitamin C retention after treatment [71].

With respect to phenolic compounds, HPP treatments (350–600 MPa; 5 min; 10–25 °C) also retained the TPC of fruit smoothies with unchanged values [69,70,77]. Furthermore, the TPC of fruit juice treated with HPP (600 MPa; 5 min; 25 °C) was only slightly reduced (≈10%), correlated with its TAC, after 30 days at 4 °C [77]. Such good phenolic retention during storage may be explained by the high PPO inactivation rates that are achieved, as previously observed [71]. Interestingly, more intense HPP treatments (500 MPa; 50 °C; 20 min) induced a TPC enhancement of 20% in gooseberry juice [76]. Such increases can be explained by several aspects: (i) enhanced solubility of phenolic compounds due to pressurization, resulting in increased antioxidant activity; and (ii) structural changes due to HPP leading to modifications in the product constituents (proteins, lipids and carbohydrates), which may allow greater access of the enzymes and promote the release of antioxidant compounds such as bioactive peptides, fat-soluble vitamins, carotenoids and polyamines [76,77]. Furthermore, phenyl ammonia lyase activity, the key enzyme in the biosynthesis pathway of polyphenols, was six-fold higher in an HPP-treated (300 MPa; 5 min; 23 °C) orange smoothie after seven days at 4 °C compared with an untreated smoothie [74]. Kouniaki et al. [113] hypothesized that the high ascorbic acid content common in fruits and vegetables may negatively impact anthocyanin content. Nevertheless, HPP treatment (600 MPa; 5 min; 25 °C) of juçara–mango juice aimed to preserve the anthocyanin content with unchanged values [77].

Carotenoid contents (α- and β-carotene) of carrot juice were better preserved with HPP (550 MPa; 6 min; <38 °C), with values 1.4–1.7-fold higher compared with heat-treated juice (110 °C; 8.6 s), which was also observed during subsequent conservation at 4 °C up to 20 days [73]. HPP treatment of whole peeled oranges (200 MPa; 1 min; 25 °C) followed by HPP of the juice (400 MPa; 1 min; 40 °C) induced 12-fold higher contents of phytoene and phytofluene, which was explained by the authors due as to the highly hydrophobic nature of these carotenoids, leading to a higher extractability after HPP and enhanced release from the cellular structures to the juice [78].

HPP has been also proposed as a green technology to reduce aflatoxins [75]. Hence, HPP treatment (500 MPa; 5 min; 45 °C) reduced aflatoxins B1/B2, G1 and G2 by 14–17%, 19%, and 29%, respectively, in grape juice [75]. The authors also stated that the effectiveness of HPP in reducing the aflatoxin content in fruit beverages may depend on the studied mycotoxin, the food matrix, and the applied conditions.

5.3. Pulsed Electric Fields

Treatment of fruit juice by PEF did not affect anthocyanin content [31]. Nevertheless, it is known that electroporation due to PEF treatments may enhance the extractability of bioactive compounds such as anthocyanins, although PEF may also promote reactions that reduce the content of these phytochemicals. Accordingly, when the previous authors supplemented fruit juice with an antioxidant (stevia) the anthocyanin content was hypothesized to be protected, and higher contents were observed [31]. A similar beneficial effect of stevia to preserve the anthocyanin content of fruit juice after HPP was observed by the same authors [114]. In addition, they found that the electric field, more so than treatment time, was the factor that most influenced anthocyanin content.

Both electric field and treatment time influenced the carotenoid content of fruit juice after PEF, showing lower electric fields can lead to enhancement of carotenoids, explained by higher extractability after PEF with low generation of ROS (which may promote the oxidation of the carotenoid chain) [31].

Nevertheless, PEF (20–40 kV/cm; 100–360 µs; <50 °C) of fruit juice led to ascorbic acid reductions, with higher electric fields strengths inducing larger decreases in this bioactive compound independently of the treatment time [31]. The authors explained such negative effects as being due to the higher extractability of intracellular contents after PEF due to electroporation, and the consequent higher oxidation reactions of ascorbic acid.

The aflatoxin content of beverages can also be reduced by PEF treatment, as previously reported [75]. These authors achieved 24–25%, 72% and 84% reductions of AFB1/AFG2, AFB2 and AFG1 reductions, respectively, after PEF treatment (3 kV/cm, 238 pulses; <75 °C) in grape juice. PEF induces the permeation of the cell membrane, forming membrane pores (temporarily or permanently), which may modify the structure of amino acids, proteins, and polysaccharides. Meanwhile, HPP treatment has minimal effect on the breakage of covalent bonds and is transmitted instantaneously; thus, no gradients are formed, which may explain the lower aflatoxin reduction after HPP compared with PEF found by Pallarés et al. [75].

As observed, PEF is an excellent green technology that has little effect on the content of phytochemicals in beverages while achieving good microbial reduction, extending the product shelf life. Hence, >5 log reductions of *E. coli*, *L. monocytogenes*, *Staphylococcus aureus* and *Salmonella typhimurium* were achieved in apple juice after PEF treatment (25 kV/cm; 63 µs; <55 °C) in [80], with the potential to extend the product shelf life as observed in PEF-treated fruit smoothies and juices [81,83].

5.4. Ultrasound

Ultrasonic processing of beverages generally enhances the content of most phytochemical compounds, such as phenolic compounds, flavonoids, carotenoids, etc., as compiled in Table 1. Thus, sonication is a well-known procedure in laboratories to improve the extraction of compounds from the food matrix.

Phenolic contents can be enhanced with US processing, as observed in fruit juices, whose TPC was increased up to 60% after treatments of 30–60 min (20–25 kHz) [90,91]. Nevertheless, shorter (<30 min) processing of beverages did not improve the TPC [88,91,92]. Such phenolic enhancements were highly correlated with TAC increases due to the highly antioxidant nature of these compounds [90,91]. Focusing on individual phenolic compounds, chlorogenic acid, caffeic acid and catechin were enhanced by 40%, 20% and 16%, while phlorizin and epicatechin increases reached up to 76% and 130% after US processing (60 min; 25 kHz) of the fruit juice [87]. Interestingly, contents of individual phenolic compounds in a nopal beverage were also enhanced after treatment, although without difference between treatment times from 10 to 40 min [91]. The observed phenolic enhancements of the authors with short treatments (10 and 20 min) similar to 40 min processing may be explained by the high frequency (42 kHz) used by those authors [91]. Similarly, treatment at 37 kHz for 10 min induced sinapic and gallic acid increases in a vegetable–coconut beverage [92]. Thus, each phenolic compound has a different sensitivity to ultrasonic waves, and its content may be increased or preserved [91]. Enhancement of flavonoids seems to be higher, since processing of just 15–30 min (20–24 kHz) induced increases of the total flavonoid content of 30–90% in several fruit juices [85,88,90].

The enhancement of phenolic compounds in beverages after US may be explained by several reasons: (i) increased extractability owing to the disruption of cell walls during the sudden change in pressure of liquid by the shear force exerted by cavitation, which can facilitate the liberation of bound polyphenolic contents; and (ii) attachment of hydroxyl radicals to the aromatic ring of phenolic compounds during sonochemical reactions occurring during US processing [84,89]. Furthermore, the addition of a second hydroxyl group to the ortho- or para-positions has previously been reported to enhance the antioxidant capacity of phenolic compounds [84]. At the same time, unchanged phenolic contents such as those observed with short US processing may be the result of their use as an antioxidant against the generation of free radicals (e.g., hydroxyl and hydrogen-free radicals) during sonication due to the dissociation of the water molecules in aqueous solutions as a result of the high temperature and pressure of the collapsing gas bubbles associated with cavitation [90].

Carotenoid contents of fruit juices were also increased after US processing [84,87,89]. In particular, the total carotenoid content of apple juice was increased by 27% after processing for 60 min (25 kHz) [84], with such enhancement of total carotenoid content of grapefruit juice increased up to 40% with a longer processing time of 90 min (28 kHz) [89]. Blanching (water bath at 100 °C for 4 min) carrots before US processing of the carrot juice improved carotenoid retention, with higher carotenoid content (1.9- and 1.7-fold higher lycopene and lutein, respectively) in the sonicated juice from blanched carrots compared to the juice from unblanched carrots [87]. These higher values after US processing when the raw product is previously blanched may be owing to the inactivation of degrading enzymes and/or additional disruption of plant cells, which enhances their extractability after the subsequent ultrasound treatment.

On the contrary, the ascorbic acid content of beverages is generally reduced under common US processing [91,92]. Thus, a 6–7% reduction in the ascorbic acid content of a vegetable–coconut beverage was observed after 10–15 min (37 kHz) of processing [92]. This ascorbic acid reduction during US processing is increased when the treatment time is augmented, as was observed in a nopal beverage after 20 and 40 min of processing (42 kHz) [91]. Interestingly, lower US frequency (24 kHz) and higher temperature (43–46 °C) during treatment, regardless of the treatment time, led to increases of the ascorbic acid content of ≈30% [90]. The latter finding may be explained by the elimination of dissolved oxygen (which leads to oxidation of ascorbic acid) during sonication being increased at higher temperatures. However, the observed ascorbic acid degradation in beverages under temperatures <30 °C may be explained by the generation of hydrogen ions (H^+), free radicals (O^-, OH^-) and hydrogen peroxide (H_2O_2) during the sonication of water molecules, which may degrade ascorbic acid [92]. In addition, such US treatments may not be enough to key enzymes on ascorbic acid stability as the ascorbate oxidases. However, the ascorbic acid reduction after US processing (e.g., 40 min; 42 kHz; <34 °C) were still far lower (≈2-fold lower) than those of conventional heat treatments (80 °C for 10 min) [91].

5.5. Cold Plasma

Cold plasma processing is also able to induce the enhancement of several phytochemicals. Cold plasma consists of an ionized gas (carbon dioxide, argon, nitrogen, helium, oxygen, or air) including active particles such as electrons, ions, free radicals and atoms. These reactive species have sufficient electrical energy to disrupt the covalent bonds of phytochemical compounds and cell membranes, which promotes their release, leading to higher contents of free phytochemicals. The reactive species generated, and therefore their effect on the phytochemical contents, depends on the parameters of the plasma source, such as its voltage, frequency, and plasma generation system, as well as the type of gas and its flow rate.

In particular, the optimal conditions to enhance the TPC of sour cherry juice with atmospheric plasma jet (single-electrode) at 2.5 kV were 3 min, 2.8 mL of sample volume and 0.75 L/h gas (argon) flow, reaching a TPC enhancement of 15% [103]. In the same study, increasing sample volume (3.2 mL) and gas flow (1.25 L/h) while maintaining treatment time (3 min) reached a total anthocyanin enhancement in sour cherry juice of 34% [103]. The authors attributed the higher anthocyanin increases to the presence of undefined small agglomerates or particles in the juice that could be dissociated during plasma processing. Furthermore, the anthocyanin stability might have been increased due to intra- and intermolecular association with other anthocyanins, or through pigmentation with copigments like flavonoids and hydroxycinnamic acids. In addition, the previous authors found that the TPC of pomegranate juice was increased up to ≈50% after plasma jet (single electrode) for 5 min, at 1 L/min argon for 3 mL of the sample [102]. Hydroxycinnamic acids seem to have higher stability than anthocyanins, as observed in chokeberry juice after single-electrode plasma jet processing [97]. The authors reported that hydroxycinnamic acids react less with radical species generated by the plasma, since they are less effective in reducing reagents, leading to their observed higher stability. Higher voltage (11 kV) during plasma jet (single electrode) using oxygen (1%)-argon at 1 L/min with longer treatment time (6 min) led to maximum TPC enhancements (7% higher) in blueberry juice [96]. Dielectric barrier discharge plasma (60 kV) of 100 mL sample (coconut–vegetable juice) for 10 min induced an enhancement of the TPC of only ≈5% [92]. With respect to the plasma orientation of the sample, atmospheric direct plasma (dielectric barrier discharge) application was more effective than indirect application in preserving the TPC, and consequently the TAC, than indirect application [101].

Increasing gas flow led to higher phenolic retention, leading to 14–28% higher TPC in cashew apple juice, with the maximum levels for 50 mL/min and longest treatment (15 min) [94]. Nevertheless, the authors agreed that flavonoids might require less energy to be released from their bonds than polyphenols, since lower flavonoid retention was achieved with the same plasma treatments regardless of gas flow or treatment time [94].

The highest increases of phenolic content with cold plasma were reported using spark discharge (10.5 kV) for 4–5 min with increases of 64–69% of the TPC of cloudy apple juice [95]. Such high phenolic enhancement may also be explained by the high PPO inactivation (70–80%) achieved with that cold plasma treatment [95]. The authors also stated, in agreement with previous literature, that spark discharge induces a higher concentration of some radical species (H_2O_2 and NO_3^-), which may induce higher enzyme inactivation [95]. Furthermore, such high phenolic enhancement could be attributed to this cold plasma method, which might increase cell membrane breakdown compared with other methods. Nevertheless, further work is needed to study phenolic changes in beverages using different cold plasma methods.

Although vitamin C is among the most labile phytochemicals during beverage processing, cold plasma has little effect compared with conventional heat treatments. Hence, short (2 min) cold plasma treatment (single-electrode plasma jet) with different oxygen concentrations (up to 1%) showed higher vitamin C retention than heat treatment (85 °C, 15 min) in blueberry juice [96]. The authors also found higher vitamin C retention even after longer cold plasma treatments (up to 6 min), when the oxygen concentration of the ionized gas was 0% [96]. Hence, prevention of vitamin C oxidation during cold plasma treatment was also studied in orange juice, with better vitamin C retention using air atmosphere compared with an oxygen-rich atmosphere (65% O_2, 30% N_2, 5% CO_2) [100]. As observed, oxidation reactions initiated by reactive oxygen (and nitrogen) species are the most important reactions responsible for the high microbicidal effect of cold plasma processing, although it may have a negative effect on vitamin C stability [93,96,98–100]. Hence, apple juice treated with cold plasma (indirect plasma field; 80 kHz) under nitrogen gas flow led to better vitamin C retention, with levels increased by up to 11% [94]. The authors explained this vitamin C increase after cold plasma processing as being due to dehydroascorbate reductase (the enzyme responsible for the dehydroascorbic acid conversion back to ascorbic acid) activation triggered by some RNS (mainly nitric oxide), which are usually generated during cold plasma treatment [94].

5.6. Combined Technologies

The combination of processing technologies to obtain additive or synergistic effects is an effective strategy to improve their individual effects on product quality [12,57,115]. Hence, high microbial inactivation (up to 5.8 log reduction of *S. cerevisiae*) of individual US and PEF were enhanced when a combined treatment (30 min US + 60 s PEF) was used in apple juice [106]. Similarly, combined US (120 W for 5 min) + UV-C (20.2 kJ/m^2) treatment achieved up to 5 log reductions of *Alicyclobacillus acidoterrestris* in apple juice [107].

Moreover, the application of combined technologies is an excellent strategy for enhancing the phytochemical content of beverages. Due to the high effectiveness of Ultrasound to preserve food quality and increase the extractability of phytochemicals, and to its relatively easy application in beverages, most studies on combined technologies for the treatment of beverages include US. For example, cold plasma (dielectric barrier discharge; 70 kV, 3 × 4 min) combined with US (750 W, 3 min) increased the individual effects of these technologies up to 22% and 34% for TPC and total carotenoid content, respectively, in carrot juice [109]. In particular, the lycopene and lutein contents of carrot juice were enhanced approximately four-fold and two-fold with combined cold plasma + US treatment [109].

US (18 kHz, 5 min) was applied in combination with HPP (450 MPa, 5 min) in prebiotic cranberry juice, with an increase of anthocyanins ranging from 14% to 20% being observed when the power level of 1200 W/L was used; meanwhile 600 W/L did not show the same benefits [110]. In addition, the same US-HPP treatment increased the contents of some fructooligosaccharides (DP4-nystose and DP5-1-fructofuranosylnystose), regardless of the power level (600 or 1200 W/L) [110]. A combination of HPP (300 MPa) and UV (28.7 J/mL) also reduced the pectinmethylesterase activity of cloudy apple juice from 190% relative activity (after increase due to single HPP treatment) to 60% [108]. A combination of US (20 kHz, 10 min) with UV increased the ascorbic acid, TPC and total carotenoid content of mango juice by approximately 47%, 35% and 200%, respectively [111]. Furthermore, mango juice processed with this combined treatment achieved better retention of these phytochemicals during cold storage, together with a higher bioaccessibility of the compounds due to the higher availability of these compounds in the treated mango juice [111].

As observed, the combination of these non-thermal technologies highly preserved and even increased the analysed content of several phytochemicals in fruit and vegetable beverages. This may be explained by the higher extractability of phytochemicals by these technologies, in particular US and HPP due to phenomena like cavitation, cell disruption, etc. In addition, the reduction of dissolved oxygen in the beverages due to cavitation during US treatment allowed for high preservation of these compounds from oxidation (of high interest for ascorbic acid). Thus, beverages treated with combined technologies may have better quality (e.g., lower colour differences due to enzymatic inactivation) [112] and higher contents of available phytochemicals (see Table 1), while ensuring food safety requirements.

6. Fortification of Beverages Using Natural Products

Fortification is a technique used to directly enrich the nutritional, bioactive and health-promoting compounds of beverages. However, the stability of the added components may be altered (oxidations, reactions with other components, etc.) due to the composition (organic acids, etc.) and physicochemical properties (pH, dissolved oxygen, etc.) of the beverage. Table 2 summarizes the fortification strategies of beverages by using natural products, which are also commented on in this section as follows.

Table 2. A review of the main phytochemical fortification conditions in beverages.

Beverage Type and Components	Fortification Conditions	Optimum Conditions	Shelf Life	Results	Reference
Apple and orange juices	Encapsulation of folic acid (synthesised) in mesoporous silica particles	Encapsulated	-	Improved stability Controlled release after consumption by modifying vitamin bioaccessibility	[116]
Apple and orange juices	Free and microencapsulated *L. acidophilus* (10 and 30%)	Microencapsulated	63 d at 4 °C	Extended survival	[117]
Blackberry juice	Glutathione, galacturonic acid, diethylenetriaminepentaacetic acid, and tannic acid (500 mg/L)	Glutathione	5 weeks at 30 °C	Great retention of the anthocyanin content	[118]
Blueberry juice	*Lactobacillus plantarum* strain J26	*L. plantarum*	-	Increased TPC by 43% Increased anthocyanin content by 15%	[119]
Carrot juice	Pomegranate peel extract obtained by HPP (0–2.5 mg/mL)	2.5 mg/mL	42 d at 4 °C	Higher TAC Improved microbial safety	[120]
Carrot juice	Carrot shreds under combination of UV-C (4 kJ/m^2) and 72 h at 15 °C in air or hyperoxia (80 kPa O_2) conditions	Non-UV-C + 80 kPa O_2 (72 h at 15 °C)	14 d at 5 °C	Increased TPC by 2060% Good microbial quality	[121]
Carrot juice	Carrot slices, peeled and unpeeled (48 h at 15 °C) + blanching (80 °C; 6 min)	Unpeeled carrot slices + blanching	-	Increased by 3600 and 195% in chlorogenic and TPC, respectively Increased minerals by 7–40%	[122]
Chokeberry juice	*Lacticaseibacillus paracasei* strain SP5 (10 g/mL)	*L. paracasei*	4 weeks at 4 °C	Increased TPC by 49%	[123]
Cranberry juice	0–10% (*v/v*) of pale brewers' spent grains (BSG) 0–10% (*v/v*) of black BSG	10% of pale or black BSG	-	Increased TAC by 120%	[124]
Date-puree nectar	Spirulina (0–20%)	10% of spirulina	-	Higher total sugar and carotenoid contents Better sensory attributes	[125]
Emmer-based beverage	*Lactiplantibacillus plantarum* 2035 + blueberry, aronia or grape juice	*L. plantarum* + aronia juice	4 weeks at 4 °C	Higher TPC and TAC High viable counts during storage	[126]
Grape-broccoli-cucumber smoothie	2.2% of alga (sea lettuce, kombu, wakame, thongweed, dulse, Irish moss, nori, chlorella, or spirulina)	Chlorella and spirulina	24 d at 5 °C	The highest vitamin C content High fucose content for thongweed, kombu, and wakame-based smoothies	[127]
Melon juice	(−) Epicatequin (1.25–5 g/mL)	2.5 g/mL	10 d at 4 °C	Increased TPC by 724%	[128]
Multi-fruit juice byproducts	Ginger and apple (50:50, *w/w*) and apple, carrot, beet, and ginger (50:29:20:1, *w/w*) juice obtained by fresh and freeze-dried byproducts	Ginger and apple juice obtained by freeze-dried by-products	-	Higher TPC, TAC, and flavonoids	[129]
Orange-apple-carrot-beet smoothie	Beet leaf extract (30% *v/v*)	Beet leaf extract (30%)	21 d at 5 °C	Increased TPC content by 50% High TAC	[130]
Orange-celery-carrot-lemon juice	Fermentation by *L. plantarum* strain HFC8	*L. plantarum*	-	Low mesophilic aerobic bacteria and yeast and mould counts Increased TPC, flavonoid, and anthocyanin content	[131]
Pasteurised broccoli juice	Fermentation by *Pediococcus pentosaceus* (isolated from fermented cherry juice and pickled pig's ear)	*P. pentosaceus* of animal origin	-	Increase riboflavin and β-carotene content Decrease free amino acids Both obtained high sulforaphane content	[132]
Pasteurised cucumber juice	Cinnamon, clove, mint, and ginger extracts (200–800 μg/mL)	200 μg/mL of clove extract	6 months RT	Better sensory attributes Higher TPC and flavonoid content Good retention for 6 months	[133]
Pasteurised peach juice	*Lactobacillus acidophilus* PTCC 1643 and *Lactobacillus fermentum* PTCC 1744	*L. acidophilus*	-	Inhibited Maillard reaction by 36.7%; 38.5% of anti-inflammatory activity Increased TAC by 74%	[134]

Table 2. *Cont.*

Beverage Type and Components	Fortification Conditions	Optimum Conditions	Shelf Life	Results	Reference
Pineapple-banana-apple smoothie	Moringa leaves (0–4.5%)	4.5% of moringa leaves	-	Increased vitamin C and E by 227% and 102%, respectively Highest TPC and TAC Lower sensorial quality	[135]
Pomegranate juice	0–0.1% of fish oil microcapsules by complex coacervation	0.1% fish oil microcapsules	42 d at 4 °C	16% of eicosapentaenoic acid and 11% of docosahexaenoic acid were released after 42 d Good sensory quality up to 0.07% microcapsules	[136]
Spinach-green apple-cucumber smoothie	2.5% of alga (*Chlorella vulgaris* and *Dunaliella salina*)	*D. salina*	28 d at 5 °C	Higher TPC and TAC Great sensory attributes	[137]
Strawberry-banana smoothie	Olive leaf extract (-OLE- 0–25 mg/100 g) + Sucrose (0–4 g/100 g)/sodium cyclamate (0–114.4 mg/100 g)/sodium chloride (0–40 mg/100 g)	OLE (20 mg/100 g) + sodium chloride (40 mg/100 g)	-	High TPC 40% less bitter taste perception	[138]
Tomato juice	Polyphenols from 0.5% of tomato extracts	Tomato extract (0.5%)	-	High TPC, lycopene, and β-carotene contents	[139]
Watermelon-apple-banana smoothie	Mint leaf extract (0–8%)	Mint leaf extract (8%)	-	High vitamin A, C, flavonoid, and TPC	[140]
Watermelon juice	Citric acid, malic acid, or lemon juice (pH = 3.8)	Non-centrifuged and addition of citric acid or lemon juice (pH = 3.8)	20 d at 4° C	Retention of sensory and functional qualities	[141]

TPC: total phenolic content; TAC: total antioxidant capacity; RT: room temperature.

6.1. Fortification of Beverages by Adding Plant Extracts and Other Health-Promoting Compounds

As is known, plants are a rich source of phytochemicals with great health-promoting properties. Such compounds may also be used for technological purposes, such as antimicrobials (extending microbiological shelf life and ensuring food safety), antioxidants (e.g., less quality degradation due to enzymatic systems), masking of undesired flavours (e.g., sucrose masks the bitter flavour of olive oil added into beverages), etc. In addition, the phytochemical content may be higher in the inedible plant parts (known in the food industry as by-products) compared with the edible parts. Therefore, their reuse becomes very important for the circular economy in the food industry, while minimising environmental impact and revalorization of phytochemicals for further use.

Fortification with added plant extracts leads to an almost proportional (to the added extract/compound quantity) increase in the level of beverage phytochemicals. However, interactions between different compounds (e.g., phenolic compounds) may lead to synergistic effects, as previously stated [124]. Nevertheless, special attention must be paid when adding some plant extracts (either by their sensory characteristics or the high concentrations used) since the sensory acceptance of the fortified beverage may be compromised [135,138].

In general, all studies published about beverage fortification by the addition of plant extracts have observed increases of their antioxidant properties, in most of cases related to their high phenolic compound contents (Table 2). Thus, beet leaf extract added to a fruit–vegetable smoothie (30% of total smoothie volume) increased both TPC and TAC by 50% [130]. Pale or black brewers' spent grain extract added (10%) to cranberry juice increased TAC by 120% [124]. Similarly, pomegranate peel dry extract (2.5 mg per mL of beverage) and clove extract (200 µg per mL of beverage) added to carrot juice and cucumber juice, respectively, enhanced the TPC, TAC and flavonoid contents of the beverages [120,133]. Fortification of the carotenoid (lycopene and β-carotene) level of tomato juice was achieved by the addition of a polyphenol extract (0.5% of the beverage) from tomatoes. Finally, fortification of beverages with vitamins from plant extracts prior to processing is an excellent strategy to counterbalance subsequent vitamin degradation (of high interest for very labile vitamins) during processing or subsequent storage of beverages. Thus, the vitamin C and vitamin A contents of a fruit smoothie were increased by the addition (8% of the beverage) of mint leaf extract [140].

Another strategy for beverage fortification is pre-enhancement of the phytochemical content in the plant raw materials used for the subsequent obtaining of the beverage. For this, several green (chemical-free) postharvest abiotic stresses have been used, such as wounding, UV-C, and hyperoxic atmospheres (Table 2). Thus, a carrot smoothie prepared from pre-incubated carrots under hyperoxia treatment (80 kPa O_2) increased the TPC of the smoothie by 2060% [121]. Similarly, the chlorogenic acid content of carrot juice was enhanced by 3600% when the unpeeled carrots were previously blanched (80 °C for 6 min) [122].

Fortification of beverages with pure compounds extracted from natural sources (plants and fish) have been also studied. The addition of some compounds may better preserve the phytochemicals contained in beverages. For example, the anthocyanin content of a blackberry juice was better retained for five weeks at 30 °C when glutathione was added (500 mg/L) [118]. Fortification of melon juice with (−)epicatechin (2.5 g/mL of the beverage) increased the TPC by ≈720% [128].

In this sense, Tarazona-Díaz & Aguayo [141] studied the effects of acidification, pasteurization, centrifugation, and refrigerated storage of watermelon juice (Table 2). Their results showed a minimal degradation of non-centrifuged juices stored at 4 °C, with particular richness in lycopene, polyphenols, and citrulline [141]. Indeed, the enrichment of Fashion watermelon juice by addition of L-citrulline (3.45 g per 500 mL) has shown to diminish muscle soreness perception from 24 to 72 h after a half-marathon race and to maintain lower concentrations of plasma lactate after exhausting exercise in amateur male runners [142]. Furthermore, lower (1.17 g per 500 mL) and higher (4.83 g per 500 mL) doses of L-citrulline in watermelon juices have reduced recovery heart rate and muscle soreness after 24 h muscle relief in athletes [143], which seems to be enhanced in combination with 22 mg pomegranate ellagitannins per 200 mL watermelon juice [144].

Nevertheless, special attention must be paid to the supplemented concentration since phytochemicals have limited bioavailability in the human body. For example, absorption of polyphenols may be relatively poor, ranging widely from 0.3 to 43%, leading to low circulating plasma concentrations [145]. For this reason, folic acid has been encapsulated with mesoporous silica particles (to be incorporated in fruit juices) in order to improve its stability, reduce the quantity of compound needed, and control release after consumption by modifying vitamin bioaccessibility [116]. Likewise, 16% of eicosanoid acid and 11% of docosahexaenoic acid were released in pomegranate juice when 0.1% of fish oil, microencapsulated by complex coacervation, was added to the juice [136].

6.2. Fortification of Beverages by Adding Algae

The culinary and health-promoting properties of marine algae, or seaweeds, have been known by Oriental cultures (mainly in Japan, China, and Korea) for centuries. In Western countries, algae use in pharmaceuticals, cosmetics, and food (mostly as thickening agents), together with its increasing acceptance as a culinary condiment, is growing day by day. Marine algae are rich sources of proteins, polysaccharides, minerals, vitamins and polyphenols, among other things [146,147], so their inclusion in beverage formulations may increase the phytochemical levels of those beverages [127]. More interestingly, these beverages may be fortified with unique health-promoting compounds of algae which are not found in plant products, for example, phlorotannins and fucoidans, among others [147]. However, special attention must be paid to the algae quantity added into the fortified beverages, since undesirable algae-related nuances may be detected. Thus, among nine green smoothies prepared with different marine algae, kombu- and wakame-fortified beverages showed the lowest sensory scores, mainly due to off-odours related to those algae [127].

Fortification of a fruit-beverage smoothie with microalga *Dunaliella salina* (2.5% of the beverage) increased the TPC and TAC while still being scored with great sensory quality [137]. Furthermore, the total sugar content of date nectar fortified with spirulina (10%) was increased, which is of high interest in masking some undesirable flavours, such as the bitter flavour of kale beverages [148]. The authors observed better sensory attributes in the date nectar fortified with spirulina.

6.3. Phytochemical Fortification of Beverages during Fermentation

Probiotics are living organisms whose ingestion provides several health benefits, including prevention of gastrointestinal diseases, promotion of antimicrobial activity, regulation of lactose metabolism, decrease in serum cholesterol levels, stimulation of the immune system, and anti-mutagenic and anti-carcinogenic effects, among others [131]. Lactic acid bacteria are widely used during food processing for fermentation, which gives typical flavours and other sensory aspects. Hence, much use of probiotics (e.g., *Lactobacillus, Bifidobacteria*, etc.), for fermentation and/or fortification purposes is made in the food industry to provide the consumer with beverages that have excellent quality and enhanced health-promoting properties.

The use of probiotics in beverages may also enhance their phytochemical content. For example, the phenolic and anthocyanin levels of blueberry juice were increased by 43% and 15%, respectively, after fermentation (24 h at 37 °C followed by 2 h at 4 °C) with *Lactobacillus plantarum* [119]. TPC and TAC were increased (by 6–8-fold) in different fruit juices after fermentation (2 h at 30 °C followed by 28 days at 4 °C) with *L. plantarum* [126]. Similarly, TPC as well as flavonoid and anthocyanin contents were increased in fruit–vegetable juice after fermentation (24 h at 37 °C followed by refrigeration at 4 °C) with *L. plantarum* [131]. Other *Lactobacillus* species also induced increases of phenolic (49% TPC increase after fermentation with *Lactobacillus paracasei*) [123], TAC (74% increase after fermentation with *Lactobacillus acidophilus*) [134] and other compounds like riboflavin, β-carotene and sulforaphane (broccoli juice fermented with *Pedioccoccus pentosaceus*) [132]. Probiotic survival under gastrointestinal conditions may be limited, with the strains being used having higher resistance to those conditions (acidic pH, enzymatic reactions, etc.); microencapsulation of probiotics to extend their survival may also be effective [117,131].

As observed, fermentation of beverages with probiotic bacteria may lead to large increases in their phytochemical contents, mainly of phenolic compounds. Such enhancement has been hypothesized to occur due to the disintegration of macromolecular polyphenol or anthocyanin structures into smaller phenols; this also increases their bioaccessibility through the specific metabolism (e.g., deglycosylation) of probiotic bacteria [119,131].

7. Conclusions

Fortification of fruit and vegetable beverages can be achieved by using green non-thermal technologies. While ultrasound is already a well-known technology used for cell disruption, UV and high-pressure processing may lead to similar increases in such beverage phytochemicals as phenolic compounds, anthocyanins, vitamin C, etc., ensuring high antioxidant properties. Pulsed electric fields and cold plasma are also green non-thermal technologies that are used to extend the shelf life of fruit and vegetable beverages; however, they are also capable of being used to fortify the phytochemical contents of these food products. Our analysis of published studies on this topic shows that these are excellent strategies to fortify the health-promoting compound contents of the plant products contained in these beverages. In addition, the optimized combination of the technologies used for treatment may highly enhance fortification rates, as most of the available literature relates to the combination of such new techniques with ultrasound. Based on this assessment, future studies may deepen understanding of different combinations of such technologies where optimal processing conditions may differ depending on beverage properties (rheology, composition, pH, etc.). Furthermore, a deeper elucidation should be sought as to whether the fortification observed with these technologies is due more to higher extractability after cell disruption or to biosynthetic reactions, for example, when substrate and enzymes come in contact.

Author Contributions: F.A.-H.; N.C.; L.M.-Z.; G.B.M.-H.: writing—original draft preparation, F.A.-H.; N.C.; L.M.-Z.; G.B.M.-H.: writing—review and editing, F.A.-H.; G.B.M.-H.: visualization, F.A.-H.: supervision. All authors have read and agreed to the published version of the manuscript.

Funding: This research received no external funding.

Institutional Review Board Statement: Not applicable.

Informed Consent Statement: Not applicable.

Data Availability Statement: Not applicable.

Acknowledgments: The authors are grateful to the Spanish Ministry of Economy and Competitiveness (MINECO) Project AGL2013−48830−C2−1−R, FEDER, and Sakata Seeds Ibérica S.L. for financial support in different projects.

Conflicts of Interest: The authors declare no conflict of interest.

References

1. DeFelice, S.L. The nutraceutical revolution: Its impact on food industry R&D. *Trends Food Sci. Technol.* **1995**, *6*, 59–61. [CrossRef]
2. Contor, L. Functional Food Science in Europe. *Nutr. Metab. Cardiovasc. Dis.* **2001**, *11*, 20–23. [CrossRef]
3. Guhr, G.; Lachance, P.A. Chapter 32: Role of Phytochemicals in Chronic Disease Prevention. 311-364. In *Nutraceuticals: Designer Foods III: Garlic, Soy and Licorice*; Lachance, P.A., Ed.; Wiley: Hoboken, NJ, USA, 1997. [CrossRef]
4. Télessy, I.G. Nutraceuticals. In *The Role of Functional Food Security in Global Health*; Elsevier Inc.: Amsterdam, The Netherlands, 2018; pp. 409–421. ISBN 978-0-12-813148-0. [CrossRef]
5. Infusino, F.; Marazzato, M.; Mancone, M.; Fedele, F.; Mastroianni, C.M.; Severino, P.; Ceccarelli, G.; Santinelli, L.; Cavarretta, E.; Marullo, A.G.M.; et al. in SARS-CoV-2 Infection: A Scoping Review. *Nutrients* **2020**, *12*, 1–21. [CrossRef] [PubMed]
6. Rodríguez-Verástegui, L.L.; Martínez-Hernández, G.B.; Castillejo, N.; Gómez, P.A.; Artés, F.; Artés-Hernández, F. Bioactive Compounds and Enzymatic Activity of Red Vegetable Smoothies during Storage. *Food Bioprocess Technol.* **2016**, 9. [CrossRef]
7. Morales-de la Peña, M.; Welti-Chanes, J.; Martín-Belloso, O. Application of Novel Processing Methods for Greater Retention of Functional Compounds in Fruit-Based Beverages. *Beverages* **2016**, *2*, 14. [CrossRef]

8. Shahbaz, H.M.; Kim, J.U.; Kim, S.H.; Park, J. *Advances in Nonthermal Processing Technologies for Enhanced Microbiological Safety and Quality of Fresh Fruit and Juice Products*; Grumezescu, A.M., Holban, A.M., Eds.; Elsevier Inc.: Amsterdam, The Netherlands, 2018; ISBN 9780128114476.
9. Buzrul, S.; Alpas, H.; Largeteau, A.; Demazeau, G. Inactivation of Escherichia coli and Listeria innocua in kiwifruit and pineapple juices by high hydrostatic pressure. *Int. J. Food Microbiol.* **2008**, *124*, 275–278. [CrossRef]
10. Di Cagno, R.; Minervini, G.; Rizzello, C.G.; De Angelis, M.; Gobbetti, M. Effect of lactic acid fermentation on antioxidant, texture, color and sensory properties of red and green smoothies. *Food Microbiol.* **2011**, *28*, 1062–1071. [CrossRef] [PubMed]
11. Arjmandi, M.; Otón, M.; Artés, F.; Artés-Hernández, F.; Gómez, P.A.; Aguayo, E. Semi-industrial microwave treatments positively affect the quality of orange-colored smoothies. *J. Food Sci. Technol.* **2016**, *53*, 3695–3703. [CrossRef]
12. Formica-Oliveira, A.C.; Martínez-Hernández, G.B.; Díaz-López, V.; Artés, F.; Artés-Hernández, F. Use of postharvest UV-B and UV-C radiation treatments to revalorize broccoli byproducts and edible florets. *Innov. Food Sci. Emerg. Technol.* **2017**, *43*, 77–83. [CrossRef]
13. Brewer, M.S. Natural Antioxidants: Sources, Compounds, Mechanisms of Action, and Potential Applications. *Compr. Rev. Food Sci. Food Saf.* **2011**, *10*, 221–247. [CrossRef]
14. Trichopoulou, A.; Martínez-González, M.A.; Tong, T.Y.N.; Forouhi, N.G.; Khandelwal, S.; Prabhakaran, D.; Mozaffarian, D.; de Lorgeril, M. Definitions and potential health benefits of the Mediterranean diet: Views from experts around the world. *BMC Med.* **2014**, *12*, 112. [CrossRef] [PubMed]
15. Martínez, L.; Ros, G.; Nieto, G. Hydroxytyrosol: Health Benefits and Use as Functional Ingredient in Meat. *Medicines* **2018**, *5*, 13. [CrossRef] [PubMed]
16. Vermeulen, M.; Klöpping-Ketelaars, I.W.A.A.; Van Den Berg, R.; Vaes, W.H.J. Bioavailability and kinetics of sulforaphane in humans after consumption of cooked versus raw broccoli. *J. Agric. Food Chem.* **2008**, *56*, 10505–10509. [CrossRef] [PubMed]
17. Lozanovski, V.J.; Houben, P.; Hinz, U.; Hackert, T.; Herr, I.; Schemmer, P. Pilot study evaluating broccoli sprouts in advanced pancreatic cancer (POUDER trial)—study protocol for a randomized controlled trial. *Trials* **2014**, *15*, 204. [CrossRef]
18. Parker, R.S.; Swanson, J.E.; You, C.-S.; Edwards, A.J.; Huang, T. Bioavailability of carotenoids in human subjects. *Proc. Nutr. Soc.* **1999**, *58*, 155–162. [CrossRef]
19. Desmarchelier, C.; Borel, P. Overview of carotenoid bioavailability determinants: From dietary factors to host genetic variations. *Trends Food Sci. Technol.* **2017**, *69*, 270–280. [CrossRef]
20. Martínez, L.; Ros, G.; Nieto, G. Fe, Zn and Se bioavailability in chicken meat emulsions enriched with minerals, hydroxytyrosol and extra virgin olive oil as measured by Caco-2 cell model. *Nutrients* **2018**, *10*, 969. [CrossRef]
21. Robles-Almazan, M.; Pulido-Moran, M.; Moreno-Fernandez, J.; Ramirez-Tortosa, C.; Rodriguez-Garcia, C.; Quiles, J.L.; Ramirez-Tortosa, M. Hydroxytyrosol: Bioavailability, toxicity, and clinical applications. *Food Res. Int.* **2018**, *105*, 654–667. [CrossRef]
22. Cooperstone, J.L.; Ralston, R.A.; Riedl, K.M.; Haufe, T.C.; Schweiggert, R.M.; King, S.A.; Timmers, C.D.; Francis, D.M.; Lesinski, G.B.; Clinton, S.K.; et al. Enhanced bioavailability of lycopene when consumed as cis-isomers from tangerine compared to red tomato juice, a randomized, cross-over clinical trial. *Mol. Nutr. Food Res.* **2015**, *59*, 658–669. [CrossRef]
23. Gallego-Jara, J.; de Diego, T.; del Real, Á.; Écija-Conesa, A.; Manjón, A.; Cánovas, M. Lycopene overproduction and in situ extraction in organic-aqueous culture systems using a metabolically engineered Escherichia coli. *AMB Express* **2015**, *5*, 1–11. [CrossRef]
24. Martínez-Hernández, G.B.; Gómez, P.A.; Talavera, N.V.G.; Hernández, F.A.; Saiz, T.M.; Álvarez, C.S.; Artés, F. Vitamin C, folates and antioxidant plasma status after consumption of kailan—hybrid broccoli. Consumption of kalian—hybrid broccoli cooking effects. *ACS Food Sci. Technol.* **2021**, *62*, 327–337. [CrossRef]
25. Jayachandran, L.E.; Chakraborty, S.; Rao, P.S. Effect of high pressure processing on physicochemical properties and bioactive compounds in litchi based mixed fruit beverage. *Innov. Food Sci. Emerg. Technol.* **2015**, *28*, 1–9. [CrossRef]
26. Fernandez, M.V.; Denoya, G.I.; Agüero, M.V.; Jagus, R.J.; Vaudagna, S.R. Optimization of high pressure processing parameters to preserve quality attributes of a mixed fruit and vegetable smoothie. *Innov. Food Sci. Emerg. Technol.* **2018**, *47*, 170–179. [CrossRef]
27. Arjmandi, M.; Otón, M.; Artés, F.; Artés-Hernández, F.; Gómez, P.A.; Aguayo, E. Continuous microwave pasteurization of a vegetable smoothie improves its physical quality and hinders detrimental enzyme activity. *Food Sci. Technol. Int.* **2017**, *23*, 36–45. [CrossRef] [PubMed]
28. Markowski, J.; Celejewska, K.; Rosłonek, A.; Kosmala, M. Impact of different thermal preservation technologies on the quality of apple-based smoothies. *LWT—Food Sci. Technol.* **2017**, *85*, 470–473. [CrossRef]
29. Amador-Espejo, G.G.; Chávez-Ocegueda, J.; Cruz-Cansino, N.; Suárez-Jacobo, A.; Gutiérrez-Martínez, P.; Valencia-Flores, D.; Velázquez Estrada, R. Thermosonication parameter effects on physicochemical changes, microbial and enzymatic inactivation of fruit smoothie. *J. Food Sci. Technol.* **2020**, *57*, 1680–1688. [CrossRef] [PubMed]
30. Ribeiro, L.O.; Brígida, A.I.S.; Sá, D.D.G.C.F.; Carvalho, C.W.P.; Silva, J.P.L.; Matta, V.M.; Freitas, S.P. Effect of sonication on the quality attributes of juçara, banana and strawberry smoothie. *J. Food Sci. Technol.* **2019**, *56*, 5531–5537. [CrossRef] [PubMed]
31. Carbonell-Capella, J.M.; Buniowska, M.; Barba, F.J.; Grimi, N.; Vorobiev, E.; Esteve, M.J.; Frígola, A. Changes of Antioxidant Compounds in a Fruit Juice-Stevia rebaudiana Blend Processed by Pulsed Electric Technologies and Ultrasound. *Food Bioprocess Technol.* **2016**, *9*, 1159–1168. [CrossRef]

32. Rabelo, M.C.; Bang, W.Y.; Nair, V.; Alves, R.E.; Jacobo-Velázquez, D.A.; Sreedharan, S.; de Miranda, M.R.A.; Cisneros-Zevallos, L. UVC light modulates vitamin C and phenolic biosynthesis in acerola fruit: Role of increased mitochondria activity and ROS production. *Sci. Rep.* **2020**, *10*, 1–14. [CrossRef]
33. Moreira-Rodríguez, M.; Nair, V.; Benavides, J.; Cisneros-Zevallos, L.; Jacobo-Velázquez, D.A. UVA, UVB light doses and harvesting time differentially tailor glucosinolate and phenolic profiles in broccoli sprouts. *Molecules* **2017**, *22*, 1065. [CrossRef]
34. Moreira-Rodríguez, M.; Nair, V.; Benavides, J.; Cisneros-Zevallos, L.; Jacobo-Velázquez, D.A. UVA, UVB Light, and Methyl Jasmonate, Alone or Combined, Redirect the Biosynthesis of Glucosinolates, Phenolics, Carotenoids, and Chlorophylls in Broccoli Sprouts. *Int. J. Mol. Sci.* **2017**, *18*, 2330. [CrossRef]
35. Castillejo, N.; Martínez-Zamora, L.; Artés–Hernández, F. Periodical Uv-B Radiation Hormesis in Biosynthesis of Kale Sprouts Nutraceuticals. *Plant Physiol. Biochem.* **2021**, *165*, 274–285. [CrossRef]
36. Martínez-Zamora, L.; Castillejo, N.; Artés, F. Postharvest UV-B and Photoperiod with Blue + Red LEDs as Strategies to Stimulate Carotenogenesis in Bell Peppers. *Appl. Sci.* **2021**, *11*, 3736. [CrossRef]
37. Formica-Oliveira, A.C.; Martínez-Hernández, G.B.; Díaz-López, V.; Artés, F.; Artés-Hernández, F. Effects of UV-B and UV-C combination on phenolic compounds biosynthesis in fresh-cut carrots. *Postharvest Biol. Technol.* **2017**, *127*, 99–104. [CrossRef]
38. Pennisi, G.; Blasioli, S.; Cellini, A.; Maia, L.; Crepaldi, A.; Braschi, I.; Spinelli, F.; Nicola, S.; Fernandez, J.A.; Stanghellini, C.; et al. Unraveling the role of red:Blue LED lights on resource use efficiency and nutritional properties of indoor grown sweet basil. *Front. Plant Sci.* **2019**, *10*, 1–15. [CrossRef] [PubMed]
39. De Oliveira, T.L.C.; Ramos, A.L.S.; Ramos, E.M.; Piccoli, R.H.; Cristianini, M. Natural antimicrobials as additional hurdles to preservation of foods by high pressure processing. *Trends Food Sci. Technol.* **2015**, *45*, 60–85. [CrossRef]
40. Wang, C.Y.; Huang, H.W.; Hsu, C.P.; Yang, B.B. Recent Advances in Food Processing Using High Hydrostatic Pressure Technology. *Crit. Rev. Food Sci. Nutr.* **2016**, *56*, 527–540. [CrossRef]
41. Raso, J.; Condón, S.; Álvarez, I. Non-Thermal Processing: Pulsed Electric Field. *Encycl. Food Microbiol. Second Ed.* **2014**, *2*, 966–973. [CrossRef]
42. Swamy, G.J.; Muthukumarappan, K. *Microwave and Radiofrequency Processing of Plant-Related Food Products*; Elsevier: Amsterdam, The Netherlands, 2021; ISBN 9780128157817.
43. Datta, A.K.; Davidson, P.M. Microwave and radio frequency processing. *J. Food Sci.* **2000**, *65*, 32–41. [CrossRef]
44. López, M.; Calvo, T.; Prieto, M.; Múgica-Vidal, R.; Muro-Fraguas, I.; Alba-Elías, F.; Alvarez-Ordóñez, A. A review on non-thermal atmospheric plasma for food preservation: Mode of action, determinants of effectiveness, and applications. *Front. Microbiol.* **2019**, *10*, 1–21. [CrossRef] [PubMed]
45. Tappi, S.; Berardinelli, A.; Ragni, L.; Dalla Rosa, M.; Guarnieri, A.; Rocculi, P. Atmospheric gas plasma treatment of fresh-cut apples. *Innov. Food Sci. Emerg. Technol.* **2014**, *21*, 114–122. [CrossRef]
46. Pankaj, S.K.; Keener, K.M. *Cold Plasma Processing of Fruit Juices*; Elsevier Inc.: Amsterdam, The Netherlands, 2018; ISBN 9780128024911.
47. Bourke, P.; Ziuzina, D.; Han, L.; Cullen, P.J.; Gilmore, B.F. Microbiological interactions with cold plasma. *J. Appl. Microbiol.* **2017**, *123*, 308–324. [CrossRef] [PubMed]
48. Angelova, Z.; Georgiev, S.; Roos, W. Elicitation of plants. *Biotechnol. Biotechnol. Equip.* **2006**, *20*, 72–83. [CrossRef]
49. Keen, N.T. Specific Elicitors of Plant Phytoalexin Production: Detenninants of Race Specificity in Pathogens? *Science* **1975**, *187*, 74–75. [CrossRef]
50. Cisneros-Zevallos, L. The use of controlled postharvest abiotic stresses as a tool for enhancing the nutraceutical content and adding-value of fresh fruits and vegetables. *J. Food Sci.* **2003**, *68*, 1560–1565. [CrossRef]
51. Jacobo-Velázquez, D.A.; Cisneros-Zevallos, L. An alternative use of horticultural crops: Stressed plants as biofactories of bioactive phenolic compounds. *Agriculture* **2012**, *2*, 259–271. [CrossRef]
52. Jacobo-Velázquez, D.A.; Cisneros-Zevallos, L. Correction: An alternative use of horticultural crops: Stressed plants as biofactories of bioactive phenolic compounds. *Agriculture* **2013**, *3*, 596–598. [CrossRef]
53. Surjadinata, B.B.; Cisneros-Zevallos, L. Biosynthesis of phenolic antioxidants in carrot tissue increases with wounding intensity. *Food Chem.* **2012**, *134*, 615–624. [CrossRef]
54. Surjadinata, B.B.; Jacobo-Velázquez, D.A.; Cisneros-Zevallos, L. UVA, UVB and UVC light enhances the biosynthesis of phenolic antioxidants in fresh-cut carrot through a synergistic effect with wounding. *Molecules* **2017**, *22*, 668. [CrossRef] [PubMed]
55. Ortega-Hernández, E.; Nair, V.; Welti-Chanes, J.; Cisneros-Zevallos, L.; Jacobo-Velázquez, D.A. Wounding and UVB light synergistically induce the biosynthesis of phenolic compounds and ascorbic acid in red prickly pears (Opuntia ficusindica cv. Rojo Vigor). *Int. J. Mol. Sci.* **2019**, *20*, 5327. [CrossRef] [PubMed]
56. Martínez-Zamora, L.; Castillejo, N.; Artés-Hernández, F. Postharvest UV-B and UV-C radiation enhanced the biosynthesis of glucosinolates and isothiocyanates in brassicaceae sprouts. *Postharvest Biol. Technol.* **2021**, *181*, 1–11. [CrossRef]
57. Formica-Oliveira, A.C.; Martínez-Hernández, G.B.; Aguayo, E.; Gómez, P.A.; Artés, F.; Artés-Hernández, F. UV-C and hyperoxia abiotic stresses to improve healthiness of carrots: Study of combined effects. *J. Food Sci. Technol.* **2016**, *53*, 3465–3476. [CrossRef] [PubMed]
58. Tomás-Callejas, A.; Otón, M.; Artés, F.; Artés-Hernández, F. Combined effect of UV-C pretreatment and high oxygen packaging for keeping the quality of fresh-cut Tatsoi baby leaves. *Innov. Food Sci. Emerg. Technol.* **2012**, *14*, 115–121. [CrossRef]

59. Pérez-Balibrea, S.; Moreno, D.A.; García-Viguera, C. Improving the phytochemical composition of broccoli sprouts by elicitation. *Food Chem.* **2011**, *129*, 35–44. [CrossRef]
60. Martínez-Hernández, G.B.; Huertas, J.-P.; Navarro-Rico, J.; Gómez, P.A.; Artés, F.; Palop, A.; Artés-Hernández, F. Inactivation kinetics of foodborne pathogens by UV-C radiation and its subsequent growth in fresh-cut kailan-hybrid broccoli. *Food Microbiol.* **2015**, *46*, 263–271. [CrossRef]
61. Pierscianowski, J.; Popović, V.; Biancaniello, M.; Bissonnette, S.; Zhu, Y.; Koutchma, T. Continuous-flow UV-C processing of kale juice for the inactivation of E. coli and assessment of quality parameters. *Food Res. Int.* **2021**, *140*, 110085. [CrossRef] [PubMed]
62. García Carrillo, M.; Ferrario, M.; Guerrero, S.; García Carrillo, M. Study of the inactivation of some microorganisms in turbid carrot-orange juice blend processed by ultraviolet light assisted by mild heat treatment. *J. Food Eng.* **2017**, *212*, 213–225. [CrossRef]
63. Fundo, J.F.; Miller, F.A.; Mandro, G.F.; Tremarin, A.; Brandão, T.R.S.; Silva, C.L.M. UV-C light processing of Cantaloupe melon juice: Evaluation of the impact on microbiological, and some quality characteristics, during refrigerated storage. *LWT* **2019**, *103*, 247–252. [CrossRef]
64. Amanina, A.K.Z.; Rosnah, S.; Noranizan, M.A.; Alifdalino, S. Comparison of UV-C and thermal pasteurisation for the quality preservation of pineapple-mango juice blend. *Food Res.* **2019**, *3*, 362–372. [CrossRef]
65. Türkmen, F.U.; Takci, H.A.M. Ultraviolet-C and ultraviolet-B lights effect on black carrot (Daucus carota ssp. sativus) juice. *J. Food Meas. Charact.* **2018**, *12*, 1038–1046. [CrossRef]
66. Baykuş, G.; Akgün, M.P.; Unluturk, S. Effects of ultraviolet-light emitting diodes (UV-LEDs) on microbial inactivation and quality attributes of mixed beverage made from blend of carrot, carob, ginger, grape and lemon juice. *Innov. Food Sci. Emerg. Technol.* **2021**, *67*, 102572. [CrossRef]
67. Riganakos, K.A.; Karabagias, I.K.; Gertzou, I.; Stahl, M. Comparison of UV-C and thermal treatments for the preservation of carrot juice. *Innov. Food Sci. Emerg. Technol.* **2017**, *42*, 165–172. [CrossRef]
68. Hurtado, A.; Guàrdia, M.D.; Picouet, P.; Jofré, A.; Bañón, S.; Ros, J.M. Shelf-life extension of multi-vegetables smoothies by high-pressure processing compared with thermal treatment. Part I: Microbial and enzyme inhibition, antioxidant status, and physical stability. *J. Food Process. Preserv.* **2019**, *43*, e14139. [CrossRef]
69. Hurtado, A.; Guàrdia, M.D.; Picouet, P.; Jofré, A.; Bañón, S.; Ros, J.M. Shelf-life extension of multi-vegetables smoothies by high pressure processing compared with thermal treatment. Part II: Retention of selected nutrients and sensory quality. *J. Food Process. Preserv.* **2019**, *43*, e14210. [CrossRef]
70. Hurtado, A.; Picouet, P.; Jofré, A.; Guàrdia, M.D.; Ros, J.M.; Bañón, S. Application of High Pressure Processing for Obtaining "Fresh-Like" Fruit Smoothies. *Food Bioprocess Technol.* **2015**, *8*, 2470–2482. [CrossRef]
71. Keenan, D.F.; Rößle, C.; Gormley, R.; Butler, F.; Brunton, N.P. Effect of high hydrostatic pressure and thermal processing on the nutritional quality and enzyme activity of fruit smoothies. *LWT—Food Sci. Technol.* **2012**, *45*, 50–57. [CrossRef]
72. Zacconi, C.; Giosuè, S.; Marudelli, M.; Scolari, G. Microbiological quality and safety of smoothies treated in different pressure–temperature domains: Effects on indigenous fruit microbiota and Listeria monocytogenes and their survival during storage. *Eur. Food Res. Technol.* **2015**, *241*, 317–328. [CrossRef]
73. Zhang, Y.; Liu, X.C.; Wang, Y.; Zhao, F.; Sun, Z.; Liao, X. Quality comparison of carrot juices processed by high-pressure processing and high-temperature short-time processing. *Innov. Food Sci. Emerg. Technol.* **2016**, *33*, 135–144. [CrossRef]
74. Formica-Oliveira, A.C.; Martínez-Hernández, G.B.; Díaz-López, V.; Otón, M.; Artés, F.; Artés-Hernández, F. High Hydrostatic Pressure Treatments for Keeping Quality of Orange Vegetables Smoothies. *Acta Hortic.* **2018**, *1194*, 575–580. [CrossRef]
75. Pallarés, N.; Berrada, H.; Tolosa, J.; Ferrer, E. Effect of high hydrostatic pressure (HPP) and pulsed electric field (PEF) technologies on reduction of aflatoxins in fruit juices. *LWT* **2021**, *142*, 111000. [CrossRef]
76. Raj, A.S.; Chakraborty, S.; Rao, P.S. Thermal assisted high-pressure processing of Indian gooseberry (*Embilica officinalis* L.) juice Impact on colour and nutritional attributes. *LWT—Food Sci. Technol.* **2019**, *99*, 119–127. [CrossRef]
77. Moreira, R.M.; Martins, M.L.; de Castro Leite Júnior, B.R.; Martins, E.M.F.; Ramos, A.M.; Cristianini, M.; da Rocha Campos, A.N.; Stringheta, P.C.; Silva, V.R.O.; Canuto, J.W.; et al. Development of a juçara and Ubá mango juice mixture with added Lactobacillus rhamnosus GG processed by high pressure. *LWT—Food Sci. Technol.* **2017**, *77*, 259–268. [CrossRef]
78. De Ancos, B.; Rodrigo, M.J.; Sánchez-Moreno, C.; Pilar Cano, M.; Zacarías, L. Effect of high-pressure processing applied as pretreatment on carotenoids, flavonoids and vitamin C in juice of the sweet oranges "Navel" and the red-fleshed "Cara Cara". *Food Res. Int.* **2020**, *132*, 109105. [CrossRef] [PubMed]
79. Barba, F.J.; Esteve, M.J.; Frigola, A. Ascorbic acid is the only bioactive that is better preserved by high hydrostatic pressure than by thermal treatment of a vegetable beverage. *J. Agric. Food Chem.* **2010**, *58*, 10070–10075. [CrossRef] [PubMed]
80. Saldaña, G.; Puértolas, E.; Monfort, S.; Raso, J.; Álvarez, I. Defining treatment conditions for pulsed electric field pasteurization of apple juice. *Int. J. Food Microbiol.* **2011**, *151*, 29–35. [CrossRef]
81. Timmermans, R.A.H.; Nederhoff, A.L.; Nierop Groot, M.N.; van Boekel, M.A.J.S.; Mastwijk, H.C. Effect of electrical field strength applied by PEF processing and storage temperature on the outgrowth of yeasts and moulds naturally present in a fresh fruit smoothie. *Int. J. Food Microbiol.* **2016**, *230*, 21–30. [CrossRef]
82. Aadil, R.M.; Zeng, X.A.; Sun, D.W.; Wang, M.S.; Liu, Z.W.; Zhang, Z.H. Combined effects of sonication and pulsed electric field on selected quality parameters of grapefruit juice. *LWT—Food Sci. Technol.* **2015**, *62*, 890–893. [CrossRef]

83. Carbonell-Capella, J.M.; Buniowska, M.; Cortes, C.; Zulueta, A.; Frigola, A.; Esteve, M.J. Influence of pulsed electric field processing on the quality of fruit juice beverages sweetened with Stevia rebaudiana. *Food Bioprod. Process.* **2017**, *101*, 214–222. [CrossRef]
84. Abid, M.; Jabbar, S.; Wu, T.; Hashim, M.M.; Hu, B.; Lei, S.; Zeng, X. Sonication enhances polyphenolic compounds, sugars, carotenoids and mineral elements of apple juice. *Ultrason. Sonochem.* **2014**, *21*, 93–97. [CrossRef] [PubMed]
85. Khan, M.K.; Asif, M.N.; Ahmad, M.H.; Imran, M.; Arshad, M.S.; Hassan, S.; Khan, M.I.; Nisa, M.U.; Iqbal, M.M.; Muhammad, N. Ultrasound-Assisted Optimal Development and Characterization of Stevia-Sweetened Functional Beverage. *J. Food Qual.* **2019**, *2019*. [CrossRef]
86. Keenan, D.F.; Tiwari, B.K.; Patras, A.; Gormley, R.; Butler, F.; Brunton, N.P. Effect of sonication on the bioactive, quality and rheological characteristics of fruit smoothies. *Int. J. Food Sci. Technol.* **2012**, *47*, 827–836. [CrossRef]
87. Jabbar, S.; Abid, M.; Hu, B.; Wu, T.; Hashim, M.M.; Lei, S.; Zhu, X.; Zeng, X. Quality of carrot juice as influenced by blanching and sonication treatments. *LWT—Food Sci. Technol.* **2014**, *55*, 16–21. [CrossRef]
88. Ahmad, K.; Imran, M.; Ahmad, T.; Ahmad, M.H.; Khan, M.K. Impact of ultrasound processing on physicochemical and bioactive attributes of grape based optimized fruit beverage. *Pakistan J. Agric. Sci.* **2020**, *57*, 1117–1124. [CrossRef]
89. Aadil, R.M.; Zeng, X.A.; Wang, M.S.; Liu, Z.W.; Han, Z.; Zhang, Z.H.; Hong, J.; Jabbar, S. A potential of ultrasound on minerals, micro-organisms, phenolic compounds and colouring pigments of grapefruit juice. *Int. J. Food Sci. Technol.* **2015**, *50*, 1144–1150. [CrossRef]
90. Guerrouj, K.; Sánchez-Rubio, M.; Taboada-Rodríguez, A.; Cava-Roda, R.M.; Marín-Iniesta, F. Sonication at mild temperatures enhances bioactive compounds and microbiological quality of orange juice. *Food Bioprod. Process.* **2016**, *99*, 20–28. [CrossRef]
91. De Albuquerque, J.G.; Escalona-Buendía, H.B.; de Magalhães Cordeiro, A.M.T.; dos Santos Lima, M.; de Souza Aquino, J.; da Silva Vasconcelos, M.A. Ultrasound treatment for improving the bioactive compounds and quality properties of a Brazilian nopal (Opuntia ficus-indica) beverage during shelf-life. *LWT—Food Sci. Technol.* **2021**, *149*, 111814. [CrossRef]
92. Mehta, D.; Sharma, N.; Bansal, V.; Sangwan, R.S.; Yadav, S.K. Impact of ultrasonication, ultraviolet and atmospheric cold plasma processing on quality parameters of tomato-based beverage in comparison with thermal processing. *Innov. Food Sci. Emerg. Technol.* **2019**, *52*, 343–349. [CrossRef]
93. Surowsky, B.; Fröhling, A.; Gottschalk, N.; Schlüter, O.; Knorr, D. Impact of cold plasma on Citrobacter freundii in apple juice: Inactivation kinetics and mechanisms. *Int. J. Food Microbiol.* **2014**, *174*, 63–71. [CrossRef]
94. Rodríguez, Ó.; Gomes, W.F.; Rodrigues, S.; Fernandes, F.A.N. Effect of indirect cold plasma treatment on cashew apple juice (Anacardium occidentale L.). *LWT—Food Sci. Technol.* **2017**, *84*, 457–463. [CrossRef]
95. Illera, A.E.; Chaple, S.; Sanz, M.T.; Ng, S.; Lu, P.; Jones, J.; Carey, E.; Bourke, P. Effect of cold plasma on polyphenol oxidase inactivation in cloudy apple juice and on the quality parameters of the juice during storage. *Food Chem. X* **2019**, *3*, 100049. [CrossRef]
96. Hou, Y.; Wang, R.; Gan, Z.; Shao, T.; Zhang, X.; He, M.; Sun, A. Effect of cold plasma on blueberry juice quality. *Food Chem.* **2019**, *290*, 79–86. [CrossRef]
97. Bursać Kovačević, D.; Gajdoš Kljusurić, J.; Putnik, P.; Vukušić, T.; Herceg, Z.; Dragović-Uzelac, V. Stability of polyphenols in chokeberry juice treated with gas phase plasma. *Food Chem.* **2016**, *212*, 323–331. [CrossRef] [PubMed]
98. Gabriel, A.A.; Aba, R.P.M.; Tayamora, D.J.L.; Colambo, J.C.R.; Siringan, M.A.T.; Rosario, L.M.D.; Tumlos, R.B.; Ramos, H.J. Reference organism selection for microwave atmospheric pressure plasma jet treatment of young coconut liquid endosperm. *Food Control* **2016**, *69*, 74–82. [CrossRef]
99. Shi, X.M.; Zhang, G.J.; Wu, X.L.; Li, Y.X.; Ma, Y.; Shao, X.J. Effect of low-temperature plasma on microorganism inactivation and quality of freshly squeezed orange juice. *IEEE Trans. Plasma Sci.* **2011**, *39*, 1591–1597. [CrossRef]
100. Xu, L.; Garner, A.L.; Tao, B.; Keener, K.M. Microbial Inactivation and Quality Changes in Orange Juice Treated by High Voltage Atmospheric Cold Plasma. *Food Bioprocess Technol.* **2017**, *10*, 1778–1791. [CrossRef]
101. Almeida, F.D.L.; Cavalcante, R.S.; Cullen, P.J.; Frias, J.M.; Bourke, P.; Fernandes, F.A.N.; Rodrigues, S. Effects of atmospheric cold plasma and ozone on prebiotic orange juice. *Innov. Food Sci. Emerg. Technol.* **2015**, *32*, 127–135. [CrossRef]
102. Herceg, Z.; Kovačević, D.B.; Kljusurić, J.G.; Jambrak, A.R.; Zorić, Z.; Dragović-Uzelac, V. Gas phase plasma impact on phenolic compounds in pomegranate juice. *Food Chem.* **2016**, *190*, 665–672. [CrossRef]
103. Elez Garofulić, I.; Režek Jambrak, A.; Milošević, S.; Dragović-Uzelac, V.; Zorić, Z.; Herceg, Z. The effect of gas phase plasma treatment on the anthocyanin and phenolic acid content of sour cherry Marasca (*Prunus cerasus* var. *Marasca*) juice. *LWT—Food Sci. Technol.* **2015**, *62*, 894–900. [CrossRef]
104. Ma, T.J.; Lan, W.S. Effects of non-thermal plasma sterilization on volatile components of tomato juice. *Int. J. Environ. Sci. Technol.* **2015**, *12*, 3767–3772. [CrossRef]
105. Pankaj, S.K.; Wan, Z.; Colonna, W.; Keener, K.M. Effect of high voltage atmospheric cold plasma on white grape juice quality. *J. Sci. Food Agric.* **2017**, *97*, 4016–4021. [CrossRef]
106. Ferrario, M.; Alzamora, S.M.; Guerrero, S. Study of the inactivation of spoilage microorganisms in apple juice by pulsed light and ultrasound. *Food Microbiol.* **2015**, *46*, 635–642. [CrossRef]
107. Tremarin, A.; Brandão, T.R.S.; Silva, C.L.M. Application of ultraviolet radiation and ultrasound treatments for Alicyclobacillus acidoterrestris spores inactivation in apple juice. *LWT—Food Sci. Technol.* **2017**, *78*, 138–142. [CrossRef]

108. Sauceda-Gálvez, J.N.; Codina-Torrella, I.; Martinez-Garcia, M.; Hernández-Herrero, M.M.; Gervilla, R.; Roig-Sagués, A.X. Combined effects of ultra-high pressure homogenization and short-wave ultraviolet radiation on the properties of cloudy apple juice. *LWT* **2021**, *136*, 110286. [CrossRef]
109. Umair, M.; Jabbar, S.; Senan, A.M.; Sultana, T.; Nasiru, M.M.; Shah, A.A.; Zhuang, H.; Jianhao, Z. Influence of Combined E ff ect of Ultra-Sonication and High-Voltage Cold Plasma Treatment on Quality. *Foods* **2019**, *8*, 593. [CrossRef]
110. Gomes, W.F.; Tiwari, B.K.; Rodriguez, Ó.; de Brito, E.S.; Fernandes, F.A.N.; Rodrigues, S. Effect of ultrasound followed by high pressure processing on prebiotic cranberry juice. *Food Chem.* **2017**, *218*, 261–268. [CrossRef] [PubMed]
111. Wang, J.; Xie, B.; Sun, Z. Quality parameters and bioactive compound bioaccessibility changes in probiotics fermented mango juice using ultraviolet-assisted ultrasonic pre-treatment during cold storage. *LWT* **2021**, *137*, 110438. [CrossRef]
112. Walkling-Ribeiro, M.; Noci, F.; Cronin, D.A.; Lyng, J.G.; Morgan, D.J. Shelf life and sensory evaluation of orange juice after exposure to thermosonication and pulsed electric fields. *Food Bioprod. Process.* **2009**, *87*, 102–107. [CrossRef]
113. Kouniaki, S.; Kajda, P.; Zabetakis, I. The effect of high hydrostatic pressure on anthocyanins and ascorbic acid in blackcurrants (Ribes nigrum). *Flavour Fragr. J.* **2004**, *19*, 281–286. [CrossRef]
114. Carbonell-Capella, J.M.; Barba, F.J.; Esteve, M.J.; Frígola, A. High pressure processing of fruit juice mixture sweetened with Stevia rebaudiana Bertoni: Optimal retention of physical and nutritional quality. *Innov. Food Sci. Emerg. Technol.* **2013**, *18*, 48–56. [CrossRef]
115. Artés-Hernández, F.; Martínez-Hernández, G.B.; Aguayo, E.; Gómez, P.A.; Artés, F. Fresh-cut fruit and vegetables: Emerging eco-friendly techniques for sanitation and preserving safety. In *Postharvest Handling*; Kahramanoglu, I., Ed.; InTech: London, UK, 2017; pp. 7–45.
116. Ruiz-Rico, M.; Pérez-Esteve, É.; Lerma-García, M.J.; Marcos, M.D.; Martínez-Máñez, R.; Barat, J.M. Protection of folic acid through encapsulation in mesoporous silica particles included in fruit juices. *Food Chem.* **2017**, *218*, 471–478. [CrossRef] [PubMed]
117. Da Silva, T.M.; Pinto, V.S.; Soares, V.R.F.; Marotz, D.; Cichoski, A.J.; Zepka, L.Q.; Jacob Lopes, E.; de Bona da Silva, C.; de Menezes, C.R. Viability of microencapsulated Lactobacillus acidophilus by complex coacervation associated with enzymatic crosslinking under application in different fruit juices. *Food Res. Int.* **2021**, *141*, 110190. [CrossRef]
118. Stebbins, N.B.; Howard, L.R.; Prior, R.L.; Brownmiller, C.; Mauromoustakos, A. Stabilization of anthocyanins in blackberry juice by glutathione fortification. *Food Funct.* **2017**, *8*, 3459–3468. [CrossRef] [PubMed]
119. Zhang, Y.; Liu, W.; Wei, Z.; Yin, B.; Man, C.; Jiang, Y. Enhancement of functional characteristics of blueberry juice fermented by Lactobacillus plantarum. *LWT* **2021**, *139*, 110590. [CrossRef]
120. Trigo, J.P.; Alexandre, E.M.C.; Oliveira, A.; Saraiva, J.A.; Pintado, M. Fortification of carrot juice with a high-pressure-obtained pomegranate peel extract: Chemical, safety and sensorial aspects. *Int. J. Food Sci. Technol.* **2020**, *55*, 1599–1605. [CrossRef]
121. Formica-Oliveira, A.C.; Martínez-Hernández, G.B.; Aguayo, E.; Gómez, P.A.; Artés, F.; Artés-Hernández, F. A Functional Smoothie from Carrots with Induced Enhanced Phenolic Content. *Food Bioprocess Technol.* **2017**, *10*, 491–502. [CrossRef]
122. Santana-Gálvez, J.; Santacruz, A.; Cisneros-Zevallos, L.; Jacobo-Velázquez, D.A. Postharvest Wounding Stress in Horticultural Crops as a Tool for Designing Novel Functional Foods and Beverages with Enhanced Nutraceutical Content: Carrot Juice as a Case Study. *J. Food Sci.* **2019**, *84*, 1151–1161. [CrossRef] [PubMed]
123. Bontsidis, C.; Mallouchos, A.; Terpou, A.; Nikolaou, A.; Batra, G.; Mantzourani, I.; Alexopoulos, A.; Plessas, S. Microbiological and chemical properties of chokeberry juice fermented by novel lactic acid bacteria with potential probiotic properties during fermentation at 4oc for 4 weeks. *Foods* **2021**, *10*, 768. [CrossRef] [PubMed]
124. McCarthy, A.L.; O'Callaghan, Y.C.; Neugart, S.; Piggott, C.O.; Connolly, A.; Jansen, M.A.K.; Krumbein, A.; Schreiner, M.; FitzGerald, R.J.; O'Brien, N.M. The hydroxycinnamic acid content of barley and brewers' spent grain (BSG) and the potential to incorporate phenolic extracts of BSG as antioxidants into fruit beverages. *Food Chem.* **2013**, *141*, 2567–2574. [CrossRef] [PubMed]
125. Aljobair, M.O.; Albaridi, N.A.; Alkuraieef, A.N.; AlKehayez, N.M. Physicochemical properties, nutritional value, and sensory attributes of a nectar developed using date palm puree and spirulina. *Int. J. Food Prop.* **2021**, *24*, 845–858. [CrossRef]
126. Dimitrellou, D.; Kandylis, P.; Kokkinomagoulos, E.; Hatzikamari, M.; Bekatorou, A. Emmer-based beverage fortified with fruit juices. *Appl. Sci.* **2021**, *11*, 3116. [CrossRef]
127. Castillejo, N.; Martínez-Hernández, G.B.; Goffi, V.; Gómez, P.A.; Aguayo, E.; Artés, F.; Artés-Hernández, F. Natural vitamin B12 and fucose supplementation of green smoothies with edible algae and related quality changes during their shelf life. *J. Sci. Food Agric.* **2018**, *98*, 2411–2421. [CrossRef]
128. Rúa Aller, J.; González González, S.; Sanz Gómez, J.; del Valle Fernandez, M.P.; García-Armesto, M.R. Assessment of (-) epicatechin as natural additive for improving safety and functionality in fresh "Piel de Sapo" melon juice. *Food Sci. Nutr.* **2021**, *9*, 2925–2935. [CrossRef] [PubMed]
129. Andrade, M.A.; de Oliveira Torres, L.R.; Silva, A.S.; Barbosa, C.H.; Vilarinho, F.; Ramos, F.; de Quirós, A.R.B.; Khwaldia, K.; Sendón, R. Industrial multi-fruits juices by-products: Total antioxidant capacity and phenolics profile by LC–MS/MS to ascertain their reuse potential. *Eur. Food Res. Technol.* **2020**, *246*, 2271–2282. [CrossRef]
130. Fernandez, M.V.; Bengardino, M.; Jagus, R.J.; Agüero, M.V. Enrichment and preservation of a vegetable smoothie with an antioxidant and antimicrobial extract obtained from beet by-products. *LWT* **2020**, *117*, 108622. [CrossRef]
131. Dogan, K.; Akman, P.K.; Tornuk, F. Role of non-thermal treatments and fermentation with probiotic Lactobacillus plantarum on in vitro bioaccessibility of bioactives from vegetable juice. *J. Sci. Food Agric.* **2021**, *101*, 4779–4788. [CrossRef]

132. Xu, X.; Bi, S.; Lao, F.; Chen, F.; Liao, X.; Wu, J. Induced changes in bioactive compounds of broccoli juices after fermented by animal- and plant-derived Pediococcus pentosaceus. *Food Chem.* **2021**, *357*, 129767. [CrossRef]
133. Saad, A.M.; Mohamed, A.S.; El-Saadony, M.T.; Sitohy, M.Z. Palatable functional cucumber juices supplemented with polyphenols-rich herbal extracts. *LWT* **2021**, *148*, 111668. [CrossRef]
134. Mohammad Bagher Hashemi, S.; Jafarpour, D.; Jouki, M. Improving bioactive properties of peach juice using Lactobacillus strains fermentation: Antagonistic and anti-adhesion effects, anti-inflammatory and antioxidant properties, and Maillard reaction inhibition. *Food Chem.* **2021**, *365*, 130501. [CrossRef] [PubMed]
135. Aderinola, T. Nutritional, Antioxidant and Quality Acceptability of Smoothies Supplemented with Moringa oleifera Leaves. *Beverages* **2018**, *4*, 104. [CrossRef]
136. Habibi, A.; Keramat, J.; Hojjatoleslamy, M.; Tamjidi, F. Preparation of Fish Oil Microcapsules by Complex Coacervation of Gelatin–Gum Arabic and their Utilization for Fortification of Pomegranate Juice. *J. Food Process Eng.* **2017**, *40*, 1–11. [CrossRef]
137. Sahin, O.I.; Ozturk, B. Microalgal biomass—a bio-based additive: Evaluation of green smoothies during storage. *Int. Food Res. J.* **2021**, *28*, 309–316.
138. Kranz, P.; Braun, N.; Schulze, N.; Kunz, B. Sensory quality of functional beverages: Bitterness perception and bitter masking of olive leaf extract fortified fruit smoothies. *J. Food Sci.* **2010**, *75*, S308–S311. [CrossRef]
139. Michaličková, D.; Belović, M.; Ilić, N.; Kotur-Stevuljević, J.; Slanař, O.; Šobajić, S. Comparison of Polyphenol-Enriched Tomato Juice and Standard Tomato Juice for Cardiovascular Benefits in Subjects with Stage 1 Hypertension: A Randomized Controlled Study. *Plant Foods Hum. Nutr.* **2019**, *74*, 122–127. [CrossRef]
140. Aduloju, A.T.V.-; Nwanja, N.M.; Ezegbe, C.C.; Okocha, K.S.; Aduloju, T.A. Phytochemicals and Vitamin Properties of Smoothie Flavoured with Mint Leaves Extract. *Int. J. Biochem. Res. Rev.* **2020**, *29*, 24–30. [CrossRef]
141. Tarazona-Díaz, M.P.; Aguayo, E. Influence of acidification, pasteurization, centrifugation and storage time and temperature on watermelon juice quality. *J. Sci. Food Agric.* **2013**, *93*, 3863–3869. [CrossRef] [PubMed]
142. Martínez-Sánchez, A.; Ramos-Campo, D.J.; Fernández-Lobato, B.; Rubio-Arias, J.A.; Alacid, F.; Aguayo, E. Biochemical, physiological, and performance response of a functional watermelon juice enriched in L-citrulline during a half-marathon race. *Food Nutr. Res.* **2017**, *61*, 1–12. [CrossRef] [PubMed]
143. Tarazona-Díaz, M.P.; Alacid, F.; Carrasco, M.; Martínez, I.; Aguayo, E. Watermelon juice: Potential functional drink for sore muscle relief in Athletes. *J. Agric. Food Chem.* **2013**, *61*, 7522–7528. [CrossRef]
144. Martínez-Sánchez, A.; Alacid, F.; Rubio-Arias, J.A.; Fernández-Lobato, B.; Ramos-Campo, D.J.; Aguayo, E. Consumption of Watermelon Juice Enriched in l -Citrulline and Pomegranate Ellagitannins Enhanced Metabolism during Physical Exercise. *J. Agric. Food Chem.* **2017**, *65*, 4395–4404. [CrossRef]
145. Manach, C.; Williamson, G.; Morand, C.; Scalbert, A.; Rémésy, C. Bioavailability and bioefficacy of polyphenols in humans. I. Review of 97 bioavailability studies. *Am. J. Clin. Nutr.* **2005**, *81*, 230–242. [CrossRef] [PubMed]
146. Silva, A.; Rodrigues, C.; Garcia-Oliveira, P.; Lourenço-Lopes, C.; Silva, S.A.; Garcia-Perez, P.; Carvalho, A.P.; Domingues, V.F.; Barroso, M.F.; Delerue-Matos, C.; et al. Screening of Bioactive Properties in Brown Algae from the Northwest Iberian Peninsula. *Foods* **2021**, *10*, 1915. [CrossRef]
147. Martínez-Hernández, G.B.; Castillejo, N.; Carrión-Monteagudo, M.M.; Artés, F.; Artés-Hernández, F. Nutritional and bioactive compounds of commercialized algae powders used as food supplements. *Food Sci. Technol. Int.* **2018**, *24*, 172–182. [CrossRef] [PubMed]
148. Klug, T.V.; Collado, E.; Martínez-Hernández, G.B.; Artés, F.; Artés-Hernández, F. Effect of stevia supplementation of kale juice spheres on their quality changes during refrigerated shelf life. *J. Sci. Food Agric.* **2019**, *99*, 2384–2392. [CrossRef] [PubMed]

foods

MDPI

Article

The Use of a Thermal Process to Produce Black Garlic: Differences in the Physicochemical and Sensory Characteristics Using Seven Varieties of Fresh Garlic

Jan Bedrníček [1], Ivana Laknerová [2], František Lorenc [1], Priscila Probio de Moraes [3], Markéta Jarošová [4], Eva Samková [1,*], Jan Tříska [5], Naděžda Vrchotová [5], Jaromír Kadlec [1] and Pavel Smetana [1]

1 Department of Food Biotechnologies and Agricultural Products' Quality, Faculty of Agriculture, University of South Bohemia in České Budějovice, Studentská 1668, 370 05 České Budějovice, Czech Republic; bedrnj00@zf.jcu.cz (J.B.); lorenf00@zf.jcu.cz (F.L.); kadlec@zf.jcu.cz (J.K.); smetana@zf.jcu.cz (P.S.)
2 Food Research Institute Prague, Radiová 1285/7, 102 00 Praha 10—Hostivař, Czech Republic; ivana.laknerova@vupp.cz
3 Department of Food Science, School of Food Engineering, University of Campinas, Monteiro Lobato No. 80, Campinas 13083-862, São Paulo, Brazil; priscilaprobiomoraes@gmail.com
4 Department of Plant Production, Faculty of Agriculture, University of South Bohemia in České Budějovice, Na Sádkách 1780, 370 05 České Budějovice, Czech Republic; jarosm08@zf.jcu.cz
5 Global Change Research Institute CAS, 603 00 Brno, Czech Republic; triska.j@czechglobe.cz (J.T.); vrchotova.n@czechglobe.cz (N.V.)
* Correspondence: samkova@zf.jcu.cz; Tel.: +420-389-032-618

Citation: Bedrníček, J.; Laknerová, I.; Lorenc, F.; Moraes, P.P.d.; Jarošová, M.; Samková, E.; Tříska, J.; Vrchotová, N.; Kadlec, J.; Smetana, P. The Use of a Thermal Process to Produce Black Garlic: Differences in the Physicochemical and Sensory Characteristics Using Seven Varieties of Fresh Garlic. *Foods* **2021**, *10*, 2703. https://doi.org/10.3390/foods10112703

Academic Editors: Marina Cano Lamadrid and Francisco Artés-Hernández

Received: 4 October 2021
Accepted: 3 November 2021
Published: 5 November 2021

Abstract: Black garlic (BG) is a product originating from fresh garlic (FG) and substantially differs in many aspects from FG due to the process called ageing. During this thermal process, the health-promoting properties of FG are enhanced, and the sensory traits are altered. However, very little is known about how the physicochemical properties of different FG varieties affect these properties of BG. Thus, the aim of this study was to investigate the influence of seven FG varieties subjected to the thermal process on the physicochemical parameters of BG. To prepare the BG samples, a fifteen-day ageing process involving a temperature gradient ranging from 30 to 82 °C was used. It was found that the antioxidant activity, the total polyphenol content, and the total soluble solids increased during ageing, while the pH level, moisture content, and lightness decreased in all the garlic varieties. The varieties of garlic differed in the studied traits significantly, both before (FG) and after ageing (BG). In the sensory analysis, significant differences between the BG varieties were observed only in the pleasantness of texture, while the remaining sensory descriptors (pleasantness of color, odor, taste and intensity of the garlic aroma, and overall acceptability) were not affected by variety. The correlations suggest that most of the FG's studied parameters in this study do not correlate with the properties of BG and cannot be used for the prediction of the quality of BG. Additionally, HPLC-MS/MS analysis revealed substantial changes in the composition of low molecular compounds.

Keywords: garlic; ageing process; antioxidant activity; bioactive compounds; organoleptic properties; HPLC-MS/MS analysis

1. Introduction

Black garlic (BG) is a functional food originating from Asia, specifically from China, South Korea, and Japan. Although BG is botanically *Allium sativa* L., it differs from fresh garlic (FG) in many aspects. It is processed using higher temperatures, ranging between 60–90 °C, with approximately 80% relative air humidity for a long period of time (10–90 days) without using any other ingredients or additives. This thermal process is called ageing and during this process the fresh white garlic gains many new properties, such as enhanced biological activity, especially antioxidant activity, and altered sensory properties because the FG's color turns from white to black, the texture changes to very soft or jelly-like, and the taste is very mild with sweet and sour tones [1]. Moreover, antimicrobial activity was investigated by Botas et al. [2] who found that BG had stronger antimicrobial activity than FG.

As reported previously, BG with excellent sensory and antioxidant properties is obtained using a temperature around 70 °C and 80% relative air humidity. Lower temperatures do not lead to the black color, and a pungent aroma is still present. A higher temperature than 85 °C accelerates the process but could give a bitter aroma to the BG, suggesting the presence of undesirable compounds [3].

BG is popular in Asia [4] and is now attracting more attention in Europe, including the Czech Republic. It is becoming an ingredient in luxury cuisine and part of a healthy diet [5].

BG has been intensively studied in the last two decades. The research has been, nevertheless, mostly focused on the physicochemical properties of BG [6]; its in vivo and in vitro effects; its bioavailability [7,8]; the influence of the processing conditions (temperature, time, and relative air humidity); and the quality of the BG [3], including pretreatment steps, such as freezing [9] and the use of hot steam [10].

Many factors influence the quality parameters of FG, such as origin, growing location, season, climate, and cultivation practice, as well as the variety, which plays a pivotal role [11]. Moreover, the different FG qualities could lead to different qualities of BG. Czech FG is assessed positively by experts and consumers due to its strong garlic aroma and taste [12]. In addition, the authors reported that the pungency of Czech varieties was higher than in European samples and much higher than in non-European. A typical, strong, pungent aroma should not be present in BG. Thus, it remains unclear whether the pungent varieties (e.g., those originating from the Czech Republic) are suitable for the preparation of BG.

Very little is known about the influence of garlic variety on BG quality. Up to now, only two papers have been published studying the influence of garlic variety on the properties of BG, with only one focusing on the quality of BG originating from conventional and organic cultivation [13–15]. However, these papers studied only three, six, and two varieties, respectively.

The aim of this study was to describe the influence of seven Czech FG varieties and the thermal ageing technology on the physicochemical properties and sensory qualities of the BG prepared from them. Moreover, the study targeted identification of the main biologically active compounds before and after ageing.

The results could be helpful for producers of BG to select garlic that has the highest quality and can also contribute to the knowledge of how the parameters affect the general quality of the BG.

2. Materials and Methods

2.1. Chemicals

All chemicals were purchased from Sigma-Aldrich, Prague, Czech Republic: acetic acid (purity ≥ 99%); sodium acetate (purity ≥ 99%); acetonitrile (LC/MS grade); methanol (LC/MS grade); formic acid (LC/MS grade, 98–100%); ferric chloride; hydrochloric acid (37%; HCl); sodium carbonate; 2,4,6-tris(2-pyridyl)-1,3,5-triazine (TPTZ, purity ≥ 98%); 2,2-diphenyl-1-picrylhydrazyl (DPPH); Trolox (purity ≥ 97%); and Folin–Ciocalteau's phenol reagent and gallic acid (97.9%).

2.2. Plant Material

In this study, seven varieties of Czech garlic (listed in the Czech State Variety Book) were used: Bjetin, Vekan, Lan, Havel, Rusák, Havran, and Lukan. All of them were purchased from a local garlic grower (Mihulková-Hradecký "Český česnek z Podkrkonoší", Czech Republic). The varieties were of the winter and bolting type (except for the Lukan, which was non-bolting) and were planted in the autumn (in the middle of October) of 2018 and harvested at the turn of the month of June and July 2019 in the Semily district (ca.

360 m above the sea level) in the Czech Republic. The average annual temperature in this area is around 7 °C with precipitations of 700 mm. The garlic samples were transported to the laboratory fresh and disease-free after harvesting. These FG samples were used for the manufacturing of BG or directly for physicochemical analyses: determination of the moisture content, soluble solids, pH level, total polyphenolic content, antioxidant activity, and texture and color analysis. Furthermore, the variety Rusák was used to determine the main biologically active compounds by means of high-performance liquid chromatography and tandem mass spectrometry (HPLC-MS/MS).

For each chemical analysis, five randomly selected bulbs of each variety were taken, and two cloves from each bulb were peeled (ten cloves per variety in total); these are assigned as "peeled garlic cloves" later in the text.

2.3. Black Garlic Processing

The BG was made from FG according to the patented method described by Vácha et al. [16]. Briefly, approximately 550 g of FG of each of the above-mentioned varieties was placed into a polyamide/polyethylene vacuum bag. Then, the samples were vacuum-packed using ETA Freshie 1762 (ETA a.s., Prague, Czech Republic). The purpose of using vacuum plastic bags is that the BG can be produced in an oven or a dryer without any humidity regulation because the high relative humidity needed for the ageing process is substituted by the water contained in the FG. Thus, the bags prevent or minimize moisture evaporation from the garlic bulbs.

The bags with the FG samples were placed into a laboratory oven (Memmert UFE 400, Memmert GmbH + Co. KG, Schwabach, Germany) to begin the process of ageing at the same time. The only parameter that was controlled was the temperature, as follows: the first stage took three days, and the temperature was set at 82 °C; in the second stage, the temperature was decreased to 30 °C for one day; and finally, the temperature was increased to 72 °C in the last stage for 11 days. Thus, the whole ageing process took 15 days. At the beginning of the ageing process, the highest temperature was used to rapidly accelerate ageing and inactivate some enzymes, such as alliinase (EC 4.4.1.4.). Due to the excessive water evaporation from the garlic bulbs, a low temperature was set in the second stage of the process. It caused the water to be condensed back to the surface of the bulbs. Afterwards, a temperature of around 70 °C in the third stage was set because it was optimal for obtaining BG with excellent properties [3].

Then, the samples from each variety were collected by the same sampling procedure as described for the FG samples. Afterwards, the black, aged samples were used for the same analyses as in the case of the fresh samples. The BG samples were additionally used for sensory analysis.

2.4. Chemical Analyses

2.4.1. Moisture-Content Analysis

The peeled garlic cloves were finely homogenized in a kitchen mixer, and 5 g of the sample was measured gravimetrically using a moisture analyzer (Kern RH 120-3, Fisher Scientific s.r.o., Pardubice, Czech Republic). The measurement was repeated three times ($n = 3$).

2.4.2. pH Measurement

The peeled garlic cloves were homogenized using a kitchen blender, and the pH of garlic samples was measured using a pH meter equipped with a spear-type electrode (HC 124, Fisher Scientific, Pardubice, Czech Republic). An analysis of each group was conducted three times ($n = 3$).

2.4.3. Determination of Total Soluble Solids

Approximately 10 g of the peeled garlic cloves was blended with 90 mL of deionized water in a glass tube using a laboratory homogenizer (Ultra-Turrax T25, IKA-Werke GmbH

& Co. Kg, Staufen, Germany). The mixture was then analyzed for total soluble solids using an optical refractometer (ATC 0–32%, Fisher Scientific s.r.o, Pardubice, Czech Republic), previously calibrated with deionized water. °Brix were then calculated using a dilution factor. An analysis for each group was conducted three times (n = 3).

2.4.4. Determination of Antioxidant Activity

To determine the antioxidant activity of the FG and BG samples, two spectrophotometric methods were selected: DPPH (2,2-diphenyl-1-picrylhydrazyl), to evaluate the free radical scavenging ability and FRAP (ferric reducing antioxidant power), to assess the reducing ability. The first step was the extraction of a sample. The peeled garlic cloves were homogenized, and 0.2 g was placed in a 15 mL centrifuge tube, then 9.8 mL of 90% methanol (v/v) was added to the sample. This mixture was shaken in a laboratory shaker for 10 min, and then it was centrifuged at 7000 rpm for 15 min at 5 °C. The supernatant was collected and immediately used for the determination of antioxidant activity. Two extractions per group were conducted, and each extract was measured twice (n = 4).

The DPPH method was conducted according to a slightly modified method described by Brand-Williams et al. [17]. The extracted sample (100 µL) was added to 4 mL of methanolic solution of DPPH (27.5 µg/mL). Afterwards, this mixture was kept for two hours in the dark at room temperature. Then, the absorbance at 515 nm was measured against a blank (DPPH solution without sample). Trolox was used to prepare a calibration curve. The results are expressed as Trolox equivalent (TE) per gram of sample on a dry matter (DM) basis (g TE/kg DM).

A slightly modified FRAP method was assessed according to Dudonné et al. [18]. Firstly, the FRAP reagent was prepared as follows: 100 mL of 300 mM (pH 3.6) acetate buffer was mixed with 10 mL of 10 nM TPTZ in 40 mM HCl and with 10 mL of ferric chloride (10 mM). After this, 0.1 mL of sample extract was pipetted to 4 mL of FRAP reagent. The reaction mixture was kept in the dark for 30 min at 37 °C, and then the absorbance was measured against a blank (acetate buffer) at 593 nm. Trolox was used as a standard; thus, the results are expressed as g TE/kg DM.

2.4.5. Determination of Total Polyphenol Content

Firstly, polyphenols from the garlic samples were extracted. The peeled garlic cloves (fresh and black) were homogenized, and 5 g was weighed into a glass tube. Then, 50 mL of 50% methanol was added. Additionally, the mixture was thoroughly homogenized using a laboratory mixer (Ultra-Turrax T25, IKA-Werke GmbH & Co. Kg, Staufen, Germany) for 30 s. Then, the tube was put into an ultrasound bath and was extracted for 15 min. Afterwards, an aliquot was pipetted into a centrifuge tube and was centrifuged at 4000 rpm for 15 min. The supernatant was stored at −18 °C until analysis. One extraction per group was conducted, and the extract was analyzed three times (n = 3).

The total polyphenolic content was spectrophotometrically determined according to the method of Lachman et al. [19]: 35 µL of the sample extract was pipetted into a disposable 4.5 mL plastic cuvette; then, 175 µL of Folin–Ciocâlteu reagent, 3465 µL of water, and 525 µL of 20% sodium carbonate were added. This reaction mixture was then incubated at room temperature for two hours in the dark. The absorbance was read at 765 nm, and gallic acid was used as a standard for the preparation of a calibration curve. The results are expressed as mg of gallic acid equivalent (GAE) per kg of sample DM (g GAE/kg DM).

2.4.6. HPLC-MS/MS Non-Targeted Analysis of Fresh and Black Garlic

Extraction was conducted according to Zhang et al. [20], with modifications. The peeled garlic cloves were homogenized in a kitchen mixer, and 5 g of the sample was immediately put into a glass tube; then, 50 mL of 100% methanol was added. The sample with methanol was additionally homogenized using a laboratory mixer (Ultra-Turrax T25, IKA-Werke GmbH & Co. KG, Staufen, Germany) for 30 s. Afterwards, the mixture was extracted in an ultrasound bath for 15 min. Then, an aliquot was transferred into a centrifuge tube and centrifuged at 4000 rpm for 15 min. The supernatant was collected and stored at −18 °C until analysis. Five µL of properly diluted sample was injected into the HPLC system (Dionex UltiMate 3000; Dionex, Sunnyvale, CA, USA) coupled with a triple quadrupole mass spectrometer (Agilent 6420; Agilent Technologies Inc., Santa Clara, CA, USA). The chromatographic separation was achieved by the Phenomenex Kinetex column (C18, 2.6 µm, 150 × 2.1 mm) maintained at 35 °C. The mobile phase A consisted of 5% acetonitrile with 0.5% formic acid, and B consisted of 100% acetonitrile. The flow rate was set at 0.2 mL/min and the gradient involved a linear increase in the mobile phase B from 5 to 80% within 30 min. After this, solvent B decreased to 5% within 10 min to be prepared for the next injection. Identification of the separated compounds was assessed using a mass spectrometer equipped with an electrospray ionization source (ESI) operating in positive ionization mode with the following settings: the drying gas (N_2) temperature of 300 °C was at a flow rate of 11 L/min, the nebulizing gas pressure was 35 psi, and the capillary voltage was set to +4kV. The fragmentor, cell accelerator, and collision energy were 135, 7, and 20 volts, respectively.

2.5. Physical Properties Analysis

2.5.1. Color Analysis

The surface color of the FG and BG peeled cloves was analyzed in the CIE L*a*b* system using a spectrophotometer ColorEye XTH (GretagMacbeth, New Windsor, NY, USA). For each sample, three independently chosen cloves were selected and analyzed (n = 3).

2.5.2. Texture Analysis

The hardness measurement of the FG and BG cloves was accomplished by using texture analyzer TA.XT Plus (Stable MicroSystems, Godalming, UK). The analyzer was equipped with a knife probe that cut the garlic cloves, and the maximal force needed to cut the cloves into halves was recorded. The following settings of the texture analyzer were applied: pre-test speed: 3 mm/s; test speed: 2 mm/s; and post-test speed: 10 mm/s. Four cloves from each group (n = 4) were randomly selected, and the results are expressed as newton (N).

2.6. Sensory Analysis of Black Garlic Samples

The sensory profile of the BG varieties was tested by sensory analysis. Seven panelists, who had been well trained in the principles and the concept of sensory evaluation, were involved in this experiment. One day before the sensory analysis, the panelists received commercially available BG cloves to get familiar with their sensory attributes.

At sensory analysis, approximately 5 g of BG per variety, corresponding to at least three cloves, were put into a 50 mL plastic container. The containers, randomly marked with a number, were served on a plate to each panelist to compare them simultaneously. The hedonic evaluation of taste, color, texture, overall acceptability, and intensity of the garlic aroma were recorded on a 10 cm unstructured hedonic scale (0 = dislike extremely, 10 = like extremely for taste, smell, color, and overall acceptability; 10 = very intense, 0 = not present for the intensity of the garlic aroma) by the same panel. The panelists then placed a mark on the scale to indicate the pleasantness or intensity of each characteristic. The values for all descriptors of BG obtained from each panelist were averaged (n = 7). Water was served as a taste neutralizer and was drunk between the judging of the following sample.

2.7. Statistical Analysis

The results of all the analyses are presented as mean ± standard deviation. Two-way ANOVA (analysis of variance) was used to determine statistically significant differences between varieties (factor one) and between the FG and BG samples (ageing process—factor two) in all the physicochemical analyses. To evaluate the statistically significant differences between the BG varieties in the sensory analysis, one-way ANOVA was used. After both ANOVAs, Tukey's HSD post hoc test was assessed for group comparisons. To estimate the associations between garlic sample properties (soluble solids, antioxidant activity, pH, total polyphenol content, etc.), Pearson's correlation coefficient was used. Statistical evaluation was performed in Statistica CZ, version 12 (StatSoft CR) and the differences were considered significant if $p < 0.05$.

3. Results and Discussion

3.1. The Effects of the Ageing Process and Variety on Moisture, pH, and Total Soluble Solid Content of Fresh and Aged Garlic Samples

The ageing process substantially affects the chemical properties of garlic [1]. The results of this study also prove the significant changes in the chemical parameters. The moisture, pH, and total soluble solid content of all the FG and BG varieties are presented in Table 1.

Table 1. Moisture content, pH, and total soluble solids of fresh and black garlic varieties.

Variety	Moisture Content (%)		pH		Total Soluble Solids (°Brix)	
	Fresh	Black	Fresh	Black	Fresh	Black
Bjetin	60.18 ± 0.64^{abA}	36.95 ± 0.93^{cdB}	6.06 ± 0.02^{abA}	4.45 ± 0.03^{aB}	39.67 ± 0.06^{aB}	52.00 ± 0.00^{bA}
Vekan	58.62 ± 1.60^{bcA}	34.28 ± 1.26^{dB}	6.05 ± 0.01^{abA}	4.48 ± 0.05^{aB}	39.00 ± 0.10^{aB}	50.33 ± 0.58^{bA}
Havel	57.04 ± 0.46^{cA}	35.50 ± 1.09^{dB}	6.11 ± 0.02^{aA}	4.39 ± 0.02^{abB}	37.67 ± 0.12^{abB}	60.00 ± 0.00^{bA}
Ivan	59.17 ± 2.18^{bcA}	39.52 ± 1.16^{bcB}	6.01 ± 0.03^{abA}	4.32 ± 0.09^{bcB}	36.67 ± 0.12^{bB}	50.00 ± 1.00^{bA}
Rusák	57.12 ± 1.20^{cA}	33.52 ± 0.40^{dB}	6.05 ± 0.01^{abA}	4.24 ± 0.05^{cB}	39.67 ± 0.06^{aB}	61.00 ± 0.00^{aA}
Havran	65.34 ± 2.46^{aA}	42.52 ± 0.87^{abB}	6.00 ± 0.00^{abA}	4.38 ± 0.02^{abB}	30.00 ± 0.00^{cB}	51.67 ± 1.15^{bA}
Lukan	62.25 ± 0.59^{abA}	45.55 ± 0.91^{aB}	5.99 ± 0.01^{bA}	4.48 ± 0.05^{aB}	32.00 ± 0.00^{cB}	50.67 ± 0.58^{bA}
Average	59.57 ± 2.44^{B}	38.26 ± 4.48^{A}	6.04 ± 0.04^{A}	4.39 ± 0.61^{B}	36.38 ± 3.68^{B}	53.67 ± 9.53^{A}

Results are presented as mean ± standard deviation ($n = 3$); $^{a–d}$ values with different superscripts within a column differ significantly ($p < 0.05$); ^{A,B} values with different superscripts within a row and the same analysis differ significantly ($p < 0.05$).

The moisture content of the FG varieties ranged between 57.04 and 65.34%, with an average value of 59.57% and between 33.52 and 45.55%, with an average value of 38.26% for the garlic samples after ageing (BG). Thus, the differences between varieties (both before and after ageing) were statistically significant ($p < 0.05$). These values are typical for FG originating from the Czech Republic, where the moisture content is usually between 55 and 70% [12]. When compared with garlic from other countries, samples originating from South Korea have an average moisture content of 63.43%, whereas Chinese samples have 66.76% [21]. Moreover, Shin et al. [22] reported that Chinese garlic samples had a moisture content higher than 70% on average. This indicates that garlic from East Asia has a higher moisture content. Regarding BG moisture, Sunanta et al. [14] reported that BG had an average moisture content of 30%. However, the authors also reported that garlic originating from China had moisture of 53.12% after ageing, which is much higher than our observation.

The ageing process resulted in an approximately 37% loss of water (from 59.57 to 38.26%), and the decrease was statistically significant ($p < 0.05$). The evaporation of water during the processing of BG is a common phenomenon. However, as stated by Xiong et al. [23], water evaporation during ageing is strongly dependent on the processing conditions, mainly the relative air humidity as well as the temperature.

The ageing process caused a significant ($p < 0.05$) decrease of pH in all the garlic samples, from approximately 6.04 to 4.39. Other authors [6,15] have reported a much lower pH, reaching values of 3.3. However, they processed garlic samples for more than 30 days. A low pH resulting from acidic conditions in BG can be caused by the formation of acids, such as formic, acetic, pyroglutamic, 3-hydroxypropionic, and succinic, which were not detected in FG, as reported by Liang et al. [24]. Although there were statistical differences ($p < 0.05$) between the varieties in pH, the range of values was relatively small in the FG and BG samples. The pH ranged between 5.99 and 6.11 for the FG samples and 4.24 and 4.48 for the BG samples. Our results are in line with Toledano Medina et al. [15], who also reported that the pH of garlic is affected by variety.

Another observed parameter in the garlic samples was total soluble solid content. The values in the FG samples varied between 30.00 to 39.67 °Brix (the average value was 36.38 °Brix). Much higher values were measured in the garlic samples after ageing, with an average value of 53.67 °Brix (ranging between 50.00 and 60.00). Thus, the ageing process had a significant ($p < 0.05$) effect on this parameter. Moreover, the variety significantly ($p < 0.05$) influenced the total soluble solid content. These results are in accordance with other authors [13,25]; however, they reported lower values, both in the FG and the BG samples. The increase in soluble solids is probably caused by the hydrolysis of polymeric compounds at a higher temperature into simpler compounds, e.g., proteins into peptides or amino acids and especially polysaccharides into oligosaccharides with a lower degree of polymerization and di- or monosaccharides [1,26].

3.2. The Effects of the Ageing Process and Variety on Antioxidant Activity and Total Polyphenol Content of Fresh and Aged Garlic Samples

Garlic possesses antioxidant activity due to the presence of many substances, such as polyphenols and organosulfur compounds, but the activity is strongly enhanced after the ageing process of garlic [27]. The antioxidant activity of the FG and BG samples is presented in Table 2.

Table 2. Antioxidant activity (DPPH and FRAP assays) and total polyphenol content (TPC) of fresh and black garlic varieties.

Variety	DPPH (g TE/kg DM)		FRAP (g TE/kg DM)		TPC (g GAE/kg DM)	
	Fresh	Black	Fresh	Black	Fresh	Black
Bjetin	1.62 ± 0.02^{bB}	13.12 ± 0.18^{bA}	0.38 ± 0.01^{B}	13.19 ± 0.29^{cA}	3.62 ± 0.08^{B}	16.01 ± 0.28^{aA}
Vekan	1.32 ± 0.02^{bB}	12.80 ± 0.53^{bcA}	0.66 ± 0.02^{B}	12.51 ± 0.35^{dA}	3.82 ± 0.16^{B}	11.60 ± 0.17^{cA}
Havel	2.52 ± 0.09^{aB}	14.00 ± 0.10^{aA}	0.56 ± 0.02^{B}	13.54 ± 0.20^{bcA}	3.61 ± 0.15^{B}	15.48 ± 0.25^{aA}
Ivan	1.89 ± 0.08^{abB}	12.28 ± 0.13^{cA}	0.47 ± 0.01^{B}	12.15 ± 0.17^{dA}	3.89 ± 0.22^{B}	14.90 ± 1.08^{aA}
Rusák	1.52 ± 0.07^{bB}	12.67 ± 0.18^{bcA}	0.38 ± 0.01^{B}	12.47 ± 0.49^{dA}	3.76 ± 0.08^{B}	15.59 ± 0.15^{aA}
Havran	2.53 ± 0.10^{aB}	14.25 ± 0.57^{aA}	0.62 ± 0.03^{B}	14.24 ± 0.36^{aA}	4.61 ± 0.11^{B}	13.42 ± 0.27^{bA}
Lukan	1.89 ± 0.06^{abB}	11.41 ± 0.49^{dA}	0.48 ± 0.01^{B}	12.16 ± 0.53^{dA}	3.47 ± 1.01^{B}	11.23 ± 0.22^{cA}
Average	1.90 ± 0.52^{B}	12.93 ± 4.33^{A}	0.50 ± 0.12^{B}	12.89 ± 4.79^{A}	3.82 ± 0.37^{B}	14.03 ± 4.44^{A}

Results are presented as mean ± standard deviation (n = 4 for DPPH and FRAP; n = 3 for TPC); $^{a–d}$ values with different superscripts within a column differ significantly ($p < 0.05$); ^{A,B} values with different superscripts within a row and the same analysis differ significantly ($p < 0.05$); DM: dry matter; TE: Trolox equivalent; DPPH: 2,2-diphenyl-1-picrylhydrazyl; FRAP: ferric reducing antioxidant power; TPC: total polyphenol content; GAE: gallic acid equivalent.

The BG samples exhibited much higher antioxidant activity, both free radical scavenging activity (DPPH assay) and reducing properties (FRAP assay), compared to the FG samples. Thus, the ageing process is a significant ($p < 0.05$) factor influencing antioxidant activity. There was an almost seven-fold increase in the DPPH assay, from 1.9 (average of FG) to 12.93 g TE/kg DM (average of BG). A very similar trend was observed for the FRAP assay, but the increase was even more significant, from 0.5 to 12.89 g TE/kg DM. The DPPH was also found to be variety dependent as significant ($p < 0.05$) differences were found between varieties (in the FG and also the BG samples). Nevertheless, no statistical differences were found between the FG samples in terms of the FRAP assay, but the BG samples differed significantly.

It was reported that some of the compounds possessing antioxidant activity in garlic might be polyphenols that can be determined by the spectrophotometric method using Folin–Ciocâlteu phenol reagent [28]. Although its selectivity and specificity are questionable [29,30], it is one of the most frequently used assays for the determination of such compounds, not only in FG and BG [6,28,31]. A relatively low concentration of total polyphenols (3.82 g GAE/kg DM on average) was found in the FG samples, as can be seen in Table 2. It should be noted that according to Pilluza and Bullitta [28] garlic had the lowest concentration of total polyphenols among 24 examined plants.

It was an interesting finding that there were no statistically significant differences ($p > 0.05$) between the FG varieties, although some distinctions were noted. The presented results of this study show that there was approximately a 3.6-fold increase in the total polyphenol content after ageing (14.03 g GAE/kg DM on average), as was also found in the previous literature [4]. On the other hand, statistically significant differences ($p < 0.05$) were found between the varieties after ageing. Although the result is in accordance with an earlier report, it remains a question what compounds in BG react with the Folin–Ciocâlteu phenol reagent. Thus, in our opinion, this phenomenon should be taken into account in further research.

Total polyphenol content often positively correlates with antioxidant activity [28], but the association between total polyphenols and antioxidant activity (DPPH and FRAP assays) in samples after ageing (BG) was found to be very small (r = 0.408 for DPPH and 0.256 for FRAP, respectively, $p > 0.05$). This also suggests that compounds other than polyphenols can contribute to the antioxidant activity, e.g., organosulfur compounds or products of the Maillard reaction. It should also be mentioned that the total polyphenol content could be affected by the particular temperature gradient. Xiong et al. [23] reported that if garlic is aged at 90 °C, the total polyphenol content rises faster than if the temperature is set at 70 °C. According to the author, if a higher temperature was set, then after five days, the polyphenol content decreased. Thus, it is advisable to process BG at 70 °C for at least 15 days.

3.3. HPLC-MS/MS Non-Targeted Analysis of Low Molecular Weight Compounds in Fresh and Black Garlic Samples

In addition to spectrophotometric analyses (TPC, DPPH, and FRAP), we decided to identify some of the low molecular weight compounds in FG and BG using HPLC-MS/MS apparatus to establish differences in the samples before and after ageing. The Rusák variety was chosen for this purpose. Compounds with high intensity (response) or significant changes due to the ageing process were subsequently identified. The identification of the compounds was based on the monitoring of molecular mass, production of MS/MS fragments, retention times, and comparison with the data from available literature [32–37] and with the help of online mass spectra databases [38,39].

Data were obtained only in positive ionization mode as bad results were achieved with negative ionization. The relative concentration change of the identified compounds on a dry matter basis (peak area vs. dry matter) was also calculated. It was expressed only as "increase" or "decrease" due to the uncertainty of the detector linearity; thus, fold change is not presented. The results of the MS/MS analysis showing tentatively identified compounds in the Rusák variety before (FG) and after ageing (BG) are presented in Table 3.

Twelve compounds in total were identified in the methanolic extracts, including amino acid, amino acid derivative, fructooligosaccharides, organosulfur compounds, and derivatives belonging to β-carbolines.

The compound with the highest m/z (1515 $[M + K]^+$) was identified as fructooligosaccharide residue with a degree of polymerization (DP) 9 as it showed a typical fragmentation pattern (m/z of 1353, 1191, 1029, 867, 705, 543, 381) with a mass shift of 162 Da corresponding to the loss of fructose moiety. Other fructooligosaccharides identified in the extracts were fructofuranosyl nystose (m/z 867, DP = 5), nystose (m/z 705, DP = 4), and kestose (m/z 543, DP = 3). The mass spectra obtained are consistent with the previous literature [37]. According to Zhang et al. [40], the usual degree of polymerization of garlic fructan is 58. Nevertheless, it was not possible to detect high molecular fructans with our instrument. The relative concentration of detected fructooligosaccharides was higher in aged garlic, which suggests that high molecular fructans in garlic decompose into smaller units due to the exposure of the garlic to high temperature for a long period of time. These low molecular fructooligosaccharides contribute to the characteristic sweet taste of BG. Other authors [15] also confirmed an increase of simple (reducing) sugars in BG.

Table 3. Tentatively identified compounds in fresh (FG) and black garlic (BG).

Compound	RT * (min)	Molecular Ion $[M + H]^+$ (*m/z*)	Fragments (*m/z*)	Presence in Samples	Change after Ageing ‡
Arginine	1.56	175	158, 130	FG, BG	↑
Fructosyl-arginine	1.57	337	319, 175, 158, 130	BG	↑
Fructooligosaccharide (Degree of polymerization = 9)	1.61	1515 $[M + K]^+$	1353, 1191, 1029, 867, 705, 543, 381 $[M + K]^+$	FG, BG	↑
Fructofuranosyl nystose (4× fructose, 1× glucose)	1.62	867 $[M + K]^+$	705, 543, 381 $[M + K]^+$	FG, BG	↑
Nystose (3× fructose, 1× glucose)	1.63	705 $[M + K]^+$	543, 381 $[M + K]^+$	FG, BG	↑
Kestose (2× fructose, 1× glucose)	1.63	543 $[M + K]^+$	381 $[M + K]^+$	FG, BG	↑
γ-L-glutamyl-S-allyl-L-cysteine	3.24	291	162, 145	FG, BG	↓
γ-L-glutamyl-S-(trans-1-propenyl)-L-cysteine	4.00	291	274, 162, 145	FG	↓
γ-L-glutamyl-phenylalanine	4.88	295	166, 120	FG, BG	↓
(1R, 3S)-1-methyl-1,2,3,4-tetrahydro-β-carbolin-3-carboxylic acid (MTCA)	6.91	231	158	BG	↑
(1S, 3S)-1-methyl-1,2,3,4-tetrahydro-β-carbolin-3-carboxylic acid (MTCA)	7.88	231	158	BG	↑
Diallyl thiosulfinate (allicin)	29.59	163	73	FG, BG	↓

* RT: retention time; ‡: relative changes in concentrations after ageing; peak areas were compared in FG and BG on a dry matter basis; ↑: increase in concentration; ↓: decrease in concentration.

Free arginine with *m/z* 175 $[M + H]^+$ was also identified based on the production of specific fragments, with *m/z* 158 and 130 corresponding to a loss of NH_3 (18) and NH_3 + CO groups (44). However, its relative concentration increased after ageing, which implies that proteins in garlic could hydrolyze under a higher temperature during ageing, which is followed by the release of free amino acids. As reported by Kodera et al. [41], arginine is the most abundant amino acid in garlic. On the other hand, Choi et al. [6] reported that arginine is the second most abundant amino acid after tyrosine.

Amino acids, together with some sugars, can subsequently enter a Maillard reaction, leading to the production of a wide variety of substances [1]. A compound related to the Maillard reaction with molecular ion *m/z* 337 $[M + H]^+$ producing fragments 319, 175, 158, and 130 was identified as fructosyl-arginine (Fru-Arg). This compound is formed in the early stage of the Maillard reaction and belongs to the group of Amadori compounds [20]. In addition, Fru-Arg was not detected in FG. On the other hand, it was reported that Fru-Arg is the most abundant Amadori compound in BG and has several biological properties, such as antioxidant activity, and could act as an inhibitor of the angiotensin I converting enzyme [42].

Two peaks with *m/z* at 231 $[M + H]^+$, producing the same fragment at *m/z* 158, were identified as isomers (1R, 3S and 1S, 3S) of alkaloid 1-methyl-1,2,3,4-tetrahydro-β-carboline-3-carboxylic acid (MTCA). These compounds are probably formed via condensation of L-tryptophan and some aldehydes or oxo-acids in the Maillard reaction and belong to the group of β-carbolines [1]. Both isomers were detected only in BG. These compounds possess hydrogen peroxide scavenging activity, but when compared to quercetin or catechin, the scavenging activity is much lower [34].

Concerning peptides, three γ-glutamyl were detected, namely γ-glutamyl-S-allyl-L-cysteine (*m/z* 291 $[M + H]^+$) with fragments at *m/z* 162 and 145; γ-glutamyl-S-(trans-1-propenyl)-L-cysteine (*m/z* 291 $[M + H]^+$) with mass spectra at *m/z* 274, 162 and 145; and γ-glutamyl-phenylalanin (*m/z* 295 $[M + H]^+$), producing the following spectra: *m/z* 166 and 120. These molecular ion masses and spectra match with data published previously [32,33,43,44]. The first two mentioned peptides containing sulfur serve in garlic as important storage peptides and are biosynthetic intermediates for a plethora of corresponding organosulfur compounds, including those with sensory importance, e.g., allicin [45]. These organosulfur compounds, together with γ-glutamyl-phenylalanin, have beneficial effects in human tissue, including antioxidant, anticancer, antinociceptive, antiplatelet, and anti-atherosclerotic activities [46]. Unfortunately, the relative concentration of γ-glutamyl peptides was lower in BG and γ-glutamyl-S-(trans-1-propenyl)-L-cysteine was detected only in FG.

A peak with molecular ion at m/z 163 $[M + H]^+$ eluting at the end of the gradient with a dominant product ion at m/z 73 was later identified as allicin. Its relative concentration after ageing was nearly undetectable compared to FG. Despite the fact that allicin has certain positive biological properties [47], its presence in BG is not suitable in terms of sensory properties, as allicin, together with other compounds, gives garlic a typical pungent flavor [45]. Although it is known that allicin is unstable and undergoes rapid decomposition [43], Zhang et al. [3] reported detectable amounts in BG even after 69 days of ageing at 60 °C.

It is interesting that we did not detect any polyphenolic compounds, neither in the FG nor in the BG, in spite of the fact that a significant increase in total polyphenols was measured using the Folin–Ciocâlteu phenol reagent in all varieties after ageing. The results are in line with a previous study by Molina-Calle et al. [35], who also did not identify polyphenolic compounds using a high-resolution quadrupole time-of-flight mass spectrometer. According to Corzo-Martínez et al. [45], flavonoids, which are abundant, e.g., in onion, are practically absent in garlic. In addition, Beato et al. [48] detected only very low levels of phenolic acids, reaching a maximum of 2.9 mg/kg of DM for caffeic acid. The only explanation could be that some polyphenolic compounds formed during the Maillard reaction. For example, Matsutomo et al. [34] identified several dilignols in aged garlic extract possessing antioxidant activity. However, these compounds were not detected in BG. As mentioned in Section 3.2, it is described in the literature [29] that the Folin–Ciocâlteu phenol reagent could sometimes overestimate the result because the reagent may react with other interfering compounds (e.g., amino acids).

3.4. Color and Texture Analysis of Black and Fresh Garlic and Sensory Analysis of Black Garlic

BG is interesting not only because of its chemical composition and biological activity, but also because it has highly remarkable organoleptic properties. It acquires a sweet-and-sour taste and a very soft, chewy, and jelly-like texture during the ageing process [49]. The pleasantness of the color, odor, texture, and taste and the intensity of the garlic aroma and the overall acceptability of the seven BG varieties were evaluated in a sensory analysis. The results of the evaluation are depicted in Figure 1A. Although the panelists perceived differences between all the varieties in all the sensory attributes, they were not statistically significant ($p > 0.05$), with the exception of texture, where the Rusák and Lukan varieties received a significantly ($p < 0.05$) lower sensory score than the remaining varieties. In terms of overall acceptability, the analysis did not reveal one variety that would be better than the others.

BG should not have as strong a garlic aroma as FG. Nevertheless, the panelists recorded that the garlic aroma was still present in all the BG samples; however, the intensity was relatively low. Although our ageing temperature was set between 70–80 °C, which is the temperature that is ideal for BG with good sensory properties, pungent garlic aroma is usually present in BG that was aged at 60 °C or for a shorter time period [3]. The reason why the garlic aroma was still perceived in our samples could be that the Czech garlic varieties have a slightly higher pungency than varieties originating from other countries [12].

It is essential to know that the variety of garlic may be an important factor affecting the sensory characteristics of BG. Recently, there have been two papers published regarding the influence of the garlic variety and storage time [14,15] and one focusing on the influence of the cultivation system (conventional vs. organic) [13] on selected quality parameters of BG. However, none of them involved a sensory assessment in the experiment. On the other hand, several papers have been published describing how the different ways of the ageing process can affect the sensory properties of BG, e.g., [3]. Sensory analysis results provide the evidence that the variety of garlic may not be the most important parameter (except for texture) for the sensory quality of BG. It is possible that processing conditions (temperature, humidity, and time) play a greater role in terms of the sensory properties of BG [5].

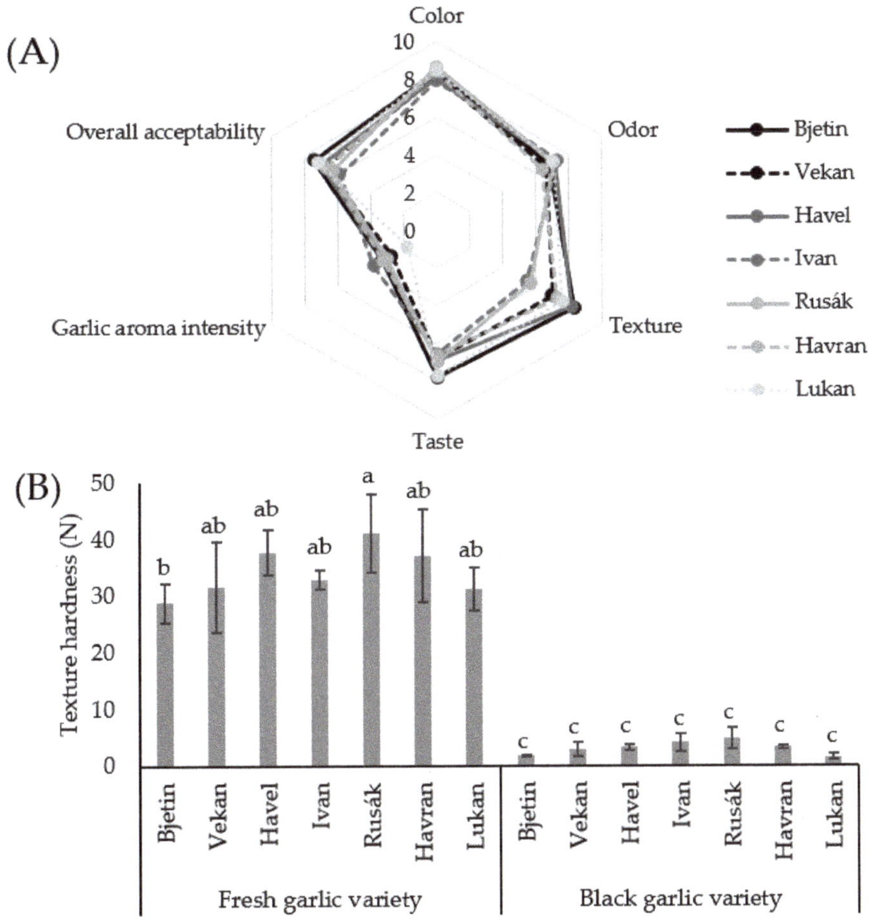

Figure 1. Sensory analysis (0 = dislike extremely, 10 = like extremely) of different black garlic varieties (**A**) and hardness of texture of white and black garlic varieties (**B**). $^{a–c}$ bars, representing means of texture hardness (n = 4) ± standard deviation, with different letters differ significantly (p < 0.05).

In addition to the sensory analysis, texture was also analyzed using a texture analyzer. The texture of BG can be characterized as soft or jelly-like and is unambiguously different from FG, whose texture is hard and stiff. The hardness of the texture of the FG and BG varieties is presented in Figure 1B. It is clearly shown that the hardness of garlic cloves was significantly affected by the ageing process as the average hardness of the FG samples was 34.29 N, but the BG samples had a much softer texture (average 3.01 N). Thus, there was approximately an 11-fold decrease in hardness.

This decrease caused by the ageing process could be related to the disruption of the cell wall polysaccharides of garlic due to a higher temperature [9]. Contrary results were presented by Karnjanapratum et al. [10], who described that the hardness of garlic cloves increased over 18 days of processing from 25.52 to 107.77 N. It is very interesting because the increase in hardness during ageing can be a sign of excessive loss of water and subsequently proof of the intensive drying of garlic cloves. However, based on our experience, values higher than 35 N reflect the firm and stiff texture of garlic cloves.

Differences between the garlic varieties (both before and after ageing) were also observed; the hardness ranged from 31.12 to 41.00 N for the FG samples and from 1.47 to 4.73 N for the FG samples. However, these differences were statistically significant ($p < 0.05$) only for the FG samples, but for the BG samples, the differences were not significant ($p > 0.05$). In other words, although there can be differences between the FG varieties, the differences are somewhat diminished after ageing. Considering the hardness and the results from the sensory analysis (Figure 1A), the pleasantness of the texture seems to be highly associated with the hardness of a BG clove ($r = -0.793$, $p < 0.05$). The panelists evaluated harder samples with a lower sensory score; on the other hand, softer samples were evaluated better.

It is, however, hard to compare our results with those of the other authors as there are no publications describing the influence of variety on the hardness of BG. According to Ríos-Ríos et al. [5], there is generally a lack of BG-texture parameters described in the literature. The authors, nevertheless, suggest that if the moisture of BG is between 40 and 50%, it would have a desirable soft texture, and when the moisture is below 40%, the texture is too hard. Results of this study show that hardness can also be a good predictor of sensory quality.

Color and appearance are one of the most important parameters of food quality that consumers evaluate as a first factor that subsequently affects the decision whether to eat the product or not. Color analysis of garlic samples, expressed in the CIE L*a*b* system, revealed that both factors (variety and ageing) significantly influence all color parameters (L*, a* and b*), as shown in Table 4.

Table 4. Color analysis of fresh and black garlic varieties in the CIE L*a*b* system.

Variety	L* (Lightness)		a* (Redness)		b* (Yellowness)	
	Fresh	Black	Fresh	Black	Fresh	Black
Bjetin	82.67 ± 0.50^{A}	10.64 ± 2.97^{dB}	1.14 ± 0.16^{B}	7.38 ± 2.86^{abA}	22.43 ± 1.45^{A}	7.36 ± 2.12^{abB}
Vekan	83.13 ± 1.68^{A}	11.35 ± 1.98^{cdB}	1.74 ± 0.09^{B}	9.50 ± 1.66^{aA}	21.04 ± 0.71^{A}	10.82 ± 2.20^{aB}
Havel	82.75 ± 1.47^{A}	13.35 ± 4.93^{cdB}	1.56 ± 0.24^{B}	3.99 ± 1.10^{cA}	22.10 ± 0.79^{A}	3.50 ± 1.49^{bcB}
Ivan	83.77 ± 1.52^{A}	23.90 ± 2.30^{aB}	0.83 ± 0.33^{B}	2.17 ± 1.43^{cA}	22.27 ± 1.88^{A}	1.66 ± 1.50^{cB}
Rusák	82.73 ± 0.58^{A}	17.70 ± 1.01^{bcB}	2.02 ± 0.20^{A}	1.26 ± 0.21^{cB}	23.18 ± 0.94^{A}	1.55 ± 0.46^{cB}
Havran	85.18 ± 1.54^{A}	23.32 ± 2.75^{aB}	0.58 ± 0.28^{B}	3.74 ± 1.69^{cA}	20.92 ± 1.10^{A}	3.17 ± 2.19^{bcB}
Lukan	87.44 ± 0.83^{A}	26.51 ± 1.71^{aB}	2.09 ± 0.28^{B}	4.40 ± 0.25^{bcA}	22.27 ± 1.03^{A}	4.76 ± 1.29^{bcB}
Average	83.95 ± 1.78^{A}	18.11 ± 6.53^{B}	1.42 ± 0.56^{B}	4.63 ± 3.04^{A}	22.03 ± 0.87^{A}	4.69 ± 7.45^{B}

Results are presented as mean ± standard deviation ($n = 3$); $^{a–d}$ values with different superscripts within a column differ significantly ($p < 0.05$); ^{A,B} values with different superscripts within a row and the same analysis differ significantly ($p < 0.05$).

During the ageing process, the color of all the samples turned from white to dark brown or black. The black color, which is a good indicator of the quality of the BG, is caused by the formation of brown color compounds usually associated with the advanced stage of the Maillard reaction [5].

Regarding lightness (L* value), values ranged between 82.67 and 87.44 in the FG samples, with an average value of 83.95 and between 10.64 and 26.51 in the BG samples (average 18.11). Thus, the lightness decreased almost five times during the ageing process, which was found to be statistically significant ($p < 0.05$). This result is in line with earlier reports [10,13].

No statistically significant differences ($p > 0.05$) were observed between the FG varieties. On the other hand, significant differences ($p < 0.05$) were observed between the BG samples.

The redness (a* value) of the garlic samples was also affected by variety and ageing ($p < 0.05$). The FG samples had lower redness (average 1.42) than the BG samples (average 4.63); thus, the value increased significantly during the ageing process. However, only minor differences (0.58–2.09) were measured between the garlic varieties before ageing (FG samples), but substantial differences were observed in the aged BG samples, where the redness ranged from 1.26 to 9.50.

As in the case of redness, yellowness (b* value) was also affected by both factors, ageing and variety, significantly ($p < 0.05$). The ageing process caused a drastic decrease of the b* value from 22.03 (FG) to 4.69 (BG) on average. Although no significant differences ($p > 0.05$) were observed between the FG varieties, after the ageing process, the BG varieties differed significantly with a very wide range (1.55 to 10.82).

3.5. The Relationships between Properties of Fresh and Black Garlic

Although two-way ANOVA can reveal the statistically significant influence of different factors (in this case, the influence of the ageing process and variety) on the quality parameters of BG, we wanted to find out the association between the parameters of FG and BG. The relationship between the moisture, pH, total soluble solids, DPPH, FRAP, TPC, hardness, L, and the a* and b* values of the FG and BG varieties is shown in Table 5.

Table 5. Correlation (r) between observed parameters of fresh garlic (FG) and black garlic (BG).

Parameter	FG × BG (r)
Moisture	0.814 ††
pH	−0.102
Total soluble solids	0.390
DPPH assay	0.588
FRAP assay	0.355
TPC	−0.051
Hardness	0.774 †
L*	0.804 ††
a*	0.071
b*	−0.481

The symbol † denotes statistical significance: † at $p < 0.05$ and †† at $p < 0.01$; $n = 7$. DPPH: 2,2-diphenyl-1-picrylhydrazyl; FRAP: ferric reducing antioxidant power; TPC: total polyphenol content.

The correlation analysis brought interesting results. Among all the examined traits, only moisture, hardness, and lightness (L* value) were found to be statistically significant ($p < 0.05$). In other words, these parameters of BG seem to be tightly related to the properties of FG before processing, e.g., the moister the FG is, the moister the BG is ($r = 0.814$, $p < 0.05$).

It was also found that there was a high positive correlation between the hardness of the FG and BG samples ($r = 0.774$, $p < 0.05$), thus it can be stated that the hardness of the BG is strongly affected by the hardness of the FG. It can be pointed out from the results that it could be recommended to choose an FG variety with a softer texture to produce BG with good texture properties.

An interesting fact is that a high correlation with statistical significance was found between the L* value of the fresh and BG samples ($r = 0.804$, $p < 0.05$). This means that the lighter the FG is, the lighter the BG is. The lightness of white garlic seems to be a relatively good predictor of the lightness of the BG.

On the other hand, the association between pH, total soluble solids, DPPH, FRAP, TPC, and the a* and b* values of FG and BG were relatively small and without any statistical significance ($p > 0.05$). Regarding the association between the DPPH values of the FG samples and the BG samples, only a small and statistically insignificant correlation ($r = 0.588$, $p > 0.05$) was found, which means that the antioxidant activity of BG is not related to the antioxidant activity of FG. Thus, it seems that the antioxidant activity of finished BG is affected by other parameters, and the antioxidant activity of FG is not a suitable parameter for the prediction of the antioxidant activity of BG. This statement can be also applied to the FRAP assay ($r = 0.355$, $p > 0.05$). A similar trend can be seen in a paper published by Toledano Medina et al. [15]. It could be hypothesized that other parameters, such as the concentration of organosulfur compounds and amino acid composition and the content of reducing sugars, could act as a predictor for the increase in antioxidant activity in BG. As mentioned by Yilmaz and Toledo [50], reducing sugars and amino acids enter the Maillard reaction, which could result in the production of compounds with high antioxidant activity.

Thus pH, total soluble solids, DPPH, FRAP, TPC, and the a* and b* values of BG are probably independent from FG, which implies that other factors play a more important role and influence these parameters more. Sunanta et al. [14] reported similar results. According to the above-mentioned authors, changes in the physicochemical traits were independent from the genotype of the garlic sample.

Additional correlation analysis, showing interesting associations, is presented in the Supplementary Material (Table S1). For instance, we found a significant correlation between the yellowness (b*) of FG and the pleasantness of odor of BG ($r = 0.76$, $p < 0.05$). However, we have not found any relevant explanation in the current literature.

In general, more research should be done to estimate the associations between the FG and BG parameters. Moreover, a higher number of varieties from different growing locations (countries or continents) should be examined to fully understand the relationships between all the evaluated parameters.

4. Conclusions

In this study, seven varieties of FG (namely Bjetin, Vekan, Havel, Ivan, Rusák, Havran, and Lukan) were examined and used to manufacture BG to evaluate the effects of different FG varieties and the ageing process on the properties of BG. Based on the results obtained, the following conclusions can be drawn. It is evident that the studied parameters were positively affected by the ageing process in all varieties, mainly the antioxidant activity, and the samples of BG obtained favorable attributes. There is a significant influence of variety on the physicochemical parameters of BG, namely on moisture, total soluble solids and total polyphenol content, pH, antioxidant activity, texture, and color. The ageing process also significantly affected all the studied traits of garlic, which is in line with previous reports [14,15].

In addition to the physicochemical parameters, sensory properties were also evaluated in all seven BG varieties, and the results show that the varieties of FG used in this study could have only a minor impact on the sensory properties of FG, except for the texture, where significant differences were found.

Moreover, in the correlation analysis, it was found that most of the studied parameters of BG seem to be independent of the properties of FG, with the exception of the moisture content, lightness, and hardness of the garlic cloves. Thus, it implies that most of the studied traits of FG cannot be used as a predictor of BG quality. For example, the FG with the highest antioxidant activity compared to other varieties does not have to have the highest antioxidant activity after ageing.

This suggests that further research is still needed to find out what parameters affect the quality of BG (e.g., amino acid composition, reducing sugars concentration, or organosulfur compounds) as the current knowledge is insufficient in this aspect. Moreover, experiments involving a higher number of varieties originating from European or Asian countries are needed for a deeper understanding of the influence of the variety of FG on the quality of BG.

5. Patents

The patent resulting from this work is Czech Patent No. 308 653: Method of preparing black garlic with antioxidant activity and black garlic prepared in this way.

Supplementary Materials: The following are available online at https://www.mdpi.com/article/10.3390/foods10112703/s1, Table S1: correlation analysis of properties of fresh garlic and black garlic.

Author Contributions: Conceptualization J.B.; methodology, J.B.; formal analysis, J.B.; investigation, J.B., I.L., P.P.d.M., F.L. and M.J.; writing—original draft preparation, J.B.; writing—review and editing, J.B., F.L., E.S., J.T. and N.V.; visualization, J.B.; supervision, J.B. and P.S.; project administration, J.B., E.S. and P.S.; funding acquisition, J.B., E.S., P.S. and J.K. All authors have read and agreed to the published version of the manuscript.

Funding: This research was funded by (1) the University of South Bohemia in České Budějovice, grant numbers GAJU057/2019/T and GAJU028/2019/Z, (2) the Ministry of Agriculture of the Czech Republic, grant number QK1910302, (3) the Ministry of Education, Youth and Sports of the Czech Republic, project SustES–Adaptation strategies for sustainable ecosystem services and food security under adverse environmental conditions (CZ.02.1.01/0.0/0.0/16_019/0000797) and (4) METROFOOD-CZ, project MEYS, Grant No: LM2018100.

Data Availability Statement: The datasets generated for this study are available on reasonable request to the corresponding author.

Acknowledgments: The authors would like to thank Justin Calvin Schaefer for the proofreading of the manuscript.

Conflicts of Interest: The authors declare no conflict of interest.

References

1. Qiu, Z.; Zheng, Z.; Zhang, B.; Sun-Waterhouse, D.; Qiao, X. Formation, nutritional value, and enhancement of characteristic components in black garlic: A review for maximizing the goodness to humans. *Compr. Rev. Food Sci. Food Saf.* **2020**, *19*, 801–834. [CrossRef]
2. Botas, J.; Fernandes, Â.; Barros, L.; Alves, M.J.; Carvalho, A.M.; Ferreira, I.C.F.R. A comparative study of black and white *Allium sativum* L.: Nutritional composition and bioactive properties. *Molecules* **2019**, *24*, 2194. [CrossRef]
3. Zhang, X.; Li, N.; Lu, X.; Liu, P.; Qiao, X. Effects of temperature on the quality of black garlic. *J. Sci. Food Agric.* **2016**, *96*, 2366–2372. [CrossRef]
4. Ryu, J.H.; Kang, D. Physicochemical properties, biological activity, health benefits, and general limitations of aged black garlic: A review. *Molecules* **2017**, *22*, 919. [CrossRef]
5. Ríos-Ríos, K.L.; Montilla, A.; Olano, A.; Villamiel, M. Physicochemical changes and sensorial properties during black garlic elaboration: A review. *Trends Food Sci. Tech.* **2019**, *88*, 459–467. [CrossRef]
6. Choi, I.S.; Cha, H.S.; Lee, Y.S. Physicochemical and antioxidant properties of black garlic. *Molecules* **2014**, *19*, 16811–16823. [CrossRef] [PubMed]
7. Moreno-Ortega, A.; Pereira-Caro, G.; Ordóñez, J.L.; Moreno-Rojas, R.; Ortíz-Somovilla, V.; Moreno-Rojas, J.M. Bioaccessibility of bioactive compounds of 'fresh garlic' and 'black garlic' through in vitro gastrointestinal digestion. *Foods* **2020**, *9*, 1582. [CrossRef] [PubMed]
8. Toledano Medina, M.A.; Merinas-Amo, T.; Fernández-Bedmar, Z.; Font, R.; del Río-Celestino, M.; Pérez-Aparicio, J.; Moreno-Ortega, A.; Alonso-Moraga, A.; Moreno-Rojas, R. Physicochemical characterization and biological activities of black and white garlic: In vivo and in vitro assays. *Foods* **2019**, *8*, 220. [CrossRef] [PubMed]
9. Li, N.Y.; Lu, X.M.; Pei, H.B.; Qiao, X.G. Effect of freezing pretreatment on the processing time and quality of black garlic. *J. Food Process Eng.* **2015**, *38*, 329–335. [CrossRef]
10. Karnjanapratum, S.; Supapvanich, S.; Kaewthong, P.; Takeungwongtrakul, S. Impact of steaming pretreatment process on characteristics and antioxidant activities of black garlic (*Allium sativum* L.). *J. Food Sci. Technol.* **2021**, *58*, 1869–1876. [CrossRef]
11. Martins, N.; Petropoulos, S.; Ferreira, I.C.F.R. Chemical composition and bioactive compounds of garlic (*Allium sativum* L.) as affected by pre- and post-harvest conditions: A review. *Food Chem.* **2016**, *211*, 41–50. [CrossRef] [PubMed]
12. Grégrová, A.; Čížková, H.; Bulantová, I.; Rajchl, A.; Voldřich, M. Characteristics of garlic of the Czech origin. *Czech J. Food Sci.* **2013**, *31*, 581–588. [CrossRef]
13. Najman, K.; Sadowska, A.; Hallmann, E. Evaluation of bioactive and physicochemical properties of white and black garlic (*Allium sativum* L.) from conventional and organic cultivation. *Appl. Sci. Basel* **2021**, *11*, 874. [CrossRef]
14. Sunanta, P.; Chung, H.H.; Kunasakdakul, K.; Ruksiriwanich, W.; Jantrawut, P.; Hongsibsong, S.; Sommano, S.R. Genomic relationship and physiochemical properties among raw materials used for Thai black garlic processing. *Food Sci. Nutr.* **2020**, *8*, 4534–4545. [CrossRef] [PubMed]

15. Toledano Medina, M.A.; Pérez-Aparicio, J.; Moreno-Ortega, A.; Moreno-Rojas, R. Influence of variety and storage time of fresh garlic on the physicochemical and antioxidant properties of black garlic. *Foods* **2019**, *8*, 314. [CrossRef]
16. Vácha, F.; Bedrníček, J.; Smetana, P.; Samková, E.; Kadlec, J.; Jirotková, D.; Tůma, K. Method of Preparing Black Garlic with Antioxidant Activity and Black Garlic Prepared in This Way. CZ PV 2020-44 No. 308653. 17 December 2020. Available online: https://isdv.upv.cz/webapp/!resdb.pta.frm (accessed on 20 August 2021).
17. Brand-Williams, W.; Cuvelier, M.E.; Berset, C. Use of a free-radical method to evaluate antioxidant activity. *Food Sci. Technol.-Lebensm.-Wiss. Technol.* **1995**, *28*, 25–30. [CrossRef]
18. Dudonné, S.; Vitrac, X.; Coutière, P.; Woillez, M.; Mérillon, J.M. Comparative study of antioxidant properties and total phenolic content of 30 plant extracts of industrial interest using DPPH, ABTS, FRAP, SOD, and ORAC assays. *J. Agric. Food Chem.* **2009**, *57*, 1768–1774. [CrossRef]
19. Lachman, J.; Hosnedl, V.; Pivec, V. Polyphenols in cereals and their positive and negative role in human and animal nutrition. In *Cereals for Human Health and Preventive Nutrition*; Vaculová, K., Ehrenbergerová, J., Eds.; MZLU: Brno, Czech Republic, 1998; pp. 118–124.
20. Zhang, X.; Shi, Y.; Wang, L.; Li, X.; Zhang, S.; Wang, X.; Jin, M.; Hsiao, C.D.; Lin, H.; Han, L.; et al. Metabolomics for biomarker discovery in fermented black garlic and potential bioprotective responses against cardiovascular diseases. *J. Agric. Food Chem.* **2019**, *67*, 12191–12198. [CrossRef]
21. Ahn, S.J.; Lee, A.; Min, S.S.; In, S.; Kim, E.; Kim, H.J. Comparison of physicochemical characteristics of garlic produced from South Korea and China. *J. Food Sci.* **2019**, *84*, 1806–1811. [CrossRef]
22. Shin, J.-H.; Lee, S.-J.; Jung, W.-J.; Kang, M.-J.; Sung, N.-J. Physicochemical characteristics of garlic (*Allium sativum* L.) on collected from the different regions. *J. Agric. Life Sci.* **2011**, *45*, 103–114.
23. Xiong, F.; Dai, C.H.; Hou, F.R.; Zhu, P.P.; He, R.H.; Ma, H.L. Study on the ageing method and antioxidant activity of black garlic residues. *Czech. J. Food Sci.* **2018**, *36*, 88–97. [CrossRef]
24. Liang, T.; Wei, F.; Lu, Y.; Kodani, Y.; Nakada, M.; Miyakawa, T.; Tanokura, M. Comprehensive NMR analysis of compositional changes of black garlic during thermal processing. *J. Agric. Food Chem.* **2015**, *63*, 683–691. [CrossRef]
25. Toledano-Medina, M.A.; Pérez-Aparicio, J.; Moreno-Rojas, R.; Merinas-Amo, T. Evolution of some physicochemical and antioxidant properties of black garlic whole bulbs and peeled cloves. *Food Chem.* **2016**, *199*, 135–139. [CrossRef] [PubMed]
26. Kimura, S.; Tung, Y.C.; Pan, M.H.; Su, N.W.; Lai, Y.J.; Cheng, K.C. Black garlic: A critical review of its production, bioactivity, and application. *J. Food Drug Anal.* **2017**, *25*, 62–70. [CrossRef] [PubMed]
27. Moreno, F.J.; Corzo-Martínez, M.; del Castillo, M.D.; Villamiel, M. Changes in antioxidant activity of dehydrated onion and garlic during storage. *Food Res. Int.* **2006**, *39*, 891–897. [CrossRef]
28. Piluzza, G.; Bullitta, S. Correlations between phenolic content and antioxidant properties in twenty-four plant species of traditional ethnoveterinary use in the Mediterranean area. *Pharm. Biol.* **2011**, *49*, 240–247. [CrossRef]
29. Everette, J.D.; Bryant, Q.M.; Green, A.M.; Abbey, Y.A.; Wangila, G.W.; Walker, R.B. Thorough study of reactivity of various compound classes toward the Folin-Ciocalteu reagent. *J. Agric. Food Chem.* **2010**, *58*, 8139–8144. [CrossRef]
30. Huang, D.J.; Ou, B.X.; Prior, R.L. The chemistry behind antioxidant capacity assays. *J. Agric. Food Chem.* **2005**, *53*, 1841–1856. [CrossRef]
31. Kim, J.S.; Kang, O.J.; Gweon, O.C. Comparison of phenolic acids and flavonoids in black garlic at different thermal processing steps. *J. Funct. Foods* **2013**, *5*, 80–86. [CrossRef]
32. Chang, W.C.; Chen, Y.T.; Chen, H.J.; Hsieh, C.W.; Liao, P.C. Comparative UHPLC-Q-Orbitrap HRMS-based metabolomics unveils biochemical changes of black garlic during aging process. *J. Agric. Food Chem.* **2020**, *68*, 14049–14058. [CrossRef]
33. Chi, M.-C.; Lo, H.-F.; Lin, M.-G.; Chen, Y.-Y.; Lin, L.-L.; Wang, T.-F. Application of *Bacillus licheniformis* γ-glutamyltranspeptidase to the biocatalytic synthesis of γ-glutamyl-phenylalanine. *Biocatal. Agric. Biotechnol.* **2017**, *10*, 278–284. [CrossRef]
34. Matsutomo, T.; Stark, T.D.; Hofmann, T. In vitro activity-guided identification of antioxidants in aged garlic extract. *J. Agric. Food Chem.* **2013**, *61*, 3059–3067. [CrossRef] [PubMed]
35. Molina-Calle, M.; de Medina, V.S.; Priego-Capote, F.; de Castro, M.D.L. Establishing compositional differences between fresh and black garlic by a metabolomics approach based on LC-QTOF MS/MS analysis. *J. Food Compos. Anal.* **2017**, *62*, 155–163. [CrossRef]
36. Wang, X.; Yang, Y.; Liu, R.; Zhou, Z.; Zhang, M. Identification of antioxidants in aged garlic extract by gas chromatography-mass spectrometry and liquid chromatography-mass spectrometry. *Int. J. Food Prop.* **2016**, *19*, 474–483. [CrossRef]
37. Zhang, H.; Zhu, L.; Chen, H.W. Direct molecular analysis of garlic using internal extractive electrospray ionization mass spectrometry. *Chin. J. Anal. Chem.* **2014**, *42*, 1634–1638. [CrossRef]
38. Metlin. Available online: http://metlin.scripps.edu (accessed on 20 August 2021).
39. MassBank. Available online: http://massbank.eu (accessed on 20 August 2021).
40. Zhang, N.; Huang, X.; Zeng, Y.; Wu, X.; Peng, X. Study on prebiotic effectiveness of neutral garlic fructan in vitro. *Food Sci. Human Wellness* **2013**, *2*, 119–123. [CrossRef]
41. Kodera, Y.; Kurita, M.; Nakamoto, M.; Matsutomo, T. Chemistry of aged garlic: Diversity of constituents in aged garlic extract and their production mechanisms via the combination of chemical and enzymatic reactions. *Exp. Ther. Med.* **2020**, *19*, 1574–1584. [CrossRef]
42. Yu, J.H.; Shan, Y.; Li, S.; Zhang, L.F. Potential contribution of Amadori compounds to antioxidant and angiotensin I converting enzyme inhibitory activities of raw and black garlic. *LWT-Food Sci. Technol.* **2020**, *129*. [CrossRef]

43. Ho, S.C.; Su, M.S. Evaluating the anti-neuroinflammatory capacity of raw and steamed garlic as well as five organosulfur compounds. *Molecules* **2014**, *19*, 17697–17714. [CrossRef]
44. Kim, S.; Park, S.L.; Lee, S.; Lee, S.Y.; Ko, S.; Yoo, M. UPLC/ESI-MS/MS analysis of compositional changes for organosulfur compounds in garlic (*Allium sativum* L.) during fermentation. *Food Chem.* **2016**, *211*, 555–559. [CrossRef]
45. Corzo-Martínez, M.; Corzo, N.; Villamiel, M. Biological properties of onions and garlic. *Trends Food Sci. Tech.* **2007**, *18*, 609–625. [CrossRef]
46. Yang, J.; Bai, W.D.; Zeng, X.F.; Cui, C. Gamma glutamyl peptides: The food source, enzymatic synthesis, kokumi-active and the potential functional properties—A review. *Trends Food Sci. Tech.* **2019**, *91*, 339–346. [CrossRef]
47. Chung, L.Y. The antioxidant properties of garlic compounds: Allyl cysteine, alliin, allicin, and allyl disulfide. *J. Med. Food* **2006**, *9*, 205–213. [CrossRef]
48. Beato, V.M.; Orgaz, F.; Mansilla, F.; Montaño, A. Changes in phenolic compounds in garlic (*Allium sativum* L.) owing to the cultivar and location of growth. *Plant. Food Hum. Nutr.* **2011**, *66*, 218–223. [CrossRef] [PubMed]
49. Martínez-Casas, L.; Lage-Yusty, M.; López-Hernández, J. Changes in the aromatic profile, sugars, and bioactive compounds when purple garlic is transformed into black garlic. *J. Agric. Food Chem.* **2017**, *65*, 10804–10811. [CrossRef]
50. Yilmaz, Y.; Toledo, R. Antioxidant activity of water-soluble Maillard reaction products. *Food Chem.* **2005**, *93*, 273–278. [CrossRef]

foods

Review

By-Products Revalorization with Non-Thermal Treatments to Enhance Phytochemical Compounds of Fruit and Vegetables Derived Products: A Review

Marina Cano-Lamadrid [1,*] and Francisco Artés-Hernández [2]

[1] Food Quality and Safety Group, Department of Agrofood Technology, Universidad Miguel Hernández, Ctra. Beniel, Km 3.2, Orihuela, 03312 Alicante, Spain
[2] Postharvest and Refrigeration Group, Department of Agronomical Engineering and Institute of Plant Biotechnology, Universidad Politécnica de Cartagena, Cartagena, 30203 Murcia, Spain; fr.artes-hdez@upct.es
* Correspondence: marina.canol@umh.es

Abstract: The aim of this review is to provide comprehensive information about non-thermal technologies applied in fruit and vegetables (F&V) by-products to enhance their phytochemicals and to obtain pectin. Moreover, the potential use of such compounds for food supplementation will also be of particular interest as a relevant and sustainable strategy to increase functional properties. The thermal instability of bioactive compounds, which induces a reduction of the content, has led to research and development during recent decades of non-thermal innovative technologies to preserve such nutraceuticals. Therefore, ultrasounds, light stresses, enzyme assisted treatment, fermentation, electro-technologies and high pressure, among others, have been developed and improved. Scientific evidence of F&V by-products application in food, pharmacologic and cosmetic products, and packaging materials were also found. Among food applications, it could be mentioned as enriched minimally processed fruits, beverages and purees fortification, healthier and "clean label" bakery and confectionary products, intelligent food packaging, and edible coatings. Future investigations should be focused on the optimization of 'green' non-thermal and sustainable-technologies on the F&V by-products' key compounds for the full-utilization of raw material in the food industry.

Keywords: zero waste; bioactive compounds; green technologies; nutraceuticals; circular economy

Citation: Cano-Lamadrid, M.; Artés-Hernández, F. By-Products Revalorization with Non-Thermal Treatments to Enhance Phytochemical Compounds of Fruit and Vegetables Derived Products: A Review. *Foods* **2022**, *11*, 59. https://doi.org/10.3390/foods11010059

Academic Editor: Urszula Tylewicz

Received: 12 November 2021
Accepted: 24 December 2021
Published: 27 December 2021

1. Introduction

The Food and Agriculture Organization (FAO) of the United Nations indicates that around a third of all food production is globally lost or wasted at some point in the food chain [1,2]. Losses vary a lot depending on the chain considered and in the case of fruit and vegetables (F&V) can reach up to 50%. Within the F&V processing operations about 25% to 30% of waste is produced [3]. The most important causes of losses on farms include inappropriate timing for harvesting, overproduction, underutilized products, climatic conditions, harvesting and handling practices, and inadequate postharvest technology [4]. At the World Food Summit held in 2017 organized by FAO, the challenges needed to achieve food stability and food availability were identified and a roadmap was proposed to reduce 50% of food waste by 2050. The principles of eco-innovation are the industrial ecology and the circular economy ("zero waste" and the use of wastes as raw materials) [5]. Among the challenges that arise different actions stand out, such as the revaluation of waste in the various stages of the production process and logistics, and/or the use of waste products (by-products) as starting raw material for the production of products with greater added value [6] and then called co-products.

The handling and processing of these raw materials generates a large number of commodity by-products being undervalued and underused, and although there are some minor uses such as the production of biomass and animal feed, these strategies do not guarantee an efficient use of this material that could offer interesting possibilities for the agri-food industry and the reduction of this environmental problem [2,7]. Horticultural by-products mainly are peels, pomace and seeds, which could be a potential good source of bioactive compounds with high added-value such as pectins, proteins, polysaccharides, flavor compounds, dietary fibers, and phytochemicals compounds [8]. To continue being relevant, it is necessary to further strengthen and dynamize the sector through the development of appropriate postharvest strategies to increase shelf life, and a model for the enhancement of horticultural by-products through the incorporation of emerging and sustainable 'Green Technologies' to its revalorization [9]. The strategies to revalue horticultural by-products can lead to a change in the productive model of the sector and evolve towards a more diversified and sustainable circular economy, giving more added value and competitiveness. These strategies can be focused on obtaining potential ingredients for the food industry, cosmetics, and/or the pharmaceutical industry. The use of plant by-products supports the low-carbon economy by using renewable resources, offering environmental and economic benefits and improving efficiency in the food industry [7,10].

Nowadays, the tendency in the food market is driven by different reasons such as health and sustainability. This phenomenon is expressed in the consumer's interest in healthy natural foods based on plant products. Food producers are increasingly striving to meet these trends by offering "Clean label" foods or ingredients. Currently, there is no legislation related to the aforementioned concept, but the growing demand for this type of food reflects the desire of consumers for food to be more "natural", wholesome, premium, and use environmentally friendly technologies [11,12]. The extracts obtained from F&V by-products can fulfill a series of technological functions such as being colorants, antioxidants, flavors or antimicrobial agents, or act directly as ingredients to enrich or improve the functional properties of some food becoming a supplemented or fortified commodity [8,13–16].

In order to obtain value-added compounds with functional (nutraceuticals) and techno-functional (pigments) properties, technologies have been developed for each side-product generated from agro-food industries [17]. Conventional and traditional thermal methods are still in use, although high energy consumption, the degradation of thermolabile nutritional compounds, and sensory quality changes occur, which require the adoption of sustainable preservation techniques without altering the sensory and nutritional quality of foods [18]. The stability of nutraceuticals is affected by different factors (temperature, pH, light stress, presence or absence of oxygen, and enzymatic activity). Focusing on temperature factor, there is increased interest in improving and optimizing non-thermal technologies to avoid degradation of key compounds, jointed with sustainable methods [19]. Among non-thermal technologies, the most common are ultrasound-assisted extraction, high-pressure processing, light stresses, fermentation technology, electro-technologies, and enzyme-assisted extraction. More detailed information is described in Section 3.

Therefore, this review is focused on generating comprehensive information about non-thermal technologies applied in F&V by-products to enhance phytochemical compounds such as polyphenols, pigments and nitrogen/organosulfur derivates, and to obtain pectin. Moreover, the potential use of such compounds will also be of a particular interest to this review.

2. Fruit and Vegetables By-Products as a Source of Valuable Compounds

Scientific research and development have been greatly increased in the last decades in the field of extraction and the application of bioactive compounds re-valorized from F&V processing by-products [20]. A large number of molecules with added-value (simple sugars, carbohydrates, polysaccharides, pectin, fibers, phenolic acids, carotenoids, tocopherols, flavonoids, vitamins and aromatic compounds) from F&V by-products can be used in the food, cosmetic, or pharma industry (co-products) [19,21,22]. This review will be focused on phytochemical compounds such as polyphenols, pigments, sulfur compounds, and pectins.

Nowadays, manufactures are focused on reducing the environmental impact of industrial by-products (zero waste and circular economy) and recovering bioactive compounds from agricultural by-products.

2.1. Phytochemical Compounds

Phytochemicals are defined as compounds obtained from plants, naturally biosynthesized in their secondary metabolism without any essential nutritional values. However, they present lots of health promoting properties according to their biological activity [23]. They are used for several purposes such as drugs, agrochemicals (biopesticides), and food additives (aroma, colorant agents). Phytochemicals were divided into different groups such as terpenoids (carotenoids and chlorophylls), polyphenols, alkaloids, nitrogen compounds, and organosulfur compounds (Figure 1). Three main subsections were made focusing on the greatest relevance groups for this review work: polyphenols (Section 2.1.1), pigments (Section 2.1.2) and organosulfur compounds (Section 2.1.3).

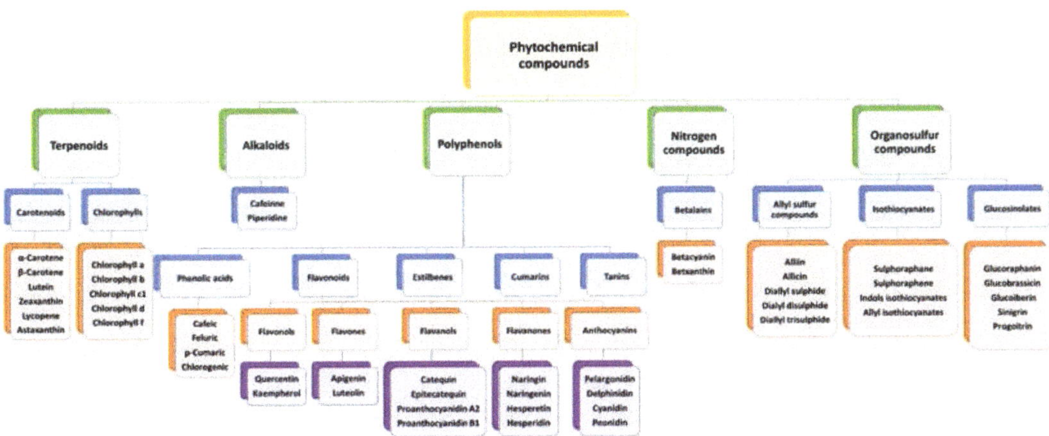

Figure 1. Classification of the main phytochemical compounds in fruit and vegetables (F&V) by-products.

2.1.1. Polyphenols

The most common key compounds from F&V by-products are polyphenols (Figure 1: phenolic acids and their polymeric derivatives, such as lignans, stilbenes, tannins, and flavonoids) in skins, pulp, seeds, or pomace [24]. Phenolic acids are common in F&V by-products such as apple pomace (chlorogenic acid, and cryptochlorogenic acid) [25], artichoke (bracts, leaves and stems) (chlorogenic acid) [26], mango kernel and leaves (gallic acid, and ellagic acid) [27], pomegranate peel (caffeic acid, chlorogenic acid, ellagic acid, and gallic acid) [28], potato peel (chlorogenic acid, ferulic, gallic, protocatechuic and caffeic acid) [29], tomato peel (3-caffeoylquinic acid, 5-caffeoylquinic acid, 3,4-di-O-caffeoylquinic acid, and 3,4,5-tri-caffeoylquinic acid) [30] and blueberry pomace (cinnamic acid derivatives) [22,31–33]. Among flavonoids, flavones, flavanones, anthocyanidins, and flavonols can be found in grape pomace (catechins, anthocyanins, stilbenes, and flavonol glycosides), onion skin (quercetin 3,40-O-diglucoside and quercetin 4-o-monoglucoside and isorhamnetin-3-glucoside) [34], tomato peel (lycopene, naringenin chalcone and naringenin) [30], apple pomace (hydroxycinnamates, phloretin glycosides, quercetin glycosides, catechins, procyanidins, and epicatechin) [25], figs peel (cyanidin-3-rutinoside, cyanidin-3,5-diglucoside, cyanidin-3-O-diglucoside, epitecatechin, catechin and quercetin-rutinoside) [32], blueberry pomace (anthocyanins and flavonol-glycosides) [33], and citrus peel (eriocitrin, hesperidin, and naringin) [22,31,32,35]. Some of them are pigments such as anthocyanins, and are explained in Section 2.1.2. jointed with other pigments (betalains, carotenoids, and chlorophylls).

2.1.2. Bioactive Pigments

Plant pigments are colored substances produced by plants and are important in controlling photosynthesis, growth, and development [36]. The market for natural colorants is experiencing a boom related to the "clean label" trend. It is worth mentioning that some of the main drivers for the increased demand of natural colorants are the health-promoting benefits of natural food colorants [37]. Researchers and the food industry are exploring stable natural colorants and new natural extracts from F&V by-products [37]. These by-products tissues are rich in betalains, anthocyanins, carotenoids, or/and chlorophylls. Figure 2 shows the classification of bioactive pigments and some examples of commodity by-products rich in these pigments.

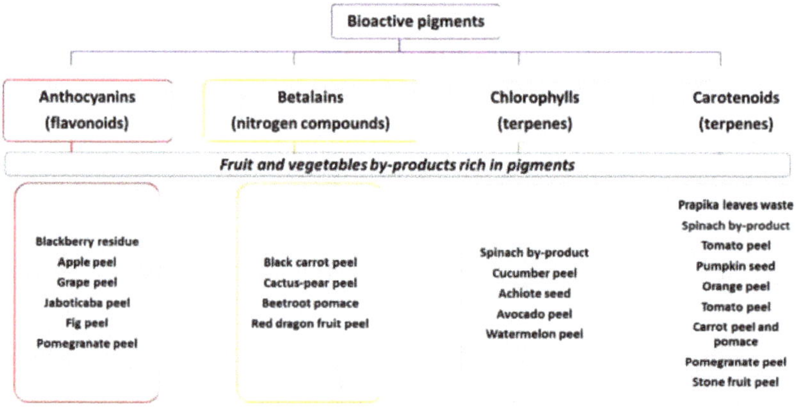

Figure 2. Types of the main bioactive pigments and examples of some fruits and vegetable (F&V) by-products rich in these pigments [8,32,38,39].

Water-Soluble Compounds: Anthocyanins and Betalains

Flavonoids are a group of secondary metabolites which belong to the class of phenylpropanoid and present a wide color range, from pale-yellow to blue. Among them, anthocyanins are responsible for the orange-to-blue colors; different parts of the plant present these compounds such as leaves, fruits, and seeds, among others. Wineries and juice manufacturer by-products are enriched sources for anthocyanin pigments that can be used as natural colorants for various food applications [38]. The use of anthocyanins as pigments (E-163) is accepted by the European Community [40].

Betalains are yellow-to-red nitrogen-containing compounds, derived from tyrosine. The use of betalains as pigments (E-162) is also accepted by the European Community and they are used in the production of jellies, jams, strawberry yogurt, among other products [41]. Betalains come from the underutilized biomass of red beetroot processing and from beetroot leaves [38,42].

Fat-Soluble Compounds: Carotenoids and Chlorophylls

Carotenoids are isoprenoids, and essential compounds of the photosystems in plants. They are responsible for the yellow-red coloration. Up to now, commercially available carotenoids synthesized chemically are being used as coloring compounds [43]. However, currently, these pigments can be obtained from F&V by-products. Carotenoids are often located in the same plant organs as anthocyanins, increasing color variety when they combine [43].

On the other hand, although chlorophylls can be used for coloring food products, there are limited available scientific reports on the use of F&V by-products for the extraction of chlorophylls and their further application as a colorant in food formulations [38].

2.1.3. Sulfur Compounds

Sulfur is an essential compound for the biosynthesis of phytoalexins, sulfur-containing glycosides (glucosinolates), and alliins, among others. Alliaceous (onion, garlic) and cruciferous vegetables (broccoli, cauliflower, radish, cabbage) are the main sources of sulfur compounds (contributed up to 42% of total sulphur intake) [44]. These metabolic compounds play a vital role in the physiology and protection of plants against several environmental stresses [45]. The alliaceous and brassica by-products (for example Bimi leaves [13] and broccoli by-products [46]) contain mainly glucosinolates as sulfur compounds. Glucosinolates can be found as not biologically active unhydrolysed compounds. However, these by-products present the myrosinase enzyme which produces several biologically active isothiocyanates and indoles, with health potential properties such as chemopreventive activity against cancer. Among them, sulforaphane is the most researched isothiocyanate from the degradation of glucoraphanin [47]. Bioactive sulfur compounds are degraded during processing, mainly by conventional thermal techniques. Even some of these compounds could not be formed by inactivation of the myrosinase enzyme.

2.2. Pectins

Pectin is a structural hetero-polysaccharide contained in the cell walls and abundant in the non-woody parts of plants, including by-products such as peel or pomace. Pectin presents beneficial properties for humans such as moderating the glycemic index and slowing gastric transit. The interaction of pectin and polyphenolic compounds contributes to systemic anti-inflammation [48]. Pectin is widely used in the food industry as a gelling agent, emulsifier, and carrier polymer for the encapsulation of food ingredients (it is an effective delivery vehicle for exogenous nutraceuticals), helping protect and promote the controlled release of biomolecules [48]. Pectin quality can be characterized by galacturonic acid content, degree of esterification and degree of methylation, affecting gelling properties [49]. Recent research summarized the characterization of the pectin composition of several F&V waste, especially form plant processing industry. One of the conclusions was that the pectin structures and recovery vary depending on the source and the applied extraction pectin as it can be observed in the Section 7 (focus on non-thermal technologies). Moreover, the information about changes in pectic polysaccharide composition after processing is essential for the industry, including the amount of uronic acid due to the requirement of the minimum of 65%. In addition, although more studies are needed, the rest of pectin below this limit could be useful in other applications [50].

3. Potential and Innovative Non-Thermal Techniques for Revalorization of Fruit & Vegetables By-Products

Due to the thermal instability of compounds (which means a reduction of their concentration level), non-thermal innovative technologies have been increasing during last decades [19], such as ultrasound-assisted extraction, high-pressure processing, light stresses, fermentation technology, electro-technologies, and enzyme-assisted extraction, among others [51,52]. Most of them are focused on the recovery of the above-mentioned compounds related to revalorization of F&V by-products [9,24]. Recovering of bioactive phytochemicals from F&V waste by non-thermal processes could improve the efficient production of potential bioactive ingredients [53].

3.1. Electro-Technologies: Pulses Electric Fields

Pulses Electric Fields (PEF) consist of subjecting the selected material to the intermittent application (<300 Hz) of electric fields at moderate-high intensity (0.1–20 kV/cm) and short duration (μs to ms) [54]. The main characteristic is the application of electric field pulsing on plant matrices that induces electro permeabilization (formation of located pores in cell membranes of cells), and the effect mainly depends on medium composition (conductivity) [55]. PEF technology has been defined as technology which requires fewer resources to produce nutritional with optimal sensory characteristics and longer shelf lives of products such as hummus, smoothies and juices [56]. Related to recovery bioactive compounds from F&V by-products, it enhances the specific recovery of bioactive intracellular compounds without increasing temperature or/and damaging the structure of the matrix. The obtained result depends on treatment intensity, physicochemical properties of the matrix and the tissues and cells composition. If the combination of the variables is optimized, reversible electroporation could occur (the membrane can return to its original state once the electric field application has finished) [54,57,58]. It is important to highlight that recent study indicated that pulsed electric field (PEF) treatment needs an optimization for more selective, quicker, and sustainable bio-active compounds extraction in the food industry [58]. Therefore, recent information about the optimal conditions of PEF were included in Tables 1–3.

3.2. Enzyme-Assisted Extraction

A novel green and non-thermal technology, enzyme-assisted technology, for bioactive compounds extraction such as phenolics and pectin has been developed during last decades for cosmetic, pharmaceutical and food applications. It is essential to highlight that enzyme-assisted extraction allows the use of F&V by-products providing a novel chance to give added-value to F&V waste [59]. The fundamental mechanism of the pectin, polyphenols, and pigments enzyme-assisted extraction from F&V by-products is based on the cell-wall degrading enzymes. These enzymes weaken, degrade partially or/and break down the cell wall polysaccharides, enhancing the possibility of the extraction of those compounds [60].

3.3. Fermentation

Fermentative processes can be classified according to different criteria. One of the most common is the group of batch fermentations which is based on the addition of the substrate and the key microorganism in the system at time zero. The produced key compounds cannot be obtained until the process is complete [61]. On the other hand, continuous and fed-batch fermentations microorganisms present another mechanism. The system can be reutilized for several batches, increasing its efficiency. In general, the industrial fermentations take place in liquid media, but sometimes solid-state fermentations microorganisms are applied. Related to fermentations and revalorizations of F&V by-products (fermentation-based valorization strategies), it has been recently developed the fermentation of date palm waste to produce lactic acid [62,63] and bioconversion of cocoa by-products using different microorganisms to obtain key enzymes, among other bioactive compounds [63,64].

3.4. High Hydrostatic Pressure

High Hydrostatic Pressure (HPP) is one of the non-thermal pasteurization processing technologies which is widely applied in the food industry [51,52]. It is a processing technique that uses a range of pressure from 100 to 900 MPa to increase shelf-life of the products due to the inactivation and elimination of microorganisms. The pressure can be applied through direct pressure and indirect pressure. HPP induces high pressure which causes severe damage to plant cells and leads to the diffusion of solvents and enhances the mass transfer and release of the extracts [65]. The uniformity of the pressure application is maintained during the process and it does not depend on the product size and geometry. It has been reported that this technique avoids no-desirable effects on texture characteristics. Moreover, this technique does not reach high temperatures, then protect characteristic flavor notes, color pigments nutrients, and antioxidant bioactive compounds which are degraded at high temperatures [66].

3.5. Light Stress

Plant by-products have been proposed as bio-factories of bioactive compounds through different induced postharvest abiotic stress mechanisms. Among them, one of the most promising techniques appears to be UV radiation, the spectrum is divided into three regions: UV-A (wavelength 320 to 400 nm), UV-B (wavelength 280 to 320 nm) and UV- C (wavelength 220–280 nm). The use of UV technology during post-harvest is an emerging technology to enhance the biosynthesis of bioactive compounds in the F&V industry, respectful with the environment, without generating waste [67,68]. The application of UV-B, alone or in combination with UV-C, has not been widely studied as a revalorization tool for maintaining and/or increasing the main key compounds in F&V by-products [69]. Although light-emitting diodes (LEDs) are increasingly adopted for the production of several vegetable modalities and for quality preservation during storage [70], influencing the metabolic pathways (biosynthesis of several bioactive compounds) [71–73]. No published information is already available concerning the effect of this light stress in F&V by-products. Recently, it has been concluded that a combination of different light stress techniques (UV-B + LEDs) could be a good strategy to enhance the bioactive compounds in commodities, being a potential tool for by-products revalorization [71].

3.6. Supercritical Fluid Extraction

Supercritical fluid extraction (SFE) is a recent extraction technique, and it is based on the use of the critical point of the solvent during the extraction. The combination of gas mass transfer and liquid solvation properties allows a high transfer mass (diffusion coefficients) than working below critical point. The majority of SFE studies have focused on the use of CO_2 due to its characteristics (non-toxic and cheap and can be easily removed after extraction) [74].

3.7. Ultrasound-Assisted Extraction (UAE)

Ultrasonication is an emerging non-thermal and green technology in the food sector, although it has been previously established in other sectors such as pharmacological. The fundamentals are based on the mechanical impact of the ultrasound waves, allowing deeper penetration of the solvent into the matrix ("sponge effect") [55]. Ultrasonication can be used with different doses (frequencies and time), which are classified as: (i) low-frequency (20 kHz–100 kHz); (ii) medium-frequency (100 kHz–1 MHz); and high-frequency ultrasonication (1 MHz–100 MHz) [75,76]. In food processing, the most common frequency range for the extraction of bioactive compounds and intensified synthesis is 20 kHz–100 kHz [51,52,77].

4. Scientific Literature Review about Non-Thermal Technologies Used for Revalorization of Fruits & Vegetable By-Products

The review is organized as a research paper. A scoping review was used to synthesize the evidence and assess the scope of the 71 studies on the topic. PRISMA Extension (PRISMA-ScR) approach was used for Scoping Reviews [78]. A comprehensive literature search using Scopus and ScienceDirect was performed in October 2021. Text words and controlled vocabulary for several concepts (Non-thermal, technologies, by-products, fruit, vegetable) within the titles, abstracts, and keywords were used. Only studies published in journals included in Journal Citation Reports (JCR) have been included. Only original research papers (Re) and reviews (Rw) including experimental design and data treatment were selected (Figure 3). This review is structured as follows: (i) the effect of non-thermal treatments on F&V by-products polyphenols; (ii) the effect of non-thermal treatments on F&V by-products pigments; and (iii) the effect of non-thermal treatments on F&V by-products pectin and sulfur compounds.

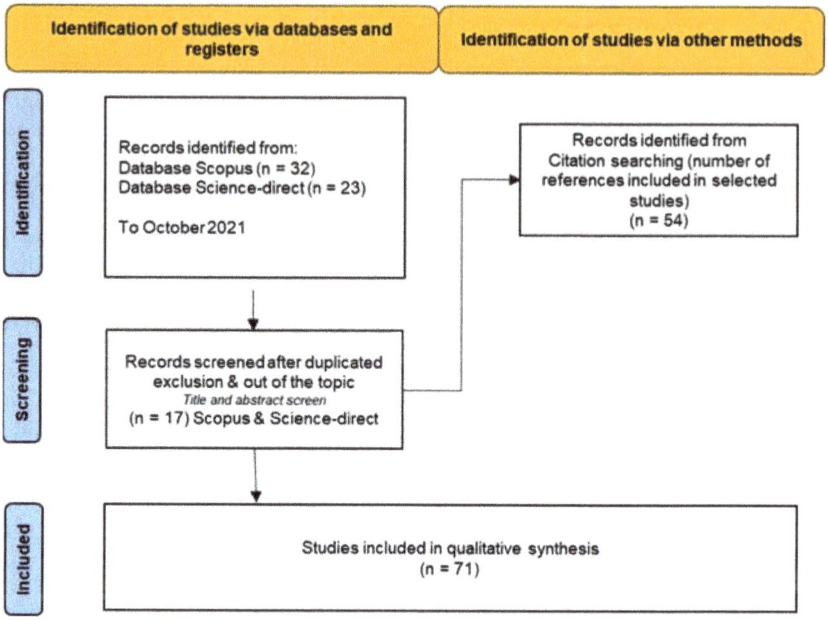

Figure 3. Flow diagram describing the study selection process of the scientific literature.

5. Effect of Non-Thermal Technologies on Fruit and Vegetable By-Products Polyphenols

Table 1 shows the non-thermal technologies applied in F&V by-products focusing on the main findings related to polyphenols. The table is divided in three parts: flavonols, polyphenols and flavonoids. The non-thermal technologies found were: solid-state fermentation, supercritical fluid extraction, ultrasounds, high pressure, hydrostatic pressure, electro-technologies, enzyme-assisted extraction, and light stress. Most of the investigations found (70%) were on fruit by-products or the wine and distillate industry, with only two studies related to vegetables (onion and broccoli) [79–81].

The recovery of bioactive compounds are mainly affected by varying the solvent concentration (ratio solvent:by-product), applied dose (wavelength, intensity, pressure, frequency), temperature and time (Table 1, Table 2 and Table 3) [82].

Table 1. Effect of non-thermal technologies on F&V b-products polyphenols (flavonols, total polyphenols, flavonoids).

	Non-Thermal Technology	By-Product	Findings	Reference
Flavonols	Solid-state Fermentation (*A. niger* and *R. oligosporus*)	Plum pomace	Increase of quercentin-3-glucoside (23 to 34 mg/100 g dry matter by *A. niger*; 22 to 24 mg/100 g dry matter by *R. oligosporus*), and quercentin-3-rutinoside (21 to 25 mg/100 g dry matter by *A. niger*) when 2×10^7 spores/g of solid was inoculated and fermentation took place during 14 days at 30 °C	[83]
	Solid-state Fermentation (*A. niger* and *R. oligosporus*)	Plum brandy distilleries waste	Increase of quercentin-3-glucoside (92 to 120 mg/100 g dry matter by *A. niger*; 92 to 110 mg/100 g dry matter by *R. oligosporus*), quercentin-3-rutinoside (42 to 64 mg/100 g dry matter by *A. niger*; 42 to 74 mg/100 g dry matter by *R. oligosporus*) and quercentin-3-galactoside (26 to 36 mg/100 g dry matter by *R. oligosporus*) when 2×10^7 spores/g of solid was inoculated and fermentation took place during 14 days at 30 °C	[83]
	Ultrasound assisted solid liquid extraction	Skins of red and yellow onions	Recovery of quercetin aglycona (118%) after extracted eight times with 20 mL Ethanol (85% *v*/*v*) for 15 min at 25 °C	[34,81]
	High hydrostatic pressure	Orange and lemon peels	More intense HPP conditions (500 MPa, 10 min), polyphenols decrease (lemon: 291.08 to 211.95 mg GAE/100 g fresh peel extracts; orange: 400 to 215.31 mg GAE/100 g fresh peel extracts).	[84,85]
	High hydrostatic pressure	Pineapple by-products	Accumulation of bromelain (increase of 350%) and TPC (increase of 36%) at 225 MPa, 8.5 min	[86,87]
	Electro-technologies	Mango peel	Recovery of polyphenols (+400%) at E = 13.3 kV/cm (160 kJ/kg); V = 40 kV (160 kJ/kg)	[54,88]
	Electro-technologies	Olive kernel	Recovery of polyphenols E = 13.3 kV/cm (0–141 kJ/kg), V = 40 kV (0–141 kJ/kg)	[54,89]
	Electro-technologies	Orange peel	Up to 159% in polyphenol extraction recovery after PEF pre-treatment at an electric field densities 1 kV/cm and 7 kV/cm (60 µs, 20 pulses, f = 1 Hz). Recovery of naringin and hesperidin increased ≈2- and 3-fold, respectively.	[84]
	Electro-technologies	Orange peel	Recovery of polyphenols (from 20%, to 159%) for orange peel PEF treated at E = 1–7 kV/cm (0.06–3.77 kJ/kg) + Pressing 5 bars	[54,90]
	Electro-technologies	Papaya peel and seeds	Recovery of polyphenols (>50%) at E = 13.3 kV/cm (160 kJ/kg); V = 40 kV (160 kJ/kg)	[54,91]
	Electro-technologies	Raspseeds stems and leaves	Recovery polyphenols (36–42%) at E = 0.2–5 kV/cm (0–700 kJ/kg)	[54,92]
	Electro-technologies	Raspseeds seeds	Recovery polyphenols (around 50%) at V = 40 kV (0–400 kJ/kg)	[54,93]
	Electro-technologies	Winery wastes and by-products (peel)	Recovery of polyphenols (42%) E = 5–10 kV/cm (1.8–6.7 kJ/kg)	[54,94]
	Electro-technologies	Winery wastes and by-products (pomace)	Recovery of polyphenols (>40%) at E = 13.3 kV/cm (0–564 kJ/kg) V = 40 kV (0–218 kJ/kg)	[54,95]
	Electro-technologies	Winery wastes and by-products (seed)	Recovery of polyphenols (>40%) at E = 8–20 kV/cm (0–53 kJ/kg) V = 40 kV (0–53 kJ/kg)	[54,96]
	Electro-technologies	Fermented grapes pomace	Increase of recovery by 1.2 kV/cm 18 kJ/kg 20 °C (the ratio of total anthocyanins to total flavan-3-ols was increased from 7.1 in non-treated to 9.0 in PEF-treated samples)	[97,98]

Table 1. *Cont.*

	Non-Thermal Technology	By-Product	Findings	Reference
Polyphenols	Electro-technologies	Winery wastes and by-products (grapes)	Increase 13% at 0.5 kV/cm, 50 pulses, 0.1 kJ/kg Increase 28% at 2.4 kV/cm, 50 pulses 2.3 kJ/kg	[99,100]
	Electro-technologies	Winery wastes and by-products (grapes)	Increase 34% at 0.7 kV/cm, 200 ms, 31 Wh/kg	[99,100]
	Electro-technologies	Winery wastes and by-products (vine shoots)	Up to 2-fold increase in TPC (Kaempferol, epicatechin, resveratrol) at 13.3 kV/cm, 0–1500 pulses, 50–762 kJ/kg/3 h compared to untreated	[99,101]
	Electro-tecnologies	Citrus peel (orange and pomelo)	Increase of polyphenols recovery, 16 mg/g dry matter for skins (for albedo + flavedo) (E = 10 kV/cm and 50% ethanol solution)	[102]
	Enzymed-assisted extraction	Grape residues	Novoferm® (1:10, 12 h and 40 °C) had the strongest effect on phenolic release (90%) from grape waste (100 mg of dry material was suspended in 1.4 mL of 0.2 M acetates buffer (pH 3.5)).	[60]
	Light stress (UV-B and UV-C; single and combined)	Bimi broccoli leaves and stalks	UV increased initial TPC of leaves/stalks up to 31–97/30–75%, 10 kJ/m^2 UV-B (UV-B10) + C induced the highest TPC increase (110%) in leaves while UV-B10 and UV-B10 + C led to the highest TPC of stalks after 48 h	[80]
	Optimized supercritical Fluid Extraction	Broccoli by-products	Decrease of polyphenols (<20%) at 400 bars, 40 °C, 5% of ethanol compared with conventional treatment	[79]
	Solid-state Fermentation (*A. niger* and *R. oligosporus*)	Plum brandy distilleries waste	Increase of 3-Caffeoylquinic acid (33 to 53 mg/100 g dry matter by *A. niger*; 33 to 46 mg/100 g dry matter by *R. oligosporus*) when 2×10^7 spores/g of solid was inoculated and fermentation took place during 14 days at 30 °C	[83]
	Solid-state Fermentation (*A. niger* and *R. oligosporus*)	Plum pomace	Increase of 5-Caffeoylquinic acid (22 to 24 mg/100 g dry matter by *A. niger*; 22 to 24 mg/100 g dry matter by *R. oligosporus*) when 2×10^7 spores/g of solid was inoculated and fermentation took place during 14 days at 30 °C	[83]
	Subcritical/critical Fluid Extraction	White grape seeds	Improved recovery of gallic acid, catechin, and epicatechin (>70%) at 1 mL/min CO_2 flow rate, 20 min extraction, 35 °C, organic modifier density (0.85–0.95 g/mL), modifier (ethanol-methanol: 10–40).	[99,103]
	Ultrasounds	Grape marc	Increase of 11–35% at 24 kHz, 20–75 W/mL	[99,104]
	Ultrasounds	Orange peel (50:50 EtOH:Water)	Recovery of caffeic (207%), p-coumaric (180%), ferulic (192%), sinapic acid (66%), and p-hydroxybenzoic (94%) at 25 KHz, 150 W, 15 min	[84]
	Ultrasounds	Orange peel (20:80; 80:20, EtOH:Water)	Recovery of naringin (38%), hesperidin (42%), TPC (31%) at 25 kHz, 50–150 W, 60 min	[84]
	Ultrasounds	Winery wastes and by-products (grapes)	Increase of 7% (sum of anthocyanins and tannins) at 24 kHz, 5–15 min, 121–363 kJ/kg	[99,104]
Flavonoids	Pulsed electric fields	Orange peel	Increase at 5 kV/cm and 20 pulses	[97]

6. Effect of Non-Thermal Technologies on Fruit & Vegetable By-Products Pigments

Non-thermal technologies applied in F&V by-products focusing on the main findings related to bioactive pigments are shown in Table 2. The table is divided in four parts: anthocyanins, betalains, carotenoids and chlorophylls. The non-thermal technologies found were: electro-technologies, high pressure, supercritical fluid extraction, ultra [105] sounds, high pressure, and combined techniques (e.g., ultrasounds + enzyme-assisted extraction). It is striking that almost 50% of the research findings are related to vegetables, mainly tomato by-products and others such as eggplant and broccoli. Related to fruits by-products, as expected, the main findings were focused on berries, olive extraction and wineries. It has been recently concluded that berry by-products from processing steps are a cheap and available source for isolating anthocyanins-rich extracts using non-thermal processing technologies as can be observed in Table 2. These technologies have been demonstrated to have unique characteristics such as being effective, rapid, low-cost, and eco-friendly [106,107]. It is essential to highlight that the accuracy of the technique depends on not only the conditions or the matrix of the F&V by-products. For example, the highest recovery of anthocyanin in plum peels was observed after US while in grapes, PEF was the most effective technology [108].

Table 2. Effect of non-thermal technologies on F&V by-products pigments (anthocyanins, betalains, carotenoids and chlorophylls).

	Non-Thermal Technology	By-Product	Findings	Reference
Anthocyanins	Electro-technologies	Winery wastes and by-products (pomace)	Recovery of polyphenols (>20%) at E = 13.3 kV/cm (0–564 kJ/kg) V = 40 kV (0–218 kJ/kg)	[95,99]
	High pressure	Wine by-products	Recovery of 41% at 600 MPa, 60 min/solvent (50–50% ethanol in water)	[99,109]
	High pressure	Wine by-products	Recovery of 22–83% at 200–600 MPa, 30–90 min, solvent (20–80%; 100–0% ethanol in water)	[99,110]
	Pulsed electric fields	Blueberry pomace (press cake)	Increase of Delphinidin, Cyanidin, Petunidin, Peonidin, and Malvidin. 51%, 71% and 95% at 1 kV/cm, 3 kV/cm, and 5 kV/cm, respectively	[107]
	Pulsed electric fields	Blueberry by-product	Anthocyanin extraction increased (>30%) with PEF process intensification (1–35 kV/cm; 1–10–41 kJ/kg; 10 Hz, 2–100 pulses, 2 µs	[105,106,111,112]
	Pulsed electric fields	Grape by-product (pomace and peel)	Improved anthocyanin extraction (up to 18.9%) at 1.2, 1.8, and 3.0 kV/cm, 18 kJ/kg, 200–2000 pulses, 100 µs	[106,108]
	Pulsed electric fields	Plum by-product	No increase anthocyanins at 37.8–289.8 W, 0.7–25.2 pulses, 10 Hz, 6 µs	[106,108]

Table 2. *Cont.*

	Non-Thermal Technology	By-Product	Findings	Reference
Anthocyanins	Pulsed electric fields	Peach by-product	Improved anthocyanin extraction (up to 11.8-fold) at 0.8 kV/cm, 0.2 kJ/kg; 0.1 Hz 4 µs	[106,113]
	Pulsed electric fields	Raspberry by-product	Increase 27.5% at 1 kV/cm, 6 kJ/kg, 20 Hz and 20 µs	[106,114]
	Pulsed electric fields	Sour cherry by-product	Improved anthocyanin extraction (up to 54%); 1 kV/cm, 10 kJ/kg, 10 Hz, 20 µs	[106,115]
	Pulsed electric fields	Sweet cherry by-product	Improved anthocyanin extraction (up to 38.4%) at 0.5 kV/cm, 10 kJ/kg, 5 Hz, 20 µs	[106,116]
	Pulsed electric fields	Winery wastes and by-products (grapes)	Increase of anthocyanins: 3-fold at 3 kV/cm 50 pulses; 1.6 and 2-fold ↑ 5 kV/cm 1 ms	[99,117]
	Pulsed electric fields	Winery wastes and by-products (grapes)	Increase of 51–62% at 0.8–5 kV/cm, 1–100 ms, 42–53 kJ/kg	[99,117]
	Pulsed electric fields	Winery wastes and by-products (grapes)	Increased anthocyanin content (1.6–1.9 fold more) at 5 kV/cm, 1 ms, 48 kJ/kg	[99,118]
	Pulsed electric fields	Winery wastes and by-products (pomace)	Increase of Anthocyanins (2-fold more) at 13.3 kV/cm, 0–564 kJ/kg	[99,101]
	Pulsed electric fields + ultrasounds	Blueberry by-products	Increase of anthoycanin extraction (3 fold more) (PEF: 60% ethanol 1:6 and 20 kV/cm; Ultrasounds: 1:6, 40 °C, 60 min at 125 W)	[119]
	Subcritical/critical Fluid Extraction	Grape skin	Recovery of 85% at 100–130 bar, pH of 2–4, 25–30% ethanol, 25–50 mL/min CO_2 flow, and 3–10% extract flow ratio	[99,120]
	Ultrasounds	Eggplant by-product	US-assisted extraction (15–45 min) was preferable to conventional solid-liquid extraction due to the lower temperature (25 °C) used and higher delphinidin 3-O-rutinoside content (1.5 fold more).	[82]
	Ultrasounds	Jabuticaba by-products	The highest concentration at 1.1 W/cm^2, 3 min, 10 KHz	[106,121,122]

Table 2. *Cont.*

	Non-Thermal Technology	By-Product	Findings	Reference
	Ultrasounds	Pomegranate peel	116 W sonication power with 80% duty cycle for 6 min for extraction of 22.51 mg cyanidin-3-glucosides/100 g pomegranate peel.	[106,123]
Betalains	Pulsed electric fields	*Opuntia stricta* peels	Total colorants to ≈80 mg/100 g FW (20 kV, frequency of 0.5 Hz, number of pulses of 50)	[124]
	Pulsed electric fields	Red pricky pear peels	Increase of 2.4 fold colorants (betanin and isobetanin) at 8–23 kV/cm 50–300 pulses + aqueous extraction	[125]
	Ultrasound	*Opuntia stricta* peels	Total colorants to ≈80 mg/100 g FW (400 W power at 24 kHz frequency for 5–15 min)	[124]
Ca-rote-noids	Electro-technologies	Olive kernel	Recovery of polyphenols (2-fold more) E = 13.3 kV/cm (0–141 kJ/kg), V = 40 kV (0–141 kJ/kg)	[54,89]
	Microemulsion (Ultrasounds + enzyme)	Tomato pomace	Recovery of lycopene (>20%). The optimal conditions (tomato pomace: double distilled water 1:6): combined ultrasound (20–37 W, amplitude 90% and sonication temperature of 10 °C for 15 min) and enzyme pretreatments (0.2 mL/kg, 30 min, pH 4, 35 °C), saponin as a natural surfactant, and glycerol as a co-surfactant.	[55,126]
	Pulsed Electric Fields	Tomato waste	Recovery of 12–18% of lycopene in acetone and ethyl lactate extracts at 5 kJ/Kg and 5 kV/cm (20 °C).	[55,105,127]
	Supercritical Fluid Extraction	Broccoli by-products	Decrease of beta-carotene (>10%) compared with conventional treatment at 400 bars, 5% of ethanol	[79]
	Supercritical fluid extraction	F&V waste: -sweet potato, tomato, apricot, pumpkin and peach peels -green, yellow and red peppers	Total carotenoid recovery values were greater than 90% *w*/*w*, with β-carotene being the most successfully extracted compound (TCRs 88–100% *w*/*w*), at 350 bar, 15 g/min CO_2, 15.5% (*v*/*v*) ethanol as co-solvent, 30 min of extraction time)	[128,129]
	Ultrasound	Orange processing waste	Optimization of β-carotene extraction with enzyme assisted technology at 20 kHz, 500 W and 25 °C	[129,130]
	Ultrasound	Red pricky pear peels	Increase of 2.6 fold colorants (betanin and isobetanin) at 400 W 5–15 min + aqueous extraction	[125]
	Ultrasound	Tomato pomace	Lycopene increase (>10%) at 25–40 °C, 0–10 min, 0–100 kPa; 58–94 µm; Hexane %: 25–75	[55,131,132]
	Ultrasounds	Tomato peel	5-fold lower all-trans lycopene content by ultrasounds (30 min 0 °C) compared to thermal extraction (75 °C, 1–2 h).	[14]
Chlorophylls	Electro-technologies	Olive kernel	Recovery of polyphenols (>30%) E = 13.3 kV/cm (0–141 kJ/kg), V = 40 kV (0–141 kJ/kg)	[54]
	Supercritical Fluid Extraction	Broccoli by-products	Increase of chlorophylls (>10%) at 400 bars, and 5% of ethanol	[79]

7. Effect of Non-Thermal Technologies on Fruit and Vegetable By-Products Pectins and Sulfur Compounds

The main findings about the effect of non-thermal technologies on pectins and sulfur compounds from F&V by-products are detailed in Table 3. The table is divided in two parts: pectin and sulfur compounds. The non-thermal technologies found were: enzyme-assisted extraction, high pressure, electro-technologies, ultrasounds, combined technologies (e.g., ultrasounds + enzyme-assisted extraction) and light stress.

Table 3. Effect of non-thermal technologies on F&V by-products pectin and sulfur components content.

	Non-Thermal Technology	By-Product	Findings	Reference
Pectin	Enzymes	Apple Pomace	Recovery of 14% by Celluclast 18 h	[133,134]
	Enzymes	Kiwi pomace	Recovery of 4% by celluclast 25 °C 0.5 h	[133,135]
	Enzymes	Passion fruit pomace	Recovery of 2.6–9.2% by Cellyclast 0.5–2 h	[133,136]
	Enzymes	Lime peel	Recovery of 26% by Validase TRL 4 h	[133,137]
	High pressure	Cactus pear peel	Increase of 22% soluble pectin at 600 MPa 10 min	[138,139]
	High pressure	Cactus pear peel	Increase of 9% insoluble pectin at 600 MPa 10 min	[138,139]
	High pressure	Mango peel	Increase of 15% soluble pectin at 600 MPa 10 min	[138,139]
	High pressure	Orange peel	Increase of 59% soluble pectin at 600 MPa 10 min	[138,139]
	High pressure	Passion fruit peel	Recovery of pectin was increased from 7.4 to 14.3% due to HPP pre-treatment. D-GalA of pectin was 65% higher than conventional treatment	[65,140]
	High pressure	Tomato peel	300 MPa pressure performed at 10, 20, 30, and 45 min. 14–15% of pectin recovery at 30 and 45 min	[65,141]
	Moderate electric field	Passion fruit peel	Increase of galacturonic acid (GA) (recovery and content) at 40 min; 100 V; pH 1 (GA); pH 3 (Recovery)	[65,140]
	Ultrasounds	Grapefruit peel	Recovery of 18.2% by 30 °C 10–60 min, 0.2–0.53 W/mL	[131,142]
	Ultrasounds-enzyme assisted extraction without or with hemicellulase or cellulase	Discarded carrots	The pectin was rich in α- and β-carotenes, lutein and α-tocopherol. US-hemicellulase led to the highest pectin recovery (27.1%) at 12.27 W/cm^2: 20 kHz, 80% amplitude, 20 min	[143]
Sulfur compounds	Electro-technologies	Raspseeds seeds	Recovery of isothyocyanates (>15%) at V = 40 kV (0–400 kJ/kg)	[54,93]
	Light stress (UV-B and UV-C; single and combined)	Bimi broccoli leaves and stalks	UV-B (10 kJ/m^2) + C increased 34% of glucobrassicin levels of leaves	[80]

8. Trends and Challenges for Fruit & Vegetable By-Products Application in Food Systems

The incorporation of F&V by-products, their compounds, and/or their extracts/powders can be a relevant strategy for the re-formulation of "Clean Label" ingredients and fortification products. When the non-thermal technologies mentioned above are optimized and applied to F&V by-products, phytochemical bioactive compounds are maintained and/or increased, and then incorporated to food systems. In the following sections, scientific evidence and opportunities for F&V by-product application in food systems are explained: minimally processed fruits, beverages and purees, bakery and confectionary products, food packaging, and cosmetics.

8.1. Minimally Processed Fresh Fruit and Vegetables

Minimally processed fresh F&V are commonly defined as any commodity that has been subjected to different processing steps to obtain a fully edible product [144]. Nowadays, the number of emerging technologies using F&V by-products is increasing. For example, a recent study concluded that the fresh-cut apples fortified with lycopene microspheres obtained from tomato peel from the food industry controlled the enzymatic browning after 9 d at 5 °C, enhancing phenolic compounds up to 56% (for chlorogenic acid) after 9 d at 5 °C [14]. In addition, broccoli by-product was incorporated for the enrichment of kale pesto sauce, increasing functional, techno-functional and sensory characteristics [13].

8.2. Fruit- & Vegetable-Based Beverages and Purees

A recompilation of the evidences of fortification of beverages by bioactive compounds from F&V by-products was recently published [9]. For instance, the incorporation of the beetroot leaves extract into a veggie smoothie was a potential tool to enrich (50%) phenolic content on the final product. Not only were functional properties enhanced, nutritional and techno-functional properties were also increased [6,89]. Another example is the fortification of coconut water by microparticles of encapsulated grape pomace extract rich in polyphenols [93]. In addition, there is a growing interest in using F&V by-products in fermented beverages for the development of novel functional foods when combining their nutritional and functional characteristics with the enzymatic mechanisms of selected lactic acid bacteria [145]. Enhancing bioactive compounds and other quality parameters could carry out adding F&V by-products to vegetable purée such as tomato puree enriched with grape skin fibers from winemaking by-products [146]. This trend is accompanied by green and non-thermal technologies such as enzyme-assisted extraction, pulsed electric field, ultrasounds, and supercritical fluid extractions [145].

8.3. Pasta, Bakery and Confectionary Products

The development of pasta, bakery and confectionary products based on natural ingredients/compounds with antioxidant properties and/or with a reduction of sugars and lipid content is a current trend to obtain new and healthier products. Therefore, the incorporation of F&V by-products is researched such as cookies fortified with purple passion fruit epicarp flour [147], candies fortified with watermelon by-products [148], nutritionally enhanced maize complementary porridges with mango seed and kernel [149], cereal-based foods fortified with by-products from the olive oil industry [150], and spaghetti enriched by persimmon and other vegetal by-products [93,151], among others.

8.4. Food Packaging and Edible Coatings

Active packaging presents several options but all of them focus on the addition of additives into the packaging system. The main purpose is to increase food quality and shelf-life. The most common additives are moisture absorbers, gas scavengers, carbon dioxide emitters, antioxidant, and antimicrobial compounds [152]. Related to food packaging obtained from F&V by-products, companies present a special interest due to the interest of circular economy and zero waste strategies around the world during last years. The unique characteristics can be described as follows: (i) to increase antioxidant and antimicrobial activity, (ii) to improve mechanical properties, and, (iii) to protect food products (to increase shelf-life) [153]. Nowadays, F&V by-product components have been proposed to improve the properties of synthetic or bio-based plastic materials [22]. F&V by-products powders and extracts are a good strategy for obtaining packaging with renewable and biodegradable biopolymers, composite films with food stability and barrier properties, active films as carriers of antioxidant and antimicrobial compounds and edible and functional food packaging [22,153].

On the other hand, the colorimetric pH indicator films can be a potential tool for obtaining smart packaging, showing alterations of the food pH by food deterioration and environmental changes. Then, consumers receive authentic information regarding the food's quality and its edibility (fresh, spoiling, and spoiled product such as milk) [152]. For example, an interesting way to use anthocyanins is building an active use by-date indicator for milk. The development of an anthocyanin-agarose film capable of changing its color in the presence of lactic acid from microbial metabolism has been reported [60].

In addition, a novel technology in which F&V by-products could be revalorized is the use of edible coating, specially containing potent antioxidants and other bioactive compounds from F&V by-products. Up to now, the most common edible coating is chitosan-based edible/biodegradable films because they can extend the shelf life of postharvest fruits. Recent studies have been indicated that coatings enriched with F&V by-products (for example with grape, blueberry and parsley pomace extracts) did not lead to a disruption of the protective function [154]. Natural antioxidants of F&V by-product extracts often contain a high amount of phenolic substances and have been used as active ingredients in the manufacture of active films [152].

8.5. Pharmacologic and Cosmetic Uses

The market for natural cosmetics is growing due to the importance of sustainable development and protecting the environment. Manufacturers present an interest in recovering bioactive compounds from F&V by-products for reducing the environmental impact of waste and for converting them into particularly valuable sources of extracts for cosmetic usage [79,155]. In addition, the potential of food and agricultural residues (rich sources of different classes of compounds with valuable active principles) for the preparation of pharmaceutical and bioactive compounds is gaining importance, taking the environmental impact of the overall production process into account [19,156]. For example, the use of broccoli by-products wasted during the preharvest stage were classified as potential ingredients for the cosmetic and pharmaceutical industries, mainly due to the antioxidant effect of its phytochemicals compounds [79]. These findings have been transferred to the industry, and several companies have been recently created, such as https://biodiversocosmetic.com/ (accessed on 25 December 2021).

9. Conclusions

Although there is an increase of research focused on the effect of non-thermal treatments on F&V by-products for enhancing phytochemicals and other compounds such as pectin, more scientific evidence is needed to establish the optimum treatments and conditions (extraction, addition, processing, storing, shelf life) for each F&V by-product. Most of the studies were focused on fruit by-products, finding a lack of clear evidence related to vegetable commodities. Even though novel extraction technologies showed a better potential to retain bioactive compounds, the use of improved sustainable methods needs further investigation towards industrial viability (energy consumption, time, equipment, value, cost, etc.). Future investigations should be focused on the effect of 'green' technologies in improving the F&V by-products extraction and incorporation for the full utilization of raw materials to preserve a circular economy while enhancing bioactive quality. In this sense, it would be of high interest to optimize nanotechnology for encapsulating extracted bioactive compounds/ingredients, preserving their degradation and optimizing their use efficacy.

Author Contributions: M.C.-L.; F.A.-H., writing—original draft preparation, review and editing, M.C.-L.; F.A.-H.: visualization, M.C.-L.; F.A.-H.: supervision. All authors have read and agreed to the published version of the manuscript.

Funding: This research received no external funding.

Institutional Review Board Statement: Not applicable.

Informed Consent Statement: Not applicable.

Data Availability Statement: Not applicable.

Acknowledgments: Not applicable.

Conflicts of Interest: The authors declare no conflict of interest.

References

1. FAO. Organización de las Naciones Unidas para la Alimentación y la Agricultura. Plataforma Técnica Sobre la Medición y la Reducción de Las Pérdidas y el Desperdicio de Alimentos. Available online: https://www.fao.org/platform-food-loss-waste/es/ (accessed on 26 November 2021).
2. Bhat, R. Sustainability challenges in the valorization of agri-food wastes and by-products. In *Valorization of Agri-Food Wastes and By-Products*; Bhat, R., Ed.; Academic Press: Cambridge, MA, USA, 2021; Chapter 1; pp. 1–27. [CrossRef]
3. Sagar, N.A.; Pareek, S.; Sharma, S.; Yahia, E.M.; Lobo, M.G. Fruit and Vegetable Waste: Bioactive Compounds, Their Extraction, and Possible Utilization. *Compr. Rev. Food Sci. Food Saf.* **2018**, *17*, 512–531. [CrossRef]
4. FAO. Organización de las Naciones Unidas para la Alimentación y la Agricultura. El Estado Mundial de la Agricultura y la Alimentación. Progresos en la Lucha Contra la Pérdida y el Desperdicio de Alimentos. Available online: https://www.fao.org/3/ca6030es/ca6030es.pdf (accessed on 26 November 2021).
5. Mirabella, N.; Castellani, V.; Sala, S. Current options for the valorization of food manufacturing waste: A review. *J. Clean. Prod.* **2014**, *65*, 28–41. [CrossRef]
6. FAO. Organización de las Naciones Unidas para la Alimentación y la Agricultura. El futuro de la Alimentación y la Agricultura: Tendencias y Desafíos. Available online: https://www.fao.org/3/i6881s/i6881s.pdf (accessed on 26 November 2021).
7. Simões, S.; Costa, A.; Faria-Silva, A.C.; Ascenso, A.; Marto, J.; Carvalheiro, M.; Gonçalves, L.M.; Marques, M.; Paiva, A.; Bento, M.; et al. Sustainable valorization of food-processing industry by-products: Challenges and opportunities to obtain bioactive compounds. In *Valorization of Agri-Food Wastes and By-Products*; Bhat, R., Ed.; Academic Press: Cambridge, MA, USA, 2021; Chapter 32; pp. 619–644. [CrossRef]
8. Coman, V.; Teleky, B.-E.; Mitrea, L.; Martău, G.A.; Szabo, K.; Călinoiu, L.-F.; Vodnar, D.C. Bioactive potential of fruit and vegetable wastes. In *Advances in Food and Nutrition Research*; Toldrá, F., Ed.; Academic Press: Cambridge, MA, USA, 2020; Volume 91, Chapter 5; pp. 157–225.
9. Artés-Hernández, F.; Castillejo, N.; Martínez-Zamora, L.; Martínez-Hernández, G.B. Phytochemical Fortification in Fruit and Vegetable Beverages with Green Technologies. *Foods* **2021**, *10*, 2534. [CrossRef] [PubMed]
10. Majumder, P.; Annegowda, H.V. Fruit and vegetable by-products: Novel ingredients for a sustainable society. In *Valorization of Agri-Food Wastes and By-Products*; Bhat, R., Ed.; Academic Press: Cambridge, MA, USA, 2021; Chapter 7; pp. 133–156. [CrossRef]
11. Rose, N.; Reynolds, T.; Kolodinsky, J. P90 Consumer Use of Food Labels Increases as "Clean Label" Trend Continues. *J. Nutr. Educ. Behav.* **2020**, *52*, S58–S59. [CrossRef]
12. Roobab, U.; Shabbir, M.A.; Khan, A.W.; Arshad, R.N.; Bekhit, A.E.-D.; Zeng, X.-A.; Inam-Ur-Raheem, M.; Aadil, R.M. High-pressure treatments for better quality clean-label juices and beverages: Overview and advances. *LWT* **2021**, *149*, 111828. [CrossRef]
13. Castillejo, N.; Martínez-Hernández, G.B.; Artés-Hernández, F. Revalorized broccoli by-products and mustard improved quality during shelf life of a kale pesto sauce. *Food Sci. Technol. Int.* **2021**, *27*, 734–745. [CrossRef]
14. Martínez-Hernández, G.B.; Castillejo, N.; Artés-Hernández, F. Effect of fresh–cut apples fortification with lycopene microspheres, revalorized from tomato by-products, during shelf life. *Postharvest Biol. Technol.* **2019**, *156*, 110925. [CrossRef]
15. Tarazona-Díaz, M.P.; Viegas, J.; Moldao-Martins, M.; Aguayo, E. Bioactive compounds from flesh and by-product of fresh-cut watermelon cultivars. *J. Sci. Food Agric.* **2011**, *91*, 805–812. [CrossRef]
16. Martínez-Sánchez, A.; López-Cañavate, M.E.; Guirao-Martínez, J.; Roca, M.J.; Aguayo, E. Aloe vera Flowers, a Byproduct with Great Potential and Wide Application, Depending on Maturity Stage. *Foods* **2020**, *9*, 1542. [CrossRef] [PubMed]
17. Bustamante, M.A.; Moral, R.; Paredes, C.; Pérez-Espinosa, A.; Moreno-Caselles, J.; Pérez-Murcia, M.D. Agrochemical characterisation of the solid by-products and residues from the winery and distillery industry. *Waste Manag.* **2008**, *28*, 372–380. [CrossRef]
18. Anticona, M.; Blesa, J.; Frigola, A.; Esteve, M.J. High Biological Value Compounds Extraction from Citrus Waste with Non-Conventional Methods. *Foods* **2020**, *9*, 811. [CrossRef] [PubMed]
19. Fierascu, R.C.; Sieniawska, E.; Ortan, A.; Fierascu, I.; Xiao, J. Fruits By-Products—A Source of Valuable Active Principles. A Short Review. *Front. Bioeng. Biotechnol.* **2020**, *8*, 319. [CrossRef]

20. Bhat, R. *Valorization of Agri-Food Wastes and By-Products. Recent Trends, Innovations and Sustainability Challenges*; Elsevier: Amsterdam, The Netherlands, 2021; Chapter 2; pp. 23–44. [CrossRef]
21. Mourtzinos, I.; Goula, A. Polyphenols in Agricultural Byproducts and Food Waste. In *Polyphenols in Plants*, 2nd ed.; Watson, R.R., Ed.; Academic Press: Cambridge, MA, USA, 2019. [CrossRef]
22. Dilucia, F.; Lacivita, V.; Conte, A.; Del Nobile, M.A. Sustainable Use of Fruit and Vegetable By-Products to Enhance Food Packaging Performance. *Foods* **2020**, *9*, 857. [CrossRef]
23. Swallah, M.S.; Sun, H.; Affoh, R.; Fu, H.; Yu, H. Antioxidant Potential Overviews of Secondary Metabolites (Polyphenols) in Fruits. *Int. J. Food Sci.* **2020**, *2020*, 9081686. [CrossRef]
24. Kelly, N.P.; Kelly, A.L.; O'Mahony, J.A. Strategies for enrichment and purification of polyphenols from fruit-based materials. *Trends Food Sci. Technol.* **2019**, *83*, 248–258. [CrossRef]
25. Li, W.; Yang, R.; Ying, D.; Yu, J.; Sanguansri, L.; Augustin, M.A. Analysis of polyphenols in apple pomace: A comparative study of different extraction and hydrolysis procedures. *Ind. Crops Prod.* **2020**, *147*, 112250. [CrossRef]
26. Jiménez-Moreno, N.; Cimminelli, M.J.; Volpe, F.; Ansó, R.; Esparza, I.; Mármol, I.; Rodríguez-Yoldi, M.J.; Ancín-Azpilicueta, C. Phenolic Composition of Artichoke Waste and its Antioxidant Capacity on Differentiated Caco-2 Cells. *Nutrients* **2019**, *11*, 1723. [CrossRef]
27. Melo, P.E.F.; Silva, A.P.M.; Marques, F.P.; Ribeiro, P.R.V.; Souza Filho, M.d.s.M.; Brito, E.S.; Lima, J.R.; Azeredo, H.M.C. Antioxidant films from mango kernel components. *Food Hydrocoll.* **2019**, *95*, 487–495. [CrossRef]
28. Hanani, Z.A.N.; Yee, F.C.; Nor-Khaizura, M.A.R. Effect of pomegranate (*Punica granatum* L.) peel powder on the antioxidant and antimicrobial properties of fish gelatin films as active packaging. *Food Hydrocoll.* **2019**, *89*, 253–259. [CrossRef]
29. Zhao, Y.; Saldaña, M.D.A. Use of potato by-products and gallic acid for development of bioactive film packaging by subcritical water technology. *J. Supercrit. Fluids* **2019**, *143*, 97–106. [CrossRef]
30. Szabo, K.; Dulf, F.V.; Diaconeasa, Z.; Vodnar, D.C. Antimicrobial and antioxidant properties of tomato processing byproducts and their correlation with the biochemical composition. *LWT* **2019**, *116*, 108558. [CrossRef]
31. Chaouch, M.A.; Benvenuti, S. The Role of Fruit by-Products as Bioactive Compounds for Intestinal Health. *Foods* **2020**, *9*, 1716. [CrossRef] [PubMed]
32. Teruel-Andreu, C.; Andreu-Coll, L.; López-Lluch, D.; Sendra, E.; Hernández, F.; Cano-Lamadrid, M. Ficus carica Fruits, by-Products and Based Products as Potential Sources of Bioactive Compounds: A Review. *Agronomy* **2021**, *11*, 1834. [CrossRef]
33. Avram, A.M.; Morin, P.; Brownmiller, C.; Howard, L.R.; Sengupta, A.; Wickramasinghe, S.R. Concentrations of polyphenols from blueberry pomace extract using nanofiltration. *Food Bioprod. Processing* **2017**, *106*, 91–101. [CrossRef]
34. Campone, L.; Celano, R.; Lisa Piccinelli, A.; Pagano, I.; Carabetta, S.; Sanzo, R.D.; Russo, M.; Ibañez, E.; Cifuentes, A.; Rastrelli, L. Response surface methodology to optimize supercritical carbon dioxide/co-solvent extraction of brown onion skin by-product as source of nutraceutical compounds. *Food Chem.* **2018**, *269*, 495–502. [CrossRef]
35. Singh, B.; Singh, J.P.; Kaur, A.; Singh, N. Phenolic composition, antioxidant potential and health benefits of citrus peel. *Food Res. Int.* **2020**, *132*, 109114. [CrossRef] [PubMed]
36. Sudhakar, P.; Latha, P.; Reddy, P.V. Plant pigments. In *Phenotyping Crop Plants for Physiological and Biochemical Traits*; Sudhakar, P., Latha, P., Reddy, P.V., Eds.; Academic Press: Cambridge, MA, USA, 2016; Chapter 15; pp. 121–127. [CrossRef]
37. Dikshit, R.; Tallapragada, P. Comparative Study of Natural and Artificial Flavoring Agents and Dyes. In *Natural and Artificial Flavoring Agents and Food Dyes*; Grumezescu, A.M., Holban, A.M., Eds.; Academic Press: Cambridge, MA, USA, 2018; Chapter 3; pp. 83–111. [CrossRef]
38. Sharma, M.; Usmani, Z.; Gupta, V.K.; Bhat, R. Valorization of fruits and vegetable wastes and by-products to produce natural pigments. *Crit. Rev. Biotechnol.* **2021**, *41*, 535–563. [CrossRef]
39. Choo, W.S.; Saik, A.Y.H. Valorization of fruit and vegetable waste for bioactive pigments: Extraction and utilization. In *Valorization of Agri-Food Wastes and By-Products*; Bhat, R., Ed.; Academic Press: Cambridge, MA, USA, 2021; Chapter 4; pp. 61–81. [CrossRef]
40. Additives, E.P.O.F.; Food, N.S.A.T. Scientific Opinion on the re-evaluation of anthocyanins (E 163) as a food additive. *EFSA J.* **2013**, *11*, 3145. [CrossRef]
41. Additives, E.P.O.F.; Food, N.S.A.T. Scientific Opinion on the re-evaluation of beetroot red (E 162) as a food additive. *EFSA J.* **2015**, *13*, 4318. [CrossRef]
42. Otalora, C.M.; Bonifazi, E.; Fissore, E.N.; Basanta, F.; Gerschenson, L.N. Thermal Stability of Betalains in By-Products of the Blanching and Cutting of Beta vulgaris L. var conditiva. *Pol. J. Food Nutr. Sci.* **2020**, *70*, 15–24. [CrossRef]
43. Tanaka, Y.; Sasaki, N.; Ohmiya, A. Biosynthesis of plant pigments: Anthocyanins, betalains and carotenoids. *Plant J.* **2008**, *54*, 733–749. [CrossRef]
44. Doleman, J.F.; Grisar, K.; Van Liedekerke, L.; Saha, S.; Roe, M.; Tapp, H.S.; Mithen, R.F. The contribution of alliaceous and cruciferous vegetables to dietary sulphur intake. *Food Chem.* **2017**, *234*, 38–45. [CrossRef]
45. Międkus, N.; Marszałek, K.; Podlacha, M.; Iqbal, A.; Puchalski, C.; Świergiel, A.H. Health Benefits of Plant-Derived Sulfur Compounds, Glucosinolates, and Organosulfur Compounds. *Molecules* **2020**, *25*, 3804. [CrossRef] [PubMed]
46. Alvarez-Jubete, L.; Valverde, J.; Kehoe, K.; Reilly, K.; Rai, D.K.; Barry-Ryan, C. Development of a Novel Functional Soup Rich in Bioactive Sulforaphane Using Broccoli (*Brassica oleracea* L. ssp. italica) Florets and Byproducts. *Food Bioprocess Technol.* **2014**, *7*, 1310–1321. [CrossRef]

47. Drabińska, N.; Ciska, E.; Szmatowicz, B.; Krupa-Kozak, U. Broccoli by-products improve the nutraceutical potential of gluten-free mini sponge cakes. *Food Chem.* **2018**, *267*, 170–177. [CrossRef]
48. Wicker, L.; Kim, Y. Pectin and Health. In *Encyclopedia of Food and Health*; Caballero, B., Finglas, P.M., Toldrá, F., Eds.; Academic Press: Oxford, UK, 2016; pp. 289–293. [CrossRef]
49. Perussello, C.A.; Zhang, Z.; Marzocchella, A.; Tiwari, B.K. Valorization of Apple Pomace by Extraction of Valuable Compounds. *Compr. Rev. Food Sci. Food Saf.* **2017**, *16*, 776–796. [CrossRef] [PubMed]
50. Müller-Maatsch, J.; Bencivenni, M.; Caligiani, A.; Tedeschi, T.; Bruggeman, G.; Bosch, M.; Petrusan, J.; Van Droogenbroeck, B.; Elst, K.; Sforza, S. Pectin content and composition from different food waste streams. *Food Chem.* **2016**, *201*, 37–45. [CrossRef]
51. Garcia-Vaquero, M.; Ravindran, R.; Walsh, O.; O'Doherty, J.; Jaiswal, A.K.; Tiwari, B.K.; Rajauria, G. Evaluation of Ultrasound, Microwave, Ultrasound–Microwave, Hydrothermal and High Pressure Assisted Extraction Technologies for the Recovery of Phytochemicals and Antioxidants from Brown Macroalgae. *Mar. Drugs* **2021**, *19*, 309. [CrossRef]
52. Garcia-Vaquero, M.; Ummat, V.; Tiwari, B.; Rajauria, G. Exploring Ultrasound, Microwave and Ultrasound-Microwave Assisted Extraction Technologies to Increase the Extraction of Bioactive Compounds and Antioxidants from Brown Macroalgae. *Mar. Drugs* **2020**, *18*, 172. [CrossRef] [PubMed]
53. Dimou, C.; Karantonis, C.H.; Skalkos, D.; Koutelidakis, E.A. Valorization of Fruits by-products to Unconventional Sources of Additives, Oil, Biomolecules and Innovative Functional Foods. *Curr. Pharm. Biotechnol.* **2019**, *20*, 776–786. [CrossRef]
54. Puértolas, E.; Barba, F.J. Electrotechnologies applied to valorization of by-products from food industry: Main findings, energy and economic cost of their industrialization. *Food Bioprod. Processing* **2016**, *100*, 172–184. [CrossRef]
55. Madia, V.N.; De Vita, D.; Ialongo, D.; Tudino, V.; De Leo, A.; Scipione, L.; Di Santo, R.; Costi, R.; Messore, A. Recent Advances in Recovery of Lycopene from Tomato Waste: A Potent Antioxidant with Endless Benefits. *Molecules* **2021**, *26*, 4495. [CrossRef] [PubMed]
56. Arshad, R.N.; Abdul-Malek, Z.; Roobab, U.; Munir, M.A.; Naderipour, A.; Qureshi, M.I.; El-Din Bekhit, A.; Liu, Z.-W.; Aadil, R.M. Pulsed electric field: A potential alternative towards a sustainable food processing. *Trends Food Sci. Technol.* **2021**, *111*, 43–54. [CrossRef]
57. Ben Ammar, J.; Lanoisellé, J.-L.; Lebovka, N.I.; Van Hecke, E.; Vorobiev, E. Impact of a Pulsed Electric Field on Damage of Plant Tissues: Effects of Cell Size and Tissue Electrical Conductivity. *J. Food Sci.* **2011**, *76*, E90–E97. [CrossRef]
58. Rai Naveed, A. Effective valorization of food wastes and by-products through pulsed electric field: A systematic review. *J. Food Process Eng.* **2021**, *44*, e13629. [CrossRef]
59. Costa, J.R.; Tonon, R.V.; Cabral, L.; Gottschalk, L.; Pastrana, L.; Pintado, M.E. Valorization of Agricultural Lignocellulosic Plant Byproducts through Enzymatic and Enzyme-Assisted Extraction of High-Value-Added Compounds: A Review. *ACS Sustain. Chem. Eng.* **2020**, *8*, 13112–13125. [CrossRef]
60. Gómez-García, R.; Martínez-Ávila, G.C.G.; Aguilar, C.N. Enzyme-assisted extraction of antioxidative phenolics from grape (*Vitis vinifera* L.) residues. *3 Biotech* **2012**, *2*, 297–300. [CrossRef]
61. Voidarou, C.; Antoniadou, M.; Rozos, G.; Tzora, A.; Skoufos, I.; Varzakas, T.; Lagiou, A.; Bezirtzoglou, E. Fermentative Foods: Microbiology, Biochemistry, Potential Human Health Benefits and Public Health Issues. *Foods* **2021**, *10*, 69. [CrossRef]
62. Azam, M.T.; Ahmad, A. Date Palm Waste: An Efficient Source for Production of Glucose and Lactic Acid. In *Sustainable Agriculture Reviews 34: Date Palm for Food, Medicine and the Environment*; Naushad, M., Lichtfouse, E., Eds.; Springer International Publishing: Cham, Switzerland, 2019; pp. 155–178. [CrossRef]
63. Sabater, C.; Ruiz, L.; Delgado, S.; Ruas-Madiedo, P.; Margolles, A. Valorization of Vegetable Food Waste and By-Products Through Fermentation Processes. *Front. Microbiol.* **2020**, *11*, 2604. [CrossRef]
64. Vásquez, Z.S.; de Carvalho Neto, D.P.; Pereira, G.V.M.; Vandenberghe, L.P.S.; de Oliveira, P.Z.; Tiburcio, P.B.; Rogez, H.L.G.; Góes Neto, A.; Soccol, C.R. Biotechnological approaches for cocoa waste management: A review. *Waste Manag.* **2019**, *90*, 72–83. [CrossRef]
65. Gavahian, M.; Mathad, G.N.; Pandiselvam, R.; Lin, J.; Sun, D.-W. Emerging technologies to obtain pectin from food processing by-products: A strategy for enhancing resource efficiency. *Trends Food Sci. Technol.* **2021**, *115*, 42–54. [CrossRef]
66. Tewari, S.; Sehrawat, R.; Nema, P.K.; Kaur, B.P. Preservation effect of high pressure processing on ascorbic acid of fruits and vegetables: A review. *J. Food Biochem.* **2017**, *41*, e12319. [CrossRef]
67. Martínez-Zamora, L.; Castillejo, N.; Artés-Hernández, F. Postharvest UV-B and UV-C radiation enhanced the biosynthesis of glucosinolates and isothiocyanates in Brassicaceae sprouts. *Postharvest Biol. Technol.* **2021**, *181*, 111650. [CrossRef]
68. Castillejo, N.; Martínez-Zamora, L.; Artés-Hernández, F. Periodical UV-B radiation hormesis in biosynthesis of kale sprouts nutraceuticals. *Plant Physiol. Biochem.* **2021**, *165*, 274–285. [CrossRef]
69. Zhang, W.; Jiang, W. UV treatment improved the quality of postharvest fruits and vegetables by inducing resistance. *Trends Food Sci. Technol.* **2019**, *92*, 71–80. [CrossRef]
70. Giménez, A.; Martínez-Ballesta, M.D.C.; Egea-Gilabert, C.; Gómez, P.A.; Artés-Hernández, F.; Pennisi, G.; Orsini, F.; Crepaldi, A.; Fernández, J.A. Combined Effect of Salinity and LED Lights on the Yield and Quality of Purslane (*Portulaca oleracea* L.) Microgreens. *Horticulturae* **2021**, *7*, 180. [CrossRef]
71. Martínez-Zamora, L.; Castillejo, N.; Artés-Hernández, F. Postharvest UV-B and Photoperiod with Blue + Red LEDs as Strategies to Stimulate Carotenogenesis in Bell Peppers. *Appl. Sci.* **2021**, *11*, 3736. [CrossRef]

72. Martínez-Zamora, L.; Castillejo, N.; Gómez, P.A.; Artés-Hernández, F. Amelioration Effect of LED Lighting in the Bioactive Compounds Synthesis during Carrot Sprouting. *Agronomy* **2021**, *11*, 304. [CrossRef]
73. Castillejo, N.; Martínez-Zamora, L.; Gómez, P.A.; Pennisi, G.; Crepaldi, A.; Fernández, J.A.; Orsini, F.; Artés-Hernández, F. Postharvest yellow LED lighting affects phenolics and glucosinolates biosynthesis in broccoli sprouts. *J. Food Compos. Anal.* **2021**, *103*, 104101. [CrossRef]
74. Wrona, O.; Rafińska, K.; Możeński, C.; Buszewski, B. Supercritical Fluid Extraction of Bioactive Compounds from Plant Materials. *J. AOAC Int.* **2017**, *100*, 1624–1635. [CrossRef]
75. Jadhav, H.B.; Annapure, U.S.; Deshmukh, R.R. Non-thermal Technologies for Food Processing. *Front. Nutr.* **2021**, *8*, 657090. [CrossRef] [PubMed]
76. Zhang, Z.-H.; Wang, L.-H.; Zeng, X.-A.; Han, Z.; Brennan, C.S. Non-thermal technologies and its current and future application in the food industry: A review. *Int. J. Food Sci. Technol.* **2019**, *54*, 1–13. [CrossRef]
77. Chemat, F.; Rombaut, N.; Sicaire, A.-G.; Meullemiestre, A.; Fabiano-Tixier, A.-S.; Abert-Vian, M. Ultrasound assisted extraction of food and natural products. Mechanisms, techniques, combinations, protocols and applications. A review. *Ultrason. Sonochem.* **2017**, *34*, 540–560. [CrossRef] [PubMed]
78. Page, M.J.; Moher, D.; Bossuyt, P.M.; Boutron, I.; Hoffmann, T.C.; Mulrow, C.D.; Shamseer, L.; Tetzlaff, J.M.; Akl, E.A.; Brennan, S.E.; et al. PRISMA 2020 explanation and elaboration: Updated guidance and exemplars for reporting systematic reviews. *BMJ* **2021**, *372*, n160. [CrossRef] [PubMed]
79. Borja-Martínez, M.; Lozano-Sánchez, J.; Borrás-Linares, I.; Pedreño, M.A.; Sabater-Jara, A.B. Revalorization of Broccoli By-Products for Cosmetic Uses Using Supercritical Fluid Extraction. *Antioxidants* **2020**, *9*, 1195. [CrossRef] [PubMed]
80. Formica-Oliveira, A.C.; Martínez-Hernández, G.B.; Díaz-López, V.; Artés, F.; Artés-Hernández, F. Use of postharvest UV-B and UV-C radiation treatments to revalorize broccoli byproducts and edible florets. *Innov. Food Sci. Emerg. Technol.* **2017**, *43*, 77–83. [CrossRef]
81. Ren, F.; Nian, Y.; Perussello, C.A. Effect of storage, food processing and novel extraction technologies on onions flavonoid content: A review. *Food Res. Int.* **2020**, *132*, 108953. [CrossRef]
82. Dranca, F.; Oroian, M. Optimization of ultrasound-assisted extraction of total monomeric anthocyanin (TMA) and total phenolic content (TPC) from eggplant (*Solanum melongena* L.) peel. *Ultrason. Sonochem.* **2016**, *31*, 637–646. [CrossRef]
83. Dulf, F.V.; Vodnar, D.C.; Socaciu, C. Effects of solid-state fermentation with two filamentous fungi on the total phenolic contents, flavonoids, antioxidant activities and lipid fractions of plum fruit (*Prunus domestica* L.) by-products. *Food Chem.* **2016**, *209*, 27–36. [CrossRef]
84. Putnik, P.; Bursać Kovačević, D.; Režek Jambrak, A.; Barba, F.J.; Cravotto, G.; Binello, A.; Lorenzo, J.M.; Shpigelman, A. Innovative "Green" and Novel Strategies for the Extraction of Bioactive Added Value Compounds from Citrus Wastes—A Review. *Molecules* **2017**, *22*, 680. [CrossRef]
85. Casquete, R.; Castro, S.M.; Villalobos, M.C.; Serradilla, M.J.; Queirós, R.P.; Saraiva, J.A.; Córdoba, M.G.; Teixeira, P. High pressure extraction of phenolic compounds from citrus peels†. *High Press. Res.* **2014**, *34*, 447–451. [CrossRef]
86. Santos, D.I.; Martins, C.F.; Amaral, R.A.; Brito, L.; Saraiva, J.A.; Vicente, A.A.; Moldão-Martins, M. Pineapple (*Ananas comosus* L.) By-Products Valorization: Novel Bio Ingredients for Functional Foods. *Molecules* **2021**, *26*, 3216. [CrossRef] [PubMed]
87. Santos, D.I.; Pinto, C.A.; Corrêa-Filho, L.C.; Saraiva, J.A.; Vicente, A.A.; Moldão-Martins, M. Effect of moderate hydrostatic pressures on the enzymatic activity and bioactive composition of pineapple by-products. *J. Food Process Eng.* **2020**, e13537. [CrossRef]
88. Parniakov, O.; Barba, F.J.; Grimi, N.; Lebovka, N.; Vorobiev, E. Extraction assisted by pulsed electric energy as a potential tool for green and sustainable recovery of nutritionally valuable compounds from mango peels. *Food Chem.* **2016**, *192*, 842–848. [CrossRef]
89. Roselló-Soto, E.; Barba, F.J.; Parniakov, O.; Galanakis, C.M.; Lebovka, N.; Grimi, N.; Vorobiev, E. High Voltage Electrical Discharges, Pulsed Electric Field, and Ultrasound Assisted Extraction of Protein and Phenolic Compounds from Olive Kernel. *Food Bioprocess Technol.* **2015**, *8*, 885–894. [CrossRef]
90. Luengo, E.; Álvarez, I.; Raso, J. Improving the pressing extraction of polyphenols of orange peel by pulsed electric fields. *Innov. Food Sci. Emerg. Technol.* **2013**, *17*, 79–84. [CrossRef]
91. Parniakov, O.; Barba, F.J.; Grimi, N.; Lebovka, N.; Vorobiev, E. Impact of pulsed electric fields and high voltage electrical discharges on extraction of high-added value compounds from papaya peels. *Food Res. Int.* **2014**, *65*, 337–343. [CrossRef]
92. Yu, X.; Bals, O.; Grimi, N.; Vorobiev, E. A new way for the oil plant biomass valorization: Polyphenols and proteins extraction from rapeseed stems and leaves assisted by pulsed electric fields. *Ind. Crops Prod.* **2015**, *74*, 309–318. [CrossRef]
93. Barba, F.J.; Boussetta, N.; Vorobiev, E. Emerging technologies for the recovery of isothiocyanates, protein and phenolic compounds from rapeseed and rapeseed press-cake: Effect of high voltage electrical discharges. *Innov. Food Sci. Emerg. Technol.* **2015**, *31*, 67–72. [CrossRef]
94. López, N.; Puértolas, E.; Condón, S.; Álvarez, I.; Raso, J. Effects of pulsed electric fields on the extraction of phenolic compounds during the fermentation of must of Tempranillo grapes. *Innov. Food Sci. Emerg. Technol.* **2008**, *9*, 477–482. [CrossRef]
95. Barba, F.J.; Brianceau, S.; Turk, M.; Boussetta, N.; Vorobiev, E. Effect of Alternative Physical Treatments (Ultrasounds, Pulsed Electric Fields, and High-Voltage Electrical Discharges) on Selective Recovery of Bio-compounds from Fermented Grape Pomace. *Food Bioprocess Technol.* **2015**, *8*, 1139–1148. [CrossRef]

96. Boussetta, N.; Vorobiev, E.; Le, L.H.; Cordin-Falcimaigne, A.; Lanoisellé, J.L. Application of electrical treatments in alcoholic solvent for polyphenols extraction from grape seeds. *LWT Food Sci. Technol.* **2012**, *46*, 127–134. [CrossRef]
97. Zia, S.; Khan, M.R.; Shabbir, M.A.; Aslam Maan, A.; Khan, M.K.I.; Nadeem, M.; Khalil, A.A.; Din, A.; Aadil, R.M. An Inclusive Overview of Advanced Thermal and Nonthermal Extraction Techniques for Bioactive Compounds in Food and Food-related Matrices. *Food Rev. Int.* **2020**, 1–31. [CrossRef]
98. Brianceau, S.; Turk, M.; Vitrac, X.; Vorobiev, E. Combined densification and pulsed electric field treatment for selective polyphenols recovery from fermented grape pomace. *Innov. Food Sci. Emerg. Technol.* **2015**, *29*, 2–8. [CrossRef]
99. Barba, F.J.; Zhu, Z.; Koubaa, M.; Sant'Ana, A.S.; Orlien, V. Green alternative methods for the extraction of antioxidant bioactive compounds from winery wastes and by-products: A review. *Trends Food Sci. Technol.* **2016**, *49*, 96–109. [CrossRef]
100. Delsart, C.; Cholet, C.; Ghidossi, R.; Grimi, N.; Gontier, E.; Gény, L.; Vorobiev, E.; Mietton-Peuchot, M. Effects of Pulsed Electric Fields on Cabernet Sauvignon Grape Berries and on the Characteristics of Wines. *Food Bioprocess Technol.* **2014**, *7*, 424–436. [CrossRef]
101. Rajha, H.N.; Boussetta, N.; Louka, N.; Maroun, R.G.; Vorobiev, E. A comparative study of physical pretreatments for the extraction of polyphenols and proteins from vine shoots. *Food Res. Int.* **2014**, *65*, 462–468. [CrossRef]
102. El Kantar, S.; Boussetta, N.; Lebovka, N.; Foucart, F.; Rajha, H.N.; Maroun, R.G.; Louka, N.; Vorobiev, E. Pulsed electric field treatment of citrus fruits: Improvement of juice and polyphenols extraction. *Innov. Food Sci. Emerg. Technol.* **2018**, *46*, 153–161. [CrossRef]
103. Palma, M.; Taylor, L.T. Extraction of polyphenolic compounds from grape seeds with near critical carbon dioxide. *J. Chromatogr. A* **1999**, *849*, 117–124. [CrossRef]
104. Vilkhu, K.; Mawson, R.; Simons, L.; Bates, D. Applications and opportunities for ultrasound assisted extraction in the food industry—A review. *Innov. Food Sci. Emerg. Technol.* **2008**, *9*, 161–169. [CrossRef]
105. Pataro, G.; Bobinaitė, R.; Bobinas, Č.; Šatkauskas, S.; Raudonis, R.; Visockis, M.; Ferrari, G.; Viškelis, P. Improving the Extraction of Juice and Anthocyanins from Blueberry Fruits and Their By-products by Application of Pulsed Electric Fields. *Food Bioprocess Technol.* **2017**, *10*, 1595–1605. [CrossRef]
106. Arruda, H.S.; Silva, E.K.; Peixoto Araujo, N.M.; Pereira, G.A.; Pastore, G.M.; Marostica Junior, M.R. Anthocyanins Recovered from Agri-Food By-Products Using Innovative Processes: Trends, Challenges, and Perspectives for Their Application in Food Systems. *Molecules* **2021**, *26*, 2632. [CrossRef]
107. Bobinaitė, R.; Pataro, G.; Raudonis, R.; Vškelis, P.; Bobinas, Č.; Šatkauskas, S.; Ferrari, G. *Improving the Extraction of Juice and Anthocyanin Compounds from Blueberry Fruits and Their by-Products by Pulsed Electric Fields*; Springer: Singapore, 2016; pp. 363–366.
108. Medina-Meza, I.G.; Barbosa-Cánovas, G.V. Assisted extraction of bioactive compounds from plum and grape peels by ultrasonics and pulsed electric fields. *J. Food Eng.* **2015**, *166*, 268–275. [CrossRef]
109. Corrales, M.; Toepfl, S.; Butz, P.; Knorr, D.; Tauscher, B. Extraction of anthocyanins from grape by-products assisted by ultrasonics, high hydrostatic pressure or pulsed electric fields: A comparison. *Innov. Food Sci. Emerg. Technol.* **2008**, *9*, 85–91. [CrossRef]
110. Corrales, M.; García, A.F.; Butz, P.; Tauscher, B. Extraction of anthocyanins from grape skins assisted by high hydrostatic pressure. *J. Food Eng.* **2009**, *90*, 415–421. [CrossRef]
111. Bobinaitė, R.; Pataro, G.; Lamanauskas, N.; Šatkauskas, S.; Viškelis, P.; Ferrari, G. Application of pulsed electric field in the production of juice and extraction of bioactive compounds from blueberry fruits and their by-products. *J. Food Sci. Technol.* **2015**, *52*, 5898–5905. [CrossRef]
112. Lončarić, A.; Celeiro, M.; Jozinović, A.; Jelinić, J.; Kovač, T.; Jokić, S.; Babić, J.; Moslavac, T.; Zavadlav, S.; Lores, M. Green Extraction Methods for Extraction of Polyphenolic Compounds from Blueberry Pomace. *Foods* **2020**, *9*, 1521. [CrossRef] [PubMed]
113. Plazzotta, S.; Ibarz, R.; Manzocco, L.; Martín-Belloso, O. Modelling the recovery of biocompounds from peach waste assisted by pulsed electric fields or thermal treatment. *J. Food Eng.* **2021**, *290*, 110196. [CrossRef]
114. Lamanauskas, N.; Pataro, G.; Bobinas, Č.; Šatkauskas, S.; Viškelis, P.; Bobinaitė, R.; Ferrari, G. Impact of pulsed electric field treatment on juice yield and recovery of bioactive compounds from raspberries and their by-products. *Zemdirbyste* **2016**, *103*, 83–90. [CrossRef]
115. Kurtulbaş Şahin, E.; Bilgin, M.; Şahin, S. Recovery of anthocyanins from sour cherry (*Prunus cerasus* L.) peels via microwave assisted extraction: Monitoring the storage stability. *Prep. Biochem. Biotechnol.* **2021**, *51*, 686–696. [CrossRef]
116. Pataro, G.; Carullo, D.; Bobinaiu, R.; Donsì, G.; Ferrari, G. Improving the Extraction Yield of Juice and Bioactive Compounds from Sweet Cherries and Their By-products by Pulsed Electric Fields. *Chem. Eng. Trans.* **2017**, *57*, 1717–1722.
117. El Darra, N.; Grimi, N.; Maroun, R.G.; Louka, N.; Vorobiev, E. Pulsed electric field, ultrasound, and thermal pretreatments for better phenolic extraction during red fermentation. *Eur. Food Res. Technol.* **2013**, *236*, 47–56. [CrossRef]
118. El Darra, N.; Grimi, N.; Vorobiev, E.; Maroun, R.G.; Louka, N. Pulsed Electric Field-Assisted Cold Maceration of Cabernet franc and Cabernet Sauvignon Grapes. *Am. J. Enol. Vitic.* **2013**, *64*, 476–484. [CrossRef]
119. Zhou, Y.; Zhao, X.; Huang, H. Effects of Pulsed Electric Fields on Anthocyanin Extraction Yield of Blueberry Processing By-Products. *J. Food Processing Preserv.* **2015**, *39*, 1898–1904. [CrossRef]
120. Bleve, M.; Ciurlia, L.; Erroi, E.; Lionetto, G.; Longo, L.; Rescio, L.; Schettino, T.; Vasapollo, G. An innovative method for the purification of anthocyanins from grape skin extracts by using liquid and sub-critical carbon dioxide. *Sep. Purif. Technol.* **2008**, *64*, 192–197. [CrossRef]

121. Gadioli Tarone, A.; Keven Silva, E.; Dias de Freitas Queiroz Barros, H.; Baú Betim Cazarin, C.; Roberto Marostica Junior, M. High-intensity ultrasound-assisted recovery of anthocyanins from jabuticaba by-products using green solvents: Effects of ultrasound intensity and solvent composition on the extraction of phenolic compounds. *Food Res. Int.* **2021**, *140*, 110048. [CrossRef] [PubMed]
122. Albuquerque, B.R.; Pinela, J.; Barros, L.; Oliveira, M.B.P.P.; Ferreira, I.C.F.R. Anthocyanin-rich extract of jabuticaba epicarp as a natural colorant: Optimization of heat- and ultrasound-assisted extractions and application in a bakery product. *Food Chem.* **2020**, *316*, 126364. [CrossRef] [PubMed]
123. More, P.R.; Arya, S.S. Intensification of bio-actives extraction from pomegranate peel using pulsed ultrasound: Effect of factors, correlation, optimization and antioxidant bioactivities. *Ultrason. Sonochem.* **2021**, *72*, 105423. [CrossRef] [PubMed]
124. Barba, F.J.; Putnik, P.; Bursać Kovačević, D.; Poojary, M.M.; Roohinejad, S.; Lorenzo, J.M.; Koubaa, M. Impact of conventional and non-conventional processing on prickly pear (Opuntia spp.) and their derived products: From preservation of beverages to valorization of by-products. *Trends Food Sci. Technol.* **2017**, *67*, 260–270. [CrossRef]
125. Koubaa, M.; Barba, F.J.; Grimi, N.; Mhemdi, H.; Koubaa, W.; Boussetta, N.; Vorobiev, E. Recovery of colorants from red prickly pear peels and pulps enhanced by pulsed electric field and ultrasound. *Innov. Food Sci. Emerg. Technol.* **2016**, *37*, 336–344. [CrossRef]
126. Amiri-Rigi, A.; Abbasi, S. Microemulsion-based lycopene extraction: Effect of surfactants, co-surfactants and pretreatments. *Food Chem.* **2016**, *197*, 1002–1007. [CrossRef]
127. Pataro, G.; Carullo, D.; Falcone, M.; Ferrari, G. Recovery of lycopene from industrially derived tomato processing by-products by pulsed electric fields-assisted extraction. *Innov. Food Sci. Emerg. Technol.* **2020**, *63*, 102369. [CrossRef]
128. de Andrade Lima, M.; Kestekoglou, I.; Charalampopoulos, D.; Chatzifragkou, A. Supercritical Fluid Extraction of Carotenoids from Vegetable Waste Matrices. *Molecules* **2019**, *24*, 466. [CrossRef]
129. Wani, F.A.; Rashid, R.; Jabeen, A.; Brochier, B.; Yadav, S.; Aijaz, T.; Makroo, H.A.; Dar, B.N. Valorisation of food wastes to produce natural pigments using non-thermal novel extraction methods: A review. *Int. J. Food Sci. Technol.* **2021**, *56*, 4823–4833. [CrossRef]
130. Shahram, H.; Dinani, S.T. Optimization of ultrasonic-assisted enzymatic extraction of β-carotene from orange processing waste. *J. Food Process Eng.* **2019**, *42*, e13042. [CrossRef]
131. Kumar, K.; Srivastav, S.; Sharanagat, V.S. Ultrasound assisted extraction (UAE) of bioactive compounds from fruit and vegetable processing by-products: A review. *Ultrason. Sonochem.* **2021**, *70*, 105325. [CrossRef] [PubMed]
132. Luengo, E.; Condón-Abanto, S.; Condón, S.; Álvarez, I.; Raso, J. Improving the extraction of carotenoids from tomato waste by application of ultrasound under pressure. *Sep. Purif. Technol.* **2014**, *136*, 130–136. [CrossRef]
133. Marić, M.; Grassino, A.N.; Zhu, Z.; Barba, F.J.; Brnčić, M.; Rimac Brnčić, S. An overview of the traditional and innovative approaches for pectin extraction from plant food wastes and by-products: Ultrasound-, microwaves-, and enzyme-assisted extraction. *Trends Food Sci. Technol.* **2018**, *76*, 28–37. [CrossRef]
134. Wikiera, A.; Mika, M.; Grabacka, M. Multicatalytic enzyme preparations as effective alternative to acid in pectin extraction. *Food Hydrocoll.* **2015**, *44*, 156–161. [CrossRef]
135. Yuliarti, O.; Goh, K.K.; Matia-Merino, L.; Mawson, J.; Brennan, C. Extraction and characterisation of pomace pectin from gold kiwifruit (Actinidia chinensis). *Food Chem.* **2015**, *187*, 290–296. [CrossRef]
136. Liew, S.Q.; Chin, N.L.; Yusof, Y.A.; Sowndhararajan, K. Comparison of Acidic and Enzymatic Pectin Extraction from Passion Fruit Peels and Its Gel Properties. *J. Food Process Eng.* **2016**, *39*, 501–511. [CrossRef]
137. Dominiak, M.; Søndergaard, K.M.; Wichmann, J.; Vidal-Melgosa, S.; Willats, W.G.T.; Meyer, A.S.; Mikkelsen, J.D. Application of enzymes for efficient extraction, modification, and development of functional properties of lime pectin. *Food Hydrocoll.* **2014**, *40*, 273–282. [CrossRef]
138. Garcia-Amezquita, L.E.; Tejada-Ortigoza, V.; Serna-Saldivar, S.O.; Welti-Chanes, J. Dietary Fiber Concentrates from Fruit and Vegetable By-products: Processing, Modification, and Application as Functional Ingredients. *Food Bioprocess Technol.* **2018**, *11*, 1439–1463. [CrossRef]
139. Tejada-Ortigoza, V.; Garcia-Amezquita, L.E.; Serna-Saldívar, S.O.; Martín-Belloso, O.; Welti-Chanes, J. High Hydrostatic Pressure and Mild Heat Treatments for the Modification of Orange Peel Dietary Fiber: Effects on Hygroscopic Properties and Functionality. *Food Bioprocess Technol.* **2018**, *11*, 110–121. [CrossRef]
140. de Oliveira, C.F.; Giordani, D.; Gurak, P.D.; Cladera-Olivera, F.; Marczak, L.D.F. Extraction of pectin from passion fruit peel using moderate electric field and conventional heating extraction methods. *Innov. Food Sci. Emerg. Technol.* **2015**, *29*, 201–208. [CrossRef]
141. Ninčević Grassino, A.; Ostojić, J.; Miletić, V.; Djaković, S.; Bosiljkov, T.; Zorić, Z.; Ježek, D.; Rimac Brnčić, S.; Brnčić, M. Application of high hydrostatic pressure and ultrasound-assisted extractions as a novel approach for pectin and polyphenols recovery from tomato peel waste. *Innov. Food Sci. Emerg. Technol.* **2020**, *64*, 102424. [CrossRef]
142. Wang, W.; Ma, X.; Xu, Y.; Cao, Y.; Jiang, Z.; Ding, T.; Ye, X.; Liu, D. Ultrasound-assisted heating extraction of pectin from grapefruit peel: Optimization and comparison with the conventional method. *Food Chem.* **2015**, *178*, 106–114. [CrossRef]
143. Encalada, A.M.I.; Pérez, C.D.; Flores, S.K.; Rossetti, L.; Fissore, E.N.; Rojas, A.M. Antioxidant pectin enriched fractions obtained from discarded carrots (Daucus carota L.) by ultrasound-enzyme assisted extraction. *Food Chem.* **2019**, *289*, 453–460. [CrossRef]
144. Artés, F.; Allende, A. Minimal Processing of Fresh Fruit, Vegetables, and Juices. In *Emerging Technologies for Food Processing*, 2nd ed.; Sun, D.-W., Ed.; Academic Press: San Diego, CA, USA, 2014; Chapter 31; pp. 583–597. [CrossRef]

145. Ruiz Rodríguez, L.G.; Zamora Gasga, V.M.; Pescuma, M.; Van Nieuwenhove, C.; Mozzi, F.; Sánchez Burgos, J.A. Fruits and fruit by-products as sources of bioactive compounds. Benefits and trends of lactic acid fermentation in the development of novel fruit-based functional beverages. *Food Res. Int.* **2021**, *140*, 109854. [CrossRef]
146. Lavelli, V.; Sri Harsha, P.S.C.; Torri, L.; Zeppa, G. Use of winemaking by-products as an ingredient for tomato puree: The effect of particle size on product quality. *Food Chem.* **2014**, *152*, 162–168. [CrossRef]
147. Ning, X.; Wu, J.; Luo, Z.; Chen, Y.; Mo, Z.; Luo, R.; Bai, C.; Du, W.; Wang, L. Cookies fortified with purple passion fruit epicarp flour: Impact on physical properties, nutrition, in vitro starch digestibility, and antioxidant activity. *Cereal Chem.* **2021**, *98*, 328–336. [CrossRef]
148. Marinelli, V.; Lucera, A.; Incoronato, A.L.; Morcavallo, L.; Del Nobile, M.A.; Conte, A. Strategies for fortified sustainable food: The case of watermelon-based candy. *J. Food Sci. Technol.* **2021**, *58*, 894–901. [CrossRef]
149. Mandha, J.; Shumoy, H.; Matemu, A.O.; Raes, K. Valorization of Mango By-Products to Enhance the Nutritional Content of Maize Complementary Porridges. *Foods* **2021**, *10*, 1635. [CrossRef] [PubMed]
150. Cedola, A.; Cardinali, A.; D'Antuono, I.; Conte, A.; Del Nobile, M.A. Cereal foods fortified with by-products from the olive oil industry. *Food Biosci.* **2020**, *33*, 100490. [CrossRef]
151. Wang, J.; Brennan, M.A.; Brennan, C.S.; Serventi, L. Effect of Vegetable Juice, Puree, and Pomace on Chemical and Technological Quality of Fresh Pasta. *Foods* **2021**, *10*, 1931. [CrossRef]
152. Mohd Basri, M.S.; Abdul Karim Shah, N.N.; Sulaiman, A.; Mohamed Amin Tawakkal, I.S.; Mohd Nor, M.Z.; Ariffin, S.H.; Abdul Ghani, N.H.; Mohd Salleh, F.S. Progress in the Valorization of Fruit and Vegetable Wastes: Active Packaging, Biocomposites, By-Products, and Innovative Technologies Used for Bioactive Compound Extraction. *Polymers* **2021**, *13*, 3503. [CrossRef]
153. Bayram, B.; Ozkan, G.; Kostka, T.; Capanoglu, E.; Esatbeyoglu, T. Valorization and Application of Fruit and Vegetable Wastes and By-Products for Food Packaging Materials. *Molecules* **2021**, *26*, 4031. [CrossRef] [PubMed]
154. Tauferova, A.; Pospiech, M.; Javurkova, Z.; Tremlova, B.; Dordevic, D.; Jancikova, S.; Tesikova, K.; Zdarsky, M.; Vitez, T.; Vitezova, M. Plant Byproducts as Part of Edible Coatings: A Case Study with Parsley, Grape and Blueberry Pomace. *Polymers* **2021**, *13*, 2578. [CrossRef] [PubMed]
155. Liu, M.; Zhang, L.; Ser, S.L.; Cumming, J.R.; Ku, K.-M. Comparative Phytonutrient Analysis of Broccoli By-Products: The Potentials for Broccoli By-Product Utilization. *Molecules* **2018**, *23*, 900. [CrossRef]
156. Espro, C.; Paone, E.; Mauriello, F.; Gotti, R.; Uliassi, E.; Bolognesi, M.L.; Rodríguez-Padrón, D.; Luque, R. Sustainable production of pharmaceutical, nutraceutical and bioactive compounds from biomass and waste. *Chem. Soc. Rev.* **2021**, *50*, 11191–11207. [CrossRef]

foods

MDPI

Review

From Pomegranate Byproducts Waste to Worth: A Review of Extraction Techniques and Potential Applications for Their Revalorization

Marina Cano-Lamadrid [1], Lorena Martínez-Zamora [1,2], Noelia Castillejo [1] and Francisco Artés-Hernández [1,*]

1 Postharvest and Refrigeration Group, Department of Agronomical Engineering and Institute of Plant Biotechnology, Universidad Politécnica de Cartagena, 30203 Cartagena, Spain
2 Department of Food Technology, Nutrition, and Food Science, Faculty of Veterinary Sciences, University of Murcia, Espinardo, 30071 Murcia, Spain
* Correspondence: fr.artes-hdez@upct.es; Tel.: +34-968325509

Abstract: The food industry is quite interested in the use of (techno)-functional bioactive compounds from byproducts to develop 'clean label' foods in a circular economy. The aim of this review is to evaluate the state of the knowledge and scientific evidence on the use of green extraction technologies (ultrasound-, microwave-, and enzymatic-assisted) of bioactive compounds from pomegranate peel byproducts, and their potential application via the supplementation/fortification of vegetal matrixes to improve their quality, functional properties, and safety. Most studies are mainly focused on ultrasound extraction, which has been widely developed compared to microwave or enzymatic extractions, which should be studied in depth, including their combinations. After extraction, pomegranate peel byproducts (in the form of powders, liquid extracts, and/or encapsulated, among others) have been incorporated into several food matrixes, as a good tool to preserve 'clean label' foods without altering their composition and improving their functional properties. Future studies must clearly evaluate the energy efficiency/consumption, the cost, and the environmental impact leading to the sustainable extraction of the key bio-compounds. Moreover, predictive models are needed to optimize the phytochemical extraction and to help in decision-making along the supply chain.

Keywords: *Punica granatum*; circular economy; sustainability; antioxidants; phenolics; encapsulation; green-technology; minimally processed; food losses; clean label

Citation: Cano-Lamadrid, M.; Martínez-Zamora, L.; Castillejo, N.; Artés-Hernández, F. From Pomegranate Byproducts Waste to Worth: A Review of Extraction Techniques and Potential Applications for Their Revalorization. *Foods* **2022**, *11*, 2596. https://doi.org/10.3390/foods11172596

Academic Editor: E. S. Brito

Received: 29 July 2022
Accepted: 20 August 2022
Published: 26 August 2022

1. Background—Food Losses and Food Waste

In accordance with the Food and Agriculture Organization of the United Nations (FAO) definition, 'food waste' is the decrease in the quantity and/or quality of food obtaining from decisions and/or actions of retailers, food service providers, and consumers, while 'food loss' refers to any food that is discarded along the food supply chain, from harvest up to retail sale [1]. FAO indicates that around one third of global food production is lost or wasted at some step in the food chain. The degree of loss greatly varies depending on the state and the basket item.

In the case of fruit and vegetables (F&V), losses over the whole supply chain could reach up to ~50% (Figure 1). FAO's future challenge is to reduce ~50% of food waste by 2050, as one of the objectives for sustainable development (OSD). The circular economy has been considered as the principle for eco-innovation, being focused on a 'zero waste' society and economy, using wastes as raw materials.

Between 2016 and 2018, FAO Statistics Division developed a food loss estimation model called '*The Food Loss and Waste database*', an online collection of data including food loss and food waste. Figure 1 shows the percentage loss of F&V (food loss + food waste) worldwide in each value chain step for the first 20 years of the twenty-first century [2]. The boxes show where ~50% of the collected data falls into, and the mid-value of the percentage loss at every stage in the supply chain is shown by a line. In this sense, postharvest and retailing are the steps in the food chain where the F&V losses represent the highest mean percentages. The mean percentage during processing is less than 10%, but in some cases, it reaches ~40%. Moreover, although the mean percentage during distribution represents less than 10%, the range is from <5% to >30%. Therefore, several strategies have been developed around the creation of active packaging with encapsulated key compounds, to avoid the high percentage of food waste/loss [3]. The range of loss percentages at each step is wide since the value depends on the type of F&V, the country, and the year.

Although this review is focused on pomegranate byproducts, the percentage of food loss related to this fruit is not available in the mentioned official database. Nevertheless, knowing that the total production of pomegranate worldwide is three million tons, and its peel and seeds represent ~54% of the fruit, this results in ~1.62 million tons of waste [4,5]. Therefore, it is a huge amount of waste produced, so it is crucial to find suitable methods to revalorize it by optimizing the bioactive compounds extraction of pomegranate residues, and then converting them into value-added products. Consequently, savings can also be made on other resources involved during production, harvesting, preservation and distribution, such as energy, water, and land, as well as contributing to the environment [5]

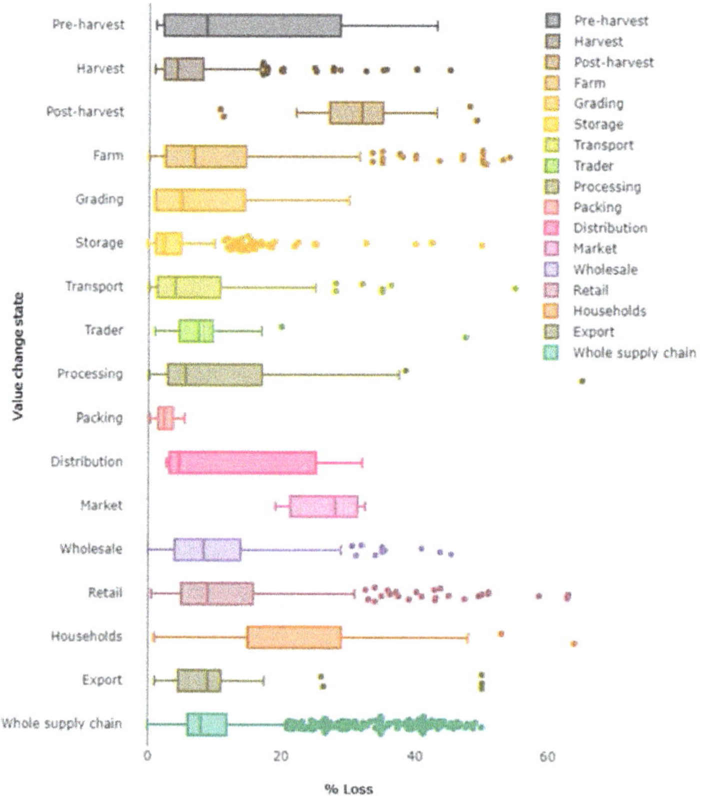

Figure 1. Food loss (%) of F&V in the world by Food Loss and Waste Database (FAO). Year range: 2000–2020; Aggregation: World; Basket items: Fruit and Vegetable; Country: All; Method data collection: all. Reprinted/adapted with permission from Ref. [2]. Copyright year: 2022; copyright owner's name: FAO.

Health, well-being, and sustainability are the current trends in the food market. Consumers and food producers are interested in 'clean label' foods or ingredients [6,7]. It means that they are interested in foods or ingredients obtained by green processing technologies (non-thermal, green solvent), and bioactive compounds with health promoting properties (nutraceuticals), among others. The bioactive compounds obtained from F&V byproducts present technological and functional features that can be incorporated within other food matrixes to enhance their nutritional, functional, and sensory quality [6,8]. Moreover, bioactive compounds from F&V byproducts have previously been classified as potential green ingredients for the cosmetic and pharmaceutical industries, and used in developing different products intended for specific populations, such as sportspeople [9].

The present review aims to evaluate the scientific evidence and knowledge on the use of green technologies for the extraction of phenolic compounds from pomegranate byproducts, and the incorporation techniques and potential applications via the supplementation/fortification of F&V matrixes to improve their quality and safety in a circular economy. For this purpose, a literature review was conducted, focusing on ultrasound-, microwave-, and enzymatic-assisted technology to enhance phenolic compounds extraction from pomegranate peel byproducts. Moreover, different incorporation techniques and applications have been reviewed. The results may provide the scientific community with an overview of the state of the art in pomegranate peel revalorization. The study may also help scientists and the food industry to develop solutions to better suit society's demands.

2. Nutritional Composition of Pomegranate Byproducts

Both primary (sugars, pectins, proteins, and fats) and secondary (polyphenols, pigments, and sulfur compounds) metabolites have been found in F&V byproducts [10]. The food industry and researchers are interested in reducing the environmental impact, and then focus on the recovery of the target compounds [6]. Carbohydrates (around 60%) [11], pectin (yield range from 6 to 25%) [12,13], proteins (around 3%) [14,15], and fats (<1%) [15] have been previously identified in pomegranate peel. Since this review is focused on the extraction of secondary metabolites from pomegranate peel, especially phenolic compounds, Figure 2 shows the classification of the main ones found [5,15].

Among them, the top ten have recently been identified and quantified [16], being punicalagin (28,000–104,000 μg/g) the major compound found, followed by ellagic acid (1580–4514 μg/g), and others such as punicalin (203–840 μg/g), catechin (115–613 μg/g), corilagin (71–418 μg/g), gallic acid (10–73 μg/g), gallocatechin (69–1429 μg/g), epigallocatechin (5–106 μg/g), epigallocatechin gallate (4–70 μg/g), and kaempferol-3-O-glucoside (16–99 μg/g) [16].

Apart from pomegranate peel, seeds (wooden part) are generated after juice processing as a byproduct. Although this review is not focused on pomegranate seeds revalorization, previous studies have indicated that pomegranate seeds are rich in polyunsaturated fatty acids (88–92%), the most abundant being linolenic acid, especially punicic acid which ranges in terms of percentage of total fatty acid profile from 59.7 to 74.3% [17,18].

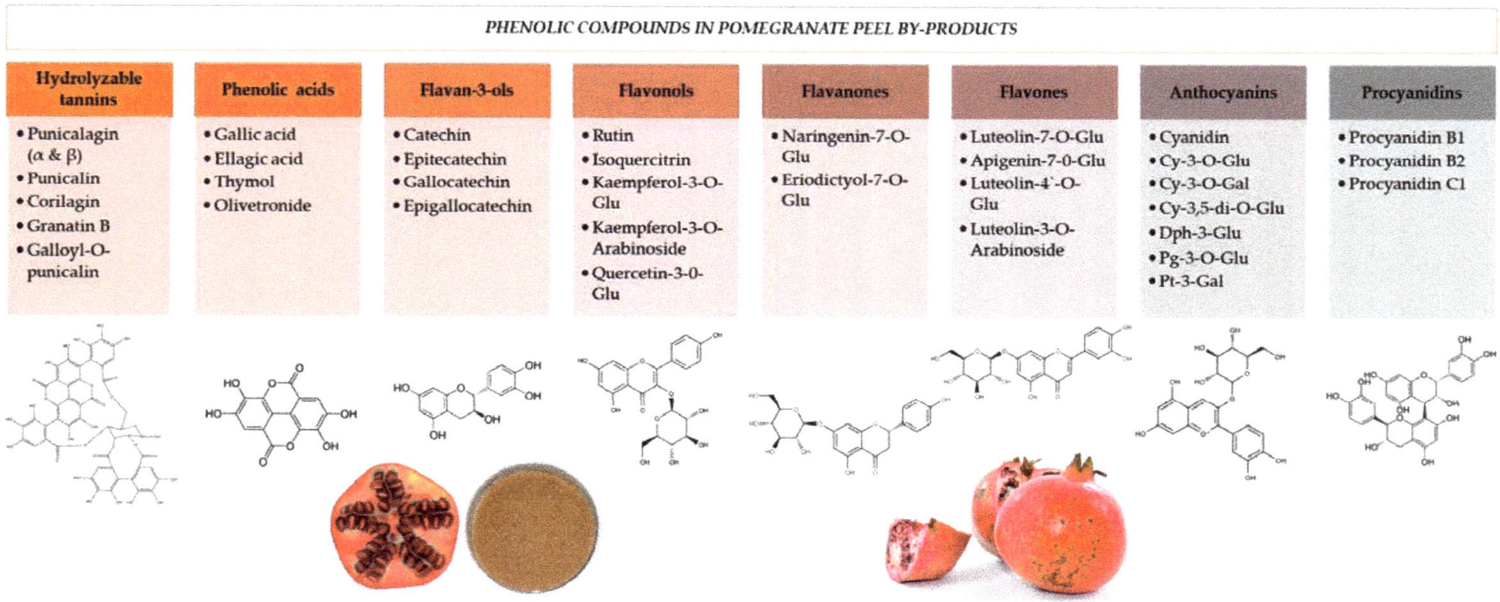

Figure 2. Classification of the main phenolic compounds in pomegranate peel [5,15]. Glu: glucoside; Cy: cyanidin; Dph: delphinidin; Pg: pelargonidin; Pt: petunidin; Gal: galactoside.

3. Scientific Literature Review

This review has been written as a research paper. Thirty-seven studies related to ultrasound-, microwave- and enzymatic-assisted extraction of phenolic compounds from pomegranate peel were collected using the PRISMA Extension (PRISMA-ScR) approach for scoping reviews. In a similar way, 21 and 28 studies were included on incorporation techniques and potential applications, respectively.

A comprehensive literature search using Web of Science and Scopus was performed in June–July 2022. Text words (pomegranate, peel, byproduct, application, ultrasound-, microwave- and enzymatic-assisted extraction) within the titles, abstracts, and keywords, were used. Original research papers and reviews with experimental design and data treatment in journals included in Journal Citation Reports (JCR) were selected.

4. Pomegranate Peel Phenolic Compounds Extraction Techniques

Conventional technologies are still in use, although these entail high energy consumption, and thermolabile nutritional compounds degradation during the process. Green extraction technologies have recently been developed using current technologies. These technologies use fewer non-green solvents, minimizing environmental and health impacts. Moreover, selective extraction is important for the bioactive compounds yield. Industry and research are focused on green extraction methods such as ultrasound-, microwave-, pulsed electric field- and enzyme-assisted extractions, among others [19].

Additionally, it is worth mentioning that processing, including drying (i.e., convective or freeze drying), homogenization, and/or grinding into powder are used as pre-treatments of extraction techniques. Even enzymatic treatment is classified as a pre-treatment of extraction processing. The drying method used for byproducts as a pre-treatment for extraction also needs to be optimized, as many of the bioactive compounds are degraded during drying. The technique, the time, and the temperature should be selected to avoid the degradation of the compounds and to have a stable material (dry byproduct) for storage until the extraction. Therefore, this process is of great importance for obtaining the best quality extracts. Depending on the drying process, the moisture content of the sample varies and influences the extraction step. Previous studies have indicated that particle size is one of the critical parameters affecting the extraction. The reduction by grinding could increase the diffusivity of the bioactive compounds, and promotes the rupture of the cell walls. Moreover, several authors indicate that blanching F&V byproducts as a pre-treatment could be a good strategy to enhance the recovery of phenolics during pomegranate peel extraction [4].

In this review, we are focused on ultrasound-assisted extraction (UAE), microwave-assisted extraction (MAE), and enzyme-assisted extraction (EAE) technologies. In the following section, the most important parameters of each technique are defined.

4.1. Ultrasound-Assisted Extraction

4.1.1. Fundamentals

Ultrasound (US) means mechanical waves propagated in an elastic medium through the transfer of energy and not of particles, to induce the longitudinal displacement of particles [20,21]. This consists in a succession of phases: (i) compression, and (ii) rarefaction, into the medium. If the strength of the rarefaction cycle is sufficient, the critical molecular distance of the liquid can be reached, and cavitation bubbles are created, creating the effect of US. The cavities increase and decrease in size during the contraction and compression phases, respectively. The bubbles generated could reach a great size, collapsing and generating large amounts of energy. The temperature and pressure at the collapse moment have been calculated to be up to 4727 °C and 5000 atm in an ultrasonic bath (25 °C) [22]. These bubbles collapse onto the surface of a solid material, and the high pressure and temperature reached create microjets directed towards the solid surface. These microjets are used in the food industry for the key bioactive compounds extraction, destroying the cell walls of the plant matrix, and its content can be released into the medium. The main parameters influencing the US technique are described below.

Type

There are two main US types: bath and probe. The first one consists of a stainless-steel tank with one or more ultrasonic transducers. The US intensity distribution is heterogeneous; therefore, the container must be located at the position where the highest intensity of sonication is achieved. The US probe is a great tool for the solid–liquid extraction of bioactive compounds. The shape and the diameter of the probe are the main characteristics that have an impact on bioactive extraction. Both US types can be applied in different modes: continuous, sweep, and pulsed mode. The main differences between the types are:

i Contact with the solution: an ultrasonic probe is submerged directly into the solution (minimum energy losses), while in a US bath, the vessel container is immersed.
ii Intensity of US: it is higher in the US probe than in the bath.
iii Maximum power achieved: in a US probe, it is the nominal power, while in a US bath, the nominal power is the minimum that can be increased due to the modulators.

Frequency

US frequency is expressed in Hertz (1 Hz $\approx$ 1 cycle/s). In a US process, the use of ultrasonic waves in the range from 20 to 100 kHz is common, and the concept time of one cycle means s/Hz.

Power/Energy Intensity/Density (Dose)

US power is expressed in watts (W), being a key parameter to express the efficiency of the process. The amount of energy applied in the system could be expressed as ultrasonic intensity (energy per second and per square meter of emitting surface, expressed in W/s o W/cm^2) or acoustic energy density (the amount of US energy per unit volume of sample, expressed in W/cm^3 or W/mL).

Amplitude level (A)

The amplitude of a wave is the height of the wave and is expressed in μm. It is important to clarify that in a US probe, the term amplitude level is commonly used. The amplitude control of the processor allows to set the ultrasonic vibrations at any desired level in the 10–100% range of the nominal power.

Pulse duration/interval ratio (Duty cycle)

This parameter is used in the pulsed US process. Pulse duration is the time for which the ultrasonic probe is *on*; pulse interval indicates the time for which ultrasonic probe is *off*; and cycle time is the sum of pulse duration and pulse interval. Duty cycle is the main way to express this parameter and can be expressed as a ratio (pulse duration/cycle time) or percentage ((pulse duration/cycle time) $\times$ 100).

Temperature

This parameter is important for the efficiency of bioactive compounds extraction. Although previous studies have indicated that an increase of temperature means an increase of extraction yield, it is essential to select an extraction temperature. The main reason is that some possible key bioactive compounds are thermolabile. Therefore, temperature optimization is needed to obtain the highest extraction yield of the key bioactive compound.

Extraction time

As with the temperature and power parameters, increasing the time in the early stages of the US process increases the extraction yield, whereas a decrease in the yield is usually observed as the extraction time increases. At the beginning, the cavitation effect of the US increases the swelling and hydration. Both swelling and hydration could be achieved by shaking. Later, the fragmentation and pore formation of the plant tissue matrix occurs, extracting the key bioactive compounds. Excessive US exposure causes structural damage to the solute and decreases the extraction yield, and even the degradation of the extracted bioactive compounds.

Solvent

The selection of the solvent for US extraction depends on the target bioactive compound. It is essential to consider the physicochemical properties of the solvent and the bioactive compounds such as viscosity, pH, surface tension, and vapor pressure of the solvent, as well as the solubility of the key bioactive compounds. The most common solvents used during US extraction are water, ethanol, alcohols, and acetone in different concentrations. Also, the concentration and the solid–liquid ratio are important factors affecting the extraction yield and properties of the bioactive compound during UAE.

4.1.2. Ultrasound-Assisted Extraction from Pomegranate Peel

Apart from the variables described above, there are other variables specific to the raw material that should be considered, such as cultivar, drying, moisture content, and particle size. Although pomegranate peel is a large reservoir of bioactive compounds, the variability of the amount and profile depends on the cultivar selected [23]. However, no cultivar information is available in 45% of the published studies, and just one of them compared two cultivars (both sour cultivars: Wonderful and Akko) with the same methodology of drying and extraction (Table 1).

Table 1 shows the different conditions of drying in all studies related to US, except one in which no information is available and one in which fresh pomegranate peel was used. Taking all the studies into account, the temperature range used is from 25 °C (room temperature) to 70 °C. On the other hand, particle size was not indicated in more than 10% of the studies showed in Table 1, and the range is from 800 μm to 25.4 mm. Moreover, one study indicated that a paste of pomegranate peel was used, and two studies described a combination of particle size in which the distribution was indicated.

The frequency range is between 20 and 80 kHz, with 20 kHz being the most common frequency used (30% of the studies included in Table 1). The power parameters were not unanimous due to the different information described: units (power, power density), the equipment used (bath, probe), and the mode (continuous, pulsed). The range of power was from 50 to 1050 W, while power density was 0–1600 expressed in W/L, and from 2.4 to 59.2 in W/cm^2. More details related to the US probe or US bath are included in "other information", such as probe diameter, submerged distance, and duty cycle. Although ethanolic solvents, with different percentages of ethanol, were mostly used (>50% of the studies included in Table 1), other authors selected solvents such as acetone, water, and methanol. It is essential to clarify that polar solvents (mainly water) extract more hydrophilic compounds (which only participate in reaction with oxygen–hydrogen bonds as sugars), and ethanolic solvents (which participate in reaction with oxygen–hydrogen and polar carbon–oxygen as ethanol) are more effective in extracting phenolic compounds [24]. The solvent changes depending on the target bioactive compounds; for instance vegetal oil was used as a solvent to extract carotenoids. The solid–liquid ratio was included in all the studies, and was highly variable among them. Information on the time and the temperature used during US extraction was lacking in about 20 and 30% of the studies, respectively. The range of time and temperature included was 0–240 min and 20–93 °C, respectively.

The target individual compounds found were punicalagin (Pn), individual phenolics (Ph), individual tannins (Tn), ellagic acid (EA), chlorogenic acid (ChlA), gallic acid (GA), individual flavonoids (Fvs), and hydroxybenzoic acids (HbA). The importance of other bioactive compounds from pomegranate as targets to optimize the extraction process, such as anthocyanins and alkaloids, should be noted as being of interest for the food, pharmacological, and cosmetic industries. In addition, spectrophotometric techniques were used to determine the yield of the extraction, the total phenolic content (TPC), total flavonoid content (TFC), antioxidant activity (AOX), total anthocyanin content (TAC), and total carotenoids content (TCC). Undesirable compounds formed during the treatment, such as hydroxymethylfurfural as a furan derivative, could be a good strategy to optimize the process.

Regarding the best conditions for extracting bioactive compounds from pomegranate peel, different optimal processing conditions can be found in the published scientific manuscripts. Following the literature review, some of the optimized conditions are presented below. The highest (506 mg g^{-1} dw) punicalagin content was obtained by UAE with 53% EtOH, s solid–liquid ratio of 1:25 w/v, and power at 757 W for 25 min [25]. More and Arya (2021) [26] concluded that the optimum processing conditions were 2:100 solid–liquid ratio, 116 W (80% duty cycle) for 6 min, obtaining a 0.48 g/g yield, and a TPC of 178 mg GAE/g dw. Pan et al. [27] observed that pulsed US extraction with 59 W/cm^2, and 5/5 of pulse duration/interval duration increased the antioxidant yield (22%) and reduced the extraction time (87%) compared with conventional extraction. Furthermore, when the US extraction was continuous with the same conditions, the antioxidant yield increased by 24% and reduced the extraction time by 90% [27]. Other authors have reported that US increased the extraction yield reducing by 20-fold the time required [28]. Moreover, the extraction yield increased with increasing extraction temperature from 25 to 35 °C [28]. Other research has proposed a mathematical model for multi-criteria optimization to enable the prediction of bioactive compounds extraction for any temperature (20–60 °C), solvent (0–100% ethanol), and US power density (0–100 W/L), at any time (0–240 min). This model reveals the optimal conditions for obtaining the best yield of the target compound with the minimum time and/or energy consumption [29]. For the recovery of ellagic acid, Muñiz-Márquez et al. [30] indicated that the best extraction conditions were at 94 °C for 55 min using ethanol 75% and 1:3 solid–liquid ratio [30]. To recover carotenoids, the most efficient extraction period to achieve the maximum yield from pomegranate peel was about 30 min with the following conditions: 52 °C, 0.10 solid–liquid ratio, amplitude of 58.8%, and sunflower oil as a solvent [31].

Table 1. Ultrasound conditions (frequency, power parameters, solvent, time, temperature) for the extraction of bioactive compounds from pomegranate peel.

Byproduct Characteristics	F (kHz)	Power Related Parameters	Solvent	Solid:Liquid Ratio	t (min)	T (°C)	Other Information	Extract Characterization	Ref.
Freeze-drying Powders < 254 µm cv information NA	NA	500–1050 W	EtOH (40–80%)	1:10/1:50	10–50	NA	NA	Pn	[25]
Vacuum oven (45 °C, 36 h) Powders < 500 µm Bhagwa cv	NA	350 W (Pulsed mode: A 10–100%)	EtOH (50%)	0.1:10/0.3:10 0.5:10/0.6:10 0.8:10/1:10	NA	NA	Duyt cycle: 10–90%	TPC, TFC, AOX (DPPH and ABTS), TAC	[26]
Cabinet hot drier (40 °C) Powders < 635 µm cv information NA	20	2.4 to 59.2 W/cm^2 (continuous) 59.2 W/cm^2 (pulsed)	NA	1:50	2–90	25	Area probe: 1.267 cm^2 Pulse duration/Interval: 2/1, 3/1, 4/1, 5/1, 6/1, 7/1, 9/1, 12/1, 2/5, 3/5,4/5, 5/5, 6/5, 7/5, 9/5, 12/5, 2/15, 3/15, 4/15, 5/15, 6/15, 7/15, 9/15, 12/15 Number of pulse repetition: 30, 60, 90, 120, 180, 270, 360, 540, 720	TPC, AOX (DPPH)	[27]
Drier (40 °C, 48 h) cv and particle size information NA	20	130 W (Pulsed mode: A 20 and 60% NP)	MetOH, EtOH, EtAc, MeOH (50%)	1:10/1:50	4–60	25–45	Ti–Al–V sonoprobe (13 mm) Pulse duration/pulse interval ratio 5/15 and 2/1	Ph	[28]
Vacuum oven (40 °C) Powder 125–150 µm cv information NA	30	50 W (A 30–70% NP)	EtOH (80%)	1:40/1:120	5–50	NA	Duty cycle (60–100%) Probe diameter 3 mm Length 80 mm	Pn	[32]
Air oven (40 °C, 18 h) Powders 3 mm cv information NA	30.8	0–100 W/L	EtOH (0–100%)	1:40	0–240	20–60	NA	TPC, AOX (DPPH)	[29]
Microwave vacuum oven (45 °C, 36 h) Powders < 508 µm Kabuli cv	20	700 W (A 40–80%)	Ac (30–90%)	1:10/1:30	10–20	45	Probe half-inch diameter Pulse-on and pulse-off time of 10 s.	Tn	[33]

Table 1. *Cont.*

Byproduct Characteristics	F (kHz)	Power Related Parameters	Solvent	Solid:Liquid Ratio	t (min)	T (°C)	Other Information	Extract Characterization	Ref.
Air oven (60 °C, 48 h) Powder 600–800 µm cv information NA	40	NA	EtOH (24–75%)	1:3/1:16	4–55	26–93	NA	EA	[30]
Fresh peel (more information NA) Small-grinded pieces (fine peel paste, 4 °C) cv information NA	20	400–1600 W/L	NA	2:10	5–50	30–70	Pulsed mode: 'on' time (5 s) Pulse interval 'off' time (3 s) Probe: 3 cm submergence of sonicator	TPC	[34]
Dried (more information NA) Powders < 0.5 mm cv information NA	40	500 W (A 50–80%)	MetOH, (30–70%)	1:15	5–15	40	Probe: 6 mm diameter, dipped up to 2 cm Duty cycle: 0.2–0.8	TPC, TFC, AOX (DPPH and FRAP)	[35]
Laminar airflow drying oven (50 °C, 24 h) Powder particle size and cv information NA	24	NA	NA	1:10	5–20	50	Titanium probe: 14-mm diameter	TPC, EA, ChlA, GA	[36]
Hot air oven (40 °C, 48 h) Fine powder (more information NA) cv information NA	45	360 W (A 40–100%)	NA	0.1:1/1:10	5–45	40–80	pH 3.5 to 6.0	TPC	[37]
Technique information NA Powder 140–425 µm Sishe Kape- Ferdos cv	20	400 W (A 20, 60 and 100%)	NA	1:4	5–15	NA	NA	TPC	[38]

Table 1. *Cont.*

Byproduct Characteristics	F (kHz)	Power Related Parameters	Solvent	Solid:Liquid Ratio	t (min)	T (°C)	Other Information	Extract Characterization	Ref.
Forced air oven (70 °C, 48 h) Powder particle size distribution: 25.4–0.105 mm (56%); 0.105–0.075 mm (17.9%); 0.075–0.037 mm (14.5%); <0.037 mm (11.6%)	37–80	180 W (continuous, pulse, and sweep modes)	EtOH (70%)	1:25	20	40–70	NA	Pn, EA, TPC	[39]
Brazilian Molar cv Drier (40 °C, 48 h) Powder < 0.2 mm cv information NA	20	130 W (Pulsed mode; 20–60%)	Sunflower oil Soy oil	1:10/3:10	NA	10–60	Ti-Al-V sonoprobe (13 mm)	TCC	[31]
Air-drier (7 days, 20 °C) Powder < 180 µm Malas cv	24	53, 79, and 105 W/cm^2 (pulse mode)	EtOH (70%)	1:10	2–10	NA	Area probe: 1.53 cm^2 Duty cycle: 50%, 70%, and 90%	Pn, HbA	[40]
Traditional heating oven (40 °C, 48 h). Microwave drying (T < 100 °C, <5 min). Powder < 150 µm. Wonderful and Akko cv	26	200 W (pulsed mode; A 50%)	EtOH (70%)	1:40	10	45	Duty cycle: 80%	EA, Pn	[41]
Hot air oven (50 °C, 48 h) Fine powder (more information NA) cv information NA	NA	NA	Ac, MetOH, EtOH (50–75%)	1:20	NA	45	NA	TPC, TFC	[42]
Air dried Room Temperature Particle size 0.3 mm cv information NA	20	400 W A 70%	EtOH (70%)	1:30	30	40	Probe tip 2 cm 22.5% duty cycle	TPC, TFC, AOX (DPPH and FRAP)	[42]

Table 1. *Cont.*

Byproduct Characteristics	F (kHz)	Power Related Parameters	Solvent	Solid:Liquid Ratio	t (min)	T (°C)	Other Information	Extract Characterization	Ref.
Ventilated oven (42 °C, 3–4 days) Particle size 0.5 mm cv information NA	NA	NA	H_2O; EtOH (70%); EtOH (100%); Ac (70%); Ac (100%);	NA	23	45	NA	TPC, EA, Pn, Individual Fvs	[43]
Convective oven 60 °C 22 h Particle size 420 µm Mollar de Elche cv	20	750 W	H_2O	4:40	NA	NA	Probe diameter 13 mm titanium	TPC, AOX (DPPH and ABTS)	[44]
Blanching (80 °C 3 m) + Tray drier 40 °C Particle size < 1 mm Wonderful cv	40	700 W	EtOH (70%)	1:15	60	40 °C	Ultrasound bath	TPC, TFC, TAC, Vit C, AOX (DPPH, FRAP and ABTS)	[45]
Tray drier 40 °C Particle size < 0.25 mm Bhagwa cv	20	20–40% A	EtOH (70%)	1:20	10–20	40–60	3 mm of probe diameter	Pn, EA, GA	[46]
Oven drier (more information NA) Particle size: size distribution using sieves: 0.85, 0.425, 0.25 and 0.18 mm cv information NA	35	140 W	EtOH (30–50–70%)	0.2:10	10–30	30–60	Ultrasound bath	TPC, AOX (FRAP and DPPH)	[47]
Oven drier (45 °C 48 h) Particle size and cv information NA	40	NA	NA	1:10	0–60	35–45	Enzymatic pre-treatment	AOX (DPPH), TPC	[48]
Hot air in cabinet drier (40 °C) Particle size < 0.420 mm Wonderful cv	20	Continuous intensity: 2.4, 4.7, 7.1, 18.9, 23.7, 30.8, 37.9, 45.0, 52.1, and 59.2 W/cm^2	H_2O	1:50	2–90 min	25	Probe with area of 1.267 cm (continuous)	AOX (DPPH)	[27]

Table 1. *Cont.*

Byproduct Characteristics	F (kHz)	Power Related Parameters	Solvent	Solid:Liquid Ratio	t (min)	T (°C)	Other Information	Extract Characterization	Ref.
Hot air in cabinet drier (40 °C) Particle size < 0.420 mm Wonderful cv	20	Pulsed mode: 2.4, 4.7, 7.1, 18.9, 23.7, 30.8, 37.9, 45.0, 52.1, and 59.2 W/cm^2	H_2O	1:50		25	Probe with area of 1.267 cm Pulsed duration/interval: 2/1, 3/1, 4/1, 5/1, 6/1, 7/1, 9/1, 12/1, 2/5, 3/5, 4/5, 5/5, 6/5, 7/5, 9/5, 12/5, 2/15, 3/15, 4/15, 5/15, 6/15, 7/15, 9/15, 12/15 Number pulse repetition: 30, 60, 90, 120, 180, 270, 360, 540, 720	AOX (DPPH)	[27]
Air-dried at 25 °C Particle size 0.75–2 mm cv information NA	NA	NA	EtOH (10–30–50–70–90%)	1:10; 1:20; 1:30; 1:40; 1:50	5–65	25–80		Pn, EA, GA, TPC	[49]
Forced air oven 70 °C 48 h Large particle size: 0.297–1.410 mm, mean: 1.05 mm Small particle size: 0.177–1 mm, mean: 0.68 mm Wonderful cv	19	0–800 W at the generator, or 0–38.5 W at the tip of the probe	H_2O; EtOH (30–50–70%)	NA	10	50–100	13 mm diameter probe Number of Cycles: 5 Assisted by pressurized liquid	Pn, EA, P	[50]

NA: Data not available; cv: cultivar; A: amplitude; F: frequency; NP: nominal power; Ac: acetone; Pn: punicalagina; P: punicalin; TPC: total polyphenolic content; TFC: total flavonoid content; AOX: total antioxidant capacity; TAC: total anthocyanin content; Ph: phenolics; Tn: Tannins; EA: ellagic acid; ChlA: chlorogenic acid; Fvs: flavonoids, GA: gallic acid; TCC: carotenoids; hydroxybenzoic acids (HbA); Fvs: flavonoids.

US treatments are also combined with other green technologies to increase the extraction effectiveness. With regard to pomegranate peel, US treatment was jointly applied with different combinations of expansion gas initial pressure [51] and system pressure [52]. These results suggest the great potential of expansion gas in pressurized liquids assisted by US using green solvents for the extraction of polyphenols [51]. Another technology used in combination with US for bioactive compounds extraction from the pomegranate peel was the extraction at the cloud point, the combination studied being the one that gave the maximum extraction of polyphenol and flavonoid content [52]. More information on the combination with EAE technology is given in Section 4.3.

4.2. Microwave-Assisted Extraction

4.2.1. Fundamentals

Microwave (MW) energy is based on electric and magnetic fields, obtaining electromagnetic waves. This energy is non-ionizing, facilitating molecular movement by ion migration and dipole rotation without altering the molecular structure, generating friction and then heat. MAE is based on the disruption or changes in the structure of cells when applying non-ionizing electromagnetic waves to a sample matrix [53]. Performance in MW-assisted processes is highly influenced by several variables. Therefore, the main characteristics and parameters are described below [54,55].

Pressure

For MAE, the most common instrument is a closed-vessel system in which the pressure and the temperature can be modulated, and then optimized to accelerate the mass transfer of target compounds from the sample matrix, avoiding degradation [56].

Power Intensity/Density (Dose)

In a similar way to the US technique, one of the most important factors to be considered in the MAE is the MW power density, expressed as the power to be applied to the product per unit weight or volume. An increase of the MW power enhances the penetration of solvent into the solid, and then the extraction and recovery of bioactive compounds. Power should be selected to optimize yields and for the selectivity of the desired components, without affecting their stability.

Temperature

An increase in temperature during extraction promotes an increase of the diffusivity of the solvent into the solid, and then a desorption of the target compounds occurs. Temperature control is carried out by a probe. Focusing on avoiding undesirable changes, the extraction temperature should be selected considering the stability and extraction yield of the desired active compounds.

Time

As the extraction time increases, the yield increases, but high power during long application times is associated with thermal degradation. Therefore, a combination of low/moderate power with longer exposure may be considered. Depending on the matrix and the target bioactive compounds, the optimal condition changes.

Solvent

Considering what has been stated in relation to solvents in the section on US, the viscosity of the solvent affects molecular rotation, and therefore the ability of samples to absorb MW.

4.2.2. Microwave-Assisted Extraction from Pomegranate Peel

Table 2 shows the scientific evidence on MAE of bioactive compounds from pomegranate peel. The drying technique, particle size, and cultivar information used during pomegranate peel from 0.5 mm to 2 mm. The power, temperature, and time used during MW treatment was from 100 to 6000 W, from 40 to 72 °C, and, from 0.5 to 40 min, respectively. Ethanol, methanol, and water were the solvents used during MAE. Different solid–solvent ratios were studied to optimize the process, from 1:10 to 1:60. Apart from the bioactive compounds, it is crucial to focus on other compounds such as hydroxymethylfurfural, when high temperatures are reached during food processing.

In a previous review related to MW extraction of pomegranate peel, several manuscripts were included that are omitted here [4] because they are not in the JCR list. Related to the optimal conditions, a previous study indicated that the optimum parameters of vacuum MAE were 10–12 min, 61–79 °C, 3797–3577 W, and 38–39% ratio of water to raw material (39.92% and 38.2%) to obtain the highest values of TPC (5.5 mg Gallic Acid Equivalent/g fresh pomegranate peel) [57]. Other authors have reported that the optimum operating conditions were extraction with ethanol 50%, 1:60 solid–liquid ratio, and 600 W [58]. The results were compared with those obtained by UAE studied in a previous work by the same research team, concluding that the MW method presented a 1.7-fold higher yield after 4 min than after 10 min by UAE [58]. Regarding phenolic extraction, another study indicated that MAE (low MW power and 50% ethanol) was useful for phytochemical extraction [59]. Although there are relevant and promising results, they are nonunanimous and scarce. Therefore, more research on MAE and comparison with other green techniques is required.

4.3. Enzymatic-Assisted Extraction

4.3.1. Fundamentals

Enzyme assisted extraction (EAE) is also classified as a green technique. The purpose of this technique is the addition of enzymes in the extraction medium, usually as a pretreatment of other techniques to enhance the yield, breaking, and/or softening the cell walls. Therefore, thanks to the digestion of the cell walls, bioactive compounds (bound or dispersed inside the cells and on cell walls) can be directed out of the cell to the solvent [19]. EAE extraction depends on several variables that are described below. Among enzymes, pectinases, proteases, and cellulases (and their combinations) are the most used for the extraction of bioactive compounds from F&V byproducts. Pectinases degrade the pectin present in cell walls, and are mainly used in food industries for the clarification and extraction of fruit juices, emerging as a new tool for the extraction of bioactive compounds [60,61]. Proteases are hydrolase enzymes that digest proteins and peptides, and even hydrolyze peptide bonds present in cell walls [62]. Cellulases are key enzymes in the food industry, as they play an important role in the overall carbon cycle. This is due to the degrading of insoluble cellulose into soluble sugars. It is important to highlight that cellulases are the most diverse type of enzymes, catalyzing the single substrate hydrolysis [63]. According to van Oort [64], the main limiting factors in the reaction speed and enzymatic activity are:

i. Solute concentration
ii. Enzyme concentration: when the enzyme concentration increases, the reaction rate will increase until a point when there is no positive or negative effect of the continued increase in enzyme concentration.
iii. Temperature: the optimum temperature for maximum enzyme activity depends on the type of enzyme since most of them are proteins which are denaturized and inactivated by heat.
iv. pH: enzymes have a characteristic pH value for their optimum activity, being acidophilic (optimal pH values are <7) and alkalophilic enzymes (pH > 7). At pH values greater or lower than the optimal, the enzymatic activity—and therefore the reaction rate—decreases. Furthermore, the optimal value of enzyme concentration, temperature, and pH depends on the conditions of the medium/matrix in which it is found.
v. The presence of inhibitors: molecules that temporarily or permanently interact with enzymes to reduce their activity and/or reaction rate. The inhibition can be classified as:
    a. *Competitive*: the inhibitor structure is like the substrate. The key bioactive compound and the complex substrate-enzyme is not formed.
    b. *Acompetitive*: the inhibitor structure is attached to the complex-enzyme.

c. *Non-competitive*: the inhibitor attached to the active center of the enzyme and enzyme is misshapen.
d. *Mixed*.

Apart from the limiting factors of the speed and enzymatic activity, the inactivation protocol is also required to optimize the extraction time. Therefore, all mentioned key parameters should be detailed and optimized. At the end of the EAE, enzyme inactivation is necessary. The inactivation conditions (temperature, time) depend on the enzymes used.

4.3.2. Enzyme-Assisted Extraction from Pomegranate Peel

The detailed information related to the EAE of bioactive compounds from pomegranate peel is shown in Table 3. The enzymes used in the literature were pectinase, protease, and cellulase, while the temperature ranged from room temperature to 45 °C. After enzymatic pre-treatment, hydrolyzed pomegranate peel continues the extraction with other green technologies (high pressure, supercritical carbon dioxide, and ultrasound). Comparing with a previous review [4] which encompassed the enzymatic extraction of pomegranate peel, it can be observed that the scientific evidence has increased over the last two years.

A previous study indicated that the combination of green technologies (US, MW, high pressure, and supercritical carbon dioxide) with enzymatic pre-treatment could be a good tool to enhance polyphenols extraction from pomegranate peel. Recent research has indicated that a higher phenolic compounds yield was obtained from pomegranate peel using an enzymatic pre-treatment (Viscozyme®) followed by MAE than when conventional solvent extraction, EAE, or MAE was used [65]. Other authors have indicated that the optimum conditions of enzymatic pre-treatment and US technology was 41 min, 1.3% Viscozyme concentrationtion, and incubation for 1.8 h at 45 °C, obtaining extracts with a TPC of 20 mg GAE/g [48]. On the other hand, the pre-treatment enzymatic extraction did not improve the extraction yield when high pressure technology was applied to obtain punicalagin rich extracts [66]. With regard to enzyme-assisted supercritical fluid extraction process, it can be said that a high content of individual phenolic acids such as vanillic, ferulic, and syringic (108, 75 and 88 µg/g of extract, respectively) were found in the extracts. These phenolic acids were extracted thanks to enzymatic-assisted tecnology using pectinase, protease, cellulase, alcalase, and viscozyme [67]. Not only have promising results been obtained in the enzymatic extraction of pomegranate peel, but there is also evidence that it works as a pre-treatment in the extraction of bioactive compounds from other F&V byproducts [68]. More studies are needed to obtain the optimum conditions, depending on the raw material and the target compounds.

Table 2. Microwave conditions (power parameters, solvent, time, temperature) for the extraction of bioactive compounds from pomegranate peel.

Byproduct Characteristics	Power (W)	Pressure	Solvent	Solid/Liquid Ratio	T (min)	T (°C)	Other Information	Extract Characterization	Ref.
Milled frozen (more information NA) Wonderful cv	2000, 4000, 6000	Vacuum 355 mbar	NA	1:10; 1:25; 1:40	10, 50, 90	40, 50, 60	Industrial-type MAC-75 multimode MW extractor	TPC, AOX (DPPH)	[57]
Drier (40 °C, 48 h) Particle size ~0.1 mm cv information NA	100, 173, 350, 527, 600	75 bar	H_2O EtOH (50–70%) MetOH (50–70%)	1:10; 1:17.3; 1:35; 1:52.7; 1:60	0.5–15	NA	Multiwave closed MW system	TPC, AOX (DPPH), Pn, EA	[58]
Air-dried (4–5 days) Particle size 0.75–2 mm cv information NA	470–800	NA	EtOH (50%)	1:10	20	41–72	Home-made setup consisting of MW oven	TPC, GA, EA, Pn, HMF	[59]

NA: Data not available; Pn: punicalagin; TPC: total polyphenolic content; AOX: total antioxidant capacity; EA: ellagic acid; GA: gallic acid; HMF: hydroxymethylfurfural.

Table 3. Enzymatic assisted conditions (enzyme, pressure, time, temperature) for the extraction of bioactive compounds from pomegranate peel.

Byproduct Characteristics	Combined with	P (MPa)	Enzymes	Inactivation Enzymes	Solid/Liqui Ratio	t (min)	T (°C)	Extract Characterization	Ref.
Drier (40 °C) cv and particle size information NA	High pressure	300	4% (vol) of pectinase 4% (vol) of cellulase	Water bath 90 °C 5 min Later ice bath	1.6:100	15	NA	TPC Individual Ph	[66]
Dried Room Temperature Particle size < 0.178 mm cv information NA	Supercritical carbon dioxide	NA	Pectinase, protease, and cellulase (25:25:50)	Water bath 90 °C 5 min Later ice bath	2.8–3.9% of enzymes in the solvent	60–120	35–60	TPC	[67]
Drier (45 °C 48 h) cv and particle size information NA	Ultrasounds (Bath 40 kHz)	NA	Viscozyme® concentration 0–2 mL/100 mL solvent	NA	1:10	0–60	30–50	TPC, TFC, AOX (DPPH)	[48]
Dried (more information NA) cv and particle size information NA	Microwave (300, 400 and 600 W)	NA	Viscozyme® concentration 0.6% (*v*/*w*) pH 4.5 and 40 °C	NA	1:20, 1:30, 1:40 g/mL EtOH 30% acidified	90–150	NA	AOX (FRAP and CUPRAC), TPC	[65]

NA: Data not available; cv: cultivar; Ph: individual phenolics compounds; TPC: total polyphenolic content; AOX: total antioxidant capacity; TFC: total flavonoid content.

5. Pomegranate Peel Byproducts Incorporation Techniques

5.1. Powders/Flours

Pomegranate peel powder/flour is commonly acquired by drying and grinding until obtaining the desired particle size. Similar drying technology applied to edible fruit and plant material could be used in F&V byproducts to avoid undesirable bioactive compound changes [69]. The most common drying technologies are convective drying, sun-drying, MW drying, and freeze-drying in which key variables should be optimized (for instance, temperature and time). Moreover, spray-drying is commonly catalogued as a good tool for byproducts drying. This powder could be applied as a solid ingredient for the fortification of different products such as meat-based, F&V-based, and bakery products (Section 6) since this material presented high dietary fiber and techno-functional properties (high water- and oil-holding capacity, and low water absorption) in previous studies [70]. Similarly, powders can be obtained from liquid extracts after bioactive compounds extraction using different technologies such as freeze-drying or spray-drying [71]. Such technologies are included in the section on encapsulation due to the need for different processes to be carried out (Section 5.3).

5.2. Liquid Extracts

With pomegranate peel powders obtained as previously detailed, extraction techniques with different solvents can be used, including those reported in this review. These liquid extracts are not suitable for direct incorporation into the different food matrixes, except when the solvents may be classified as a food ingredient (e.g., water). Therefore, these solvents must be removed through evaporation. Once they have been evaporated, drying should be carried out (for instance convective or freeze-drying) to later redissolve it in water, as the most common liquid. In this way, the liquid extract is ready to be incorporated into the matrixes at different solid–liquid ratio, as observed in Section 6. In addition, liquid extracts can be used to obtain coatings, and can be encapsulated by different carriers and techniques, as outlined in Section 5.3.

5.3. Encapsulation

Encapsulation is a means to protect sensitive key bioactive compounds found in the food industry byproducts against undesirable heat, oxygen, light, and pH conditions [72]. The process needs a carrier agent and a technique to create the protective capsules. Different techniques may be used for the encapsulation of target compounds from F&V byproducts, such as spray-drying, freeze-drying, complex coacervation, and ion gelation [73], among others. Spray-drying is the liquid food drying method and has been widely used to obtain powders from F&V juices [69,74–76]. Currently, the transformation of F&V byproduct extracts (liquid) into powders using a spray-drier (the extracts are sprayed into a hot air chamber) has garnered attention because the process is complex, although this technique is one of the fastest, cheapest, and more reproducible, despite its complexity. In lyophilization as well as in spray-drying, a solution, dispersion, or emulsion is first obtained depending on the encapsulating agent and the active compound. The first step of freeze-drying-based encapsulation consists in creating an emulsion between the carriers and the target compounds, followed by a conversion into microcapsules by applying the freeze-drying technique [77], which consists of water removal by sublimation (primary drying) and secondary drying. Table 4 shows the main technologies (spray-drying, freeze-drying, double emulsion, and ion gelation) and the carriers used to encapsulate target bioactive compounds from pomegranate peel. It can be seen that there is an interest in using novel carriers such as citrus byproducts.

Table 4. Main technologies used to encapsulate target compounds from pomegranate peel.

Technology	Carriers	Target Compound/Activity	Ref.
Spray-drying	Maltrodextrin	F-TPC, UPLC-TPC, Pn, EA, P, GA	[78,79]
	Maltrodextrin + others: Tween 80 (99:1); Skimmed milk powder (50:50); Whey protein isolate (50:50); Gum arabic (50:50)	NA (Yield/Stability)	[28,80]
	Skimmed milk power	NA (Yield/Stability)	[28,80]
	Orange juice byproduct	F-TPC, DPPH	[81,82]
	Maltodextrin/Pectin	TPC, Pn, EA	[83]
	Whey protein	Pn, EA, P, GA	[79]
	Arabic gum	Pn, EA	[84]
	Chitosan	Pn, EA	[84,85]
	Pectin	Pn, EA	[84]
	Modified starch	Pn, TPC, HTC, DPPH	[86]
	Alginate	NA (Yield/Stability)	[85]
Freeze-drying	Soy phosphatidylcholine liposomes	Pn, EA, rutin, epifallocatechin, TPC	[87]
	Maltodextrin (5 and 10%) and b-cyclodextrin (5 and 10%).	F-TPC, FRAP	[38]
	Prunus armeniaca gum exudates	FRAP, DPPH	[88]
	Chitosan	FRAP, DPPH	[88]
	Maltrodextrin	TPC, TFC, Pn, EA, FRAP, DPPH	[89]
	Maltodextrin and calcium alginate	ANCs, FRAP, DPPH	[90]
	Maltodextrin and soy lecitin	NA (Yield/Stability)	[91]
Double emulsion	Water[1] in Oil in Water[2]: Water[1] (ethanolic solutions) in Oil (castor, soybean, sunflower, medium chain triglyceride and orange) in Water[2] (aqueous solution with Tween[80])	NA (Yield/Stability)	[92]
Ion gelation	Chitosan gel (1%):gelatin 2:1	F-TPC, DPPH	[93]
	Spirulina	TPC, DPPH	[72]
	Microalgae	EA	[94]
	Chitosan + others: Dialdehyde guar gum Gelatin-based materials	F-TPC, DPPH	[95]

NA: Data not available; cv: cultivar; EA: ellagic acid; F-TPC: total polyphenolic content by Folin assay; UPLC-TPC: total polyphenolic content by UPCL; TFC: total flavonoid content; Pn: punicalagin; P: punicalin; GA: gallic acid; HTC: hydrolysable tanin content; ANCs: anthocyanins.

In addition, other technologies were applied for other pomegranate byproducts, such as complex coacervation, to obtain encapsulated pomegranate oil rich in punicic acid [96]. Complex coacervation is a liquid–liquid phase separation phenomenon that consists between oppositely charged biopolymers through electrostatic interaction, and this technique is increasingly used in the food industry due to its high encapsulation efficiency and optimal processing conditions [97]. After encapsulation processing, the encapsulated material presents the characteristics to be incorporated in other matrixes.

6. Potential Applications in the Food Industry

Pomegranate peel (in powders, liquid extract, and/or encapsulated, among others) have been reported in several food matrixes [98] such as F&V-based (Table 5), meat-based [15], fish-based [99,100], oil [101], dairy-based [102], confectionary [103], and baking products [82,104,105], among others. Packaging evidence have been reported by other authors, which has proven to be a good tool to preserve foods without altering their composition [106].

Since the bibliography on the incorporation of pomegranate byproducts into different food matrixes is extensive, this review has been focused on the scientific evidence related to the use of pomegranate peel byproducts during F&V handling and processing in the form of fresh whole, fresh-cut, minimally processed F&V, and beverages. Table 5 includes information about the characteristics of pomegranate peel byproducts (drying technique, particle size, and cultivar), extraction technique (US, maceration), incorporation method (liquid extracts, coating, dipping), and benefits tested after its incorporation (shelf life, bioactive compounds fortification). In the following sections, more specifications related to F&V based products are detailed.

6.1. Fresh Whole F&V

In this case, more than 15 types of evidence have been found, in which pomegranate peel extracts were incorporated in different F&V (Table 5), being >25% incorporated into citrus fruits. The incorporation of pomegranate peel extract as a postharvest technique in fresh whole F&V has been reported in ~90% of the included studies. A coating enriched with pomegranate peel extract is described in 42% of them, the control formulation in which the extracts were added being chitosan and alginate solutions. Additionally, scientific evidence related to preharvest application is reported (pomegranate peel atomization in tomato leaves and the incorporation of the soil in a sage herb field). Table 5 shows specific information related to the drying technique, particle size, and cultivar of pomegranate; the extraction technique; the extracts formulation and incorporation method (atomization, liquid extracts, coating, dipping); and the main results obtained by the authors.

6.2. Minimally Processed, or Fresh-Cut F&V

Since fresh-cut F&V usually present a short shelf life mainly due to enzymatic browning, dehydration, and microbial growth, it is necessary to look for innovative tools to preserve its quality and safety. Table 5 shows the scientific evidence in which pomegranate peel extracts were used in minimally processed or fresh-cut F&V. There is a need to focus on the different ways of incorporating extracts into other fresh-cut F&V, and salads (for instance, baby leaves and younger plants such as sprouts or microgreens). There is a lack of knowledge on the effect of pomegranate peel extracts on vegetable commodities.

6.3. F&V Based Beverages

The fortification of F&V based beverages with bioactive compounds has been recently reviewed and reported [8]. The goal of the fortification with target compounds could be to enhance functionality (high content of polyphenols and other compounds) and/or techno-functional properties (color maintenance, sensory quality, inhibition of microbial growth). Moreover, if the key biocompounds have been extracted by green technologies from F&V byproducts, their incorporation replaces or reduces synthetic additives. Table 5 shows the incorporation of pomegranate peel extracts in F&V juices as an alternative to enhance quality parameters. Future research should be focused on the fortification of other F&V-based matrixes such as cold/hot/dried soups and culinary sauces with pomegranate peel. For instance, a previous study indicated that the incorporation of horticultural byproducts improved the quality and shelf life of a kale pesto sauce [107].

Table 5. Application of pomegranate peel in fresh fruit and vegetable, minimally processed fruit and vegetable, and beverages.

	Matrix	Pomegranate Peel Byproduct	Extraction	Incorporation Method	Benefit	Ref.
Fresh whole F&V (pre- and postharvest)	Tomato	Drier (50–60 °C, 72 h) Fine powder (more information NA) cv information NA	Ratio 3:10 EtOH 48 h + evaporator (65 °C) + re-dissolved in sterile distilled water (0.05%, 0.5%, 1% and 5% *w*/*v*)	Preharvest. Tomato plants were sprayed in the leaves (bacteria inoculation) with the aqueous extract + 24 h drying	Antibacterial activity at least 15 days Replacing, reducing, or even alternating treatments involving copper compounds	[108]
	Sage herb	Air dried (more information NA) Grinder (more information NA) cv information NA	1:10 solid–liquid ratio in water or EtOH 80% 24 h + evaporator + water dilution	Preharvest. Added in the soil (2, 4, and 6 g per plot)	Higher dry mass and essential oils Inhibition of free radical scavenging	[109]
	Olive	Oven drier (40 °C) Powder home grinder (more information NA) Wonderful cv	120 g/L EtOH solvent (50 and 80%) + 1% Citric acid	Postharvest. Treatment of 1 × 1-mm injuries and inoculated (*C. acutatum*) by 10 µL of pomegranate peel extract (12, 1.2, or 0.12 g/L)	Reduction of fungal and bacterial population	[110]
	Potato tubers	Air drier (28 °C, 10–15 days) Fine powder (more information NA) Baladi cv	1:10 solid–liquid (MetOH) 48 h 28 °C + evaporator + oven 50 °C 48 h	Postharvest. Wound (3 × 3 mm φ and deep) + inoculation (*F. sambucinum*) (24 h) + dipping (1.25, 2.5, 5, 10, and 20 mg/mL water) + air dried (2 h at 28 °C).	Antifungal activity on the mycelial growth and spore germination	[111]
	Strawberry	Drying and particle size information NA Dente di caballo cv	US 40 °C 80% A 3 min (3 on, 8 off) Ratio 1:10 (H_2O 25%, propanol 25%, ethanol 25% and methanol 25%) + evaporator + Freeze-drier + re-dissolved in water	Postharvest. Immersion (30 s in a 2 L solution of pomegranate peel extract) + air-drying (1 h)	Extension of shelf life Substitution of synthetic pesticides	[112]
	Sweet cherry	Oven drier (40 °C) Particle size NA Mollar de Elche cv	EtOH solvent (50 and 80%) + 1% Citric acid +evaporator + Water dilution	Postharvest. Dipping (2 min) in the pomegranate extract (12, 2.4 or 1.2 g/L) + air drying (2 h, 28 °C) + storage at 1 °C	Inhibition of all fungal spore germination	[113]

Table 5. *Cont.*

	Matrix	Pomegranate Peel Byproduct	Extraction	Incorporation Method	Benefit	Ref.
Fresh whole F&V (pre- and postharvest)	Sweet cherry	Oven drier (40 °C) Fine powder < 470 µm cv information NA	1:8 solid–liquid ratio (Water 28 °C 24 h)	Postharvest. Immersion in pomegranate peel extracts (3 min 20 °C) + room temperature drying	Pomegranate peel extracts and calcium sulphate coatings, alone or in combination, decreased weight loss, decay, respiration rate, and increased acidity, firmness, ascorbic acid, DPPH, TPC, and TAC	[114]
	Apple	Oven drier (40 °C) Particle size NA Mollar de Elche cv	EtOH solvent (50 and 80%) + 1% citric acid + evaporator + water dilution	Postharvest. Wounds treated with 10 µL of pomegranate peel extract (12, 1.2 or 0.12 g/L) + inoculation (10 µL *P. expansum*)	Inhibition of fungal spore germination and decay of artificial inoculations	[113]
	Mango	Freeze drying (−45 °C, 94 h) Particle size and cv information NA	MetOH 45 °C 30 min + Bath US + evaporator + water dilution	Postharvest. Chitosan (2%) in 0.5% citric acid solution + Pullulan (2%) in water (50:50 ratios). During stiring: 1% glycerol + 5% of pomegranate peel extract (0.02 g/mL). Dipping for 2 min	Increase of firmness, TPC and AOX. Prolonged the shelf life	[115]
	Apricot	Drier (60 °C, 48 h) Particle size < 0.251 mm cv information NA	80% EtOH 25 °C + evaporator	Postharvest. Chitosan coating solution (1% chitosan in glacial acetic 1% + 0.8% glycerol + Tween 80 + 0.50, 0.75, and 1% pomegranate peel extract)	Reduction of % decay and weight loss. Maintenance of DPPH radical scavenging activity, ascorbic acid content, titratable acidity and firmness.	[116]
	Figs	Air dried few days (more information NA) Pulverized (more information NA) cv information NA	Alcoholic buffer (EtOH 50%)	Postharvest. Alginic acid: agar (70:30) + 0.25 and 0.5% pomegranate peel extract Dipping in the coating solution + coating gelation	Prolonged the shelf life	[117]
	Dates	Drier (48 °C, 52 h) Ground peels (more information NA)cv information NA	EtOH 70% + evaporator + Water dilution	Postharvest. 1% Chitosan, 1% nanochitosan or 1% pomegranate peel extract in 1% glacial acetic	Growth inhibition of any fungal spore after 48 h of coating.	[118]

Table 5. *Cont.*

	Matrix	Pomegranate Peel Byproduct	Extraction	Incorporation Method	Benefit	Ref.
Fresh whole F&V (pre- and postharvest)	Citrus	Hot air drier (50 °C, 48 h) Particle size 0.250 mm cv information NA	2.5:10 Solid–liquid ratio (Ac, EtOH, MetOH, H_2O, DMSO) + shaking (6 h) + re-extracted with water evaporation	Postharvest. Immersion of wounded lemons (2 × 1 mm long and wide tip) in pomegranate peel extract (pre-infection and post-infection with *P. digitatum*) + air drying	Prevention and control of *P. digitatum*	[119]
	Grapefruit	Oven drier (40 °C) Particle size information NA Mollar de Elche cv	EtOH solvent (50 and 80%) + 1% citric acid evaporator + water dilution	Postharvest. Wounds treated with 10 µL of pomegranate peel extract (12, 1.2 or 0.12 g/L) + inoculation 10 µL *P. digitatum* and *P. italicum*	Inhibition of all fungal spore germination and decay of artificial inoculations	[120]
	Lemon	Oven drier (40 °C) Particle size information NA Mollar de Elche cv	EtOH solvent (50 and 80%) + 1% citric acid + evaporator + water dilution	Postharvest. Wounds treated with 10 µL of pomegranate peel extract (12, 1.2 or 0.12 g/L) + inoculation 10 µL *P. digitatum* and *P. italicum*	Inhibition of all fungal spore germination and decay of artificial inoculations	[113,120]
	Mandarin	Drier (70 °C, 48 h) Ground peels (more information NA) Shirine Shahvar cv	0.25:10 solid–liquid ratio (60% EtOH + 0.1% citric acid)	Postharvest. Wounded (1 × 2 mm φ and depth) + dipping 1 min in pomegranate peel extract concentrations (25, 50, 75, 100%) + inoculation (*P. italicum* and *P. digitatum*) + drying	Reduction of % infected wound and lesion φ (75% or/and 100% extract). Increase of TPC, TFC, and PAL activity (75% or/and 100% extract)	[121]
	Orange	Drier (35 °C, 2 days) Particle size NA Gabsi cv	1:10, 0.6:10, 0.3:10 solid–liquid ratio (MetOH or Water) + evaporated + drying (40 °C or freeze-drying) + re-dissolved in water	Postharvest. Chitosan coating solution (1% chitosan in glacial acetic 1% + 0.5% Locust bean gum + 20% glycerol + 7, 18, and 36% dry waster/MetOH pomegranate peel extract). Wounded oranges (4 times: 3 × 3 mm φ × deep) + Inoculation (20 µL of a *P. digitatum*) + drying + dipping in different coating solutions (2 min)	Controlled growth of *Penicillium digitatum* Reduction of postharvest decay	[122]

Table 5. *Cont.*

	Matrix	Pomegranate Peel Byproduct	Extraction	Incorporation Method	Benefit	Ref.
Fresh whole F&V (pre- and postharvest)	Orange	Oven drier (40 °C) Particle size NA Mollar de Elche cv	EtOH solvent (50 and 80%) + 1% citric acid + evaporator + water dilution	Postharvest. Wounded oranges (3 times 2 × 2 mm φ and deep) + 20 μL pomegranate peel extract (12 g/L) + Inoculation (20 μL of a *P. digitatum)* + 1% citric acid + drying	Enhanced defense pathways (antibiotic biosynthesis)	[123]
	Guava	Drier (60 °C, 72 h) Particle size 0.420 mm Bhagwa cv	1:10 solid–liquid ratio (80% EtOH) + evaporation	Postharvest. Chitosan (1% chitosan in glacial acetic 1% + 0.75% glycerol) and alginate solution (2% alginate + 10% glycerol + 2% calcium chloride) with 1% pomegranate peel extract	Preserved quality for 20 d under refrigeration	[124]
	Capsicum	Drier (60 °C, 72 h) Particle size 0.420 mm Bhagwa cv	1:10 solid–liquid ratio (80% EtOH) + evaporation	Postharvest. Chitosan (1% chitosan in glacial acetic 1% + 0.75% glycerol) and alginate solution (2% alginate + 10% glycerol + 2% calcium chloride) with 1% pomegranate peel extract	Inhibition of microbial growth. Preserved sensory quality. Extension of shelf life up to 25 d at 10 °C	[125]
	Pear	Drier (60 °C, 72 h) Particle size 0.420 mm Bhagwa cv	1:10 solid–liquid ratio (80% EtOH) + evaporation	Postharvest. Chitosan (1% chitosan in glacial acetic 1% + 0.75% glycerol) and alginate solution (2% alginate + 10% glycerol + 2% calcium chloride) with 2% pomegranate peel extract	Lowered the cell wall degrading enzymes activity (firmness preservation)	[126]

Table 5. *Cont.*

	Matrix	Pomegranate Peel Byproduct	Extraction	Incorporation Method	Benefit	Ref.
Fresh-cut/Minimally processed F&V	Fruit salad: nectarine and pineapple in cubes covered with fructose syrup	Oven drier (38 °C, 48 h) Particle size 500 mm Wonderful cv	Powder	2.5–5% (*w*/*v*) of pomegranate peel powder at the container bottom	Inhibition of mesophilic bacteria, total psychrotrophic microorganisms, yeasts, and lactic acid bacteria No negative effect on sensory characteristics	[127]
	Fresh-cut pear, apple and melon (plugs)	Oven drier (40 °C) Particle size NA Mollar de Elche cv	EtOH solvent (50 and 80%) + 1% citric acid + evaporator + water dilution	Inoculated plugs were dipped (10 min, 150 rpm) + dried (25 °C 30 min)	Reduction of *Listeria monocytogenes*	[128]
	Fresh-cut Golden apple wedges: thickness 30-mm and 30 g	Drying and particle size information NA Dente di cavallo cv	Pulsed UAE (10 min, <50 °C, 1:40, 26 kHz, 200 W, 40% A, 50% duty cycle) + encapsulation with pectin from citrus peel by spray drying	Enrichment with microencapsulates reconstituted in water 1:1	Reduction of enzymatic browning. Color preservation	[129]
Beverages	Carrot juice	Oven drier (40 °C) Grounded in a colloid mill (more information NA) cv information NA	High pressure-assisted extraction	5 mg pomegranate peel extract per mL of carrot juice	Improvement of microbiological safety and AOX during storage. Color preservation	[130,131]
	Apple juice	Oven drier (55 °C, 12 h) Particle size and cv information NA	Maceration extraction (1:50, 80% EtOH 1 h shaking)	Different% of pomegranate peel extract (0–2%)	Enhancing sensory quality and AOX. Low toxicity with 1% of pomegranate peel extract	[132]
	Kiwi juice	Information NA	Commercial pomegranate extract (PureBulk, Roseburg)	Extract incorporation (180 µg/mL kiwi juice) + US bath (40 kHz, 180 W, 20 °C, 10–30 min)	US and pomegranate extract combined treatment: higher reductions on yeast and molds	[133]
	Red wine	Green decoction: Boiled in water 60 min (1:40) Freeze-drying of the extract Wonderful cv	Powder	Purification to obtain the tannins. 8 analyzed tannins (1 g L^{-1} wine solution)	Increase of protein stability Increase of color stability Reduction of sulfites	[134]

Table 5. *Cont.*

Matrix	Pomegranate Peel Byproduct	Extraction	Incorporation Method	Benefit	Ref.
Symbiotic drink powder	Hot oven (40 °C, 48 h) Particle size Kitchen-miller (more information NA) cv information NA	Ethanolic extract (80%; 1:15) + evaporator + Freeze-drier	Formulation: beetroot peel extract powder (3%), pomegranate peel extract powder (1%), grape pomace extract powder (1.5%), quince seed gum (0.5%), stevia (4%), mint (0.1%) and water (89.9%). Pasteurization: 72 °C, 90 s	Maintenance of *L. casei* viability of the recommended level of 10^{-7} CFU/g	[135]

NA: Data no available; A: amplitude; cv: cultivar; TPC: total polyphenolic content; TFC: total flavonoid content; AOX: total antioxidant capacity; TAC: total anthocyanin content; PAL: phenylalanine ammonia-lyase.

7. Conclusions and Future Perspectives

The research community and the food industry are quite interested in the use of (techno)-functional bioactive compounds from pomegranate byproducts in different food matrixes to reduce the use of synthetic additives and to develop 'clean label' products. However, the optimal extraction technique greatly depends on the raw material and conditions (cultivar, moisture, drying technology, particle size, etc.), so specific parameters should be recommended after a proper evaluation, on which more studies are needed. There is a lack of important information about the main characteristics of pomegranate peel, making it more difficult to have a more definitive conclusion on the optimal conditions of their bioactive compound extraction. Considering the three green extraction technologies included in this review, more than 80% of the evidence is focused on ultrasound-assisted technology. Therefore, more research on enzymatic and microwave-assisted methods, and their combinations, should be carried out. The combination of enzyme-assisted treatment with other green technologies usually increases the yield, shortening the extraction time. However, further research is still needed to optimize such combined treatments. In addition, a review of other green technologies for the extraction of bioactive compounds from pomegranate byproducts should be of interest to the research community, as well as other pomegranate byproducts such as seeds or arils. In future studies, the energy efficiency/consumption, the cost, and the environmental impact leading to a sustainable extraction of the key biocompounds must be evaluated. Additionally, predictive models are needed to optimize the phytochemical extraction and help in decision-making.

Author Contributions: Conceptualization, M.C.-L. and F.A.-H.; methodology, formal analysis, investigation, M.C.-L.; resources, M.C.-L. and N.C.; data curation, M.C.-L., N.C. and L.M.-Z.; writing—original draft preparation, M.C.-L.; writing—review and editing, M.C.-L., L.M.-Z., N.C. and F.A.-H.; visualization, M.C.-L., L.M.-Z. and F.A.-H.; supervision, F.A.-H.; project administration, F.A.-H.; funding acquisition, F.A.-H. All authors have read and agreed to the published version of the manuscript.

Funding: This research was funded by the Spanish Ministry of Science and Innovation, Knowledge Generation Projects 2021, Type B Oriented Research Modality, grant number PID2021-123857OB-I00, REVALFOOD PROJECT. L.M.-Z. contract has been financed by Programme for the Re-qualification of the Spanish University System, Margarita Salas modality, by the University of Murcia. M.C.-L. contract has been co-financed by Juan de la Cierva-Formación (FJC2020-043764-I) from the Spanish Ministry of Education.

Data Availability Statement: No new data were created or analyzed in this study. Data sharing is not applicable to this article.

Conflicts of Interest: The authors declare no conflict of interest.

References

1. FAO. Technical Platform on the Measurement and Reduction of Food Loss and Waste. 2022. Available online: https://www.fao.org/platform-food-loss-waste/en/ (accessed on 20 July 2022).
2. FAO. Food Loss and Waste Database. 2022. Available online: https://www.fao.org/platform-food-loss-waste/flw-data/en/ (accessed on 20 July 2022).
3. López–Gómez, A.; Ros–Chumillas, M.; Buendía-Moreno, L.; Martínez–Hernández, G.B. Active Cardboard Packaging with Encapsulated Essential Oils for Enhancing the Shelf Life of Fruit and Vegetables. *Front. Nutr.* **2020**, *7*, 559978. [CrossRef] [PubMed]
4. Magangana, T.P.; Makunga, N.P.; Fawole, O.A.; Opara, U.L. Processing Factors Affecting the Phytochemical and Nutritional Properties of Pomegranate (*Punica granatum* L.) Peel Waste: A Review. *Molecules* **2020**, *25*, 4690. [CrossRef] [PubMed]
5. Ko, K.; Dadmohammadi, Y.; Abbaspourrad, A. Nutritional and Bioactive Components of Pomegranate Waste Used in Food and Cosmetic Applications: A Review. *Foods* **2021**, *10*, 657. [CrossRef] [PubMed]
6. Cano-Lamadrid, M.; Artés-Hernández, F. By-Products Revalorization with Non-Thermal Treatments to Enhance Phytochemical Compounds of Fruit and Vegetables Derived Products: A Review. *Foods* **2022**, *11*, 59. [CrossRef] [PubMed]
7. Cano-Lamadrid, M.; Valverde, J.M.; Lipan, L.; Carbonell-Barrachina, Á.A.; Sendra, E. Introduction. The Age of Clean Label Foods. In *The Age of Clean Label Foods*; Galanakis, C.M., Ed.; Springer International Publishing: Cham, Switzerland, 2022; pp. 1–35. ISBN 978-3-030-96698-0.

8. Artés-Hernández, F.; Castillejo, N.; Martínez-Zamora, L.; Martínez-Hernández, G.B. Phytochemical Fortification in Fruit and Vegetable Beverages with Green Technologies. *Foods* **2021**, *10*, 2534. [CrossRef] [PubMed]
9. Martínez-Sánchez, A.; Alacid, F.; Rubio-Arias, J.A.; Fernández-Lobato, B.; Ramos-Campo, D.J.; Aguayo, E. Consumption of Watermelon Juice Enriched in L-Citrulline and Pomegranate Ellagitannins Enhanced Metabolism during Physical Exercise. *J. Agric. Food Chem.* **2017**, *65*, 4395–4404. [CrossRef] [PubMed]
10. Carrillo, C.; Nieto, G.; Martínez-Zamora, L.; Ros, G.; Kamiloglu, S.; Munekata, P.E.S.; Pateiro, M.; Lorenzo, J.M.; Fernández-López, J.; Viuda-Martos, M.; et al. Novel Approaches for the Recovery of Natural Pigments with Potential Health Effects. *J. Agric. Food Chem.* **2022**, *70*, 6864–6883. [CrossRef] [PubMed]
11. Turker, S.; Polat, A.A.; Bindak, R. Seasonal Changes of Carbohydrates in Fruit Peels, Leaves and Shoots of Three Pomegranate (*Punica granatum* L.) Cultivars Grown in Upper Euphrates Basin. *Sci. Hortic.* **2022**, *304*, 111315. [CrossRef]
12. Abid, M.; Renard, C.M.G.C.; Watrelot, A.A.; Fendri, I.; Attia, H.; Ayadi, M.A. Yield and Composition of Pectin Extracted from Tunisian Pomegranate Peel. *Int. J. Biol. Macromol.* **2016**, *93*, 186–194. [CrossRef]
13. Talekar, S.; Patti, A.F.; Vijayraghavan, R.; Arora, A. An Integrated Green Biorefinery Approach towards Simultaneous Recovery of Pectin and Polyphenols Coupled with Bioethanol Production from Waste Pomegranate Peels. *Bioresour. Technol.* **2018**, *266*, 322–334. [CrossRef]
14. Viuda-Martos, M.; Ruiz-Navajas, Y.; Martin-Sánchez, A.; Sánchez-Zapata, E.; Fernández-López, J.; Sendra, E.; Sayas-Barberá, E.; Navarro, C.; Pérez-Álvarez, J.A. Chemical, Physico-Chemical and Functional Properties of Pomegranate (*Punica granatum* L.) Bagasses Powder Co-Product. *J. Food Eng.* **2012**, *110*, 220–224. [CrossRef]
15. Gullón, P.; Astray, G.; Gullón, B.; Tomasevic, I.; Lorenzo, J.M. Pomegranate Peel as Suitable Source of High-Added Value Bioactives: Tailored Functionalized Meat Products. *Molecules* **2020**, *25*, 2859. [CrossRef]
16. Man, G.; Xu, L.; Wang, Y.; Liao, X.; Xu, Z. Profiling Phenolic Composition in Pomegranate Peel From Nine Selected Cultivars Using UHPLC-QTOF-MS and UPLC-QQQ-MS. *Front. Nutr.* **2022**, *8*. [CrossRef]
17. Cano-Lamadrid, M.; Lech, K.; Calín-Sánchez, Á.; Rosas-Burgos, E.C.; Figiel, A.; Wojdyło, A.; Wasilewska, M.; Carbonell-Barrachina, Á.A. Quality of Pomegranate Pomace as Affected by Drying Method. *J. Food Sci. Technol.* **2018**, *55*, 1074–1082. [CrossRef]
18. Alcaraz-Mármol, F.; Nuncio-Jáuregui, N.; Calín-Sánchez, Á.; Carbonell-Barrachina, Á.A.; Martínez, J.J.; Hernández, F. Determination of Fatty Acid Composition in Arils of 20 Pomegranates Cultivars Grown in Spain. *Sci. Hortic.* **2015**, *197*, 712–718. [CrossRef]
19. Panja, P. Green Extraction Methods of Food Polyphenols from Vegetable Materials. *Curr. Opin. Food Sci.* **2018**, *23*, 173–182. [CrossRef]
20. Gallo, M.; Ferrara, L.; Naviglio, D. Application of Ultrasound in Food Science and Technology: A Perspective. *Foods* **2018**, *7*, 164. [CrossRef]
21. Chemat, F.; Zill-E-Huma; Khan, M.K. Applications of Ultrasound in Food Technology: Processing, Preservation and Extraction. *Ultrason. Sonochemistry* **2011**, *18*, 813–835. [CrossRef]
22. Khadhraoui, B.; Fabiano-Tixier, A.S.; Robinet, P.; Imbert, R.; Chemat, F. Ultrasound Technology for Food Processing, Preservation, and Extraction. *Green Food Process. Tech. Preserv. Transform. Extr.* **2019**, *24*, 23–56. [CrossRef]
23. Montefusco, A.; Durante, M.; Migoni, D.; de Caroli, M.; Ilahy, R.; Pék, Z.; Helyes, L.; Fanizzi, F.P.; Mita, G.; Piro, G.; et al. Analysis of the Phytochemical Composition of Pomegranate Fruit Juices, Peels and Kernels: A Comparative Study on Four Cultivars Grown in Southern Italy. *Plants* **2021**, *10*, 2521. [CrossRef]
24. Yusoff, I.M.; Mat Taher, Z.; Rahmat, Z.; Chua, L.S. A Review of Ultrasound-Assisted Extraction for Plant Bioactive Compounds: Phenolics, Flavonoids, Thymols, Saponins and Proteins. *Food Res. Int.* **2022**, *157*, 111268. [CrossRef]
25. Liu, Y.; Kong, K.W.; Wu, D.T.; Liu, H.Y.; Li, H.B.; Zhang, J.R.; Gan, R.Y. Pomegranate Peel-Derived Punicalagin: Ultrasonic-Assisted Extraction, Purification, and Its α-Glucosidase Inhibitory Mechanism. *Food Chem.* **2022**, *374*, 131635. [CrossRef]
26. More, P.R.; Arya, S.S. Intensification of Bio-Actives Extraction from Pomegranate Peel Using Pulsed Ultrasound: Effect of Factors, Correlation, Optimization and Antioxidant Bioactivities. *Ultrason. Sonochem.* **2021**, *72*, 105423. [CrossRef]
27. Pan, Z.; Qu, W.; Ma, H.; Atungulu, G.G.; McHugh, T.H. Continuous and Pulsed Ultrasound-Assisted Extractions of Antioxidants from Pomegranate Peel. *Ultrason. Sonochem.* **2011**, *18*, 1249–1257. [CrossRef] [PubMed]
28. Kaderides, K.; Goula, A.M.; Adamopoulos, K.G. A Process for Turning Pomegranate Peels into a Valuable Food Ingredient Using Ultrasound-Assisted Extraction and Encapsulation. *Innov. Food Sci. Emerg. Technol.* **2015**, *31*, 204–215. [CrossRef]
29. Dimitrov, K.; Pradal, D.; Vauchel, P.; Baouche, B.; Nikov, I.; Dhulster, P. Modeling and Optimization of Extraction and Energy Consumption during Ultrasound-Assisted Extraction of Antioxidant Polyphenols from Pomegranate Peels. *Environ. Prog. Sustain. Energy* **2019**, *38*, 13148. [CrossRef]
30. Muñiz-Márquez, D.B.; Wong-Paz, J.E.; Aguilar-Zárate, P.; Sepúlveda, L.; Buenrostro-Figueroa, J.; Ascacio-Valdés, J.A.; Aguilar, C.N. Effect of Ultrasound on the Extraction of Ellagic Acid and Hydrolysis of Ellagitannins from Pomegranate Husk. *Environ. Technol. Innov.* **2021**, *24*, 102063. [CrossRef]
31. Goula, A.M.; Ververi, M.; Adamopoulou, A.; Kaderides, K. Green Ultrasound-Assisted Extraction of Carotenoids from Pomegranate Wastes Using Vegetable Oils. *Ultrason. Sonochem.* **2017**, *34*, 821–830. [CrossRef]
32. Rakshit, M.; Srivastav, P.P.; Bhunia, K. Kinetic Modeling of Ultrasonic-Assisted Extraction of Punicalagin from Pomegranate Peel. *J. Food Process Eng.* **2020**, *43*, e13533. [CrossRef]

33. Rifna, E.J.; Dwivedi, M. Effect of Pulsed Ultrasound Assisted Extraction and Aqueous Acetone Mixture on Total Hydrolysable Tannins from Pomegranate Peel. *Food Biosci.* **2022**, *45*, 101496. [CrossRef]
34. Selvakumar, P.; Karthik, V.; Kumar, P.S.; Asaithambi, P.; Kavitha, S.; Sivashanmugam, P. Enhancement of Ultrasound Assisted Aqueous Extraction of Polyphenols from Waste Fruit Peel Using Dimethyl Sulfoxide as Surfactant: Assessment of Kinetic Models. *Chemosphere* **2021**, *263*, 128071. [CrossRef]
35. Kumar, A.; Srinivasa Rao, P. Optimization of Pulsed-Mode Ultrasound Assisted Extraction of Bioactive Compounds from Pomegranate Peel Using Response Surface Methodology. *J. Food Meas. Charact.* **2020**, *14*, 3493–3507. [CrossRef]
36. Rajha, H.N.; Koubaa, M.; Boussetta, N.; Maroun, R.G.; Louka, N.; Lebovka, N.; Vorobiev, E. Selective Ultrasound-Assisted Aqueous Extraction of Polyphenols from Pomegranate Peels and Seeds. *J. Food Process. Preserv.* **2020**, *44*, e14545. [CrossRef]
37. Siriangkhawut, W.; Didpinrum, P.; Khanhuathon, Y.; Ponhong, K.; Grudpan, K. Small-Scale Ultrasound-Assisted Extraction of Phenolics from Pomegranate Peels and Their Application as a Natural Reagent for the Colorimetric Assay of Iron. *Anal. Lett.* **2020**, *53*, 887–904. [CrossRef]
38. Sharayei, P.; Azarpazhooh, E.; Zomorodi, S.; Ramaswamy, H.S. Ultrasound Assisted Extraction of Bioactive Compounds from Pomegranate (*Punica granatum* L.) Peel. *LWT* **2019**, *101*, 342–350. [CrossRef]
39. Machado, A.P.D.F.; Sumere, B.R.; Mekaru, C.; Martinez, J.; Bezerra, R.M.N.; Rostagno, M.A. Extraction of Polyphenols and Antioxidants from Pomegranate Peel Using Ultrasound: Influence of Temperature, Frequency and Operation Mode. *Int. J. Food Sci. Technol.* **2019**, *54*, 2792–2801. [CrossRef]
40. Kazemi, M.; Karim, R.; Mirhosseini, H.; Abdul Hamid, A. Optimization of Pulsed Ultrasound-Assisted Technique for Extraction of Phenolics from Pomegranate Peel of Malas Variety: Punicalagin and Hydroxybenzoic Acids. *Food Chem.* **2016**, *206*, 156–166. [CrossRef]
41. Turrini, F.; Zunin, P.; Catena, S.; Villa, C.; Alfei, S.; Boggia, R. Traditional or Hydro-Diffusion and Gravity Microwave Coupled with Ultrasound as Green Technologies for the Valorization of Pomegranate External Peels. *Food Bioprod. Process.* **2019**, *117*, 30–37. [CrossRef]
42. Ranjha, M.M.A.N.; Amjad, S.; Ashraf, S.; Khawar, L.; Safdar, M.N.; Jabbar, S.; Nadeem, M.; Mahmood, S.; Murtaza, M.A. Extraction of Polyphenols from Apple and Pomegranate Peels Employing Different Extraction Techniques for the Development of Functional Date Bars. *Int. J. Fruit Sci.* **2020**, *20*, S1201–S1221. [CrossRef]
43. Hasnaoui, N. Pomegranate Peel Phenolics: One Step Water Extraction, in Vitro Antiproliferative Activity, and Effect on Sensory Attributes of Muffins and Cheese. *Food Chem.* **2022**, *392*, 133297. [CrossRef]
44. Ordaz-Rodríguez, S.B.; Abadía-García, L.; Huerta-Manzanilla, E.L.; Mendoza-Sánchez, M. Ultrasound-Assisted Extraction of Pomegranate Peel Antioxidants: A Green Process to Obtain a Meat Preservative. *J. Food Process. Preserv.* **2022**, *46*, e16712. [CrossRef]
45. Magangana, T.P.; Makunga, N.P.; la Grange, C.; Stander, M.A.; Fawole, O.A.; Opara, U.L. Blanching Pre-Treatment Promotes High Yields, Bioactive Compounds, Antioxidants, Enzyme Inactivation and Antibacterial Activity of 'Wonderful' Pomegranate Peel Extracts at Three Different Harvest Maturities. *Antioxidants* **2021**, *10*, 1119. [CrossRef] [PubMed]
46. Foujdar, R.; Bera, M.B.; Chopra, H.K. Optimization of Process Variables of Probe Ultrasonic-Assisted Extraction of Phenolic Compounds from the Peel of *Punica granatum* Var. Bhagwa and It's Chemical and Bioactivity Characterization. *J. Food Process. Preserv.* **2020**, *44*, e14317. [CrossRef]
47. Tabaraki, R.; Heidarizadi, E.; Benvidi, A. Optimization of Ultrasonic-Assisted Extraction of Pomegranate (*Punica granatum* L.) Peel Antioxidants by Response Surface Methodology. *Sep. Purif. Technol.* **2012**, *98*, 16–23. [CrossRef]
48. Nag, S.; Sit, N. Optimization of Ultrasound Assisted Enzymatic Extraction of Polyphenols from Pomegranate Peels Based on Phytochemical Content and Antioxidant Property. *J. Food Meas. Charact.* **2018**, *12*, 1734–1743. [CrossRef]
49. Živković, J.; Šavikin, K.; Janković, T.; Ćujić, N.; Menković, N. Optimization of Ultrasound-Assisted Extraction of Polyphenolic Compounds from Pomegranate Peel Using Response Surface Methodology. *Sep. Purif. Technol.* **2018**, *194*, 40–47. [CrossRef]
50. Sumere, B.R.; de Souza, M.C.; dos Santos, M.P.; Bezerra, R.M.N.; da Cunha, D.T.; Martinez, J.; Rostagno, M.A. Combining Pressurized Liquids with Ultrasound to Improve the Extraction of Phenolic Compounds from Pomegranate Peel (*Punica granatum* L.). *Ultrason. Sonochem.* **2018**, *48*, 151–162. [CrossRef]
51. Santos, M.P.; Souza, M.C.; Sumere, B.R.; da Silva, L.C.; Cunha, D.T.; Bezerra, R.M.N.; Rostagno, M.A. Extraction of Bioactive Compounds from Pomegranate Peel (*Punica granatum* L.) with Pressurized Liquids Assisted by Ultrasound Combined with an Expansion Gas. *Ultrason. Sonochem.* **2019**, *54*, 11–17. [CrossRef]
52. Motikar, P.D.; More, P.R.; Arya, S.S. A Novel, Green Environment-Friendly Cloud Point Extraction of Polyphenols from Pomegranate Peels: A Comparative Assessment with Ultrasound and Microwave-Assisted Extraction. *Sep. Sci. Technol.* **2021**, *56*, 1014–1025. [CrossRef]
53. Gomez, L.; Tiwari, B.; Garcia-Vaquero, M. Emerging Extraction Techniques: Microwave-Assisted Extraction. *Sustain. Seaweed Technol.* **2020**, 207–224. [CrossRef]
54. Llompart, M.; Garcia-Jares, C.; Celeiro, M.; Dagnac, T. Extraction | Microwave-Assisted Extraction. *Encycl. Anal. Sci.* **2019**, 67–77. [CrossRef]
55. de Castro, M.D.L.; Castillo-Peinado, L.S. Microwave-Assisted Extraction of Food Components. *Innov. Food Process. Technol. Extr. Sep. Compon. Modif. Process Intensif.* **2016**, 57–110. [CrossRef]
56. Sparr Eskilsson, C.; Björklund, E. Analytical-Scale Microwave-Assisted Extraction. *J. Chromatogr. A* **2000**, *902*, 227–250. [CrossRef]

57. Skenderidis, P.; Leontopoulos, S.; Petrotos, K.; Giavasis, I. Optimization of Vacuum Microwave-Assisted Extraction of Pomegranate Fruits Peels by the Evaluation of Extracts' Phenolic Content and Antioxidant Activity. *Foods* **2020**, *9*, 1655. [CrossRef]
58. Kaderides, K.; Papaoikonomou, L.; Serafim, M.; Goula, A.M. Microwave-Assisted Extraction of Phenolics from Pomegranate Peels: Optimization, Kinetics, and Comparison with Ultrasounds Extraction. *Chem. Eng. Process.-Process Intensif.* **2019**, *137*, 1–11. [CrossRef]
59. Vladić, J.; Janković, T.; Živković, J.; Tomić, M.; Zdunić, G.; Šavikin, K.; Vidović, S. Comparative Study of Subcritical Water and Microwave-Assisted Extraction Techniques Impact on the Phenolic Compounds and 5-Hydroxymethylfurfural Content in Pomegranate Peel. *Plant Foods Hum. Nutr.* **2020**, *75*, 553–560. [CrossRef]
60. Xia, J.L.; Li, P.J. Pectic Enzymes. *Encycl. Food Chem.* **2019**, 270–276. [CrossRef]
61. Díaz, A.B.; Bolívar, J.; de Ory, I.; Caro, I.; Blandino, A. Applicability of Enzymatic Extracts Obtained by Solid State Fermentation on Grape Pomace and Orange Peels Mixtures in Must Clarification. *LWT-Food Sci. Technol.* **2011**, *44*, 840–846. [CrossRef]
62. Philipps-Wiemann, P. Proteases—General Aspects. *Enzym. Hum. Anim. Nutr. Princ. Perspect.* **2018**, 257–266. [CrossRef]
63. Wilson, D.B. Cellulases. *Encycl. Microbiol.* **2009**, 252–258. [CrossRef]
64. van Oort, M. Enzymes in Food Technology–Introduction. In *Enzymes in Food Technology*; John Wiley & Sons, Ltd.: Hoboken, NJ, USA, 2009; pp. 1–17. ISBN 9781444309935.
65. Kumar, M.; Tomar, M.; Punia, S.; Amarowicz, R.; Kaur, C. Evaluation of Cellulolytic Enzyme-Assisted Microwave Extraction of *Punica granatum* Peel Phenolics and Antioxidant Activity. *Plant Foods Hum. Nutr.* **2020**, *75*, 614–620. [CrossRef] [PubMed]
66. Alexandre, E.M.C.; Silva, S.; Santos, S.A.O.; Silvestre, A.J.D.; Duarte, M.F.; Saraiva, J.A.; Pintado, M. Antimicrobial Activity of Pomegranate Peel Extracts Performed by High Pressure and Enzymatic Assisted Extraction. *Food Res. Int.* **2019**, *115*, 167–176. [CrossRef] [PubMed]
67. Mushtaq, M.; Sultana, B.; Anwar, F.; Adnan, A.; Rizvi, S.S.H. Enzyme-Assisted Supercritical Fluid Extraction of Phenolic Antioxidants from Pomegranate Peel. *J. Supercrit. Fluids* **2015**, *104*, 122–131. [CrossRef]
68. Costa, J.R.; Tonon, R.V.; Cabral, L.; Gottschalk, L.; Pastrana, L.; Pintado, M.E. Valorization of Agricultural Lignocellulosic Plant Byproducts through Enzymatic and Enzyme-Assisted Extraction of High-Value-Added Compounds: A Review. *ACS Sustain. Chem. Eng.* **2020**, *8*, 13112–13125. [CrossRef]
69. Calín-Sánchez, Á.; Lipan, L.; Cano-Lamadrid, M.; Kharaghani, A.; Masztalerz, K.; Carbonell-Barrachina, Á.A.; Figiel, A. Comparison of Traditional and Novel Drying Techniques and Its Effect on Quality of Fruits, Vegetables and Aromatic Herbs. *Foods* **2020**, *9*, 1261. [CrossRef]
70. Viuda-Martos, M.; Pérez-Álvarez, J.A.; Sendra, E.; Fernández-López, J. In vitro antioxidant properties of pomegranate (*Punica granatum*) peel powder extract obtained as coproduct in the juice extraction process. *J. Food Process. Preserv.* **2013**, *37*, 772–776. [CrossRef]
71. Bhandari, B. 1-Introduction to Food Powders. In *Handbook of Food Powders*; Bhandari, B., Bansal, N., Zhang, M., Schuck, P., Eds.; Woodhead Publishing Series in Food Science, Technology and Nutrition; Woodhead Publishing: Shaston, UK, 2013; pp. 1–25. ISBN 978-0-85709-513-8.
72. Yağmur, N.; Şahin, S. Encapsulation of Ellagic Acid from Pomegranate Peels in Microalgae Optimized by Response Surface Methodology and an Investigation of Its Controlled Released under Simulated Gastrointestinal Studies. *J. Food Sci.* **2020**, *85*, 998–1006. [CrossRef]
73. Marcillo-Parra, V.; Tupuna-Yerovi, D.S.; González, Z.; Ruales, J. Encapsulation of Bioactive Compounds from Fruit and Vegetable By-Products for Food Application–A Review. *Trends Food Sci. Technol.* **2021**, *116*, 11–23. [CrossRef]
74. Michalska, A.; Wojdyło, A.; Honke, J.; Ciska, E.; Andlauer, W. Drying-Induced Physico-Chemical Changes in Cranberry Products. *Food Chem.* **2018**, *240*, 448–455. [CrossRef]
75. Michalska, A.; Lech, K. The Effect of Carrier Quantity and Drying Method on the Physical Properties of Apple Juice Powders. *Beverages* **2018**, *4*, 2. [CrossRef]
76. Watson, M.A.; Lea, J.M.; Bett-Garber, K.L. Spray Drying of Pomegranate Juice Using Maltodextrin/Cyclodextrin Blends as the Wall Material. *Food Sci. Nutr.* **2017**, *5*, 820–826. [CrossRef]
77. Carpena, M.; Garcia-Oliveira, P.; Lourenco-Lopez, C.; Gonzalez Pereira, A.; Fraga-Corral, M.; Angel Prieto, M.; Simal-Gandara, J. Freeze-Drying Encapsulation as a Mechanism of Choice in Oils: Methods and Mechanism. In *Basic Protocols in Encapsulation of Food Ingredients*; Gomez-Zavaglia, A., Ed.; Springer: New York, NY, USA, 2021; pp. 91–101; ISBN 978-1-0716-1649-9.
78. Çam, M.; İçyer, N.C.; Erdoğan, F. Pomegranate Peel Phenolics: Microencapsulation, Storage Stability and Potential Ingredient for Functional Food Development. *LWT-Food Sci. Technol.* **2014**, *55*, 117–123. [CrossRef]
79. Šavikin, K.; Nastić, N.; Janković, T.; Bigović, D.; Miličević, B.; Vidović, S.; Menković, N.; Vladić, J. Effect of Type and Concentration of Carrier Material on the Encapsulation of Pomegranate Peel Using Spray Drying Method. *Foods* **2021**, *10*, 1968. [CrossRef]
80. Goula, A.M.; Lazarides, H.N. Integrated Processes Can Turn Industrial Food Waste into Valuable Food By-Products and/or Ingredients: The Cases of Olive Mill and Pomegranate Wastes. *J. Food Eng.* **2015**, *167*, 45–50. [CrossRef]
81. Kaderides, K.; Goula, A.M. Encapsulation of Pomegranate Peel Extract with a New Carrier Material from Orange Juice By-Products. *J. Food Eng.* **2019**, *253*, 1–13. [CrossRef]
82. Kaderides, K.; Mourtzinos, I.; Goula, A.M. Stability of Pomegranate Peel Polyphenols Encapsulated in Orange Juice Industry By-Product and Their Incorporation in Cookies. *Food Chem.* **2020**, *310*, 125849. [CrossRef]

83. Yang, B.; Kealey, K.; Chen, J.; Solval, K.M. Developing Microencapsulated Powders Containing Polyphenols and Pectin Extracted from Georgia-Grown Pomegranate Peels. *LWT* **2022**, *154*, 112644. [CrossRef]
84. Fathi, F.; Ebrahimi, S.N.; Pereira, D.M.; Estevinho, B.N.; Rocha, F. Preliminary Studies of Microencapsulation and Anticancer Activity of Polyphenols Extract from *Punica granatum* Peels. *Can. J. Chem. Eng.* **2021**. [CrossRef]
85. Endo, E.H.; Ueda-Nakamura, T.; Nakamura, C.V.; Filho, B.P.D. Activity of Spray-Dried Microparticles Containing Pomegranate Peel Extract against Candida Albicans. *Molecules* **2012**, *17*, 10094–10107. [CrossRef]
86. Bustamante, A.; Hinojosa, A.; Robert, P.; Escalona, V. Extraction and Microencapsulation of Bioactive Compounds from Pomegranate (*Punica granatum* Var. Wonderful) Residues. *Int. J. Food Sci. Technol.* **2017**, *52*, 1452–1462. [CrossRef]
87. Marín, D.; Alemán, A.; Montero, P.; Gómez-Guillén, M.C. Encapsulation of Food Waste Compounds in Soy Phosphatidylcholine Liposomes: Effect of Freeze-Drying, Storage Stability and Functional Aptitude. *J. Food Eng.* **2018**, *223*, 132–143. [CrossRef]
88. Jafarian, M.; Taghinia, P.; Baeghbali, S.; Samadi Ghorbani Dozdozani, N. Application of Prunus Armeniaca Gum Exudates and Chitosan for Encapsulation of Pomegranate Peel Extract as a Natural Antioxidant. *J. Food Meas. Charact.* **2021**, *15*, 4992–4999. [CrossRef]
89. Hamid; Thakur, N.S.; Thakur, A. Microencapsulation of Wild Pomegranate Flavedo Phenolics by Lyophilization: Effect of Maltodextrin Concentration, Structural Morphology, Functional Properties, Elemental Composition and Ingredient for Development of Functional Beverage. *LWT* **2020**, *133*, 110077. [CrossRef]
90. Azarpazhooh, E.; Sharayei, P.; Zomorodi, S.; Ramaswamy, H.S. Physicochemical and Phytochemical Characterization and Storage Stability of Freeze-Dried Encapsulated Pomegranate Peel Anthocyanin and In Vitro Evaluation of Its Antioxidant Activity. *Food Bioprocess Technol.* **2019**, *12*, 199–210. [CrossRef]
91. Okumuş, E.; Bakkalbaşı, E.; Javidipour, I.; Meral, R.; Ceylan, Z. A Novel Coating Material: Ellagitannins-Loaded Maltodextrin and Lecithin-Based Nanomaterials. *Food Biosci.* **2021**, *42*, 101158. [CrossRef]
92. Sanhueza, L.; García, P.; Giménez, B.; Benito, J.M.; Matos, M.; Gutiérrez, G. Encapsulation of Pomegranate Peel Extract (*Punica granatum* L.) by Double Emulsions: Effect of the Encapsulation Method and Oil Phase. *Foods* **2022**, *11*, 310. [CrossRef]
93. Bertolo, M.R.V.; Martins, V.C.A.; Horn, M.M.; Brenelli, L.B.; Plepis, A.M.G. Rheological and Antioxidant Properties of Chitosan/Gelatin-Based Materials Functionalized by Pomegranate Peel Extract. *Carbohydr. Polym.* **2020**, *228*, 115386. [CrossRef]
94. Yağmur, N.; Şahin, S.; Korkmaz, E. Microencapsulation of Ellagic Acid Extracted from Pomegranate Peel onto Spirulina: Characterization, Loading, and Storage Stability Properties. *J. Food Process. Preserv.* **2021**, *45*, e15086. [CrossRef]
95. Maroufi, L.Y.; Tabibiazar, M.; Ghorbani, M.; Jahanban-Esfahlan, A. Fabrication and Characterization of Novel Antibacterial Chitosan/Dialdehyde Guar Gum Hydrogels Containing Pomegranate Peel Extract for Active Food Packaging Application. *Int. J. Biol. Macromol.* **2021**, *187*, 179–188. [CrossRef]
96. Costa, A.M.M.; Moretti, L.K.; Simões, G.; Silva, K.A.; Calado, V.; Tonon, R.V.; Torres, A.G. Microencapsulation of Pomegranate (*Punica granatum* L.) Seed Oil by Complex Coacervation: Development of a Potential Functional Ingredient for Food Application. *LWT* **2020**, *131*, 109519. [CrossRef]
97. Eghbal, N.; Choudhary, R. Complex Coacervation: Encapsulation and Controlled Release of Active Agents in Food Systems. *LWT* **2018**, *90*, 254–264. [CrossRef]
98. Kaderides, K.; Kyriakoudi, A.; Mourtzinos, I.; Goula, A.M. Potential of Pomegranate Peel Extract as a Natural Additive in Foods. *Trends Food Sci. Technol.* **2021**, *115*, 380–390. [CrossRef]
99. Martínez-Zamora, L.; Ros, G.; Nieto, G. Designing a Clean Label Fish Patty with Olive, Citric, Pomegranate, or Rosemary Extracts. *Plants* **2020**, *9*, 659. [CrossRef]
100. Martínez, L.; Castillo, J.; Ros, G.; Nieto, G. Antioxidant and Antimicrobial Activity of Rosemary, Pomegranate and Olive Extracts in Fish Patties. *Antioxidants* **2019**, *8*, 86. [CrossRef]
101. Mekawi, E.M.; Sharoba, A.M.; Ramadan, M.F. Reduction of Acrylamide Formation in Potato Chips during Deep-Frying in Sunflower Oil Using Pomegranate Peel Nanoparticles Extract. *J. Food Meas. Charact.* **2019**, *13*, 3298–3306. [CrossRef]
102. Chan, C.-L.; Gan, R.; Shah, N.P.; Corke, H. Enhancing Antioxidant Capacity of Lactobacillus Acidophilus-Fermented Milk Fortified with Pomegranate Peel Extracts. *Food Biosci.* **2018**, *26*, 185–192. [CrossRef]
103. Palabiyik, I.; Toker, O.S.; Konar, N.; Gunes, R.; Güleri, T.; Alaşalvar, H.; Çam, M. Phenolics Release Kinetics in Sugared and Sugar-Free Chewing Gums: Microencapsulated Pomegranate Peel Extract Usage. *Int. J. Food Sci. Technol.* **2018**, *53*, 2657–2663. [CrossRef]
104. da Silva Veloso, F.; Caleja, C.; Calhelha, R.C.; Pires, T.C.S.; Alves, M.J.; Barros, L.; Genena, A.K.; Barreira, J.C.M.; Ferreira, I.C.F.R. Characterization and Application of Pomegranate Epicarp Extracts as Functional Ingredients in a Typical Brazilian Pastry Product. *Molecules* **2020**, *25*, 1481. [CrossRef]
105. Martins, Z.E.; Pinho, O.; Ferreira, I.M.P.L.V.O.; Jekle, M.; Becker, T. Development of Fibre-Enriched Wheat Breads: Impact of Recovered Agroindustrial by-Products on Physicochemical Properties of Dough and Bread Characteristics. *Eur. Food Res. Technol.* **2017**, *243*, 1973–1988. [CrossRef]
106. Kumar, N.; Daniloski, D.; Pratibha; Neeraj; D'Cunha, N.M.; Naumovski, N.; Petkoska, A.T. Pomegranate Peel Extract–A Natural Bioactive Addition to Novel Active Edible Packaging. *Food Res. Int.* **2022**, *156*, 111378. [CrossRef]
107. Castillejo, N.; Martínez-Hernández, G.B.; Artés-Hernández, F. Revalorized Broccoli By-Products and Mustard Improved Quality during Shelf Life of a Kale Pesto Sauce. *Food Sci. Technol. Int.* **2021**, *27*, 734–745. [CrossRef]

108. Quattrucci, A.; Ovidi, E.; Tiezzi, A.; Vinciguerra, V.; Balestra, G.M. Biological Control of Tomato Bacterial Speck Using *Punica granatum* Fruit Peel Extract. *Crop Prot.* **2013**, *46*, 18–22. [CrossRef]
109. Abd-Rabbu, H.S.; Wahba, H.E.; Khalid, K.A. Pomegranate Peel Modifies Growth, Essential Oil and Certain Chemicals of Sage (*Salvia officinalis* L.) Herb. *Biocatal. Agric. Biotechnol.* **2021**, *33*, 101978. [CrossRef]
110. Pangallo, S.; Nicosia, M.G.L.D.; Agosteo, G.E.; Abdelfattah, A.; Romeo, F.V.; Cacciola, S.O.; Rapisarda, P.; Schena, L. Evaluation of a Pomegranate Peel Extract as an Alternative Means to Control Olive Anthracnose. *Phytopathology* **2017**, *107*, 1462–1467. [CrossRef]
111. Elsherbiny, E.A.; Amin, B.H.; Baka, Z.A. Efficiency of Pomegranate (*Punica granatum* L.) Peels Extract as a High Potential Natural Tool towards Fusarium Dry Rot on Potato Tubers. *Postharvest Biol. Technol.* **2016**, *111*, 256–263. [CrossRef]
112. Rongai, D.; Sabatini, N.; Pulcini, P.; di Marco, C.; Storchi, L.; Marrone, A. Effect of Pomegranate Peel Extract on Shelf Life of Strawberries: Computational Chemistry Approaches to Assess Antifungal Mechanisms Involved. *J. Food Sci. Technol.* **2018**, *55*, 2702–2711. [CrossRef] [PubMed]
113. Li Destri Nicosia, M.G.; Pangallo, S.; Raphael, G.; Romeo, F.V.; Strano, M.C.; Rapisarda, P.; Droby, S.; Schena, L. Control of Postharvest Fungal Rots on Citrus Fruit and Sweet Cherries Using a Pomegranate Peel Extract. *Postharvest Biol. Technol.* **2016**, *114*, 54–61. [CrossRef]
114. Parsa, Z.; Roozbehi, S.; Hosseinifarahi, M.; Radi, M.; Amiri, S. Integration of Pomegranate Peel Extract (PPE) with Calcium Sulphate ($CaSO_4$): A Friendly Treatment for Extending Shelf-Life and Maintaining Postharvest Quality of Sweet Cherry Fruit. *J. Food Process. Preserv.* **2021**, *45*, e15089. [CrossRef]
115. Kumar, N.; Pratibha; Neeraj; Petkoska, A.T.; AL-Hilifi, S.A.; Fawole, O.A. Effect of Chitosan–Pullulan Composite Edible Coating Functionalized with Pomegranate Peel Extract on the Shelf Life of Mango (*Mangifera indica*). *Coatings* **2021**, *11*, 764. [CrossRef]
116. Gull, A.; Bhat, N.; Wani, S.M.; Masoodi, F.A.; Amin, T.; Ganai, S.A. Shelf Life Extension of Apricot Fruit by Application of Nanochitosan Emulsion Coatings Containing Pomegranate Peel Extract. *Food Chem.* **2021**, *349*, 129149. [CrossRef]
117. Paolucci, M.; di Stasio, M.; Sorrentino, A.; la Cara, F.; Volpe, M.G. Active Edible Polysaccharide-Based Coating for Preservation of Fresh Figs (*Ficus carica* L.). *Foods* **2020**, *9*, 1793. [CrossRef]
118. Alotaibi, M.A.; Tayel, A.A.; Zidan, N.S.; el Rabey, H.A. Bioactive Coatings from Nano-Biopolymers/Plant Extract Composites for Complete Protection from Mycotoxigenic Fungi in Dates. *J. Sci. Food Agric.* **2019**, *99*, 4338–4343. [CrossRef]
119. Tayel, A.A.; El-Baz, A.F.; Salem, M.F.; El-Hadary, M.H. Potential Applications of Pomegranate Peel Extract for the Control of Citrus Green Mould. *J. Plant Dis. Prot.* **2009**, *116*, 252–256. [CrossRef]
120. Pangallo, S.; Li Destri Nicosia, M.G.; Raphael, G.; Levin, E.; Ballistreri, G.; Cacciola, S.O.; Rapisarda, P.; Droby, S.; Schena, L. Elicitation of Resistance Responses in Grapefruit and Lemon Fruits Treated with a Pomegranate Peel Extract. *Plant Pathol.* **2017**, *66*, 633–640. [CrossRef]
121. Givi, F.; Gholami, M.; Massah, A. Application of Pomegranate Peel Extract and Essential Oil as a Safe Botanical Preservative for the Control of Postharvest Decay Caused by Penicillium Italicum and Penicillium Digitatum on "Satsuma" Mandarin. *J. Food Saf.* **2019**, *39*, e12639. [CrossRef]
122. Kharchoufi, S.; Parafati, L.; Licciardello, F.; Muratore, G.; Hamdi, M.; Cirvilleri, G.; Restuccia, C. Edible Coatings Incorporating Pomegranate Peel Extract and Biocontrol Yeast to Reduce Penicillium Digitatum Postharvest Decay of Oranges. *Food Microbiol.* **2018**, *74*, 107–112. [CrossRef]
123. Belgacem, I.; Pangallo, S.; Abdelfattah, A.; Romeo, F.V.; Cacciola, S.O.; Li Destri Nicosia, M.G.; Ballistreri, G.; Schena, L. Transcriptomic Analysis of Orange Fruit Treated with Pomegranate Peel Extract (PGE). *Plants* **2019**, *8*, 101. [CrossRef]
124. Nair, M.S.; Saxena, A.; Kaur, C. Effect of Chitosan and Alginate Based Coatings Enriched with Pomegranate Peel Extract to Extend the Postharvest Quality of Guava (*Psidium guajava* L.). *Food Chem.* **2018**, *240*, 245–252. [CrossRef]
125. Nair, M.S.; Saxena, A.; Kaur, C. Characterization and Antifungal Activity of Pomegranate Peel Extract and Its Use in Polysaccharide-Based Edible Coatings to Extend the Shelf-Life of Capsicum (*Capsicum annuum* L.). *Food Bioprocess Technol.* **2018**, *11*, 1317–1327. [CrossRef]
126. Megha, M.; Gill, P.S.; Jawandha, S.K.; Kaur, N.; Gill, M.S. Effect of Chitosan Coating Incorporated with Pomegranate Peel Extract on Pear Fruit Softening, Quality, and Cell Wall Degrading Enzymes during Cold Storage. *J. Food Process. Preserv.* **2021**, *45*, e15984. [CrossRef]
127. Lacivita, V.; Incoronato, A.L.; Conte, A.; del Nobile, M.A. Pomegranate Peel Powder as a Food Preservative in Fruit Salad: A Sustainable Approach. *Foods* **2021**, *10*, 1359. [CrossRef] [PubMed]
128. Belgacem, I.; Schena, L.; Teixidó, N.; Romeo, F.V.; Ballistreri, G.; Abadias, M. Effectiveness of a Pomegranate Peel Extract (PGE) in Reducing Listeria Monocytogenes in Vitro and on Fresh-Cut Pear, Apple and Melon. *Eur. Food Res. Technol.* **2020**, *246*, 1765–1772. [CrossRef]
129. Turrini, F.; Boggia, R.; Donno, D.; Parodi, B.; Beccaro, G.; Baldassari, S.; Signorello, M.G.; Catena, S.; Alfei, S.; Zunin, P. From Pomegranate Marcs to a Potential Bioactive Ingredient: A Recycling Proposal for Pomegranate-Squeezed Marcs. *Eur. Food Res. Technol.* **2020**, *246*, 273–285. [CrossRef]
130. Trigo, J.P.; Alexandre, E.M.C.; Oliveira, A.; Saraiva, J.A.; Pintado, M. Fortification of Carrot Juice with a High-Pressure-Obtained Pomegranate Peel Extract: Chemical, Safety and Sensorial Aspects. *Int. J. Food Sci. Technol.* **2020**, *55*, 1599–1605. [CrossRef]
131. Trigo, J.P.; Alexandre, E.M.C.; Silva, S.; Costa, E.; Saraiva, J.A.; Pintado, M. Study of Viability of High Pressure Extract from Pomegranate Peel to Improve Carrot Juice Characteristics. *Food Funct.* **2020**, *11*, 3410–3419. [CrossRef] [PubMed]

132. Altunkaya, A.; Hedegaard, R.V.; Harholt, J.; Brimer, L.; Gökmen, V.; Skibsted, L.H. Palatability and Chemical Safety of Apple Juice Fortified with Pomegranate Peel Extract. *Food Funct.* **2013**, *4*, 1468–1473. [CrossRef] [PubMed]
133. Tomadoni, B.; Moreira, M.D.R.; Espinosa, J.P.; Ponce, A. Individual and Combined Effects of Pomegranate Extract and Ultrasonic Treatments on Kiwifruit Juice Quality Parameters. *J. Food Process Eng.* **2017**, *40*, e12339. [CrossRef]
134. Canuti, V.; Cecchi, L.; Khatib, M.; Guerrini, L.; Mulinacci, N.; Zanoni, B. A New Extract from Pomegranate (*Punica granatum* L.) By-Products as a Potential Oenological Tannin: Preliminary Characterization and Comparison with Existing Commercial Products. *Molecules* **2020**, *25*, 4460. [CrossRef]
135. Jouki, M.; Khazaei, N.; Rashidi-Alavijeh, S.; Ahmadi, S. Encapsulation of Lactobacillus Casei in Quince Seed Gum-Alginate Beads to Produce a Functional Synbiotic Drink Powder by Agro-Industrial by-Products and Freeze-Drying. *Food Hydrocoll.* **2021**, *120*, 106895. [CrossRef]

foods

MDPI

Review

Postharvest Ultraviolet Radiation in Fruit and Vegetables: Applications and Factors Modulating Its Efficacy on Bioactive Compounds and Microbial Growth

Magalí Darré [1], Ariel Roberto Vicente [1,*], Luis Cisneros-Zevallos [2] and Francisco Artés-Hernández [3]

[1] LIPA—Laboratorio de Investigación en Productos Agroindustriales, Universidad Nacional de La Plata, Calle 60 y 119 s/n, La Plata CP 1900, Argentina; lipa@agro.unlp.edu.ar
[2] Department of Horticultural Sciences, Texas A&M University, College Station, TX 77843, USA; lcisnero@tamu.edu
[3] Postharvest and Refrigeration Group, Department of Agronomical Engineering & Institute of Plant Biotechnology, Universidad Politécnica de Cartagena, 30203 Murcia, Spain; fr.artes-hdez@upct.es
* Correspondence: arielvicente@agro.unlp.edu.ar

Abstract: Ultraviolet (UV) radiation has been considered a deleterious agent that living organisms must avoid. However, many of the acclimation changes elicited by UV induce a wide range of positive effects in plant physiology through the elicitation of secondary antioxidant metabolites and natural defenses. Therefore, this fact has changed the original UV conception as a germicide and potentially damaging agent, leading to the concept that it is worthy of application in harvested commodities to take advantage of its beneficial responses. Four decades have already passed since postharvest UV radiation applications began to be studied. During this time, UV treatments have been successfully evaluated for different purposes, including the selection of raw materials, the control of postharvest diseases and human pathogens, the elicitation of nutraceutical compounds, the modulation of ripening and senescence, and the induction of cross-stress tolerance. Besides the microbicide use of UV radiation, the effect that has received most attention is the elicitation of bioactive compounds as a defense mechanism. UV treatments have been shown to induce the accumulation of phytochemicals, including ascorbic acid, carotenoids, glucosinolates, and, more frequently, phenolic compounds. The nature and extent of this elicitation have been reported to depend on several factors, including the product type, maturity, cultivar, UV spectral region, dose, intensity, and radiation exposure pattern. Even though in recent years we have greatly increased our understanding of UV technology, some major issues still need to be addressed. These include defining the operational conditions to maximize UV radiation efficacy, reducing treatment times, and ensuring even radiation exposure, especially under realistic processing conditions. This will make UV treatments move beyond their status as an emerging technology and boost their adoption by industry.

Keywords: UV; UVB; UVC; UV illumination; photochemical treatments; abiotic stress; antioxidants; phytochemicals; quality; food safety

Citation: Darré, M.; Vicente, A.R.; Cisneros-Zevallos, L.; Artés-Hernández, F. Postharvest Ultraviolet Radiation in Fruit and Vegetables: Applications and Factors Modulating Its Efficacy on Bioactive Compounds and Microbial Growth. *Foods* **2022**, *11*, 653. https://doi.org/10.3390/foods11050653

Academic Editor: Filipa V. M. Silva

Received: 27 December 2021
Accepted: 7 February 2022
Published: 23 February 2022

1. Introduction

Ultraviolet (UV) radiation comprises the region of the electromagnetic spectrum (EM) between visible light and X-rays (100–400 nm) [1]. It was discovered in 1801 by Johann Wilhelm Ritter, who observed that radiation outside the violet end of the visible solar spectrum could decompose silver chloride [2]. Seven decades later, it was discovered that UV light could prevent microbial growth [3].

The UV region is frequently divided into three sub-regions, UVA (315–400 nm), UVB (280–315 nm), and UVC (100—280 nm) (Figure 1), which are used for CIE and ISO standards [4]. Further sub categorization has been performed by some authors to discriminate within the UVC region vacuum UV (100 and 200 nm) with stronger ionizing power, but less penetration [5–7]. Based on the known mechanisms of plant photoreception, the UVA region has been split into short-UVA (315–350 nm) and long-UVA (350–400) [8].

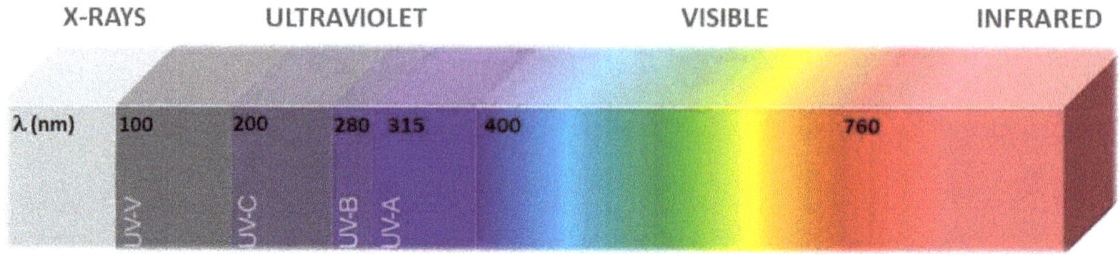

Figure 1. Subregions of the UV spectrum relevant for technological applications and plant photoreception.

2. UV Radiation Sources

Commercially available UV sources include mercury lamps (low- and medium-pressure), pulsed light (PL), and light-emitting diodes (LEDs) [9]. Although the market has become highly dynamic with the improvement of LEDs, it is still dominated by mercury lamps. These sources are based on the excitation of gas discharges and feature several pitfalls, including a relatively high voltage requirement to operate and a substantial amount of heat released [10]. However, one advantage is that, especially with medium-pressure Hg lamps, high output powers can be achieved.

Xenon inert-gas lamps were introduced during the late 1970s in Japan, leading to the development of sterilizing technology, called PureBright® [11]. PL treatments consist in the exposure of fresh produce to polychromatic light (200–1100 nm), including ultraviolet (180–400 nm), visible (400–700 nm), and near-infrared (700–1100 nm) wavelengths, in the form of intense, but short, pulses (1 μs–0.1 s) [12,13].

Light-emitting diodes (LEDs) are based on the junction of two-terminal semiconductors (p-n junction) converting electricity into radiation. Depending on the materials out of which the semiconductors are made, the LEDs emit at different wavelengths [14]. The first LEDs, in the early 1960s, emitted infrared (IR) light. Over the years, it became possible to develop LEDs of shorter wavelengths. UV LEDs have several advantages relative to mercury lamps, including their lack of requirement for warming time, their lack of mercury, their compactness, their robustness (with UV LEDs, no protection against glass breakage is necessary, and mobile use is possible), and their large wavelength diversity (210 nm to 360 nm by varying the aluminum content in the AlGaN quantum wells) [10]. In addition, they have lower electromagnetic interference, are easily adaptable for fast modulation in terms of radiation intensity and pulse duration, present narrow-band emission without spurious peaks, and require low maintenance [15]. Two important advantages of LEDs are their long lifespan (expected lifetimes of many 10,000 s of h) and low heat emission [16]. The Achilles heel of UV LEDs is their relatively low quantum efficiency [17]. However, in recent years, by reducing dislocations and defects and improving semiconductor doping and light extraction, their quantum efficiency has been increased [18].

3. Uses in the Food Industry

UV technology has been applied in the food industry for many different purposes (Figure 2).

Surface sterilization: One of the most common uses of germicide UVC lamps is as environmental sterilizers in foodstuffs filling equipment, conveyor belts, containers, and working surfaces [19,20]. Sterilizing UV lamps are frequently used for aseptic packing, a technology that is expected to continue growing in the coming years [21,22].

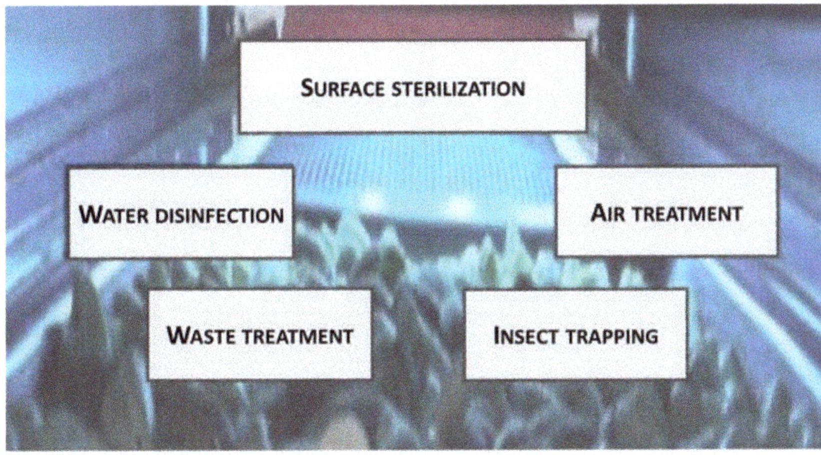

Figure 2. UV radiation applications in the food industry.

Fluid disinfection: UV radiation in the C zone has been used for water disinfection since 1909. It has been also applied for juice pasteurization [23]. UVC does not generate undesirable by-products, but on the other hand, it does not provide residual disinfection capacity [24]. It has been applied to reduce chlorine use.

Air treatment: Air disinfection can be achieved through different strategies, ranging from irradiating just the air in the upper region to treating all air, either when the room is empty or during circulation through air-conditioning systems [25]. The fact that UVC relatively low radiation doses (0.1–0.3 kJ m^{-2} for 2 log cycle reductions) can inactivate human SARS-coronaviruses has increased the recent interest in using UV radiation for air treatment [26–28].

Waste treatment: Another application of UV radiation has been the elimination of undesirable volatile organic compounds (VOCs) in industrial exhausts [29,30]. This has been achieved through advanced oxidation processes combining UV radiation with photo-catalysts, such as TiO_2 [9,31]. This strategy generates highly oxidative environments, which facilitates the degradation of unwanted molecules [32].

Insect trapping: For a long time, it has been known that UV radiation can attract insects; thus, it is used for trapping purposes [33]. The most common insect light traps use "black-light" fluorescent tubes emitting ultraviolet (UVA) as an insect attractant in both pre and postharvest [34]. Furthermore, insects may be trapped in glued materials or killed in electrically charged grids [35].

4. Uses in Fruits and Vegetables Postharvest

UV technology may be of interest for the postharvest treatment of fruits and vegetables for many different purposes [36,37] (Figure 3).

Raw material selection: The presence of skin defects or wounding is a main factor affecting consumer acceptability and purchase decisions. Consequently, one of the most intensive activities of packinghouses is to separate fruit with these defects. Normally, this is performed through visual inspection or machine optical sorting when the fruit is illuminated under proper white light [38]. In citrus, the use of UV lamps during initial classification has may facilitate the identification of physical damages. UVA "black light" illuminates fruit, showing that small peel cracks fluoresce intensely, allowing segregation at early classification steps [39]. The conveyor belts transporting the fruit cross these rooms, where the operators must wear protective glasses and gloves.

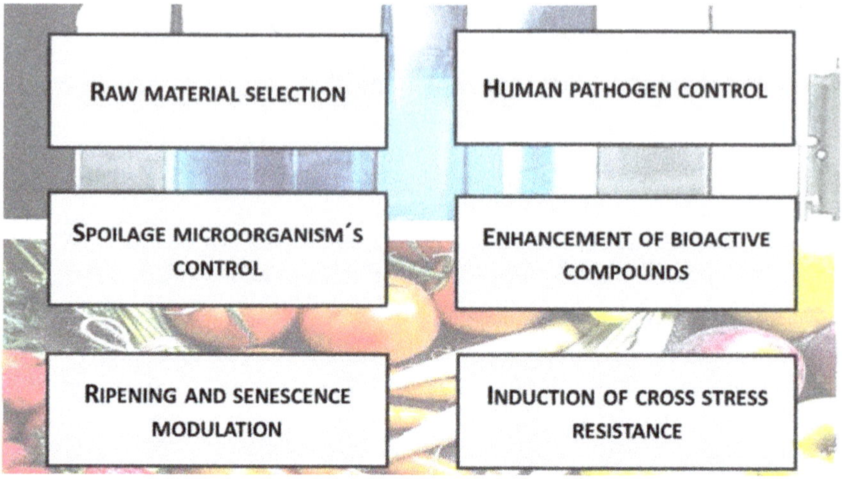

Figure 3. Main uses of postharvest UV treatments for direct application in fruit and vegetables.

Control of spoilage and pathogenic microorganisms: Relying on UVC germicide properties, a large body of research evaluated this technology in fruits and vegetables to control surface microorganisms [40]. Microbial death induced by UVC has been attributed to DNA mutations, including the formation of cyclobutyl-type dimers (pyrimidine dimers) and pyrimidine adducts [41,42]. Furthermore, the overproduction of reactive oxygen species (ROS) induced by UV radiation can oxidize membrane lipids and inhibit critical cellular enzymes [43]. Most enzymes that contain aromatic amino acids are likely to be sensitive to UV radiation to some extent due to their absorption in this region. Due to its higher energy, UVC is the most effective at killing microorganisms [44]. UV radiation is lethal to bacteria, viruses, protozoa, and fungi [45]. Successive studies showed that UV radiation is more efficient for inactivating Gram-negative than Gram-positive bacteria. This effect has been associated with the difference in the cell wall peptidoglycan structure, which can affect radiation penetration [46]. Furthermore, eukaryotic organisms are normally more resistant to UV than bacteria due to their higher cell size, complexity, and genetic redundancy [47]. The relatively high yeast resistance to UV radiation has also been associated with lower DNA pyrimidine content relative to bacteria, which may increase the likelihood of photons being absorbed by other compounds [48].

Several works have studied the impact of postharvest pre-storage single exposure on most common postharvest fungal pathogens, including *Rhizopus*, *Penicillum digitatum*, *P. expansum* and *Penicillum italicum*, *Monilinia* sp., *Botrytis cinerea*, *Colletrichcum* sp. and *Fusarium* sp. among others [49,50], with positive effects with regards to reducing disease incidence and severity. Direct germicide action compromising microbial viability has been frequently reported [51], but less severe effects, such as the reduced germination speed of viable conidia, have also been observed [52]. With regards to human pathogens, the direct UV irradiation of fresh produce reduced the viability of *E. coli*, *Salmonella*, and *Listeria* [53–57]. These studies have so far mostly been conducted on a laboratory or, in the best scenario, pilot scale. A review of the available research suggests that their widespread commercial use requires some important aspects to be solved, especially with regards to the adaptability of this treatment to continuous processing lines (where the treatment duration may require minutes at low irradiances), safety procedures for workers and even radiation exposure, while avoiding mechanical damage, especially to fruits and vegetables. Furthermore, the fact that wet cleaning methods have long been applied to fresh produce could limit the fast adoption of a different technology.

The mechanisms through which UV prevents deterioration exceed radiation germicide properties and involve host-induced physiological responses [58–60]. They include the inhibition of ripening-related genes [61] and the induction of enzymes and compounds, improving tolerance to opportunistic pathogens or other environmental effectors causing oxidative damage. An array of defensive responses has been observed in UV-stressed tissues. It includes the activation of glucanases and chitinases thought to be involved in the degradation of microbial cell walls, the induction of genes related to phenolic compound biosynthesis or oxidation, such as phenylalanine ammonia lyase (PAL), polyphenol oxidases (PPOs), and peroxidases (PODs) [62,63]. Another frequent change reported is the upregulation of gene coding for antioxidative enzymes, such as superoxide dismutases (SODs), ascorbate peroxidases (APXs), and catalases (CATs) [64]. More recently, Rabelo et al. [65] proposed that the UV response is mediated by the generation of oxidative stress as the primary signaling molecule generated through the partial ionization of water and an increase in mitochondrial activity. With regards to the induction of compounds that could contribute to increasing host tolerance to pathogen attack, there have been several reports on metabolites with direct antimicrobial activity (i.e., hydroxycinnamic acid derivatives, 6-methoxymellein, scoparone, scopoletin, rishitin), or reinforcing structural barriers [49,66]. Phenolic compounds showing in many cases antimicrobial and antioxidant properties have been the most frequently identified family of induced antimicrobial compounds.

Enhancement of bioactive compounds: Early on, in 1977, Langcake and Pryce [67] showed that UV exposure induced bioactive compounds in grapes. This raised the interest in using UV treatments to improve fruit and vegetable phytochemical profiles. A literature review shows that the number of publications on bioactive compounds and UV radiation has increased exponentially in the last three decades. Table 1 provides an overview of some relevant studies that used UV radiation on whole and fresh-cut fruit and vegetables, focusing on bioactive compounds. As a rule, the results show that the induction of antioxidants by UV radiation tends to be greatest for phenolic compounds, although increases in other antioxidant metabolites (ascorbic acid, glutathione, carotenoids) have also been reported [65,68]. The subclass of compounds induced is mostly dependent on the species considered [69,70] and even on the cultivar [71]. Increases in phenolic acids, non-anthocyanin flavonoids, anthocyanins, other flavonoids, isoflavones, and stilbenes have been frequently reported [62,72–76]. Furthermore, the subregion of the UV spectra applied may determine the type of metabolite elicited; in carrots, chlorogenic acid and isocoumarin were more inducible by UVB and UVC radiation, whereas ferulic acid was elicited by all UV regions to comparable levels [77,78]. At low irradiances, UVB and UVC regions are considered more inductive of secondary metabolites than UVA and UVB, which still have effects on antioxidants but also initiate several photomorphogenic responses [79,80]. Other factors affecting phytochemical accumulation besides the UV region and dose that have been less studied include the irradiances of the illuminating source and the mode of exposure (pulse number duration, interval between successive irradiations, etc.) [13,81,82]. Recently, UV has been reported as a suitable green strategy to enhance the nutraceutical content of fruit and vegetable beverages [83]. Considering there are receptors for UVB and UVA but no receptors for UVC, it has been proposed that ROS plays a key role in secondary metabolite biosynthesis for all three UV lights [77,78]. However, further studies are needed to confirm the role of oxidative stress through UVB and UVA receptor-mediated responses. Another proposed hypothesis is that UVB and UVC could share the same photoreceptors, since the action spectrum of UVR8 protein (as the main UVB receptor) ranges from 250 to 310 nm [8], including the UVC region; therefore, this UVB receptor could also be activated by UVC. It can explain why UVC produces similar effects to UVB in some cases according to previous research, since the energy provided by shorter wavelengths of UVC may activate the UVR8 protein [84]. Nevertheless, the UVR8 spectrum of action could slightly vary, depending on the metabolic pathway. The action spectrum on PAL induction is closer to the UVB region and thus may explain why UVB or even the combined effect of UVC + B obtained a higher increase in total phenolics than UVC [85].

Table 1. Effects of UV radiation on the phytochemical profile of fruit and vegetables and on other quality attributes.

UV Region	Product	Treatment Conditions	Main Results Found	Reference
UVA	Blueberries	Dose: 6 kJ m^{-2}	UVA had lower inductive effect than UVB or UVC	[86]
	Fresh-cut carrot	Intensity 12.73 W m^{-2} Dose: 45.8 kJ m^{-2}	Induction of PAL and increase in total antioxidant capacity and phenolics.	[77]
	Lettuce	Intensity: 3.7 W UVA, 4.2 W UVB, 7.5 W UVC.	Induction of PAL and phenolic antioxidants in all UV regions. UVA caused no growth inhibition. UVB inhibited growth and UVC caused most severe lesions.	[87]
	Tomato	λ: 353, 365 or 400 nm. Intensity:0.28–0.33 W m^{-2} Dose: 0.17–7.1 kJ m^{-2}	All wavelengths increased phenolics and carotenoids.	[88]
		Daily dose: UVA 2.9 kJ m^{-2} UVA 11.5 kJ m^{-2} UVB 0.941 kJ m^{-2} UVB 0.353 kJ m^{-2}	Increased antioxidant capacity and flavonoid accumulation. UVA was more promising than UVB with regards to firmness maintenances and antioxidant elicitation.	[89]
	Soybean sprouts	UVA 173.0 kJ m^{-2} UVA 346.0 kJm^{-2} Intensity: 2 W·m^{-2}	Treatments elicited isoflavone and flavonol accumulation.	[90]
UVB	Apples	Dose: 219 kJ m^{-2}	Increased content of flavonoids (64%) and hydroxycinnamic acids (38%) in the peel after 14 days.	[91]
	Bell pepper	Dose: 9.0 kJ m^{-2}. Storage 4 d at 20 °C under retail sale photoperiod (14 h fluorescent + 10 h Blue & Red LEDs)	Capsaicinoids increased by ~22%, ~38%, and ~27% in the content of capsanthin, capsanthin laurate, and capsanthin esters, respectively, after the UVB treatment. This effect was enhanced by ~18% after an LED-supplemented photoperiod.	[92]
	Brassicaceae sprouts	Dose: 15.0 kJ m^{-2}	Increased the total phenolics and antioxidant activity. Increased the glucosinolate content by ~30%. Sulforaphane was enhanced by 37.5% in broccoli sprouts. Sulforaphene was increased by 72% in radish sprouts.	[93]
	Broccoli	Intensity: 3.2–5.0 W m^{-2} Dose: 2–12 kJ m^{-2}	Low doses and intensities delayed chlorophyll degradation, whereas high intensity elicited antioxidant accumulation.	[81]
		Dose: 1.5–7.2 kJ m^{-2}	UVB increased glucobrassicins by 18–22%. Glucoraphanin was enhanced by 11% in florets exposed to 1.5 kJ m^{-2}, while a dose of 7.2 kJ m^{-2} by 16%. Florets exposed to 1.5 and 7.2 kJ m^{-2} UVB increased hydroxyl-cinnamic acids by 12%.	[94]
		Dose: 5–15 kJ m^{-2} alone or in combination with UVC (9 kJ m^{-2}).	Combination of moderate UVB and UVC doses reported the highest inductive effect on phenolics and total antioxidant activity. A high UVB dose (15 kJ m^{-2}), single or combined with moderate UVC, induced a higher level of glucoraphanin and sulforaphane.	[95]
	Fresh-cut carrot	Dose: 1.5 kJ m^{-2} alone or in combination with 4.0 kJ m^{-2} UVC	UVB caused the largest increase in phenolics and antioxidant accumulation after 3 days at 15 °C.	[77]

Table 1. *Cont.*

UV Region	Product	Treatment Conditions	Main Results Found	Reference
UVB		Intensity: 12.73 W m^{-2} (UVA) 10.44 W m^{-2} (UVB) 11.8 W m^{-2} (UVC) Dose: 46–275 kJ m^{-2} (UVA) 37.5–225.5 kJ m^{-2} (UVB) 42.5–255 kJ m^{-2} (UVC)	Phenolics (chlorogenic acid and its isomers, ferulic acid, and isocoumarin), antioxidant capacity, and PAL activity increments. Chlorogenic acid was induced by all UV radiations but mostly by UVB and UVC.	[77]
	Kale sprouts	Dose: 0, 5, 10, and 15 kJ m^{-2}	Enhanced the total antioxidant activity by 20%. Doses of 10 and 15 kJ m^{-2} stimulated the glucoraphanin and glucobrassicin synthesis by 30%.	[96]
	Lemon	Dose: 22 kJ m^{-2}	Increased levels of anthocyanins, flavonols and flavanones-dihydroflavonols. Increased antifungal activity of flavedo extracts against *Penicillium digitatum*.	[97]
	Mango	Dose: 5 kJ m^{-2}	Increased ascorbic acid (42%) and phenolic compound (36%).	[98]
	Red cabbage sprouts	Dose 10 kJ m^{-2} proportionally applied on germination days 3, 5, 7, and 10 days,	Phenolics were increased by 40%, while total antioxidant activity and flavonols content was increased by 35 and 30%, respectively. Carotenoids were also enhanced.	[99]
	Peach and nectarine	Dose: 73–219 kJ m^{-2}	Cultivar-dependent response: the stimulation of phenol accumulation occurred after 24 h in 'Big Top' (69%) and 36 h in 'Suncrest' (21%). Decreased phenolics in of 'Babygold 7' after 36 h.	[100]
		Dose: 1.39 and 8.33 kJ m^{-2}	Transient increase 24 h after illumination, especially for flavanols, flavonols, and flavones (+123, +70, +55, and +50%, respectively). Phenolics induced not only in the peel but also in the pulp. UVB increased the glycoside/aglycone ratio of flavonols and anthocyanins.	[74]
	Prickly pear (red)	Intensity: 6.4 W·m^{-2} Dose: 5.76 kJ m^{-2}	Highest phenolic accumulation. The main phenolics were quercetin, sinapic acid, kaempferol, rosmarinic acid, and sinapyl malate, showing increases of 709.8%, 570.2%, 442.8%, 439.9%, and 186.2%, respectively.	[101]
		Intensity 6.4 W m^{-2} Dose: 5.76–69 kJ m^{-2}	Immediate accumulation of betalains (33–40%) and ascorbic acid (54–58%) in the pulp and peel of wounded tissue.	[70]
UVC	Blueberry	Dose: 4.0 kJ m^{-2}	Increased anthocyanins (70%). Antioxidant enzymes induced (SOD, APX).	[102]
	Broccoli	Intensity: UVB s of 9.27 and UVC 25.21 W m^{-2}, Dose 5, 10 or 15 kJ m^{-2} UVB, UVC: 9 kJ m^{-2}.	UVB + UVC increased glucobrassicin (34%) at 15 °C. UVB15 + C induced the highest glucoraphanin levels of florets after 72 h at 15 °C. UVB10 + C induced the highest total phenolic content increase (110%) in leaves.	[95]
	Carambola	Dose: 13 kJ m^{-2}	UVC induced antioxidant enzymes (CAT, POX and SOD) and phenols accumulation.	[103]
	Carrot	Dose: 9 kJ m^{-2} + hyperoxia (80 kPa O_2)	Increase in phenolic compounds, which was also observed in hyperoxia for 72 h. UVC + hyperoxia showed higher accumulation of chlorogenic acid.	[104]

Table 1. *Cont.*

UV Region	Product	Treatment Conditions	Main Results Found	Reference
UVC	Fresh-cut watermelon	Dose: 1.6–7.2 kJ m^{-2}	Increase in antioxidant capacity (7%), maintenance of lycopene and ascorbic acid. Microbial growth retardation. Only the lowest doses (1.6 and 2.8 kJ m^{-2}) preserved sensory attributes.	[105]
	Fresh-cut Bimi® Broccoli	Dose: 1.5–15 kJ m^{-2}	Increased total phenolics (25%). Hydroxycinnamoyl acid derivates were immediately increased after the treatments. The higher the UVC doses, the higher total antioxidant capacity values. UVC delayed chlorophyll degradation.	[106]
	Fresh-cut tatsoi baby leaves	Dose: 4.54 kJ m^{-2} with hyperoxia (100 kPa O_2)	Improved phenolic content and total antioxidant capacity retention throughout storage. UVC and the combined UVC + O_2-controlled the epiphytic microbes.	[107]
	Fresh-cut pomegranate arils	Dose: 4.54 kJ m^{-2}	Combination of UVC and high O_2 preserved SOD and CAT and decreased POD and PPO.UVC combined with high O_2 maintained the level of anthocyanins and phenolics.Combining UVC to high O_2 enhanced the benefits of applying each treatment alone.All treatments involving high O_2 and/or UVC kept anthocyanins high, especially phenolic content.	[108]
		Dose: 4.54 kJ m^{-2}	The lowest antioxidant activity was found in hot water + UVC + superatmospheric O_2 packaging (HO) and the highest in UVC + HO and HO treatments. Hot water alone or in combination with UVC and HO inhibited mesophile, mold and yeast growth, while UVC + HO was most effective for controlling yeast and mold growth.	[109]
	Fresh-cut carrot	Intensity: UVB 9.27 W m^{-2}, UVC 25.21 W m^{-2}, Dose: UVB 1.5 kJ m^{-2}. UVC 4.0 kJ m^{-2}. Treatments alone or in combination	Combined UVC + UVB showed better results than each treatment alone.	[96]
	Fresh-cut red pepper	1.5; 3; 5; 6; 10 and 20 kJ m^{-2} in the inner (I), outer (O) or both fruit surfaces (I + O).	10 kJ m^{-2} (I + O) reduced decay and softening.UVC induced the accumulation of hydroxycinnamic acid-derivatives.Pectin solubilization and wall disassembly were delayed under UVC.UVC may control soft rots by modulating the host susceptibility.	[110]
	Garlic	Dose: 2.0 kJ m^{-2}	Increased total phenolics (11%) and reduced microbial loads.	[111]
	Grape	Dose: 0, 0.5, 1.0, 2.0, or 4.0 kJ m^{-2}	Increased activity of antioxidant enzymes (SOD and CAT) and induction of glutathione reductase and guaiacol peroxidase at longer times. Increased total thiol content by more than 2.0 kJ m^{-2}, total phenolics (20%), anthocyanin (35%) for 5d at 20 °C.	[112]

Table 1. *Cont.*

UV Region	Product	Treatment Conditions	Main Results Found	Reference
UVC	Red pepper	Dose: 10 kJ m^{-2}	UVC treatments do not cause marked modifications in DPPH radical scavenging capacity or AA content. UVC treatments increase the activity of enzymes involved in the detoxification of superoxide and hydrogen peroxide (SOD, CAT and APX) during early cold storage.	[64]
		Dose: 6 kJ m^{-2} UV (B or C) and 6 + 6 kJ m^{-2} UV (B + C)	UVC greatly enhanced the flavonoid accumulation. UVC + UVB increased by ~94% the carotenoid content and the flavonoid biosynthesis. Rutin accumulation was highly enhanced (~70%).	[92]
	Spinach	Dose: 1.5–3 kJ m^{-2}	Greatest induction of antioxidants (60%) and total phenolics (50%) with 1.5 kJ m^{-2}	[113]
		Dose: 4.54–11.35 kJ m^{-2}	Total antioxidant activity and polyphenols decreased throughout storage; this was more evident in higher UVC doses.Mesophilic and psychrophilic counts were reduced at similar level than conventional sanitization washing.	[114]
	Strawberry	Dose: 4.1 kJ m^{-2}	Induction of anthocyanin biosynthesis and related enzymes, PAL, tyrosine ammonia-lyase and cinnamate 4-hydroxylase.	[115]
	Sweet cherry	Dose: 4 kJ m^{-2} or Interactions of UVC with 2 regulated deficit irrigation (RDI)	UVC increased phenols (21–36%) after shelf-life in RDI fruit.	[116]
		Dose: 1.0–4.2 kJ m^{-2}	Induction of total phenolics, flavonoids, and anthocyanins (26%, 35% and 76% respectively). Induction of phenylpropanoid genes (PAL, C4H, 4CL).	[117]
	Tomato	Dose: 3.7 kJ m^{-2}	Increased the accumulation of phenolic compounds and lignin.	[60]

In addition to their effects on phenolics, carotenoids and ascorbic acid, UV treatments have been shown to increase vitamin D content. Exposure to sunlight and dietary foods are the most important ways for humans to obtain vitamin D [118–120]. Lifestyle changes due to the SARS-CoV-2 pandemic have substantially reduced our regular exposure to the sun, resulting in vitamin D deficiency [120]. Furthermore, patients with vitamin D deficiency were five times more likely to be positive for COVID-19 than patients with no deficiency [121]. Mushrooms are rich in ergosterol, a precursor to vitamin D2, which can be converted to vitamin D2 under proper UV exposure. The eliciting capacity of UVA, UVB, and UVC has been tested in different types of edible mushrooms, increases ranging between 25 and 8000% reported [122]. The UVB zone of radiation shows the greatest inductive effect, with other important factors being the radiation dose applied, the product's water content, and the degree of processing [123].

Retardation of ripening and senescence: In some commodities and under proper treatment conditions, UV radiation may delay ripening and senescence [60,124]. These effects could be understood by recognizing that both developmental processes are genetically regulated and require specific transcriptional programs to be induced. Under UV radiation, cells redirect their normal developmental programs to primarily respond to external stressing stimuli [125,126]. Stress acclimation favors the induction of UV-responsive genes at the expense of ripening and or senescence-related genes [61,127,128]. The efficacy of UV treatments to retard ripening is certainly dependent on the UV irradiation schedule used, and to a great extent on the initial ripening state of the commodity [129]. Ripening progression would be more effectively delayed in fruit treated at early stages. Hundreds of genes are up- and down-regulated in fruit irradiated with UV radiation. Most frequently, up-regulated genes are mainly involved in signal transduction, defense response, and metabolism. Conversely, genes related to cell wall disassembly, photosynthesis, and lipid metabolism are usually suppressed. The retardation of ripening-related changes such as softening may be one important contributor to the improved tolerance to postharvest spoilage pathogens observed in treated products [61,82,130]. In some cases, especially with long-term exposure, UVB has been shown to promote fruit ripening [131].

Induction of cross-stress resistance and synergistic responses: It is currently known that biotic and abiotic stress responses use common signals, pathways, and triggers [40]. This overlap includes common changes in cellular redox status, reactive oxygen species, hormones, protein kinase cascades, and calcium gradients as common elements [131] and helps to explain cross-tolerance phenomena, whereby exposure to one type of stress can improve tolerance to several different types of stress [132]. UV treatments preceding cold storage have been reported to improve the chilling tolerance of sensitive commodities such as peach [133], sweet pepper [134], and tomato [135]. Some of the metabolic changes behind cross-tolerance include the induction of polyamine biosynthesis in stone fruit and an increase in antioxidant enzymes in the case of pepper [64]. Another effect observed from combining stresses is a synergistic response in plant tissues, as in those reported with wounding and UV exposure. This synergism among stresses applied simultaneously is due to the activation of similar signaling molecules and signaling pathways [136]. This has been reported in the biosynthesis of polyphenols [77,78,102] and betalains [137]. In Table 2, the physiological responses could potentially be higher in fresh-cut products compared to whole tissues, where synergistic effects take place due to skin removal in whole produce, being skin-determinant in the response due to the partial blockage of UV penetration [77].

Table 2. Advantages and drawbacks of using UV radiation in foods.

Advantages	Drawbacks
• Simple. • Non-ionizing treatment. • Approved by food control agencies. • Strongly germicide (UVC) and broad microbiological control. • Able to elicit hormetic responses inducing phytochemical accumulation in metabolically active foods. • Relatively small changes in physicochemical quality attributes. • Energy-efficient and cost-effective. • Lower restriction than other irradiation methods. • Could be combined with other preservation methods. • No wastes or by-products generated. • Does not require water.	• Low penetration power in solids or turbid liquids. • Little or no residual effect. • Direct exposure required for germicide action and maximum effects. • Absorbed by commonly used polymeric packing materials. • Difficult to adapt to commercial operations/continuous processing. • Harmful to operators if not properly protected. • Consumers concerns although it is a non-ionizing radiation.

Advantages and disadvantages of UV treatments: UV technology provides several advantages over other conventional preservation methods, but also has some important drawbacks (Table 2). UV treatments can be simply applied and are able to inactivate a wide range of pathogenic and spoilage microorganisms while causing negligible changes in nutritional and sensory quality [95,96]. In contrast to other disinfection practices, it does not require water or generate wastes and leaves no residues on treated surfaces and foods [48]. In addition, it is approved with no major restrictions by the European Food Safety Authority (EFSA), the US Food and Drug Administration (FDA), and most other food regulatory agencies. When compared to electron beam or gamma irradiation, UV technology also offers several advantages (i.e., lower investment required, fewer regulations, no treatment label needed in the products) [138]. Another plus for UV treatments relates to their ability to be relatively simply combined with other preservation techniques in the search for additive or even synergistic effects [70,139]. UV irradiation is considered a very valuable tool within the hurdle technology, an integrated approach aimed in creating safe and stable foods by combining multiple physical, chemical, and/or biological preservation methods [139]. UV treatments have been applied in metabolically active fruit and vegetables not just due to their germicide action but to activate natural defense mechanisms (i.e., phytoalexins, free radical scavengers' antimicrobial and antioxidant enzymes) [140]. Compared to many conventional methods, UV treatments are also energy-efficient and cost-effective [141]. Finally, UV has historically being applied for a wide range of applications and validation data is available.

However, there are several limitations of UV technology which prevent its wide use in the food industry. It has a low penetration power, especially in solids and turbid media [142]. Moreover, the fact that it does not provide residuality may be a drawback when long-lasting effects are desired. To properly express the UV germicide properties direct and even exposure is required [143]. This may be challenging in some cases under commercial operations. Another difficulty could be to adapt UV treatments to continue processing lines, especially if long exposure times or low irradiances are required. Considering the moments in which UV could be used, UV treatments are defeated compared to ionizing irradiation, since they cannot be applied after packing [144]. UV radiation could be also harmful to operators if direct exposure occurs, but proper safety procedures and personal protective equipment can easily prevent such risk.

Factors determining UV treatment efficacy: In the last two decades, more than 500 publications have tested postharvest UV illumination strategies in fresh horticultural produce. These studies have been useful in identifying the main factors determining the efficacy of the technology on both the commodity and the treatment sides (Table 3).

Table 3. Factors affecting the efficacy of postharvest UV treatments in fruit and vegetables.

Product Variables	Process Variables
Commodity type	Radiation wavelength
Cultivar	Radiation dose (fluence)
Ripening stage	Radiation intensity (fluence rate/irradiance)
Degree of processing	Exposure pattern
Product–radiation interphase	Radiation uniformity
Product–microorganism interphase	Post irradiation illumination

The factor that has received the most attention is the type of commodity. So far, most commercially relevant fresh fruit and vegetable (>100 products) have been tested. In general, positive results have been found in one or more of the effects outlined in the previous section, depending on the species considered. In a few cases, some damages, mostly related to tissue discoloration, have also been reported [145,146]. The outcome of UV treatments has also been reported to be dependent on the cultivar [71,147]. A third relevant factor with regards to UV irradiation efficacy is the product-ripening stage. As a rule, the treatment of unripe fruit may lead to stronger phenotypes regarding ripening delay than fruit at advanced maturity stages. The ripening stage not only affects the impact of UV irradiation on ripening changes, but also its ability to control postharvest decay. In fresh-cut bell peppers, UV radiation was more effective at reducing soft rot in red ripe fruit [148,149]. The degree of commodity processing also has a substantial impact on quality maintenance and antioxidant elicitation [102,114]. This could be due to the increased surface during processing, which results in the exposure of a greater area of the product. However, the induction of synergic effects induced by simultaneous UV and wounding stresses has been suggested to be involved as well [70,100]. Finally, the nature of the interphase between UV radiation and the product can also have a relevant effect.

For instance, surface wetting would be expected to reduce UV penetration in vegetable tissues. Another effect may be the impact of interphases on microorganism arrangements. Heterogeneous microorganism distributions in liquid droplets can lead to preferential concentration in the outer layer at the liquid–air interface, which may protect the cells inside the droplet from UV bactericidal action [55].

Regarding treatment, the subregion of the UV spectra employed has a large influence on the kind of responses obtained [150]. For instance, changes in nutraceutical compounds have been reported for all three zones, but the greatest impact on disease control is clearly observed for UVC [151]. The second relevant process variable is the radiation dose (energy per unit area). A broad range of doses has been explored (i.e., 0.1–50 kJ m^{-2} for UVC region). However, a single study tried more than two or three doses and complete optimization studies are currently lacking. Working with strawberries, Cote et al. [152] showed that the radiation intensity (energy per unit time and unit area) is another key process factor determining the efficacy of treatments. Strikingly, this variable has been overlooked in several studies that do not report the radiation intensity of the illuminant used. Subsequent work by Darré et al. [81] reported that both radiation dose and intensity should be considered simultaneously when optimizing UV treatments. Low UVB doses (5 kJ m^{-2}) and intensities delay chlorophyll degradation and may be useful to complement refrigeration. Instead, high-intensity UVB exposure may be better suited for the freezing industry as a pre-treatment to increase the antioxidant capacity prior to further processing.

The irradiation pattern is another aspect to be considered when selecting a proper treatment schedule. This has been relatively well studied for pulsed treatments with xenon lamps. Bauer et al. [153] found that UV germicidal efficacy against *Bacillus* spores was a function of pulsed radiation parameter, with shorter pulses and lower frequencies being more effective. With regards to treatments with conventional single-UV radiation sources, almost no attention has been paid to the relevance of the radiation exposure pattern. Ortiz-Araque et al. [82] showed that at the same total dose (4 kJ m^{-2}) and intensity (36 W m^{-2}), fractionated (two-step and five-step) treatments were much more effective at controlling softening and decay than single pre-storage irradiation. The fractionation of the treatments during storage delayed pectin debranching and delayed the solubilization of polyuronides [130].

Although treatment uniformity and ensuring that all food surfaces are exposed to UVC light may be one of most problematic causes in commercial settings, there are very few studies attempting to overcome this limitation. Finally, some studies have reported that light exposure after treatment may affect the outcome of UV treatments. This may be due to photo repair mechanisms that may be activated in the presence of light [11,154,155]. It is not clear whether this effect would be significant in products. It is important to state that eventually, all these factors and their combinations result in the generation of signal molecules, such as ROS, and ROS levels determine the molecular and physiological response of stress-challenged produce, either staying in homeostasis, going through hormesis, or even responding to extreme stress [136].

5. Concluding Remarks

After microbial control, the elicitation of bioactive compounds is the aspect that has received the most attention with regards to the use of postharvest UV treatments in fresh produce. So far, several works have reported that appropriate exposure to UV radiation may stimulate the biosynthesis of phenolic compounds and, to a lesser extent, of ascorbic acid, carotenoids, and/or glucosinolates. Recent studies have identified the most relevant factors determining the nature and extent of such changes on both the commodity and process sides. These include the species, cultivar, ripening stage, degree of processing, radiation-product interphase, radiation wavelength, dose, intensity, exposure pattern (i.e., pulse frequency, duration), and illumination regime after treatment.

Despite the significant advances achieved, some important limitations remain. These has likely slowed down the adoption of UV technology by industrial stakeholders. For fresh products, the main challenges remaining are reducing the treatment times to facilitate their compatibility with continuous processing lines and increasing the treatment uniformity for large produce volumes without causing mechanical damage to products. Solving these challenges would likely help UV irradiation to move beyond the stage of emerging technology and translate all the knowledge accumulated about its application into production.

Funding: This research was funded by the Agencia Nacional de Promoción Científica y Tecnológica, Argentina grant number PICT 3679-2018.

Conflicts of Interest: The authors declare no conflict of interest.

References

1. Delorme, M.M.; Guimarães, J.T.; Coutinho, N.M.; Balthazar, C.F.; Rocha, R.S.; Silva, R.; Margalho, L.P.; Pimentel, T.L.; Silva, M.C.; Freitas, M.Q.; et al. Ultraviolet radiation: An interesting technology to preserve quality and safety of milk and dairy foods. *Trends Food Sci. Technol.* **2020**, *102*, 146–154. [CrossRef]
2. Shin, J.Y.; Kim, S.J.; Kim, D.K.; Kang, D.H. Fundamental characteristics of deep-UV light-emitting diodes and their application to control foodborne pathogens. *Appl. Environ. Microbiol.* **2016**, *82*, 2–10. [CrossRef] [PubMed]
3. Downes, A.; Blunt, T.P. Researches on the effect of light upon bacteria and other organisms. *Proc. R. Soc. Lond.* **1877**, *26*, 488–500. [CrossRef]
4. Bjorn, L.O. History ultraviolet-A, B, and C. *UV4Plants Bull.* **2015**, *1*, 17–18.
5. Zoschke, K.; Bornick, H.; Worch, E. Review vacuum-UV radiation at 185 nm in water treatment. A review. *Water Res.* **2014**, *52*, 134–145. [CrossRef]
6. Szeto, W.; Yam, W.C.; Huang, H.; Leung, D.Y. The efficacy of vacuum-ultraviolet light disinfection of some common environmental pathogens. *BMC Infect. Dis.* **2020**, *20*, 127. [CrossRef]
7. Kim, J.; Jang, J. Inactivation of airborne viruses using vacuum ultraviolet photocatalysis for a flow-through indoor air purifier with short irradiation time. *Aerosol Sci. Technol.* **2018**, *52*, 557–566. [CrossRef]
8. Rai, N.; Morales, L.O.; Aphalo, P.J. Perception of solar UV radiation by plants: Photoreceptors and mechanisms. *Plant Physiol.* **2021**, *186*, 1382–1396. [CrossRef]
9. Natarajan, T.S.; Natarajan, K.; Bajaj, H.C.; Tayade, R.J. Energy Efficient UV-LED Source and TiO2 Nanotube Array-Based Reactor for Photocatalytic Application. *Ind. Eng. Chem. Res.* **2011**, *50*, 7753. [CrossRef]
10. Bergman, R.S. Germicidal UV Sources and Systems. *Photochem. Photobiol.* **2021**, *97*, 466–470. [CrossRef]
11. Gómez-López, V.M.; Ragaert, P.; Debevere, J.; Devlieghere, F. Pulsed light for food decontamination: A review. *Trends Food Sci. Technol.* **2007**, *18*, 464–473. [CrossRef]
12. Oms-Oliu, G.; Martín-Belloso, O.; Soliva-Fortuny, R. Pulsed light treatments for food preservation. A review. *Food Bioprocess Technol.* **2010**, *3*, 13–23. [CrossRef]
13. Denoya, G.I.; Pataroc, G.; Ferrari, G. Effects of postharvest pulsed light treatments on the quality and antioxidant properties of persimmons during storage. *Postharvest Biol. Technol.* **2020**, *160*, 111055. [CrossRef]
14. Pulli, T.; Dönsberg, T.; Poikonen, T.; Manoocheri, F.; Kärhä, P.; Ikonen, E. Advantages of white LED lamps and new detector technology in photometry. *Light Sci. Appl.* **2015**, *4*, e332. [CrossRef]
15. Nassarawa, S.S.; Abdelshafy, A.M.; Xu, Y.; Li, L.; Luo, Z. Effect of light-emitting diodes (LEDs) on the quality of fruits and vegetables during postharvest period: A review. *Food Bioproc. Technol.* **2021**, *14*, 388–414. [CrossRef]
16. Zhang, F.; Shi, Z.F.; Ma, Z.Z.; Li, Y.; Li, S.; Wu, D.; Du, G.T. Silica coating enhances the stability of inorganic perovskite nanocrystals for efficient and stable down-conversion in white light-emitting devices. *Nanoscale* **2018**, *10*, 20131–20139. [CrossRef]
17. Guttmann, M.; Susilo, A.; Sulmoni, L.; Susilo, N.; Ziffer, E.; Wernicke, T.; Kneissl, M. Light extraction efficiency and internal quantum efficiency of fully UVC-transparent AlGaN based LEDs. *J. Phys. D Appl. Phys.* **2021**, *54*, 335101. [CrossRef]
18. Loi, M.; Villani, A.; Paciolla, F.; Mulè, G.; Paciolla, C. Challenges and Opportunities of Light-Emitting Diode (LED) as Key to Modulate Antioxidant Compounds in Plants. *A Review Antioxidants* **2021**, *10*, 42. [CrossRef]
19. Bintsis, T.; Litopoulou-Tzanetaki, E.; Robinson, R.K. Existing and potential applications of ultraviolet light in the food industry—A critical review. *J. Sci. Food Agric.* **2000**, *80*, 637–645. [CrossRef]
20. López-Gómez, A.; Castano-Villar, A.M.; Palop, A.; Marín-Iniesta, F. Hygienic design and microbial control of refrigeration and air conditioning systems for food processing and packaging plants. *Food Eng. Rev.* **2013**, *5*, 18–35. [CrossRef]
21. Koutchma, T. Advances in ultraviolet light technology for non-thermal processing of liquid foods. *Food Bioproc. Technol.* **2009**, *2*, 138–155. [CrossRef]
22. Jildeh, Z.B.; Wagner, P.H.; Schöning, M.J. Sterilization of objects, products, and packaging surfaces and their characterization in different fields of industry: The status in 2020. *Phys. Estado Solidi* **2021**, *218*, 2000732. [CrossRef]
23. Tran, H.D.M.; Boivin, S.; Kodamatani, H.; Ikehata, K.; Fujioka, T. Potential of UV-B and UV-C irradiation in disinfecting microorganisms and removing N-nitrosodimethylamine and 1, 4-dioxane for potable water reuse: A review. *Chemosphere* **2021**, *286*, 131682. [CrossRef]

24. Wang, C.P.; Chang, C.S.; Lin, W.C. Efficiency improvement of a flow-through water disinfection reactor using UV-C light emitting diodes. *J. Water Process. Eng.* **2021**, *40*, 101819. [CrossRef]
25. Memarzadeh, F. A Review of Recent Evidence for Utilizing Ultraviolet Irradiation Technology to Disinfect Both Indoor Air and Surfaces. *Appl. Biosaf.* **2021**, *26*, 52–56. [CrossRef]
26. Buonanno, M.; Welch, D.; Shuryak, I.; Brenner, D.J. Far-UVC light (222 nm) efficiently and safely inactivates airborne human coronaviruses. *Sci. Rep* **2020**, *10*, 10285. [CrossRef]
27. Quevedo, R.; Bastías-Montes, J.M.; Espinoza-Tellez, T.; Ronceros, B.; Balic, I.; Muñoz, O. Inactivation of Coronaviruses in food industry: The use of inorganic and organic disinfectants, ozone, and UV radiation. *Sci. Agropec.* **2020**, *11*, 257–266. [CrossRef]
28. Ma, B.; Gundy, P.M.; Gerba, C.P.; Sobsey, M.D.; Linden, K.G. UV Inactivation of SARS-CoV-2 across the UVC Spectrum: KrCl* Excimer, Mercury-Vapor, and Light-Emitting-Diode (LED) Sources. *Appl. Environ. Microbiol.* **2021**, *87*, e01532-21. [CrossRef]
29. Li, Y.W.; Ma, W.L. Photocatalytic oxidation technology for indoor air pollutants elimination: A. review. *Chemosphere* **2021**, *280*, 130667. [CrossRef]
30. Lee, C.S.; Shayegan, Z.; Haghighat, F.; Zhong, L.; Bahloul, A.; Huard, M. Experimental evaluation of in-duct electronic air cleaning technologies for the removal of ketones. *Build. Environ.* **2021**, *196*, 107782. [CrossRef]
31. Varma, K.S.; Tayade, R.J.; Shah, K.J.; Joshi, P.A.; Shukla, A.D.; Gandhi, V.G. Photocatalytic degradation of pharmaceutical and pesticide compounds (PPCs) using doped TiO_2 nanomaterials: A review. *Water-Energy Nexus* **2020**, *3*, 46–61. [CrossRef]
32. Matafonov, G.; Batoev, V. Recent advances in application of UV light-emitting diodes for degrading organic pollutants in water through advanced oxidation processes: A review. *Water Res.* **2018**, *132*, 177–189. [CrossRef] [PubMed]
33. Windsor, H. 'Electric chair' for insects aids farmers. *Pop. Mechan.* **1934**, *61*, 406–407.
34. González, M.; Alarcón-Elbal, P.M.; Valle-Mora, J.; Goldarazena, A. Comparison of different light sources for trapping Culicoides biting midges, mosquitoes and other dipterans. *Vet. Parasitol.* **2016**, *226*, 44–49. [CrossRef]
35. Sliney, D.H.; Gilbert, D.W.; Terry, L. Ultraviolet safety assessments of insect light traps. *J. Occup. Environ. Hyg.* **2016**, *13*, 413–424. [CrossRef]
36. Ribeiro, C.; Canadá, J.; Alvarenga, B. Prospects of UV radiation for application in postharvest technology. *Emir. J. Food Agric.* **2012**, *24*, 586–597. [CrossRef]
37. Turtoi, M. Ultraviolet light treatment of fresh fruits and vegetables surface: A review. *J. Agroalim. Proc. Technol.* **2013**, *19*, 325–337. Available online: https://hyperviolet.com.br/wp-content/uploads/2020/04/White-Paper-UV-C-Frutas-2-1.pdf (accessed on 15 November 2021).
38. Blasco, J.; Aleixos, N.; Gómez, J.; Molto, E. Citrus sorting by identification of the most common defects using multispectral computer vision. *J. Food Eng.* **2007**, *83*, 384–393. [CrossRef]
39. Firouzjaei, R.A.; Minaei, S.; Beheshti, B. Sweet lemon mechanical damage detection using image processing technique and UV radiation. *J. Food Meas. Charact.* **2018**, *12*, 1513–1518. [CrossRef]
40. Charles, F. Current challenges of physical treatments to control quality and postharvest diseases of fresh fruits and vegetables. *IOP Conf. Ser. Earth Environ. Sci.* **2018**, *309*, 012003. [CrossRef]
41. Livneh, Z.; Cohen-Fix, O.; Skaliter, R.; Elizur, T. Replication of damaged DNA and the molecular mechanism of ultraviolet light mutagenesis. *Crit. Rev. Biochem. Mol. Biol.* **1993**, *28*, 465–513. [CrossRef]
42. Leung, W.Y.; Murray, V. The influence of DNA methylation on the sequence specificity of UVB-and UVC-induced DNA damage. *J. Photochem. Photobiol. B Biol.* **2021**, *221*, 112225. [CrossRef] [PubMed]
43. Visser, P.M.; Poos, J.J.; Scheper, B.B.; Boelen, P.; van Duyl, F.C. Diurnal variations in depth profiles of UV-induced DNA damage and inhibition of bacterioplankton production in tropical coastal waters. *Marine. Ecol. Prog. Ser.* **2002**, *228*, 28–33. [CrossRef]
44. Hoerter, J.D.; Arnold, A.A.; Kuczynska, D.A.; Shibuya, A.; Ward, C.S.; Sauer, M.G.; Gizachew, A.; Hotchkiss, T.M.; Fleming, T.J.; Johnson, S. Effects of sublethal UVA irradiation on activity levels of oxidative defense enzymes and protein oxidation in *Escherichia coli*. *J. Photochem. Photobiol. B Biol.* **2005**, *81*, 171–180. [CrossRef]
45. Akhila, P.P.; Sunooj, K.V.; Aaliya, B.; Navaf, M.; Sudheesh, C.; Sabu, S.; Khaneghah, A.M. Application of electromagnetic radiations for decontamination of fungi and mycotoxins in food products: A comprehensive review. *Trends Food Sci. Technol.* **2021**, *114*, 399–409. [CrossRef]
46. Kim, D.K.; Kim, S.J.; Kang, D.H. Bactericidal effect of 266 to 279 nm wavelength UVC-LEDs for inactivation of Gram positive and Gram negative foodborne pathogenic bacteria and yeasts. *Food Res. Int.* **2017**, *97*, 280–287. [CrossRef]
47. Soler, P.; Moreno-Mesonero, L.; Zornoza, A.; Macián, V.J.; Moreno, Y. Characterization of eukaryotic microbiome and associated bacteria communities in a drinking water treatment plant. *Sci. Total Environ.* **2021**, *797*, 149070. [CrossRef] [PubMed]
48. Gayán, E.; Condón, S.; Álvarez, I. Biological aspects in food preservation by ultraviolet light: A review. *Food Bioprocess Technol.* **2014**, *7*, 1–20. [CrossRef]
49. D'hallewin, G.; Schirra, M.; Manueddu, E.; Piga, A.; Ben-Yehoshua, S. Scoparone and scopoletin accumulation and ultraviolet-C induced resistance to postharvest decay in oranges as influenced by harvest date. *J. Am. Soc. Hortic. Sci.* **1999**, *124*, 702–707. [CrossRef]
50. Filho, F.O.; Silva, E.D.O.; Lopes, M.M.D.A.; Ribeiro, P.R.V.; Oster, A.H.; Guedes, J.A.C.; Zocolo, G.J. Effect of pulsed light on postharvest disease control-related metabolomic variation in melon (*Cucumis melo*) artificially inoculated with *Fusarium pallidoroseum*. *PLoS ONE* **2020**, *15*, e0220097. [CrossRef]

51. Erkan, T.; Wang, C.Y.; Krizek, D.T. UV-C irradiation reduces microbial populations and deterioration in *Cucurbita pepo* fruit tissue. *Environ. Exp. Bot.* **2001**, *45*, 1–9. [CrossRef]
52. Pan, J.; Vicente, A.R.; Martínez, G.A.; Chaves, A.R.; Civello, P.M. Combined use of UV-C irradiation and heat treatment to improve postharvest life of strawberry fruit. *J. Sci. Food Agric.* **2004**, *84*, 1831–1838. [CrossRef]
53. Gayán, E.; Monfort, S.; Álvarez, I.; Condón, S. UV-C inactivation of *Escherichia coli* at different temperatures. *Innov. Food Sci. Emerg. Technol.* **2011**, *12*, 531–541. [CrossRef]
54. Gayán, E.; Serrano, M.J.; Raso, J.; Álvarez, I.; Condón, S. Inactivation of *Salmonella enterica* by UV-C light alone and in combination with mild temperatures. *Appl. Environ. Microbiol.* **2012**, *78*, 8353–8361. [CrossRef]
55. Cheng, Y.; Chen, H.; Sánchez Basurto, L.A.; Protasenko, V.V.; Bharadwaj, S.; Islam, M.; Moraru, C.I. Inactivation of *Listeria* and *E. coli* by Deep-UV LED: Effect of substrate conditions on inactivation kinetics. *Sci. Rep.* **2020**, *10*, 3411. [CrossRef] [PubMed]
56. Kebbi, Y.; Muhammad, A.I.; Sant'Ana, A.S.; do Prado-Silva, L.; Liu, D.; Ding, T. Recent advances on the application of UV-LED technology for microbial inactivation: Progress and mechanism. *Compr. Rev. Food Sci. Food Saf.* **2020**, *19*, 3501–3527. [CrossRef]
57. Prasad, A.; Gänzle, M.; Roopesh, M.S. Inactivation of *Escherichia coli* and *Salmonella* using 365 and 395 nm high intensity pulsed light emitting diodes. *Foods* **2019**, *8*, 679. [CrossRef]
58. Charles, M.T.; Benhamou, N.; Arul, J. Physiological basis of UV-C induced resistance to *Botrytis cinerea* in tomato fruit. III. Ultrastructural modifications and their impact on fungal colonization. *Postharvest Biol. Technol.* **2008**, *47*, 27–40. [CrossRef]
59. Charles, M.T.; Goulet, A.; Arul, J. Physiological basis of UV-C induced resistance to *Botrytis cinerea* in tomato fruit. IV. Biochemical modification of structural barriers. *Postharvest. Biol. Technol.* **2008**, *47*, 41–53. [CrossRef]
60. Charles, M.T.; Mercier, J.; Makhlouf, J.; Arul, J. Physiological basis of UV-C induced resistance to *Botrytis cinerea* in tomato fruit. I. Role of pre- and post-challenge accumulation of the phytoalexin-rishitin. *Postharvest Biol. Technol.* **2008**, *47*, 10–20. [CrossRef]
61. Pombo, M.A.; Dotto, M.C.; Martínez, G.A.; Civello, P.M. UV-C irradiation delays strawberry fruit softening and modifies the expression of genes involved in cell wall degradation. *Postharvest. Biol. Technol.* **2009**, *51*, 141–148. [CrossRef]
62. Nigro, F.; Ippolito, A.; Lattanzio, V.; Salerno, M. Effect of ultraviolet-C light on postharvest decay of strawberry. *J. Plant Pathol.* **2000**, *82*, 29–37. Available online: https://www.jstor.org/stable/41997977 (accessed on 15 November 2021).
63. Sheng, K.; Shui, S.S.; Yan, L.; Liu, C.; Zheng, L. Effect of postharvest UV-B or UV-C irradiation on phenolic compounds and their transcription of phenolic biosynthetic genes of table grapes. *J. Food Sci. Technol.* **2018**, *55*, 3292–3302. [CrossRef] [PubMed]
64. Andrade Cuvi, M.J.; Vicente, A.R.; Concellón, A.; Chaves, A.R. Changes in red pepper antioxidants as affected by UV-C treatments and storage at chilling temperatures. *Food Sci. Technol.* **2011**, *44*, 1666–1671. [CrossRef]
65. Rabelo, M.C.; Bang, W.Y.; Nair, V.; de Miranda, M.R.A.; Cisneros-Zevallos, L. UVC light modulates vitamin C and phenolic biosynthesis in acerola fruit: Role of increased mitochondria activity and ROS production. *Sci. Rep.* **2020**, *10*, 21972. [CrossRef]
66. Mercier, J.; Roussel, D.; Charles, M.T.; Arul, J. Systemic and local responses associated with UV- and pathogen-induced resistance to *Botrytis cinerea* in stored carrot. *Phytopathology* **2000**, *90*, 981–986. [CrossRef]
67. Langcake, P.; Pryce, R.J. The production of resveratrol and the viniferins by grapevines in response to ultraviolet irradiation. *Phytochemistry* **1977**, *16*, 1193–1196. [CrossRef]
68. Erkan, M.; Wang, S.Y.; Wang, C.Y. Effect of UV treatment on antioxidant capacity, antioxidant enzyme activity and decay in strawberry fruit. *Postharvest Biol. Technol.* **2008**, *48*, 163–171. [CrossRef]
69. Charles, M.T.; Arul, J. UV treatment of fresh fruits and vegetables for improved quality: A status report. *Stewart Postharvest Rev.* **2007**, *3*, 1–8. [CrossRef]
70. Ortega-Hernández, E.; Welti-Chanes, J.; Jacobo-Velázquez, D.A. Effects of UVB light, wounding stress, and storage time on the accumulation of betalains, phenolic compounds, and ascorbic acid in red prickly pear (*Opuntia ficus-indica* cv. Rojo Vigor). *Food Bioprocess Technol.* **2018**, *11*, 2265–2274. [CrossRef]
71. Xie, Z.; Charles, M.T.; Fan, J.; Deschênes, M.; Dubé, C. Effects of preharvest ultraviolet-C irradiation on fruit phytochemical profiles and antioxidant capacity in three strawberry (*Fragaria* × *ananassa* Duch.) cultivars. *J. Sci. Food Agric.* **2015**, *95*, 2996–3002. [CrossRef]
72. Cantos-Villar, E.; Espín, J.C.; Fernández, M.J.; Oliva, J. Postharvest UV-C-irradiated grapes as a potential source for producing stilbene-enriched red wines. *J. Agric. Food Chem.* **2003**, *51*, 1208–1214. [CrossRef] [PubMed]
73. Fernández-Marín, M.I.; Puertas, B.; Guerrero, R.F.; García-Parrilla, M.C.; Cantos-Villar, E. Preharvest methyl jasmonate and postharvest UVC treatments: Increasing stilbenes in wine. *J. Food Sci.* **2014**, *79*, C310. [CrossRef] [PubMed]
74. Santin, M.; Miras-Moreno, B.; Rocchetti, G.; Lucini, L.; Hauser, M.T.; Ranieri, A. Beyond the visible and below the peel: How UV-B radiation influences the phenolic profile in the pulp of peach fruit. A biochemical and molecular study. *Front. Plant Sci.* **2020**, *11*, 1707. [CrossRef] [PubMed]
75. Santin, M.; Ranieri, A.M.; Hauser, M.T.; Miras-Moreno, B.; Rocchetti, G.; Lucini, L.; Strid, A.; Castagna, A. The outer influences the inner: Postharvest UV-B irradiation modulates peach flesh metabolome although shielded by the skin. *Food Chem.* **2021**, *338*, 127782. [CrossRef] [PubMed]
76. Li, Y.; Liu, C.; Shi, Q.; Yang, F.; Wei, M. Mixed red and blue light promotes ripening and improves quality of tomato fruit by influencing melatonin content. *Environ. Exp. Bot.* **2021**, *185*, 104407. [CrossRef]
77. Surjadinata, B.B.; Jacobo-Velázquez, D.A.; Cisneros-Zevallos, L. UVA, UVB and UVC light enhances the biosynthesis of phenolic antioxidants in fresh-cut carrot through a synergistic effect with wounding. *Molecules* **2017**, *22*, 668. [CrossRef]

78. Surjadinata, B.B.; Jacobo-Velázquez, D.A.; Cisneros-Zevallos, L. Physiological role of reactive oxygen species, ethylene, and jasmonic acid on UV light induced phenolic biosynthesis in wounded carrot tissue. *Postharvest Biol. Technol.* **2021**, *172*, 111388. [CrossRef]
79. Jenkins, G.I. Photomorphogenic responses to ultraviolet-B light. *Plant Cell Environ.* **2017**, *40*, 2544–2557. [CrossRef]
80. Vanhaelewyn, L.; Van Der Straeten, D.; De Coninck, B.; Vandenbussche, F. Ultraviolet radiation from a plant perspective: The plant-microorganism context. *Front. Plant Sci.* **2020**, *11*, 597642. [CrossRef]
81. Darré, M.; Valerga, L.; Araque, L.C.O.; Lemoine, M.L.; Demkura, P.V.; Vicente, A.R.; Concellón, A. Role of UV-B irradiation dose and intensity on color retention and antioxidant elicitation in broccoli florets (*Brassica oleracea* var. Italica). *Postharvest Biol. Technol.* **2017**, *128*, 76–82. [CrossRef]
82. Ortiz Araque, L.C.; Rodoni, L.M.; Darré, M.; Ortiz, C.M.; Civello, P.M.; Vicente, A.R. Cyclic low dose UV-C treatments retain strawberry fruit quality more effectively than conventional pre-storage single high fluence applications. *LWT* **2018**, *92*, 304–311. [CrossRef]
83. Artés–Hernández, F.; Castillejo, N.; Martínez-Zamora, L.; Martínez-Hernández, G.B. Phytochemical fortification in fruit and vegetable beverages with green technologies. *Foods* **2021**, *10*, 2534. [CrossRef] [PubMed]
84. Artés-Hernández, F.; Castillejo, N.; Martínez-Zamora, L. UV and Visible Spectrum LED Lighting as Abiotic Elicitors of Bioactive Compounds in Sprouts, Microgreens, and Baby Leaves—A Comprehensive Review including Their Mode of Action. *Foods* **2022**, *11*, 265. [CrossRef]
85. Martínez-Zamora, L.; Castillejo, N.; Artés–Hernández, F. Postharvest UV-B and photoperiod with blue + red LEDs as strategies to stimulate carotenogenesis in bell peppers. *Appl. Sci. Spec. Issue Funct. Foods Human Health* **2021**, *11*, 3736. [CrossRef]
86. Nguyen, C.T.; Kim, J.; Yoo, K.S.; Lim, S.; Lee, E.J. Effect of prestorage UV-A,-B, and-C radiation on fruit quality and anthocyanin of 'Duke' blueberries during cold storage. *J. Agric. Food Chem.* **2014**, *62*, 12144–12151. [CrossRef] [PubMed]
87. Lee, M.J.; Son, J.E.; Oh, M.M. Growth and phenolic compounds of *Lactuca sativa* L. grown in a closed-type plant production system with UV-A,-B, or-C lamp. *J. Sci. Food Agric.* **2014**, *94*, 197–204. [CrossRef]
88. Dyshlyuk, L.; Babich, O.; Prosekov, A.; Ivanova, S.; Pavsky, V.; Chaplygina, T. The effect of postharvest ultraviolet irradiation on the content of antioxidant compounds and the activity of antioxidant enzymes in tomato. *Heliyon* **2020**, *6*, e03288. [CrossRef]
89. Mariz-Ponte, N.; Martins, S.; Gonçalves, A.; Correia, C.M.; Ribeiro, C.; Dias, M.C.; Santos, C. The potential use of the UV-A and UV-B to improve tomato quality and preference for consumers. *Sci. Hort.* **2019**, *246*, 777–784. [CrossRef]
90. Lim, Y.J.; Lyu, J.I.; Kwon, S.J.; Eom, S.H. Effects of UV-A radiation on organ-specific accumulation and gene expression of isoflavones and flavonols in soybean sprout. *Food Chem.* **2021**, *339*, 128080. [CrossRef]
91. Assumpção, C.F.; Hermes, V.S.; Pagno, C.; Castagna, A.; Mannucci, A.; Sgherri, C.; de Oliveira Rios, A. Phenolic enrichment in apple skin following post-harvest fruit UV-B treatment. *Postharvest Biol. Technol.* **2018**, *138*, 37–45. [CrossRef]
92. Martínez-Zamora, L.; Castillejo, N.; Artés–Hernández, F. Postharvest UV radiation enhanced biosynthesis of flavonoids and carotenes in bell peppers. *Postharvest Biol. Technol.* **2022**, *184*, 111774. [CrossRef]
93. Martínez-Zamora, L.; Castillejo, N.; Artés–Hernández, F. Postharvest UV-B and UV-C radiation enhanced the biosynthesis of glucosinolates and isothiocyanates in brassicaceae sprouts. *Postharvest Biol. Technol.* **2021**, *181*, 111650. [CrossRef]
94. Duarte-Sierra, A.; Hasan, S.M.M.; Angers, P.; Arul, J. UV-B radiation hormesis in broccoli florets: Glucosinolates and hydroxy-cinnamates are enhanced by UV-B in florets during storage. *Postharvest Biol. Technol.* **2020**, *168*, 111278. [CrossRef]
95. Formica-Oliveira, A.C.; Martínez-Hernández, G.B.; Díaz-López, V.; Artés, F.; Artés-Hernández, F. Use of postharvest UV-B and UV-C radiation treatments to revalorize broccoli byproducts and edible florets. *Innov. Food Sci. Emerg. Technol.* **2017**, *43*, 77–83. [CrossRef]
96. Martínez-Zamora, L.; Castillejo, N.; Artés–Hernández, F. Periodical UV-B radiation hormesis in biosynthesis of kale sprouts nutraceuticals. *Plant Physiol. Biochem.* **2021**, *165*, 274–285. [CrossRef]
97. Ruiz, V.E.; Cerioni, L.; Zampini, I.C.; Cuello, S.; Isla, M.I.; Hilal, M.; Rapisarda, V.A. UV-B radiation on lemons enhances antifungal activity of flavedo extracts against *Penicillium digitatum*. *LWT* **2017**, *85*, 96–103. [CrossRef]
98. Jiang, Z.; Zheng, Y.; Qiu, R.; Yang, Y.; Xu, M.; Ye, Y.; Xu, M. Short UV-B exposure stimulated enzymatic and non-enzymatic antioxidants and reduced oxidative stress of cold-stored mangoes. *J. Agric. Food Chem.* **2015**, *63*, 10965–10972. [CrossRef]
99. Martínez-Zamora, L.; Castillejo, N.; Artés–Hernández, F. UV-B radiation as abiotic elicitor to enhance phytochemicals and development of red cabbage sprouts. *Horticulturae* **2021**, *7*, 567. [CrossRef]
100. Scattino, C.; Castagna, A.; Neugart, S.; Chan, H.M.; Schreiner, M.; Crisosto, C.H.; Ranieri, A. Post-harvest UV-B irradiation induces changes of phenol contents and corresponding biosynthetic gene expression in peaches and nectarines. *Food Chem.* **2014**, *163*, 51–60. [CrossRef]
101. Ortega-Hernández, E.; Nair, V.; Welti-Chanes, J.; Cisneros-Zevallos, L.; Jacobo-Velázquez, D.A. Wounding and UVB light synergistically induce the biosynthesis of phenolic compounds and ascorbic acid in red prickly pears (*Opuntia ficus-indica* cv. Rojo Vigor). *Int. J. Mol. Sci.* **2019**, *20*, 5327. [CrossRef]
102. Xu, F.; Wang, S.; Xu, J.; Liu, S.; Li, G. Effects of combined aqueous chlorine dioxide and UV-C on shelf-life quality of blueberries. *Postharvest Biol. Technol.* **2016**, *117*, 125–131. [CrossRef]

103. Andrade Cuvi, M.J.; Moreno Guerrero, C.; Henríquez Bucheli, A.; Gómez Gordillo, A.; Concellón, A. Influencia de la radiación UV-C como tratamiento postcosecha sobre carambola (*Averroha Carambola*, L.) mínimamente procesada almacenada en refrigeración. *Rev. Iberoam. Tecnol. Postcosecha* **2010**, *11*. Available online: https://digital.cic.gba.gob.ar/handle/11746/4393 (accessed on 15 November 2021).
104. Formica-Oliveira, A.C.; Martínez-Hernández, G.B.; Aguayo, E.; Gómez, P.A.; Artés, F.; Artés-Hernández, F. UV-C and hyperoxia abiotic stresses to improve healthiness of carrots: Study of combined effects. *J. Food Sci. Technol.* **2016**, *53*, 3465–3476. [CrossRef] [PubMed]
105. Artés-Hernández, F.; Robles, P.A.; Gómez, P.A.; Tomás–Callejas, A.; Artés, F.; Martínez–Hernández, G.B. Quality changes of fresh–cut watermelon during shelf-life as affected by cut intensity and UV–C pre-treatment. *Food Bioprocess Technol.* **2021**, *14*, 505–517. [CrossRef]
106. Martínez-Hernández, G.B.; Gómez, P.A.; Pradas, I.; Artés, F.; Artés-Hernández, F. Moderate UV-C pretreatment as a quality enhancement tool in fresh-cut Bimi® broccoli. *Postharvest Biol. Technol.* **2011**, *62*, 327–337. [CrossRef]
107. Tomás-Callejas, A.; Otón, M.; Artés, F.; Artés-Hernández, F. Combined effect of UV-C pretreatment and high oxygen packaging for keeping the quality of fresh-cut tatsoi baby leaves. *Innov. Food Sci. Emerg. Technol.* **2012**, *14*, 115–121. [CrossRef]
108. Maghoumi, M.; Gómez, P.A.; Artés-Hernández, F.; Mostofi, Y.; Zamani, Z.; Artés, F. Hot water, UV-C and superatmospheric oxygen packaging as hurdle techniques for keeping overall quality of fresh-cut pomegranate arils. *J. Sci. Food Agric.* **2013**, *93*, 1162–1168. [CrossRef] [PubMed]
109. Maghoumi, M.; Gómez, P.A.; Mostofi, Y.; Zamani, Z.; Artés-Hernández, F.; Artés, F. Combined effect of heat treatment, UV-C and superatmospheric oxygen packaging on antioxidant and browning related enzymes of fresh-cut pomegranate arils. *LWT* **2013**, *54*, 389–396. [CrossRef]
110. Rodoni, L.M.; Zaro, M.J.; Hasperué, J.H.; Concellón, A.; Vicente, A.R. UV-C treatments extend the shelf life of fresh-cut peppers by delaying pectin solubilization and inducing local accumulation of phenolics. *LWT* **2015**, *63*, 408–414. [CrossRef]
111. Park, M.H.; Kim, J.G. Low-dose UV-C irradiation reduces the microbial population and preserves antioxidant levels in peeled garlic (*Allium sativum* L.) during storage. *Postharvest Biol. Technol.* **2015**, *100*, 109–112. [CrossRef]
112. Maurer, L.H.; Bersch, A.M.; Santos, R.O.; Trindade, S.C.; Costa, E.L.; Peres, M.M.; Emanuelli, T. Postharvest UV-C irradiation stimulates the non-enzymatic and enzymatic antioxidant system of 'Isabel' hybrid grapes (*Vitis labrusca* × *Vitis vinifera* L.). *Int. Food Res. J.* **2017**, *102*, 738–747. [CrossRef] [PubMed]
113. Martínez-Sánchez, A.; Lozano-Pastor, P.; Artés-Hernández, F.; Artés, F.; Aguayo, E. Preharvest UV-C treatment improves the quality of spinach primary production and postharvest storage. *Postharvest Biol. Technol.* **2019**, *155*, 130–139. [CrossRef]
114. Artés-Hernández, F.; Escalona, V.H.; Robles, P.A.; Martínez-Hernández, G.B.; Artés, F. Effect of UV-C radiation on quality of minimally processed spinach leaves. *J. Sci. Food Agric.* **2009**, *89*, 414–421. [CrossRef]
115. Li, D.; Luo, Z.; Mou, W.; Wang, Y.; Ying, T.; Mao, L. ABA and UV-C effects on quality, antioxidant capacity and anthocyanin contents of strawberry fruit (*Fragaria ananassa* Duch.). *Postharvest Biol. Technol.* **2014**, *90*, 56–62. [CrossRef]
116. Martínez-Hernández, G.B.; Blanco, V.; Blaya-Ros, P.J.; Torres, R.; Domingo, R.; Artés-Hernández, F. Effects of UV-C on bioactive compounds and quality changes during shelf life of sweet cherry grown under conventional or regulated deficit irrigation. *Sci. Hort.* **2020**, *269*, 109398. [CrossRef]
117. Zhang, Q.; Yang, W.; Liu, J.; Liu, H.; Lv, Z.; Zhang, C.; Jiao, Z. Postharvest UV-C irradiation increased the flavonoids and anthocyanins accumulation, phenylpropanoid pathway gene expression, and antioxidant activity in sweet cherries (*Prunus avium* L.). *Postharvest Biol. Technol.* **2021**, *175*, 111490. [CrossRef]
118. Cardwell, G.; Bornman, J.F.; James, A.P.; Black, L.J. A Review of mushrooms as a potential source of dietary vitamin D. *Nutrients* **2018**, *10*, 1498. [CrossRef]
119. Black, L.J.; Lucas, R.M.; Sherriff, J.L.; Björn, L.O.; Bornman, J.F. In Pursuit of Vitamin D in Plants. *Nutrients* **2017**, *9*, 136. [CrossRef]
120. Arabi, S.M.; Ranjbar, G.; Bahrami, L.S.; Vafa, M.; Norouzy, A. The effect of vitamin D supplementation on hemoglobin concentration: A systematic review and meta-analysis. *Nutr. J.* **2020**, *19*, 11. [CrossRef]
121. Katz, J.; Yue, S.; Xue, W. Increased risk for COVID-19 in patients with vitamin D deficiency. *Nutrition* **2021**, *84*, 111106. [CrossRef]
122. Salemi, S.; Saedisomeoli, S.; Azimi, F.; Zolfigol, S.; Mohajerani, E.; Mohammadi, M.; Yaseri, M. Optimizing the production of vitamin D in white button mushrooms (*Agaricus bisporus*) using ultraviolet radiation and measurement of its stability. *LWT* **2021**, *137*, 110401. [CrossRef] [PubMed]
123. Jiang, Q.; Zhang, M.; Mujumdar, A.S. UV induced conversion during drying of ergosterol to vitamin D in various mushrooms: Effect of different drying conditions. *Trends Food Sci. Technol.* **2020**, *105*, 200–210. [CrossRef] [PubMed]
124. Costa, M.L.; Vicente, A.R.; Civello, P.M.; Chaves, A.R.; Martínez, G.A. UV-C treatment delays postharvest senescence in broccoli florets. *Postharvest Biol. Technol.* **2006**, *39*, 204–210. [CrossRef]
125. Carbonell-Bejerano, P.; Diago, M.P.; Martínez-Abaigar, J.; Martínez-Zapater, J.M.; Tardáguila, J.; Núñez-Olivera, E. Solar ultraviolet radiation is necessary to enhance grapevine fruit ripening transcriptional and phenolic responses. *BMC Plant Biol.* **2014**, *14*, 183. [CrossRef]
126. Dotto, M.; Casati, P. Developmental reprogramming by UV-B radiation in plants. *Plant Sci.* **2017**, *264*, 96–101. [CrossRef]

127. Pombo, M.A.; Rosli, H.G.; Martínez, G.A.; Civello, P.M. UV-C treatment affects the expression and activity of defense genes in strawberry fruit (*Fragaria × ananassa*, Duch.). *Postharvest Biol. Technol.* **2011**, *59*, 94–102. [CrossRef]
128. Liu, C.; Cai, L.; Han, X.; Ying, T. Temporary effect of postharvest UV-C irradiation on gene expression profile in tomato fruit. *Gene* **2011**, *486*, 56–64. [CrossRef]
129. Li, T.; Yamane, H.; Tao, R. Preharvest long-term exposure to UV-B radiation promotes fruit ripening and modifies stage-specific anthocyanin metabolism in highbush blueberry. *Hort. Res.* **2021**, *8*, 67. [CrossRef]
130. Ortiz Araque, L.C.; Ortiz, C.M.; Darré, M.; Rodoni, L.M.; Civello, P.M.; Vicente, A.R. Role of UV-C irradiation scheme on cell wall disassembly and surface mechanical properties in strawberry fruit. *Postharvest Biol. Technol.* **2019**, *150*, 122–128. [CrossRef]
131. Atkinson, N.J.; Urwin, P.E. The interaction of plant biotic and abiotic stresses: From genes to field. *J. Exp. Bot.* **2012**, *63*, 3523–3544. [CrossRef]
132. Pastori, G.M.; Foyer, C.H. Common components, networks, and pathways of cross-tolerance to stress. The central role of 'redox' and abscisic acid-mediated controls. *Plant Physiol.* **2002**, *129*, 460–468. [CrossRef] [PubMed]
133. González-Aguilar, G.; Wang, C.Y.; Buta, G.J. UV-C irradiation reduces breakdown and chilling injury of peaches during cold storage. *J. Sci. Food Agric.* **2004**, *84*, 415–422. [CrossRef]
134. Vicente, A.R.; Pineda, C.; Lemoine, L.; Civello, P.M.; Martinez, G.A.; Chaves, A.R. UV-C treatments reduce decay, retain quality, and alleviate chilling injury in pepper. *Postharvest Biol. Technol.* **2005**, *35*, 69–78. [CrossRef]
135. Liu, C.; Jahangir, M.M.; Ying, T. Alleviation of chilling injury in postharvest tomato fruit by preconditioning with ultraviolet irradiation. *J. Sci. Food Agric.* **2012**, *92*, 3016–3022. [CrossRef] [PubMed]
136. Cisneros-Zevallos, L.; Jacobo-Velazquez, D.A. Controlled Abiotic Stresses Revisited: From Homeostasis through Hormesis to Extreme Stresses and the Impact on Nutraceuticals and Quality during Pre- and Postharvest Applications in Horticultural Crops. *J. Agric. Food Chem.* **2020**, *68*, 11877–11879. [CrossRef] [PubMed]
137. Ortega-Hernández, E.; Nair, V.; Serrano-Sandoval, S.N.; Welti-Chanes, J.; Cisneros-Zevallos, L.; Jacobo-Velázquez, D.A. Wounding and UVB light synergistically induce the postharvest biosynthesis of indicaxanthin and betanin in red prickly pears. *Postharvest Biol. Technol.* **2020**, *167*, 111247. [CrossRef]
138. Civello, P.M.; Villarreal, N.; Gómez Lobato, M.E.; Martínez, G.A. Physiological effects of postharvest UV treatments: Recent progress. *Stewart Postharv. Rev.* **2014**, *3*, 8.
139. Esua, O.J.; Chin, N.L.; Yusof, Y.A.; Sukor, R. A review on individual and combination technologies of UV-C radiation and ultrasound in postharvest handling of fruits and vegetables. *Processes* **2020**, *8*, 1433. [CrossRef]
140. Zhang, W.; Jiang, W. UV treatment improved the quality of postharvest fruits and vegetables by inducing resistance. *Trends Food Sci. Technol.* **2019**, *92*, 71–80. [CrossRef]
141. Guerrero-Beltrán, J.A.; Barbosa-Cánovas, G.V. Review: Advantages and limitations on processing foods by UV light. *Food Sci. Technol. Int.* **2004**, *10*, 137–147. [CrossRef]
142. Chacha, J.S.; Zhang, L.; Ofoedu, C.E.; Suleiman, R.A.; Dotto, J.M.; Roobab, O.; Agunbiade, A.O.; Duguma, H.T.; Mkojera, B.T.; Hossaini, S.M.; et al. Revisiting non-thermal food processing and preservation Methods—action mechanisms, pros, and cons: A technological update (2016–2021). *Foods* **2021**, *10*, 1430. [CrossRef] [PubMed]
143. Civello, P.M.; Vicente, A.R.; Martínez, G.A. UV-C technology to control postharvest diseases of fruits and vegetables. In *Recent Advances in Alternative Postharvest Technologies to Control Fungal Diseases in Fruits and Vegetables*; Troncoso-Rojas, R., Tiznado-Hernández, M.E., González-León, A., Eds.; Transworld Research Network: Trivandrum, India, 2006; pp. 71–102.
144. Vicente, A.R.; Repice, B.; Costa, M.L.; Martínez, G.A.; Civello, P.M.; Sozzi, G.O. Maintenance of fresh boysenberry fruit quality with UV-C light and heat treatments combined with low storage temperature. *J. Hort. Sci. Biotechnol.* **2004**, *79*, 246–251. [CrossRef]
145. Piga, A.; D'Hallewin, G.; D'Aquino, S.; Agabbio, M. Influence of film wrapping and UV irradiation on cactus pear quality after storage. *Packag. Technol. Sci.* **1997**, *10*, 59–68. [CrossRef]
146. Maharaj, R.; Arul, J.; Nadeau, P. Effect of photochemical treatment in the preservation of fresh tomato (*Lycopersicon esculentum* cv. Capello) by delaying senescence. *Postharvest Biol. Technol.* **1999**, *15*, 13–23. [CrossRef]
147. Mditshwa, A.; Magwaza, L.S.; Tesfay, S.Z.; Mbili, N.C. Effect of ultraviolet irradiation on postharvest quality and composition of tomatoes: A review. *J. Food Sci. Technol.* **2017**, *54*, 3025–3035. [CrossRef]
148. Rodoni, L.M.; Concellón, A.; Chaves, A.R.; Vicente, A.R. Use of UV-C Treatments to Maintain Quality and Extend the Shelf Life of Green Fresh-cut Bell Pepper (*Capsicum annuum* L.). *J. Food Sci.* **2012**, *77*, C632–C639. [CrossRef]
149. Rodoni, L.M. Evaluación de Metodologías No Contaminantes para Mantener la Calidad y Extender la Vida Poscosecha de Pimientos (*Capsicum annuum* L.) Frescos Cortados. Ph.D. Thesis, Universidad Nacional de La Plata, La Plata, Argentina, 2014.
150. Bisht, B.; Bhatnagar, P.; Gururani, P.; Kumar, V.; Tomar, M.S.; Sinhmar, R.; Kumar, S. Food irradiation: Effect of ionizing and non-ionizing radiations on preservation of fruits and vegetables–a review. *Trends Food Sci. Technol.* **2021**, *114*, 372–385. [CrossRef]
151. Neugart, S.; Schreiner, M. UVB and UVA as eustressors in horticultural and agricultural crops. *Sci. Hortic.* **2018**, *234*, 370–381. [CrossRef]
152. Cote, S.; Rodoni, L.M.; Miceli, E.; Concellón, A.; Civello, P.M.; Vicente, A.R. Effect of radiation intensity on the outcome of postharvest UV-C treatments. *Postharvest Biol. Technol.* **2013**, *83*, 83–89. [CrossRef]

153. Bauer, S.; Holtschmidt, H.; Otta, G. Ultraviolet germicidal efficacy as a function of pulsed radiation parameters studied by spore film dosimetry. *J. Photochem. Photobiol. B. Biol.* **2018**, *178*, 69–75. [CrossRef]
154. Marquenie, D.; Geeraerd, A.H.; Lammertyn, J.; Soontjens, C.; Van Impe, J.F.; Michiels, C.W.; Nicolaı, B.M. Combinations of pulsed white light and UV-C or mild heat treatment to inactivate conidia of *Botrytis cinerea* and *Monilia fructigena*. *Int. J. Food Microbiol.* **2003**, *85*, 185–196. [CrossRef]
155. Cleaver, J.E. Photoreactivation. *DNA Repair.* **2003**, *2*, 629–638. [CrossRef]

foods

MDPI

Perspective

Non-Thermal Technologies as Tools to Increase the Content of Health-Promoting Compounds in Whole Fruits and Vegetables While Retaining Quality Attributes

Daniel A. Jacobo-Velázquez [1,*] and Jorge Benavides [2]

[1] Tecnologico de Monterrey, Escuela de Ingeniería y Ciencias, Av. General Ramón Corona 2514, Zapopan C.P. 45201, Jalisco, Mexico
[2] Tecnologico de Monterrey, Escuela de Ingeniería y Ciencias, Av. Eugenio Garza Sada 2501, Monterrey C.P. 64849, Nuevo Leon, Mexico; jorben@tec.mx
* Correspondence: djacobov@tec.mx; Tel.: +52-312-119-1650

Abstract: Fruits and vegetables contain health-promoting compounds. However, their natural concentration in the plant tissues is low and in most cases is not sufficient to exert the expected pharmacological effects. The application of wounding stress as a tool to increase the content of bioactive compounds in fruits and vegetables has been well characterized. Nevertheless, its industrial application presents different drawbacks. For instance, during the washing and sanitizing steps post-wounding, the primary wound signal (extracellular adenosine triphosphate) that elicits the stress-induced biosynthesis of secondary metabolites is partially removed from the tissue. Furthermore, detrimental reactions that affect the quality attributes of fresh produce are also activated by wounding. Therefore, there is a need to search for technologies that emulate the wound response in whole fruits and vegetables while retaining quality attributes. Herein, the application of non-thermal technologies (NTTs) such as high hydrostatic pressure, ultrasound, and pulsed electric fields are presented as tools for increasing the content of health-promoting compounds in whole fruits and vegetables by inducing a wound-like response. The industrial implementation and economic feasibility of using NTTs as abiotic elicitors is also discussed. Whole fruits and vegetables with enhanced levels of bioactive compounds obtained by NTT treatments could be commercialized as functional foods.

Keywords: controlled abiotic stresses; nutraceuticals; elicitation; functional foods; whole fruits and vegetables; stress-induced biosynthesis; secondary metabolites; innovative technologies; non-thermal technologies; health-promoting compounds

Citation: Jacobo-Velázquez, D.A.; Benavides, J. Non-Thermal Technologies as Tools to Increase the Content of Health-Promoting Compounds in Whole Fruits and Vegetables While Retaining Quality Attributes. *Foods* **2021**, *10*, 2904. https://doi.org/10.3390/foods10122904

Academic Editors: Marina Cano Lamadrid and Francisco Artés-Hernández

Received: 20 September 2021
Accepted: 9 November 2021
Published: 23 November 2021

1. Introduction

Fruits and vegetables are essential sources of health-promoting compounds; thus, their consumption aids in the prevention of chronic and degenerative diseases [1]. However, the content of bioactive compounds in most fruits and vegetables is insufficient to exert the pharmacological effect expected of nutraceutical and functional foods [2]. Therefore, there is a need to find innovative technologies that increase the content of secondary metabolites with health-promoting properties in plant tissues. In this context, the application of controlled abiotic stresses during postharvest has been recognized as an effective strategy to elicit the secondary metabolism of fruits and vegetables, leading to the accumulation of bioactive ompounds [3,4]. Of the different postharvest abiotic stresses evaluated to induce the biosynthesis of plant bioactives, wounding stress has been recognized as one of the postharvest treatments with the highest impact on the activation of primary and secondary plant metabolism, generating significant increases in the accumulation of secondary metabolites [5–7]. However, the application of wounding stress at industrial scales presents different drawbacks. For instance, during the washing and sanitizing steps post-wounding, which are needed to ensure the safety of fresh-cut produce, the primary wound signal (extracellular adenosine triphosphate, ATP) that elicits the stress-induced biosynthesis of secondary metabolites is partially removed from the tissue [8]. Furthermore, detrimental oxidative reactions that affect the quality attributes of the fresh produce (i.e., color, flavor, and texture) are also activated by wounding [9]. Therefore, there is a need to search for technologies that emulate the wound response in whole fruits and vegetables while retaining quality attributes.

The application of non-thermal technologies (NTTs) as abiotic elicitors to induce the biosynthesis of bioactive compounds in plant foods was proposed by Jacobo-Velázquez et al. [10]. This proposal was based on previous reports where elicitation by NTTs was observed in different plant cell cultures [11–13]; thus, it was hypothesized that the results could be extrapolated to fruits and vegetables. In their initial proposal, Jacobo-Velázquez et al. [10] presented a hypothetical model indicating that NTTs could elicit the biosynthesis of secondary metabolites by inducing a wound-like response in plant tissues. In this context, it was stated that high hydrostatic pressure (HHP), ultrasound (US), and pulsed electric fields (PEF) generate cell membrane disruption by different driving forces [10]. The scientific community responded well to the concept of using NTTs as elicitors to induce the biosynthesis of bioactive compounds in plant foods, and several research groups have evaluated the potential of HHP, US, and PEF to induce the biosynthesis of secondary metabolites (i.e., carotenoids, phenolics, and glucosinolates) in different whole fruits and vegetables [14,15]. Furthermore, the hypothetical model proposed by Jacobo-Velázquez et al. [10] explaining the physiological mechanisms governing the biosynthesis of bioactive compounds induced by NTTs was recently revisited by López-Gámez et al. [14].

In the present article, the application of NTTs such as HHP, US, and PEF are presented as tools to increase the content of health-promoting compounds in whole fruits and vegetables by inducing a wound-like response. The main effects of NTTs on secondary metabolite biosynthesis and the quality attributes of whole fresh produce are discussed based on recently published scientific literature. Likewise, the industrial implementation and economic feasibility of using NTTs as abiotic elicitors is also discussed. Finally, research needs in this emerging research area are highlighted.

2. Wounding Stress as a Tool to Induce the Biosynthesis of Health-Promoting Compounds in Horticultural Crops: Mechanism and Drawbacks in Industrial Practice

As mentioned in the introduction, applying wounding stress to fruits and vegetables postharvest is an effective strategy to induce the biosynthesis and accumulation of health-promoting compounds. However, wounding stress presents different drawbacks when evaluating the feasibility of its application at industrial scale. To better explain these drawbacks and search for alternative technologies that could emulate the wound response in horticultural crops, the physiological mechanisms governing the wound-induced biosynthesis of secondary metabolites in the plant cell as well as the main drawbacks of wounding stress application as a postharvest technology are described herein and summarized in Figure 1.

As an immediate response to wounding stress, adenosine triphosphate (ATP) is released from the cytoplasm of damaged cells, serving as the primary signal to elicit the wound response [8,16,17]. Subsequently, ATP binds to receptors from intact cells, eliciting the production of secondary stress-signaling molecules such as reactive oxygen species (ROS), ethylene, and jasmonic acid (Figure 1A). These secondary signals trigger the expression of transcription factors, genes, and enzymes related to the biosynthesis of health-promoting compounds [4]. The accumulation of secondary metabolites results from a balance between their biosynthesis and their utilization rates to accomplish a specific physiological function in the plant tissue. For instance, soluble phenolic compounds are biosynthesized in wounded plant cells to serve as precursors for the biosynthesis of lignin or suberin, which prevents water loss in the wounded tissue [18–20]. Likewise, glucosinolates in wounded tissue also serve as precursors for isothiocyanate production; these are toxic for insects and protecting plant tissues from pathogen attack [21,22].

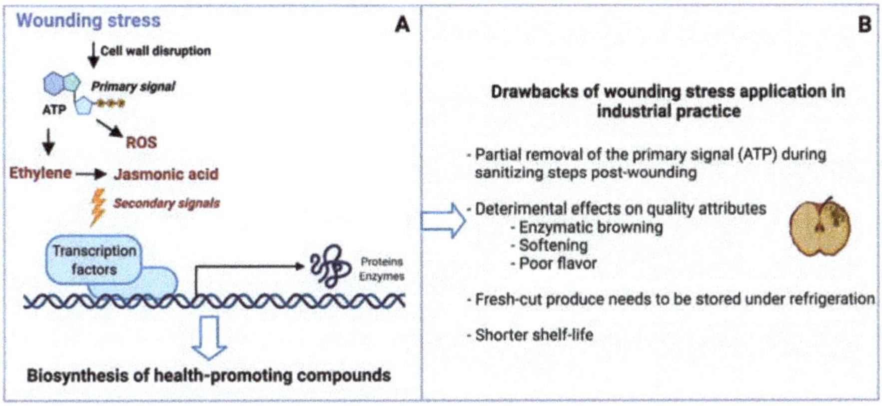

Figure 1. Wounding stress as a tool to induce the biosynthesis of health-promoting compounds in horticultural crops: physiological mechanism involved in stress response, (**A**); drawbacks in industrial practice, (**B**). Abbreviations: ATP, adenosine triphosphate; ROS, reactive oxygen species.

The industrial implementation of wounding stress as a technology to produce minimally processed fruits and vegetables with enhanced health benefits presents different drawbacks that need to be addressed (Figure 1B). For instance, as a common practice to ensure the safety of fresh-cut products, the tissue is subjected to washing and sanitizing procedures after cutting. These steps partially remove ATP released from wounded cells, which serves as the primary wound signal (Figure 1A). Therefore, the partial removal of ATP can impede the accumulation of bioactive compounds in the tissue. On this note, sanitizing procedures applied only before peeling and cutting are recommended in order to diminish this drawback [8]. Moreover, wounding stress induces changes in the physiology of horticultural crops, resulting in a loss of quality attributes of fresh produce. For example, cutting fruits and vegetables alters flavor by modifying their acid and sugar content and by inducing synthesis of volatile organic compounds. Furthermore, wounding stress favors the biosynthesis, degradation, and oxidation of secondary metabolites, and induces the depolymerization and degradation of cell membranes, resulting in color changes and texture softening [9]. All of these physiological modifications in the tissue induced by wounding result in a loss of homeostasis and shorter shelf-life of the product. Thus, minimally processed fruits and vegetables need to be commercialized under refrigeration in order to reduce metabolic activity and reduce the loss of quality attributes of the tissue. Furthermore, incorporating preservatives in fresh-cut fruits and vegetables to extend their shelf-life is also a common practice that needs to be reduced through consumer demand for clean labels.

3. Scientific Evidence for Non-Thermal Technologies (NTTs) That Emulate Wound-Induced Biosynthesis of Bioactive Compounds in Whole Fruits and Vegetables

As described in the previous section, wounding stress induces high activation of the secondary metabolism of horticultural crops, leading to the accumulation of bioactive compounds. However, the industrial application of wounding to generate fruits and vegetables with enhanced health benefits presents different drawbacks associated with a partial removal of the primary wound signal (ATP) due to sanitizing steps (decreasing elicitation), as well as an increase in oxidative reactions that results in quality loss in the fresh produce. Therefore, it is highly relevant to search for technologies that emulate the wound-induced biosynthesis of bioactive compounds while decreasing the detrimental effects of cutting. In this context, technologies that can increase the health benefits of whole fruits and vegetables are highly valuable and can make for very attractive commercialized functional food products.

The application of NTTs (HHP, US, and PEF) as abiotic elicitors to induce the biosynthesis of bioactive compounds in whole fruits and vegetables has been recently evaluated in different studies. Jacobo-Velázquez et al. [10] proposed that this stress response is elicited through a wound-like response mainly due to the cell membrane disruption induced by NTTs through different driving forces. For PEF and US, cell membrane disruption is produced by an electric potential and by a pressure/temperature gradient in the membrane, respectively. On the other hand, membrane permeability via HHP is induced by a pressure gradient in the plasma membrane [10,23]. The model presented by Jacobo-Velázquez et al. [10] hypothesizes that when membrane permeability is induced, ATP is released from damaged cells and serves as primary signal to induce a wound-like response (Figure 1A). A key aspect to consider is the NTT processing conditions selected, which should be moderate in order to elicit the stress response while preventing plant cell death. In the specific case of HHP, processing conditions should not exceed 100 MPa, while for US, high-intensity treatments with high energy and low frequency (20 to 100 kHz) are adequate for elicitation. Finally, moderate-intensity PEF (MIPEF, 0.5–5 kV cm^{-1}, 1–20 kJ kg^{-1}) generates reversible damage and elicitation in the plant cell [10].

In the following sections, recent scientific reports describing the effects of HHP, US, and PEF as elicitors to induce the biosynthesis of health-promoting compounds in different whole fruits and vegetables are summarized. Likewise, the effects of NTTs on the quality attributes of whole fresh produce are also described.

3.1. High Hydrostatic Pressure (HHP)

The results from previous studies evaluating the effects of HHP treatment on the biosynthesis of health-promoting compounds and the physiological and quality attributes of whole mangoes and carrots are summarized in Table 1.

3.1.1. Mango (*Mangifera indica*)

Hu et al. [24] evaluated the effects of HHP applied at 20, 40, 60, and 80 MPa for 10 min on the biosynthesis and accumulation of carotenoids, phenolics, and ascorbic acid in whole mangoes stored for 16 days at 13 °C. The authors reported that HHP increased the biosynthesis of carotenoids at the transcriptional level, whereas samples treated at 20 MPa presented the highest carotenoid content (43.7% higher than the non-treated samples) during storage. Furthermore, treating mango at 40 MPa resulted in higher ascorbic acid retention, whereas samples treated at 20 MPa presented higher flavonoid content. Similarly, Álvarez-Virrueta et al. [25] and Ortega et al. [26] evaluated the effects of HHP treatment (15, 30, or 60 MPa for 10 or 20 min) on the accumulation of carotenoids and phenolics in whole mangoes stored for 14 days at 25 °C. The authors found that treating whole mangoes at 30 and 60 MPa for 20 min increased the ascorbic acid content by 30.7–46.1% after storage. Likewise, HHP treatment at 15 MPa for 10 min induced a higher accumulation of total carotenoids (100%) and phenolics (41.2%) compared with the control at 14 days of storage. These results indicate that HHP under mild conditions (15 and 20 MPa for 10 min) induced the biosynthesis of carotenoids and phenolics in whole mangoes [24–26].

Table 1. Effects of high hydrostatic pressure (HHP) on the biosynthesis of health-promoting compounds and the quality and physiological attributes of whole fruits and vegetables.

Horticultural Crop	HHP Processing and Storage Conditions Evaluated	Main Findings		References
		Effects on the Biosynthesis of Health-Promoting Compounds	Effects on Quality and Physiological Attributes	
Mango (*Mangifera indica*)	Whole mangoes (cv. Tainong) were subjected to 20, 40, 60, and 80 MPa for 10 min. Samples were stored for 16 days at 13 °C and ~85% relative humidity (RH).	HHP increased carotenoid biosynthesis at the transcriptional level. Samples treated at 20 MPa showed 43.7% higher total carotene content after storage. HHP increased ascorbic acid retention during storage. Samples treated with 40 MPa showed higher ascorbic acid retention. Except for the 40 MPa treatment, HHP-treated samples showed higher flavonoid content. In general, 20 MPa treatment resulted in the highest accumulation of total phenolics and carotenoids, while 40 MPa treated mangoes showed higher levels of ascorbic acid.	HHP treatment reduced respiration rate by 26.62, 20.25, 32.72, and 41.81%, for mangoes treated at 20, 40, 60, and 80 MPa, respectively, compared with the control. HHP treated samples showed higher a * (redness) values during ripening compared with the control. HHP treatment decreased firmness at the initial storage time (1 day). However, after 7 days, no significant difference was observed in firmness values between HHP-treated samples and the control. HHP-treated mangoes showed higher titratable acidity (from days 7 to 16), higher reducing sugar content and higher moisture loss.	[24]
	Whole mangoes (cv. Ataulfo) were treated with HHP at 15, 30, or 60 MPa for 10 or 20 min. Non-treated fruit was used as the control. Samples were stored for 14 days at 25 °C and 85–90% of RH.	HHP treatments at 30 and 60 MPa for 20 min increased ascorbic acid content by 30.7–46.1% compared to the control before storage. HHP treatment at 15 MPa for 10 min induced a higher accumulation of total carotenoids (100%) and phenolics (41.2%) compared with the control at 14 days of storage. Results indicated that low-pressure treatments induced the biosynthesis of nutraceuticals in mango.	HHP did not inhibit the ripening process of mango. At the climacteric peak (9 days), the HHP-treated samples showed a lower respiration rate. Ethylene production was lower in samples treated at 30 and 60 MPa, regardless of the pressurization time (10 or 20 min). HHP-treated mangoes showed lower firmness values during storage, whereas at 14 days, no significant difference was detected between HHP samples and the control. HHP treatment increased moisture loss. At 9 days of storage, the pulp of HHP-treated mangoes showed a more intense orange color than non-treated samples.	[25,26]
Carrot (*Daucus carota*)	Whole carrots were treated at 60 or 100 MPa for the come-up time (CUT). Samples were stored for 3 days at 15 °C.	Immediately after HHP application, carrots treated at 100 MPa showed an increase of free (5-*O*-caffeoylquinic acid, 63.9% and 3,4-di-*O*-feruloylquinic acid, 228.6%) and bound (p-coumaric acid, 82.6%) phenolics. On day 1, the 60 MPa samples showed accumulation of 4,5-di-*O*-caffeoylquinic acid (60.2%) and isocoumarin (98.9%), whereas the 100 MPa samples presented increases of chlorogenic acid (291.2%) and 3,4-di-*O*-feruloylquinic acid (466.1%). On day 2, an increase in bound phenolics (rutin, 85.5% and p-coumaric acid, 214.7%) was observed in samples treated at 60 MPa. On day 3, the 100 MPa samples presented higher quercetin (371.2%) content.	On day 2, the 60 MPa and 100 MPa samples showed 380.2% and 139.7% higher phenylalanine-ammonia lyase (PAL) activity, respectively, than the control. On day 3, the 60 MPa samples presented 212% higher PAL activity than the control, whereas no significant difference was observed between the 100 MPa samples and the control. During storage, higher ethylene production and respiration rate were detected in the 60 and 100 MPa samples compared with the control.	[27]
	HHP treatments were applied for the CUT as a single pulse or multi-pulse (2P, 3P, and 4P). In addition, a single sustained treatment (5 min) was applied at 60 or 100 MPa. Samples were stored for 48 h at 15 °C.	Immediately after HHP treatment, the extractability of phenolics increased by 66.65% and 80.77% in 3P 100 MPa and 4P 60 MPa samples, respectively. After storage, CUT 60 MPa samples accumulated free (163.05%) and bound (36.95%) phenolics. Total xanthophylls increased by 27.16% after CUT 60 MPa treatment, whereas no changes were observed after storage.	The authors did not report quality characteristics and physiological measurements of the samples.	[28]

The effect of HHP treatments on the quality and physiological attributes of whole mangoes was also reported in the studies described above [24–26]. Pressurization did not inhibit the ripening process of mango, but reduced the respiration rate by 26.62, 20.25, 32.72, and 41.81% for the whole tissue treated at 20, 40, 60, and 80 MPa, respectively, as compared with the control. In addition, at the climacteric peak, whole mangos treated with HHP showed a lower respiration rate [24]. Likewise, HHP-treated mangoes showed higher a* (redness) values during ripening than the control. Moreover, pressure treatments decreased firmness at the initial storage time (1 day), whereas after 7 days of storage no significant difference was observed between HHP treated samples and the control. The immediate decrease in firmness was attributed to the loss of turgor induced by vacuolar content leakage upon pressurization [24]. Likewise, the increase in firmness of HHP-treated samples during storage was attributed to an increase in pectinmethylesterase (PME) activity, which upon contact with the substrate results in the formation of pectates and pectin gelation, promoting firmness [24]. Pressurized whole mangoes showed higher titratable acidity (from days 7 to 16), higher reducing sugar content, and higher moisture loss during storage [24]. Furthermore, ethylene production was lower in whole mangoes treated at 30 and 60 MPa for 10 or 20 min. Finally, during storage, the pulp of HHP treated mangoes showed a more intense orange color, associated with the biosynthesis of carotenoids during ripening [25,26,29].

3.1.2. Carrot (*Daucus carota*)

Regarding carrots, Viacava et al. [27] evaluated the effects of HHP treatment and storage time (15 °C for 3 days) on the biosynthesis of health-promoting compounds and physiological attributes of the whole tissue. Carrots were treated at 60 or 100 MPa for the time needed to reach the desired pressure (come-up time, CUT). Immediately after HHP processing, an increase of free (5-*O*-caffeoylquinic acid (63.9%) and 3,4-di-*O*-feruloylquinic acid (228.6%)) and bound (*p*-coumaric acid (82.6%)) phenolics was detected. Furthermore, whole carrots treated at 60 MPa and stored for 1 day showed accumulation of the free phenolic 4,5-di-*O*-caffeoylquinic acid (60.2%), whereas samples treated at 100 MPa showed increases of chlorogenic acid (291.2%) and 3,4-di-*O*-feruloylquinic acid (466.1%). On day 2, samples treated at 60 MPa showed an increase in bound phenolics (rutin (85.5%) and *p*-coumaric acid (214.7%)). Finally, at the end of the storage period (day 3), samples treated at 100 MPa showed higher quercetin content (371.2%) than the control. Similarly, Viacava et al. [28] evaluated the effect of HHP treatments (60 or 100 MPa) applied for a single sustained pulse (5 min) or the CUT as a single pulse (P) or multi-pulse (2P, 3P, and 4P) on the accumulation of phenolic compounds and carotenoids after 48 h of storage at 15 °C. The authors reported that immediately after HHP treatment, the extractability of phenolics increased by 66.65% and 80.77% in 3P 100 MPa and 4P 60 MPa samples, respectively. Likewise, whole carrots treated for the CUT 60 MPa accumulated free (163.05%) and bound (36.95%) phenolics after storage. Furthermore, the content of total xanthophylls increased by 27.16% immediately after CUT 60 MPa treatment, whereas no changes in carotene concentration were observed after storage.

Regarding the effects of HHP treatment on the physiological attributes of carrots, after 2 days of storage (15 °C), whole carrots pressurized at 60 MPa and 100 MPa showed 380.2% and 139.7% higher phenylalanine-ammonia lyase (PAL) activity, respectively, as compared with the control, whereas at day 3, samples treated at 60 MPa showed 212% higher PAL activity than the control, with no significant difference observed in PAL activity between the 100 MPa samples and the control. HHP-induced PAL activation confirmed that pressure induced the biosynthesis of phenolic compounds in whole carrots. During storage, samples treated at 60 and 100 MPa showed higher ethylene production and respiration rates than the control [27]. These results are in contrast with the results reported for mangoes, where HHP reduced respiration and ethylene production, suggesting that the HHP effect on the physiological attributes of fresh produce is tissue-dependent.

One of the main physiological differences between carrots and mangoes that could result in a different postharvest behavior as a response to HHP is that carrot is a non-climacteric vegetable, whereas mango is a climacteric fruit. Climacteric and non-climacteric classification of fruits and vegetables depends on their ripening patterns and the presence of a burst in ethylene production and respiration rate during postharvest, also known as the climacteric peak [9]. Results from the literature cited herein indicate that HHP affects climacteric and non-climacteric produce differently. Interestingly, if a wound-like response is expected due to HHP application in mango, an accelerated climacteric peak and an associated increased loss of quality characteristics should be occurring in the pressurized tissue. However, in contrast with wounding, HHP reduced ethylene production and respiration of the fruit, preserving its quality characteristics during storage and overcoming this major drawback of wounding stress application.

3.2. *Ultrasound (US)*

The effects of US treatment on the biosynthesis of health-promoting compounds and physiological attributes of whole fruits and vegetables such as lettuce, strawberry, carrot, and broccoli have been previously reported (Table 2). Furthermore, the application of US has also been explored as a pretreatment in seeds to improve sprouting indexes and improve the content of health-promoting compounds and the quality characteristics of sprouts commercialized as a ready-to-eat vegetable (Table 2).

3.2.1. Lettuce (*Lactuca sativa*)

Yu et al. [30] evaluated the effects of US treatment (25 kHz) at an acoustic power density of 26 W/L for 1–3 min on whole leaf lettuce. The US-treated samples were stored at room temperature for 150 h. Immediately after US treatment, no significant difference in total phenolics was detected between the control and the processed lettuce. At 60 h of storage, samples treated for 1 min of US showed 22.50% higher phenolics than the control, which was accompanied by an increase in PAL activity, confirming that US induced the activation of the phenylpropanoid metabolism. Samples treated with US showed higher firmness during storage, while the color was not affected [30].

3.2.2. Strawberry (*Fragaria x ananassa*)

Strawberries were treated with US (33 kHz, 60 W, 25 °C) for different times (0, 10, 20, 30, 40, and 60 min) and stored for 15 days at 4 °C [31]. US-treated strawberries showed higher ascorbic acid retention in the fruit, with decreases of 54.68, 36.68, 35.57, and 32.20% observed for treatment times of 0, 10, 20, and 30 min, respectively. US increased total phenolics on day 1. As the US time increased from 0 to 40 min, a significant increase (7.91%) in total phenolics was quantified. However, 60 min US treatment decreased phenolic content by 7.91%. Except for 60 min US treatment, total phenolics increased in both US treated samples and the control.

Regarding the effect of HHP on the physiological and quality attributes of whole strawberries, at day 15 an increase of 13.84, 4.32, 3.20, 3.50, 2.93, and 13.88% in pH value was detected in samples subjected to US for 0, 10, 20, 30, 40, and 60 min, respectively. At day 15, total soluble solids increased by 24.66, 21.43, 18.23, 17.07, 18.84, and 30.05% for US treatment of 0, 10, 20, 30, 40, and 60 min, respectively. Immediately after treatment, color change (ΔE^*) values increased by 4.61, 5.25, 6.16, 7.94, and 11.29% for US treatments of 10, 20, 30, 40, and 60 min, respectively. Furthermore, US-treated strawberries showed better color retention during storage, and fruit firmness was better retained in US (20–30 min) treated samples. Microbial load also decreased with US treatment. Interestingly, as observed for mango treated with HHP, strawberries subjected to US showed improved texture during storage. This observation suggests that US could also activate PME, likely by promoting contact between substrate and enzyme, leading to the production of pectates and pectin gelation.

Table 2. Effects of ultrasound (US) on the biosynthesis of health-promoting compounds and on the quality and physiological attributes of whole fruits and vegetables.

Horticultural Crop	US Processing and Storage Conditions Evaluated	Main Findings		References
		Effects on the Biosynthesis of Health-Promoting Compounds	Effects on Quality and Physiological Attributes	
Romaine lettuce (*Lactuca sativa*, var. *longifolia*)	Whole leaf lettuce was treated with US (25 kHz) at an acoustic power density of 26 W/L for 1–3 min. Samples were stored at room temperature for 150 h.	Immediately after treatment, no significant difference in total phenolics was detected between the control and US-treated lettuce. At 60 h of storage, samples treated for 1 min of US showed 22.50% higher phenolics than the control. Lettuce treated with US for 2 and 3 min did not show significant increases in phenolics compared with the control. After 90 h, no significant difference in phenolic content was detected between the control and US-treated samples.	In the first 30 h, no phenylalanine ammonia-lyase (PAL) activation was detected due to US treatment. At 60 h, samples treated with 2 and 3 min of US showed the highest PAL activity. US-treated lettuce showed higher firmness (maximum force, N) than the control (water washed) immediately after treatment and during storage. No significant difference was detected in the color characteristics of lettuce between treatments.	[30]
Strawberry (*Fragaria x ananassa*, cv. chandler)	US treatment (33 kHz, 60 W, 25 °C) was applied to strawberries for different times (0, 10, 20, 30, 40, and 60 min). Samples were stored for 15 days at 4 °C.	US treatments (10–30 min) increased ascorbic acid retention in the tissue, where decreases of 54.68, 36.68, 35.57, and 32.20% were observed for treatment times of 0, 10, 20, and 30 min, respectively. US treatments of 40 and 60 min resulted in a decrease in the retention of ascorbic acid. US increased total phenolics on day 1. As the US time increased from 0 to 40 min, a significant increase (7.91%) in total phenolics was quantified. However, 60 min US treatment decreased phenolic content by 7.91%. Except for 60 min US treatment, total phenolics increased in US-treated samples as well as in the control.	On day 15, an increase of 13.84, 4.32, 3.20, 3.50, 2.93, and 13.88% in pH value was detected in samples subjected to US for 0, 10, 20, 30, 40, and 60 min. respectively. At day 15, total soluble solids increased by 24.66, 21.43, 18.23, 17.07, 18.84, and 30.05% for US treatment of 0, 10, 20, 30, 40, and 60 min, respectively. Immediately after treatment, the value of ΔE^* increased by 4.61, 5.25, 6.16, 7.94, and 11.29% for 10, 20, 30, 40, and 60 min US treatments, respectively. US-treated strawberries had better color retention during storage. Fruit firmness was better retained in US (20–30 min)-treated samples. Microbial load also decreased with US treatment.	[31]
Carrot (*Daucus carota*)	Whole carrots were treated with US (frequency 24 kHz, amplitude 100 mm) for 5 min at 20 °C. Samples were stored at 20 °C for 3 days.	As an immediate response to US, carrots showed 21.1% higher levels of total carotenoids as compared with the control. After storage, carotenoid content decreased. However, US-treated samples showed a lower decrease in carotenoid (−7.6%) than the control (−16.4%). US treatment induced an immediate decrease in total phenolic content (−62.5%) compared with the control. After 3 days of storage, phenolic content in US-treated samples increased by 129.2%, whereas for control samples the phenolic content did not change. Chlorogenic acid was the main phenolic compound that increased in US-treated whole carrots, which showed 41.8% higher content at 1 day of storage than the control.	On day 3 of storage, US-treated whole carrots showed the highest isocoumarin content, 164.0% higher than the control and 240.8% higher than the control before storage. Isocoumarin exerts a bitter flavor in carrots at concentrations of 200 mg/kg [LaFuente et al. 1996]. However, the levels detected in sonicated carrots were below the threshold of sensory perception (40 mg/kg). US-treated samples showed a 2.04-fold increase in expression of the *PAL* gene immediately after treatment. During the first 0.5 days of storage, PAL activity increased in US-treated carrots, showing 64.8% higher activity than the control before storage. US induced an immediate increase in respiration rate. US-treated whole carrots showed 27.9%, 66.0%, and 162.0%, higher levels of volatile organic compounds (indicators of ethylene production) at 0.5, 1.5, and 2 days of storage, respectively.	[32]
Broccoli (*Brassica oleracea* L. var. *Italica*, cv. Tlaloc®)	Broccoli florets were treated with US (20 min, frequency 24 kHz, amplitude 100 μm). US-treated broccoli florets were subjected to exogenous methyl jasmonate (MJ, 250 ppm) and/or ethylene (ET, 1000 ppm) during storage. Samples were stored for 3 days at 15 °C.	As an immediate response to US treatment, the extractability of glucosinolates (glucoraphanin (795%), glucobrassicin (78.6%), and 4-hydroxy glucobrassicin (153%)) and phenolics (1-sinapoyl-2-feruloylgentiobiose (57.23%)) was increased. At 3 days, samples treated with US and MJ showed the highest accumulation of gluconasturtiin (755.9%), neoglucobrassicin (232.8%), 4-hydroxy glucobrassicin (187.1%), glucoerucin (111.92%), 1,2,2-trisinapoylgentiobiose (136.7%), 3-*O*-caffeoylquinic acid (73.4%), and 1-sinapoyl-2-ferulolylgentiobiose (56.0%) as compared with the control. Ascorbic acid content decreased during storage. However, US and exogenous phytohormones (MJ and ET) in combination reduced ascorbic acid degradation.	The authors did not report quality characteristics and physiological measurements of the samples.	[33]

Table 2. *Cont.*

Horticultural Crop	US Processing and Storage Conditions Evaluated	Main Findings		References
		Effects on the Biosynthesis of Health-Promoting Compounds	Effects on Quality and Physiological Attributes	
Common bean (*Phaseolus vulgaris*, cv. Kabulengeti) sprouts	Seeds were treated with US at different power (0, 180, and 360 W) and time (0, 30, 45, and 60 min) before sprouting. Seeds were sprouted in darkness (at 25 °C) for 96 h.	At 96 h of sprouting, the accumulation of total phenolic acids, flavonoids, and anthocyanins increased with the intensity of the US treatment. Total phenolic acids, flavonoids, and anthocyanins were 1065.27%, 559.8%, and 1052.9%, higher at 96 h of sprouting, respectively, in seeds treated with 360 W (60 min) as compared with the control.	US treatments decreased radicle emergence time, increased radicle length (72.37%), and induced the maximum hypocotyl growth (60.44%) when comparing 360 W (60 min) with the control at 96 h of sprouting time. US treatment enhanced the sprouting percentage, sprouting index and vigor of the samples, improving the quality characteristics of common bean sprouts. Hydrogen peroxide production and the activity of catalase, glutathione peroxidase, PAL, and tyrosine ammonia-lyase were also higher during sprouting of seeds treated with 360 W for 60 min compared with the control.	[34]
Soybean (*Glycine max* L. cv. Dongnong 48) sprouts	Soybean seeds were treated with US at different power levels (0, 100, 200, and 300 W, 40 kHz, for 30 min). The seeds were germinated at 30 °C for 5 days in darkness.	Gamma-aminobutyric acid (GABA) accumulated in higher concentrations in soybean sprouts as the US power level increased, with sprouts from seeds treated at 300 W for 30 min showing 43.5% higher GABA content compared with the control. US pretreatment in seeds (300 W, 30 min) generated soybean sprouts with lower content of daidzin (−79.62%) and genistin (−70.95%), and higher content of daidzein (39.13%) and genistein (94.91%) as compared with the control sprouts.	The germination rate and average length of sprouts increased by 18.07% and 20.41%, respectively, after US (300 W, 30 min) as compared with the control. US pretreatment (300 W, 30 min) reduced lipoxygenase (LOX) activity by 36.22 to 55.57% (depending on the LOX isomer), resulting on sprouts with improved odor and flavor. Sprouts obtained from seeds treated at 300 W for 30 min showed the highest decrease in IgE-binding potency (51.39%), and resulted in soybean sprouts with 98.78% less trypsin inhibitor.	[35]
Peanut (*Arachis hypogaea* L.) sprouts. Three cultivars were evaluated (Fuhua12, Fuhua 18, and Baisha 1016)	Peanut seeds were treated with US at 28, 45, and 100 kHz for 15, 20, and 30 min. Seeds were germinated under dark at 28 °C and 90% relative humidity for 5 days.	Resveratrol content of sprouts increased with prolonged US and soaking times, showing optimum conditions at 20 min and 6 h, respectively. Fuhua cultivar treated with US (100 kHz, 20 min) presented an increase in resveratrol content (980.1%) after sprouting compared with the control.	Germination rate was increased for seeds treated with 100 kHz ultrasonic waves. The allergic protein content completely decreased at 3 days of germination when treating the seeds with 100 kHz for 20 min before germination.	[36]

3.2.3. Carrot (*Daucus carota*)

Cuellar-Villarreal et al. [32] evaluated the effect of US (frequency 24 kHz, amplitude 100 mm) applied for 5 min at 20 °C, on the phenolic and carotenoid content in carrots immediately after treatment and after storage (20 °C for 3 days). As an immediate response to US, carrots showed 21.1% higher levels of total carotenoids as compared with the control. After storage, carotenoid content decreased in carrots. However, US-treated samples showed a lower decrease in carotenoid (−7.6%) than the control (−16.4%). US treatment induced an immediate decrease in total phenolic content (−62.5%) compared with the control. Moreover, after 3 days of storage, phenolic content in US-treated carrots increased by 129.2%, whereas for control samples the phenolic content did not change. Chlorogenic acid was the main phenolic compound that increased in US-treated whole carrots, which showed 41.8% higher content at 1 day of storage than the control. Interestingly, US-treated whole carrots at 3 days of storage showed the highest isocoumarin accumulation, 164.0% higher than the control and 240.8% higher than the control before storage. It is well known that isocoumarin exerts bitter flavor in carrots at 200 mg/kg [37]. However, the levels detected in sonicated carrots were below the threshold for sensory perception (40 mg/kg). The US-induced biosynthesis of phenolic compounds in carrots was confirmed by the immediate increase (2.04-fold increase) in *PAL* gene expression detected after treatment, which was accompanied by an increase in PAL enzymatic activity during the 0.5 days of storage, which was 64.8% higher as compared with the control before storage. US induced an immediate increase in respiration rate and higher levels of volatile organic compounds compared with the control.

3.2.4. Broccoli (*Brassica oleracea*)

The US-induced accumulation of health-promoting compounds has also been evaluated in broccoli [33]. The authors treated broccoli florets with US (20 min, frequency 24 kHz, amplitude 100 μm) and stored (3 days at 15 °C) the ultrasonicated samples in the presence of exogenous methyl jasmonate (MJ, 250 ppm) and/or ethylene (ET, 1000 ppm). As an immediate response to US treatment, the extractability of glucosinolates (glucoraphanin (795%), glucobrassicin (78.6%), and 4-hydroxy glucobrassicin (153%)) and phenolics (1-sinapoyl-2-feruloylgentiobiose (57.23%)) was increased. Furthermore, at 3 days of storage, broccoli florets treated with US and MJ showed the highest accumulation of gluconasturtiin (755.9%), neoglucobrassicin (232.8%), 4-hydroxy glucobrassicin (187.1%), glucoerucin (111.92%), 1,2,2-trisinapoylgentiobiose (136.7%), 3-*O*-caffeoylquinic acid (73.4%), and 1-sinapoyl-2-ferulolylgentiobiose (56.0%) as compared with the control. Likewise, ascorbic acid content decreased during storage of broccoli florets; however, the combined application of US, MJ, and ET reduced ascorbic acid degradation. These results indicate that the combined application of US treatment and exogenous phytohormones resulted in an effective strategy to induce the biosynthesis of glucosinolates and retain vitamin C content during storage.

3.2.5. Common Bean (*Phaseolus vulgaris*) Sprouts

The application of US has also been explored as a pretreatment in seeds to improve the content of health-promoting compounds and quality characteristics of sprouts. In this context, common bean seeds were treated with US at different power (0, 180, and 360 W) and time (0, 30, 45, and 60 min) before sprouting [34]. US-treated bean seeds and the control were sprouted in darkness (at 25 °C) for 96 h in an incubator. At 96 h of sprouting, the accumulation of total phenolic acids, flavonoids, and anthocyanins increased with the intensity of the US treatment. Total phenolic acids, flavonoids, and anthocyanins were 1065.27%, 559.8%, and 1052.9% higher at 96 h of sprouting, respectively, in seeds treated with 360 W (60 min) as compared with the control. US treatments decreased radicle emergence time, increased radicle length (72.37%), and induced the maximum hypocotyl growth (60.44%) when comparing 360 W (60 min) with the control at 96 h of sprouting time. Sprouting percentage was enhanced by US treatment, with seeds treated with 360 W for 60 min showing the highest values at 36 h of sprouting. US enhanced the sprouting index and vigor of samples, improving the quality characteristics of common bean sprouts. Hydrogen peroxide production and the enzymatic activity of catalase, glutathione peroxidase, PAL, and tyrosine ammonia-lyase were also higher during sprouting of seeds treated with 360 W for 60 min as compared with the control, suggesting that US induced the biosynthesis of phenolic compounds through a ROS-mediated mechanism.

3.2.6. Soybean (*Glycine max*) Sprouts

Soybean seeds were treated with US at different power levels (0, 100, 200, and 300 W, 40 kHz, for 30 min). The seeds were germinated at 30 °C for 5 days in darkness [35]. Gamma-aminobutyric acid (GABA) accumulated in higher concentrations in soybean sprouts as US power levels increased, with sprouts from seeds treated at 300 W for 30 min showing 43.5% higher GABA content compared with the control. US pretreatment in seeds (300 W, 30 min) generated soybean sprouts with a lower content of daidzin (−79.62%) and genistin (−70.95%), and higher content of daidzein (39.13%) and genistein (94.91%) compared with the control sprouts, indicating that US application before germination of soybean seeds induces the activation of glucosidases during sprouting, generating isoflavone aglycones.

The germination rate and average length of soybean sprouts increased by 18.07% and 20.41%, respectively, after US (300 W, 30 min) compared with the control. US pretreatment (300 W, 30 min) generated a reduction in lipoxygenase (LOX) activity by 36.22% and 55.57% (depending on the LOX isomer), resulting in sprouts with improved odor and flavor. Sprouts obtained from seeds treated at 300 W for 30 min showed the highest decrease in IgE-binding potency (51.39%), indicating that US treatment induced the degradation of allergens into peptides and amino acids. US pretreatment at 300 W resulted in soybean sprouts with 98.78% less trypsin inhibitor.

3.2.7. Peanut (*Glycine max*) Sprouts

Yu et al. [36] evaluated the effect of US (28, 45, and 100 kHz for 15, 20, and 30 min) as a pretreatment for germination in peanut seeds and determined the resveratrol content during germination for 5 days (Table 2). The authors reported that the resveratrol content of sprouts increased (980.1%) with prolonged US and soaking times, showing optimum conditions at 20 min and 6 h, respectively. Germination rate increased for seeds treated with 100 kHz ultrasonic waves. The allergic protein content completely decreased at 3 days of germination time by treating the seeds with 100 kHz for 20 min prior to germination. Thus, in addition to increasing the content of health-promoting compounds, US pretreatment decreases the natural antinutrient content in the seeds.

3.3. *Pulsed Electric Fields (PEF)*

The effect of moderate-intensity pulsed electric fields (MIPEF) on the accumulation of health-promoting compounds has been evaluated in fruits and vegetables such as apple [38], tomato [39–41], and carrot [42,43] (Table 3).

Table 3. Effects of moderate intensity pulsed electric fields (MIPEF) on the biosynthesis of health-promoting compounds and on the quality and physiological attributes of whole fruits and vegetables.

Horticultural Crop	MIPEF Processing and Storage Conditions Evaluated	Main Findings		References
		Effects on the Biosynthesis of Health-Promoting Compounds	Effects on Quality and Physiological Attributes	
Apple (*Malus domestica*, var. Golden delicious)	Whole apple fruits were treated at 0.4–2 kV cm^{-1} using 5–35 monopolar pulses of 4 µs at a frequency of 0.1 Hz (energy input of 0.008–1.3 kJ kg^{-1}). Samples were stored at 4 and 22 °C for 48 h.	MIPEF treatment (0.008 kJ kg^{-1}) induced an increase in total phenolic (13%) and flavan-3-ol (92%) contents in fruits stored at 22 °C for 24 h, and in flavonoids (58%) in samples stored at 4 °C for 24 h.	The authors did not report quality characteristics and physiological measurements of the samples.	[38]
Tomato (*Lycopersicon esculentum* Mill. cv. Daniella)	Tomato fruits were subjected to different electric field strengths (from 0.4 to 2.0 kV cm^{-1}) and number of pulses (from 5 to 30). Samples were stored at 4 °C for 24 h.	Except for 2 kV/cm treatment, MIPEF-treated tomatoes showed higher phenolic content after 24 h of treatment than the control. The increases in phenolic content of tomatoes ranged from 6.6% (five pulses at 0.4 kV cm^{-1}) to 44.6% (30 pulses at 1.2 kV cm^{-1}). Accumulation of lycopene was detected after storage of MIPEF-treated tomato fruit, which ranged from 0.6% (18 pulses at 2 kV cm^{-1}) to 31.8% (five pulses at 1.2 kV cm^{-1}) as compared with non-treated tomato.	The authors did not report quality characteristics and physiological measurements of the samples.	[39]
		MIPEF treatment at 1.2 kV cm^{-1} and 30 pulses resulted in the highest accumulation of caffeic acid-*O*-glucoside (170%), caffeic acid (140%), and chlorogenic acid (152%). MIPEF treatments at 1.2 kV cm^{-1} and five pulses presented the highest accumulation of α-carotene (93%), 9-*cis*-lycopene (94%) and 13-*cis*-lycopene (140%), respectively.	The authors did not report quality characteristics and physiological measurements of the samples.	[40]
Tomato (*Licopersicon esculentum* Mill. cv. Raf)	Tomato fruits were subjected to different electric field strengths (40, 120, and 200 kV m^{-1}) and number of pulses (5, 18 and 30 pulses). Specific energy input ranged from 0.02 to 2.31 kJ kg^{-1}. Samples were stored at 4 °C for 24 h.	The highest increase in carotenoid content (50%) was achieved in tomato fruits treated with an energy input of 2.31 kJ kg^{-1} (200 kV m^{-1}–30 pulses).	MIPEF-treated tomatoes showed a significant increase in the respiration rate of tomato fruit. Ethylene concentration was higher (53%) in fruits subjected to the lowest electric field strength. Further increase in PEF intensity inhibited ethylene production. Acetaldehyde synthesis was induced when tomatoes were treated at energy inputs higher than 0.38 kJ kg^{-1}. The higher the treatment intensity, the greater the softening effect. Total soluble solids and pH values increased with the treatment intensity.	[41]

Table 3. *Cont.*

Horticultural Crop	MIPEF Processing and Storage Conditions Evaluated	Main Findings		References
		Effects on the Biosynthesis of Health-Promoting Compounds	Effects on Quality and Physiological Attributes	
Carrot (*Daucus carota* cv. Nantes)	Whole carrots were subjected to different electric field strengths (0.8, 2, and 3.5 kV cm^{-1}) and number of pulses (5, 12, and 30). Samples were stored for 48 h at 4 °C.	The largest increase in phenolic content was observed at 24 h, after applying five pulses of 3.5 kV cm^{-1} (39.5%) and 30 pulses of 0.8 kV cm^{-1} (40.1%).	A correlation between the specific energy input and cell viability was found. After applying 3.5 kV cm^{-1}, viability decreased by 87.5–79.4%. At 24 h, whole carrots treated with five pulses of 3.5 kV cm^{-1} and 30 pulses of 0.8 kV cm^{-1} showed texture softening while preserved the color.	[42]
	Whole carrots were processed in a PEF batch system. Samples were treated with five pulses of 350 kV m^{-1} (580 ± 80 J kg^{-1}). Samples were stored for 36 h at 4 °C.	Immediately after PEF treatment, whole carrots showed a decrease in *p*-coumaric acid (−42.3%), protocatechuic acid (−78.1%), and ferulic acid (−56.3%). The maximum accumulation of phenolics in whole carrots was reached after 24 h of PEF treatment, where whole carrots presented a significant increase in total phenolics (80.2%) and chlorogenic (74.9%), ferulic (52.2%), and *p*-OH-benzoic (94.7%) acid compared to the control. At 36 h of storage a decrease in phenolic content was observed.	PEF induced an immediate increase in respiration. From 12 to 36 h, PEF-treated carrots presented a 123–164% higher respiration rate than untreated carrots. PEF did not induce an immediate increase in volatile organic compound production in whole carrots. However, at 12 h of storage, samples treated with PEF generated higher amounts of acetaldehyde (7 pg kg^{-1} s^{-1}), ethanol (68 ng kg^{-1} s^{-1}), and ethylene (50 ng kg^{-1} s^{-1}), whereas these volatiles were not detected in untreated carrots. PEF application delayed the peak of maximum peroxidase enzymatic activity for 12 h. Pectin methylesterase (PME) activity increased during the first 12 h of storage in the control. In contrast, PEF induced an immediate increase (164%) in enzyme activity, which remained stable for the following storage time. Polygalacturonase (PG) activity immediately decreased by 31%–32% after treatment. Phenylalanine ammonia-lyase (PAL) activity in untreated carrots remained stable during storage, whereas PEF-treated samples showed a constant increase in PAL activity during storage, showing the highest increase (153%) at the end of the study.	[43]

3.3.1. Apple (*Malus domestica*)

Soliva-Fontuny et al. [38] treated whole apple fruits at 0.4–2 kV cm^{-1} using 5–35 monopolar pulses of 4 μs at a frequency of 0.1 Hz, corresponding to a specific energy input of 0.008–1.3 kJ kg^{-1}. The apples were stored at 4 and 22 °C for 48 h. MIPEF treatment (0.008 kJ kg^{-1}) induced an increase in total phenolic (13%) and flavan-3-ol (92%) contents in fruits stored for 24 h at 22 °C, and in flavonoids (58%) in samples stored at 4 °C. The authors did not report quality characteristics or physiological measurements of MIPEF-treated apples.

3.3.2. Tomato (*Lycopersicon esculentum*)

Tomato fruits were subjected to different electric field strengths (from 0.4 to 2.0 kV cm^{-1}) and number of pulses (from 5 to 30), and the content of phenolics and carotenoids was detected after storage (4 °C for 24 h). Except for the 2 kV/cm treatment, MIPEF-treated tomatoes showed a higher phenolic content after 24 h of treatment compared to the control [39,40]. This increase ranged from 6.6% (5 pulses at 0.4 kV/cm) to 44.6% (30 pulses at 1.2 kV/cm). Accumulation of lycopene was detected after storage of MIPEF-treated tomato fruit, which ranged from 0.6% (18 pulses at 2 kV/cm) to 31.8% (5 pulses at 1.2 kV cm^{-1}) as compared with non-treated tomato. Likewise, MIPEF treatments at 1.2 kV cm^{-1} and 30 pulses resulted in the highest accumulation of caffeic acid-*O*-glucoside (170%), caffeic acid (140%), and chlorogenic acid (152%). MIPEF treatments at 1.2 kV cm^{-1} and five pulses resulted on the highest accumulation of α-carotene (93%), 9-*cis*-lycopene (94%) and 13-*cis*-lycopene (140%), respectively. The authors did not report quality characteristics and physiological measurements of MIPEF-treated tomato fruits [40].

In another study, González-Casado et al. [41] subjected tomato fruits at different electric field strengths (40, 120, and 200 kV m^{-1}) and number of pulses (5, 18, and 30 pulses). Specific energy input ranged from 0.02 to 2.31 kJ kg^{-1}. PEF-treated tomato fruits were stored at 4 °C for 24 h. The highest increase in carotenoid content (50%) was achieved in tomato fruits treated with an energy input of 2.31 kJ kg^{-1} (200 kV m^{-1} and 30 pulses). MIPEF-treated tomatoes showed a significant increase in the respiration rate of tomato fruit. Ethylene concentration was higher (53%) in fruits subjected to the lowest electric field strength, whereas a further increase in PEF intensity resulted in depletion of ethylene concentration. This result suggests that PEF could induce ethylene biosynthesis in even the mildest conditions. Acetaldehyde synthesis was induced when tomatoes were subjected to energy inputs higher than 0.38 kJ kg^{-1}, indicating that the tissue was undergoing plant cell death. Higher treatment intensities resulted in a more pronounced softening effect, with a reduction in the firmness values of 80% observed in the fruits treated with the most intense PEF conditions. Total soluble solids and pH values increased with treatment intensity. Accumulation of sugars in PEF-treated tomatoes was attributed to a stress response in the tissue related to osmoregulation to restore plant cell activity as well as with a PEF-induced increase in the ripening process of the fruit [41]. Furthermore, the authors associated the increase in pH with increased respiration induced by PEF, since organic acids are used as substrates.

3.3.3. Carrot (*Daucus carota*)

Regarding carrots, López-Gámez et al. [42] treated the whole tissue with different electric field strengths (0.8, 2.0, and 3.5 kV cm^{-1}) and number of pulses (5, 12, and 30). Samples were stored for 48 h at 4 °C, and the effects of PEF on phenolic content, cell viability, texture softening, and color of carrots were determined (Table 3). The most significant increase in phenolic content was observed at 24 h, after applying five pulses of 3.5 kV cm^{-1} (39.5%) and 30 pulses of 0.8 kV cm^{-1} (40.1%). A correlation between the specific energy input and cell viability was found. After applying 3.5 kV cm^{-1}, viability decreased by 87.5–79.4%. At 24 h, whole carrots treated with five pulses of 3.5 kV cm^{-1} and 30 pulses of 0.8 kV cm^{-1} showed texture softening with color preservation.

Similarly, López-Gámez et al. [43] processed whole carrots in a PEF batch system with five pulses of 350 kV m^{-1} (580 ± 80 J kg^{-1}) and evaluated the treatment effect on the phenolic profile, respiration rate, and color during storage (36 h at 4 °C) of the tissue. Immediately after PEF treatment, the whole carrots showed a decrease in *p*-coumaric acid (42.3%), protocatechuic acid (−78.1%), and ferulic acid (−56.3%). The maximum accumulation of phenolics in whole carrots was reached after 24 h of PEF treatment, showing a significant increase in total phenolics (80.2%), chlorogenic acid (74.9%), ferulic acid (52.2%), and *p*-OH-benzoic (94.7%) acid as compared with the control. At 36 h of storage, a decrease in phenolic content was observed. The application of PEF induced an immediate increase in respiration rate after treatment. From 12 to 36 h, PEF-treated carrots showed between 123–164% more CO_2 production than untreated carrots. PEF did not induce an immediate increase in production of volatile organic compounds in whole carrots. However, at 12 h of storage, samples treated with PEF generated higher amounts of acetaldehyde (7 pg kg^{-1} s^{-1}), ethanol (68 ng kg^{-1} s^{-1}) and ethylene (50 ng kg^{-1} s^{-1}), whereas these volatiles were not detected in untreated carrots. The authors attributed the presence of acetaldehyde and ethanol to PEF-induced anaerobic metabolism related to structural damage and intracellular content leakage. On the other hand, the increase of ethylene was associated with serving as a stress signaling molecule that induces the activation of secondary metabolism [43].

PEF application delayed the peak of maximum peroxidase enzymatic activity for 12 h. Pectin methylesterase (PME) activity increased during the first 12 h of storage in the control, whereas PEF induced an immediate increase (164%) in enzyme activity, which remained stable for the following storage time. Polygalacturonase (PG) activity immediately decreased by 31–32% after treatment. PAL activity in untreated carrots remained stable during storage, whereas PEF-treated samples showed a constant increase in PAL activity, showing the highest increase (153%) at the end of the storage time. As earlier described, PME strengthens vegetable tissues through cross-linking between pectin chains, whereas PG activity is related to solubilization of pectic substances and softening in many fruits and vegetables. Thus, the results indicate that a PEF-induced increase of PME activity and an induced decrease of PG generates increased cell-wall rigidity in carrots [43].

4. Industrial Implementation and Economic Feasibility of Using Non-Thermal Technologies as Tools to Enhance the Content of Health-Promoting Compounds in Whole Fruits and Vegetables

For the last 20 years, NTTs have come a long way in technological development, process optimization, industrial applicability, and consumer acceptability. HHP and PEF have become widely used processing stages on industrial scales [44]. On the other hand, US application in food and beverage processing is relatively new compared to HHP and PEF. Although there are some uses of US on a large scale, there is still work to do to achieve the same acceptability and industrial applicability as other NTTs. Nevertheless, these three NTTs have proven to achieve enzyme and microbial inactivation, enhance shelf-life, and preserve nutritional and nutraceutical properties in foods and beverages. Furthermore, they may be used to enhance extractability and bioavailability in food matrixes [14].

Table 4 presents a qualitative comparison between HHP, PEF, and US processing technology. Each technology is assessed in terms of scaling-up feasibility and cost of operation/maintenance. As mentioned before, HHP is a much more mature technology compared to PEF and US, as it has been studied and optimized for more than 35 years [45,46]. This context is relevant when comparing these technologies. It can be expected that some of the limitations or drawbacks associated with PEF and US will lessen as the technology develops further. All three technologies can process in batch or continuous/semi-continuous mode. Furthermore, as described in the previous section, all of them can process whole fruits and vegetables.

Table 4. Comparison between non-thermal technologies (high hydrostatic pressure, pulsed electric fields, and ultrasound) in terms of their scaling-up feasibility and cost of operation/maintenance.

Non-Thermal Technology	General Principle of Action	Scaling-Up Feasibility	Cost of Operation/Maintenance
High Hydrostatic Pressure (HHP)	A high-pressure pump injects liquid into a treatment chamber where the product of interest is located. High pressure generates structural changes in tissues, cells, and molecules depending on process conditions.	**High** There is industrial equipment that can process tens of tons per hour.	**High** Due to the mechanical complexity of the equipment and the constant wearing of sealing components, the cost of maintenance is relatively high.
Pulsed Electric Fields (PEF)	Electrodes generate repetitive short electric pulses in a treatment chamber designed for solids, suspensions, or liquids. The pulsed electric field generates changes in the permeability of the cell wall and cell membrane, causing changes in metabolic flux and cell viability depending on process conditions.	**Medium** There is equipment available at industrial scales, although processing capability is somewhat limited compared to HHP.	**Medium** The equipment is robust in general terms, and the configuration of the electrodes may enhance energy efficiency.
Ultrasound (US)	Ultrasound waves are generated, propagating through a transmitting medium. As the sound wave propagates, the fluctuations in pressure generate cavitation (formation and collapse of microbubbles), promoting shear stress that affects the integrity of cells and molecules depending on process conditions.	**Low-Medium** Most of the equipment is still at pilot-plant scale. Although there are efforts for application at larger scales, the reach of sound wave propagation seems to be a limiting factor.	**Medium** The equipment is robust in general terms. The sound source elements erode constantly and must be periodically replaced. A fraction of the energy dissipates as heat, decreasing efficiency.

As mentioned above (Tables 1–3), NTTs can also be used as a tool to elicit the biosynthesis of nutraceutical compounds in whole fruits and vegetables, enhancing their health-promoting properties. This elicitation is promoted using process conditions that activate the secondary plant metabolism without significantly affecting quality attributes (texture, flavor, color, etc.). HHP, PEF, and US processing conditions to induce stress responses on whole fruits and vegetables are milder than those needed for achieving microbial or enzyme inactivation in processed foods. These milder processing conditions may represent an additional advantage in terms of energy expenditure and equipment maintenance. The processing conditions shown in Tables 1–3 can be applied at pilot-plant and industrial scales to obtain whole fresh fruits (i.e., apples, mangoes, tomatoes, and strawberries) and vegetables (i.e., carrots, broccoli, and lettuce) with enhanced levels of health-promoting compounds such as carotenoids, phenolics, and glucosinolates, among others.

Consumer awareness of the benefits of consuming fruits and vegetables meeting desirable quality attributes, safety requirements, and high nutraceutical value increases daily. In such a way, having whole fresh produce with enhanced health-promoting properties is undoubtedly of interest to most consumers. Regarding the economic point of view, it is essential to reflect on how NTT processing may affect product price once in the market. How much would it cost to process whole carrots with HHP to increase their content of health-promoting compounds? Or mangoes? In order to calculate the processing cost, it is necessary to consider aspects such as the energy demand of the equipment (kWh), the cost of all inputs needed (such as the water for the treatment chamber), and the cost of maintenance (estimated to be about 5% of the total inversion cost of the equipment annually, as a rule of thumb). It is also necessary to define a scenario.

In this regard, let us suppose that the goal is to use HHP to process 30 tons of whole carrots per day in order to enhance their health-promoting properties. HHP equipment throughput at industrial scales typically ranges from 200 to 3500 kg/h. Given the processing goal, equipment with an average throughput of 1500 kg/h would be adequate (20 h processing time). Such HHP equipment would have an approximate annual maintenance cost of USD 500,000 and an approximate power requirement of 230 kWh. HHP processing of 30 tons of whole carrots per day would require about 30 m^3 (1059 ft^3) of water (the total volume of the chamber minus the volume occupied by the product); this technical information has been taken or estimated using commercial literature from an HHP equipment supplier [47].

In this way, the cost for HHP processing of 30 tons of carrots per day (20 h processing time) can be estimated, considering approximate electricity and water costs for industrial consumers in the USA, as:

- Energy cost: 230 kWh ∗ USD 0.067/kWh ∗ 20 h/day = USD 308.20/day
- Water cost: 1059 ft^3/day ∗ USD 0.011/ft^3 = USD 11.65/day
- Maintenance cost: USD 500,000/year ∗ 0.0027 years/day = USD 1350/day
- Estimated HHP processing cost for 30 tons of carrots = USD 1669.85/day

For one kg of HHP-processed whole carrot, the cost is USD 0.056. In the USA, the average price of one kg of fresh carrot is USD 1.19. Adding the HHP processing cost to the average commercial price of carrots gives USD 1.246 per kg, or a 4.71% increase. HHP-processed whole carrots with enhanced health-promoting properties could be easily commercialized at a price 15–30% higher than non-processed carrots, validating the economic feasibility of the process. This same analysis could be done with other NTT such as PEF and US. A wide variety of fruits and vegetables (those included in Tables 1–3, as well as many others) may be feasible for large-scale processing using NTT to enhance their health-promoting properties, while preserving their quality attributes. These are exciting times in the optimization and development of new NTT. This technology, used under abiotic stress conditions, may soon constitute a widely-used strategy for enhancing whole fruits and vegetables at a commercial (pilot-plant and industrial) scale.

5. Further Research Needs

As described in the previous sections, some scientific research is being performed to evaluate the feasibly of using NTTs as abiotic elicitors in order to increase the content of bioactive compounds in whole fruits and vegetables. However, most of the recent scientific reports are focused on studying the effects of NTTs on either the accumulation of specific secondary metabolites or on the quality and physiological parameters of the crop under investigation. A holistic approach should be followed, considering advances in key applied research aspects of the industrial application of NTTs as tools to increase the content of health-promoting compounds in whole fruits and vegetables. Figure 2 summarizes the holistic approach proposed herein to determine the feasibility of NTTs as tools to improve the content of bioactive compounds in whole fruits and vegetables.

When selecting the fruit or vegetable to investigate, it is essential to consider if it is climacteric or non-climacteric. As observed in the cited research (Tables 1–3), NTT effects on whole fruits and vegetables are tissue-dependent. For instance, in the case of climacteric crops, postharvest ripening could be decreased depending on the processing condition selected. Therefore, it is important to evaluate the physiological attributes of the fruit or vegetable to increase the understanding of the response of the tissue at the physiological level. The selected NTT processing conditions should be mild, in order to induce a stress response in the plant cell while maintaining its viability. Conditions proposed for evaluation are highlighted in Figure 2. The effect of NTT treatment on bioactive plant components, quality, and the physiological attributes of the whole fresh produce under investigation should be evaluated both immediately after processing and during storage. Evaluating the immediate effect of NTT processing determines the most intense processing conditions that can be applied in the tissue while retaining essential quality attributes and cell viability.

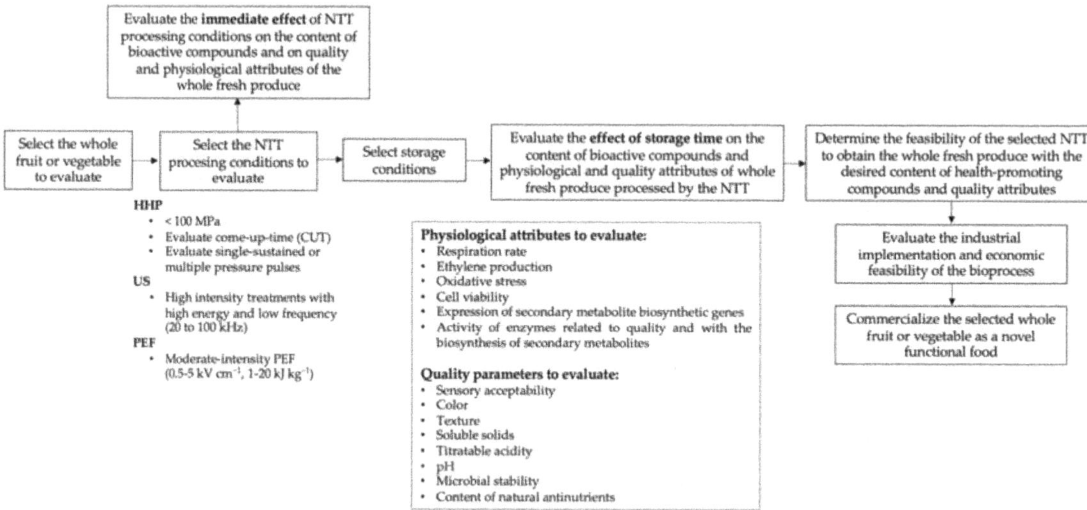

Figure 2. Key aspects to consider when determining the feasibility of an NTT to enhance the content of health-promoting compounds in whole fruits and vegetables while preserving quality attributes. More fundamental knowledge is needed to advance the industrial application of NTTs as abiotic elicitors. Abbreviations: HHP, high hydrostatic pressure; US, ultrasound; PEF, pulsed electric fields; NTT, non-thermal technology.

Once the whole tissue is treated with the NTT under investigation, the fruit or vegetable should be stored under adequate conditions used for their commercialization in order to determine the effect of the applied conditions on the biosynthesis of bioactive compounds and on the physiological and quality attributes during storage of the crop, which allow for determining its shelf-life. Physiological attributes proposed for evaluation immediately after processing and during storage include respiration rate, ethylene production, oxidative stress, cell viability, expression of secondary metabolite biosynthesis genes, and enzyme activity related to quality and biosynthesis of secondary metabolites. Likewise, quality parameters suggested for evaluation in the processed whole tissue include sensory acceptability, color, texture, soluble solids, pH, and microbial counts; in the specific case of seed treated with NTTs as pretreatments for germination, measurement of the content of natural antinutrients in the sprouts is also suggested.

Based on the results obtained regarding the effects of NTTs on secondary metabolite content and on physiological and quality attributes evaluated immediately after processing and during storage, the technological feasibility of using an NTT on the selected crop can then be determined. The economic feasibility of bioprocessing should be carefully considered when proposing the commercialization of whole fruits and vegetables treated with NTTs. The holistic approach to evaluating the technical and industrial feasibility of NTT as abiotic elicitors presented herein will allow the advancement of this research, taking these functional whole fruits and vegetables from the laboratories to the market.

6. Conclusions

Herein, an update regarding recent research reporting the use of NTTs as abiotic elicitors to induce the biosynthesis of health-promoting compounds in whole fruits and vegetables has been presented. Results from the reviewed literature demonstrate that

NTT application under certain conditions in different crops results in the development of next-generation whole fruits and vegetables with improved bioactive compound contents and enhanced shelf-life. Moreover, the analysis presented regarding the industrial implementation and economic feasibility of NTT application in obtaining functional whole fruits and vegetables is positive, providing motivation for the scientific community to continue researching this fascinating emerging research field.

Although this paper has focused on potential industrial implementation of NTT for enhancing the health-promoting properties of whole fruits and vegetables, it may also be interesting to explore scenarios in which consumers are the ones applying NTT in their homes. Would it be possible to have affordable, practical, and safe-to-use countertop NTT equipment for residential use, allowing home processing of whole fruits and vegetables in order to increase their nutraceutical value? Currently, commercial US equipment for residential use (not necessarily designed for food applications) may partially meet the processing conditions at which abiotic stress is elicited in whole fruits and vegetables. Would it be possible to have residential PEF or HHP equipment adapted to the typical home food preparation process? This "Do-It-Yourself (DIY)" approach to NTT processing should be further explored in order to assess its technical and economic feasibility.

Author Contributions: Conceptualization, D.A.J.-V.; investigation, D.A.J.-V. and J.B.; literature search, D.A.J.-V. and J.B.; writing—original draft preparation, D.A.J.-V. and J.B.; writing—review and editing, D.A.J.-V. and J.B. All authors have read and agreed to the published version of the manuscript.

Funding: This study was based upon research supported by Tecnologico de Monterrey—Bioprocess Research Group.

Institutional Review Board Statement: Not applicable.

Informed Consent Statement: Not applicable.

Data Availability Statement: Not applicable.

Conflicts of Interest: The authors declare no conflict of interest.

References

1. Cisneros-Zevallos, L. The power of plants: How fruit and vegetables work as source of nutraceuticals and supplements. *Int. J. Food Sci. Nutr.* **2021**, *72*, 660–664. [CrossRef]
2. Santini, A.; Novellino, E. Nutraceuticals—Shedding light on the grey area between pharmaceuticals and food. *Expert Rev. Clin. Pharmacol.* **2018**, *11*, 545–547. [CrossRef]
3. Cisneros-Zevallos, L. The use of controlled postharvest abiotic stresses as a tool for enhancing the nutraceutical content and adding-value of fresh fruits and vegetables. *J. Food Sci.* **2003**, *68*, 1560–1565. [CrossRef]
4. Cisneros-Zevallos, L.; Jacobo-Velázquez, D.A. Controlled abiotic stresses revisited: From homeostasis through hormesis to extreme Stresses and the impact on nutraceuticals and quality during pre- and postharvest applications in horticultural crops. *J. Agric. Food Chem.* **2020**, *68*, 11877–11879. [CrossRef]
5. Reyes, L.F.; Villarreal, J.E.; Cisneros-Zevallos, L. The increase in antioxidant capacity after wounding depends on the type of fruit or vegetable tissue. *Food Chem.* **2007**, *101*, 1254–1262. [CrossRef]
6. Jacobo-Velázquez, D.A.; Cisneros-Zevallos, L. An alternative use of horticultural crops: Stressed plants as biofactories of bioactive phenolic compounds. *Agriculture* **2012**, *2*, 259–271. [CrossRef]
7. Jacobo-Velázquez, D.A.; González-Agüero, M.; Cisneros-Zevallos, L. Cross-talk between signaling pathways: The link between plant secondary metabolite production and wounding stress response. *Sci. Rep.* **2015**, *5*, 8608. [CrossRef] [PubMed]
8. Gastélum-Estrada, A.; Hurtado-Romero, A.; Santacruz, A.; Cisneros-Zevallos, L.; Jacobo-Velázquez, D.A. Sanitizing after fresh-cutting carrots reduces the wound-induced accumulation of phenolic antioxidants compared to sanitizing before fresh-cutting. *J. Sci. Food Agric.* **2020**, *100*, 4995–4998. [CrossRef] [PubMed]
9. Cisneros-Zevallos, L.; Jacobo-Velázquez, D.A.; Pech, J.C.; Koiwa, H. Signaling Molecules Involved in the Postharvest Stress Response of Plants. In *Handbook of Plant and Crop Physiology*, 3rd ed.; CRC Press: Boca Raton, FL, USA, 2014; pp. 259–276.
10. Jacobo-Velázquez, D.A.; Cuéllar-Villarreal, M.D.R.; Welti-Chanes, J.; Cisneros-Zevallos, L.; Ramos-Parra, P.A.; Hernández-Brenes, C. Nonthermal processing technologies as elicitors to induce the biosynthesis and accumulation of nutraceuticals in plant foods. *Trends Food Sci. Technol.* **2017**, *60*, 80–87. [CrossRef]
11. Schreck, S.; Dörnenburg, H.; Knorr, D. Research report: Evaluation of hydrogen peroxide production in tomato (*Lycopersicum esculentum*) suspension cultures as a stress reaction to high pressure treatment. *Food Biotechnol.* **1996**, *10*, 163–171. [CrossRef]

12. Dörnenburg, H.; Knorr, D. Monitoring the impact of high-pressure processing on the biosynthesis of plant metabolites using plant cell cultures. *Trends Food Sci. Technol.* **1998**, *9*, 355–361. [CrossRef]
13. Lin, L.; Wu, J.; Ho, K.-P.; Qi, S. Ultrasound-induced physiological effects and secondary metabolite (saponin) production in *Panax ginseng* cell cultures. *Ultrasound Med. Biol.* **2001**, *27*, 1147–1152. [CrossRef]
14. López-Gámez, G.; Elez-Martínez, P.; Martín-Belloso, O.; Soliva-Fortuny, R. Enhancing carotenoid and phenolic contents in plant food matrices by applying non-thermal technologies: Bioproduction vs improved extractability. *Trends Food Sci. Technol.* **2021**, *112*, 622–630. [CrossRef]
15. Denoya, G.; Colletti, A.C.; Vaudagna, S.R.; Polenta, G. Application of non-thermal technologies as a stress factor to increase the content of health-promoting compounds of minimally processed fruits and vegetables. *Curr. Opin. Food Sci.* **2021**, *42*, 224–236. [CrossRef]
16. Song, C.J.; Steinebrunner, I.; Wang, X.; Stout, S.C.; Roux, S.J. Extracellular ATP induces the accumulation of superoxide via NADPH oxidases in Arabidopsis. *Plant Physiol.* **2006**, *140*, 1222–1232. [CrossRef] [PubMed]
17. Jacobo-Velázquez, D.A.; Martínez-Hernández, G.B.; Rodríguez, S.D.C.; Cao, C.-M.; Cisneros-Zevallos, L. Plants as biofactories: Physiological role of reactive oxygen species on the accumulation of phenolic antioxidants in carrot tissue under wounding and hyperoxia stress. *J. Agric. Food Chem.* **2011**, *59*, 6583–6593. [CrossRef] [PubMed]
18. Becerra-Moreno, A.; Redondo-Gil, M.; Benavides, J.; Nair, V.K.; Cisneros-Zevallos, L.; Jacobo-Velázquez, D.A. Combined effect of water loss and wounding stress on gene activation of metabolic pathways associated with phenolic biosynthesis in carrot. *Front. Plant Sci.* **2015**, *6*, 837. [CrossRef] [PubMed]
19. Viacava, F.; Santana-Gálvez, J.; Heredia-Olea, E.; Pérez-Carrillo, E.; Nair, V.; Cisneros-Zevallos, L.; Jacobo-Velázquez, D.A. Sequential application of postharvest wounding stress and extrusion as an innovative tool to increase the concentration of free and bound phenolics in carrots. *Food Chem.* **2020**, *307*, 125551. [CrossRef]
20. Torres-Contreras, A.M.; Jacobo-Velázquez, D.A. Effects of wounding stress and storage temperature on the accumulation of chlorogenic acid isomers in potatoes (*Solanum tuberosum*). *Appl. Sci.* **2021**, *11*, 8891. [CrossRef]
21. Torres-Contreras, A.M.; Nair, V.; Cisneros-Zevallos, L.; Jacobo-Velázquez, D.A. Stability of bioactive compounds in broccoli as affected by cutting styles and storage time. *Molecules* **2017**, *22*, 636. [CrossRef]
22. Chhajed, S.; Mostafa, I.; He, Y.; Abou-Hashem, M.; El-Domiaty, M.; Chen, S. Glucosinolate biosynthesis and the glucosinolate–myrosinase system in plant defense. *Agronomy* **2020**, *10*, 1786. [CrossRef]
23. Serment-Moreno, V.; Jacobo-Velázquez, D.A.; Torres, J.A.; Welti-Chanes, J. Microstructural and physiological changes in plant cell induced by pressure: Their role on the availability and pressure-temperature stability of phytochemicals. *Food Eng. Rev.* **2017**, *9*, 314–334. [CrossRef]
24. Hu, K.; Peng, D.; Wang, L.; Liu, H.; Xie, B.; Sun, Z. Effect of mild high hydrostatic pressure treatments on physiological and physicochemical characteristics and carotenoid biosynthesis in postharvest mango. *Postharvest Biol. Technol.* **2021**, *172*, 111381. [CrossRef]
25. Álvarez-Virrueta, D.; García-López, E.; Montalvo-González, E.; Ramírez, J.; Mata-Montes-De-Oca, M.; Tovar-Gómez, B. Efecto de las altas presiones hidrostáticas en la fisiología postcosecha del mango 'Ataulfo'. *CyTA-J. Food* **2012**, *10*, 173–181. [CrossRef]
26. Ortega, V.G.; Ramirez, J.; Velazquez, G.; Tovar, B.; Mata, M.; Montalvo-González, E. Effect of high hydrostatic pressure on antioxidant content of 'Ataulfo' mango during postharvest maturation. *Food Sci. Technol.* **2013**, *33*, 561–568. [CrossRef]
27. Viacava, F.; Ortega, E.; Welti-Chanes, J.; Cisneros-Zevallos, L.; Jacobo-Velázquez, D.A. Using high hydrostatic pressure processing come-up time as an innovative tool to induce the biosynthesis of free and bound phenolics in whole carrots. *Food Bioprocess Technol.* **2020**, *13*, 1717–1727. [CrossRef]
28. Viacava, F.; Ramos-Parra, P.; Welti-Chanes, J.; Jacobo-Velázquez, D. High hydrostatic pressure processing of whole carrots: Effect of static and multi-pulsed mild intensity hydrostatic pressure treatments on bioactive compounds. *Foods* **2021**, *10*, 219. [CrossRef] [PubMed]
29. Ibarra-Garza, I.P.; Ramos-Parra, P.A.; Hernández-Brenes, C.; Jacobo-Velázquez, D.A. Effects of postharvest ripening on the nutraceutical and physicochemical properties of mango (*Mangifera indica* L. cv Keitt). *Postharvest Biol. Technol.* **2015**, *103*, 45–54. [CrossRef]
30. Yu, J.; Engeseth, N.J.; Feng, H. High intensity ultrasound as an abiotic elicitor—Effects on antioxidant capacity and overall quality of romaine lettuce. *Food Bioprocess Technol.* **2016**, *9*, 262–273. [CrossRef]
31. Gani, A.; Baba, W.N.; Ahmad, M.; Shah, U.; Khan, A.A.; Wani, I.A.; Masoodi, F.; Gani, A. Effect of ultrasound treatment on physico-chemical, nutraceutical and microbial quality of strawberry. *LWT-Food Sci. Technol.* **2016**, *66*, 496–502. [CrossRef]
32. Cuéllar-Villarreal, M.D.R.; Ortega, E.; Becerra-Moreno, A.; Welti-Chanes, J.; Cisneros-Zevallos, L.; Jacobo-Velázquez, D.A. Effects of ultrasound treatment and storage time on the extractability and biosynthesis of nutraceuticals in carrot (*Daucus carota*). *Postharvest Biol. Technol.* **2016**, *119*, 18–26. [CrossRef]
33. Aguilar-Camacho, M.; Welti-Chanes, J.; Jacobo-Velázquez, D.A. Combined effect of ultrasound treatment and exogenous phytohormones on the accumulation of bioactive compounds in broccoli florets. *Ultrason. Sonochem.* **2019**, *50*, 289–301. [CrossRef] [PubMed]
34. Ampofo, J.O.; Ngadi, M. Ultrasonic assisted phenolic elicitation and antioxidant potential of common bean (*Phaseolus vulgaris*) sprouts. *Ultrason. Sonochem.* **2020**, *64*, 104974. [CrossRef] [PubMed]

35. Yang, H.; Gao, J.; Yang, A.; Chen, H. The ultrasound-treated soybean seeds improve edibility and nutritional quality of soybean sprouts. *Food Res. Int.* **2015**, *77*, 704–710. [CrossRef]
36. Yu, M.; Liu, H.; Shi, A.; Liu, L.; Wang, Q. Preparation of resveratrol-enriched and poor allergic protein peanut sprout from ultrasound treated peanut seeds. *Ultrason. Sonochem.* **2016**, *28*, 334–340. [CrossRef]
37. Lafuente, M.T.; López-Gálvez, G.; Cantwell, M.; Yang, S.F. Factors influencing ethylene-induced isocoumarin formation and increased respiration in carrots. *J. Am. Soc. Hortic. Sci.* **1996**, *121*, 537–542. [CrossRef]
38. Soliva-Fortuny, R.; Vendrell-Pacheco, M.; Martín-Belloso, O.; Elez-Martínez, P. Effect of pulsed electric fields on the antioxidant potential of apples stored at different temperatures. *Postharvest Biol. Technol.* **2017**, *132*, 195–201. [CrossRef]
39. Vallverdu-Queralt, A.; Oliu, G.O.; Odriozola-Serrano, I.; Lamuela-Raventos, R.M.; Martín-Belloso, O.; Elez-Martinez, P. Effects of pulsed electric fields on the bioactive compound content and antioxidant capacity of tomato fruit. *J. Agric. Food Chem.* **2012**, *60*, 3126–3134. [CrossRef]
40. Vallverdu-Queralt, A.; Oliu, G.O.; Odriozola-Serrano, I.; Lamuela-Raventos, R.M.; Martín-Belloso, O.; Elez-Martinez, P. Metabolite profiling of phenolic and carotenoid contents in tomatoes after moderate-intensity pulsed electric field treatments. *Food Chem.* **2013**, *136*, 199–205. [CrossRef]
41. González-Casado, S.; Martín-Belloso, O.; Elez-Martinez, P.; Soliva-Fortuny, R. Enhancing the carotenoid content of tomato fruit with pulsed electric field treatments: Effects on respiratory activity and quality attributes. *Postharvest Biol. Technol.* **2018**, *137*, 113–118. [CrossRef]
42. López-Gámez, G.; Elez-Martinez, P.; Martín-Belloso, O.; Soliva-Fortuny, R. Enhancing phenolic content in carrots by pulsed electric fields during post-treatment time: Effects on cell viability and quality attributes. *Innov. Food Sci. Emerg. Technol.* **2020**, *59*, 102252. [CrossRef]
43. López-Gámez, G.; Elez-Martinez, P.; Martín-Belloso, O.; Soliva-Fortuny, R. Pulsed electric fields affect endogenous enzyme activities, respiration and biosynthesis of phenolic compounds in carrots. *Postharvest Biol. Technol.* **2020**, *168*, 111284. [CrossRef]
44. Khouryieh, H.A. Novel and emerging technologies used by the U.S. food processing industry. *Innov. Food Sci. Emerg. Technol.* **2020**, *67*, 102559. [CrossRef]
45. Jadhav, H.B.; Annapure, U.S.; Deshmukh, R.R. Non-thermal technologies for food processing. *Front. Nutr.* **2021**, *8*, 657090. [CrossRef] [PubMed]
46. Gómez-López, V.M.; Pataro, G.; Tiwari, B.; Gozzi, M.; Meireles, M.Á.A.; Wang, S.; Guamis, B.; Pan, Z.; Ramaswamy, H.; Sastry, S.; et al. Guidelines on reporting treatment conditions for emerging technologies in food processing. *Crit. Rev. Food Sci. Nutr.* **2021**, 1–25. [CrossRef]
47. Hiperbaric Website. Available online: https://www.hiperbaric.com/en/hpp-technology/equipment/ (accessed on 4 September 2021).

foods

MDPI

Article

Applying Pulsed Electric Fields to Whole Carrots Enhances the Bioaccessibility of Carotenoid and Phenolic Compounds in Derived Products

Gloria López-Gámez, Pedro Elez-Martínez, Olga Martín-Belloso and Robert Soliva-Fortuny *

Department of Food Technology, University of Lleida, Agrotecnio Center, Av. Alcalde Rovira Roure, 191, 25198 Lleida, Spain; gloria.lopez@udl.cat (G.L.-G.); pedro.elez@udl.cat (P.E.-M.); olga.martin@udl.cat (O.M.-B.)
* Correspondence: robert.soliva@udl.cat

Abstract: We propose the application of pulsed electric fields (PEF) to carrots to obtain derived products with increased phenolic and carotenoid bioaccessibility. For this purpose, juices, purees, and oil-added purees were obtained from whole PEF-treated carrots (five pulses of 3.5 kV cm^{-1}; 0.61 kJ kg^{-1}). In order to obtain shelf-stable products, the effect of a thermal treatment (70 °C for 10 min) was also studied. Carrot juices exhibited the highest carotenoid (43.4 mg/100 g fresh weight) and phenolic (322 mg kg^{-1} dry weight) contents. However, caffeic and coumaric acid derivatives were highly sensitive to PEF. The phenolic bioaccessibility reached 100% in purees obtained from the PEF-treated carrots, whereas the further thermally treated oil-added purees exhibited the greatest carotenoid bioaccessibility (7.8%). The increase in carotenoid bioaccessibility could be related to their better release and solubilization into micelles. The results suggest that food matrix aspects apart from particle size (e.g., pectin characteristics) are involved in phenolic bioaccessibility.

Keywords: carotenoids; phenolic compounds; puree; juice; bioaccessibility; pulsed electric fields; carrot; microstructure; quality attributes

Citation: López-Gámez, G.; Elez-Martínez, P.; Martín-Belloso, O.; Soliva-Fortuny, R. Applying Pulsed Electric Fields to Whole Carrots Enhances the Bioaccessibility of Carotenoid and Phenolic Compounds in Derived Products. *Foods* **2021**, *10*, 1321. https://doi.org/10.3390/foods10061321

Academic Editors: Marina Cano Lamadrid and Francisco Artés-Hernández

Received: 26 April 2021
Accepted: 3 June 2021
Published: 8 June 2021

Publisher's Note: MDPI stays neutral with regard to jurisdictional claims in published maps and institutional affiliations.

1. Introduction

Today's life pace leads consumers to increasingly demand healthier minimally processed products that are easy to prepare and consume. Carrots are one of the most consumed vegetables worldwide, and are, thus, a significant source of antioxidants, including carotenoids and phenolic compounds. Clinical studies have demonstrated that α-carotene and β-carotene, the most abundant carotenoids in carrots, can prevent suffering atherosclerosis, cancer, or macular degeneration [1]. Likewise, chlorogenic acid, the main phenolic compound found in carrots, possesses anti-diabetic and cardioprotective properties [2]. Therefore, due to their health-promoting properties, carrots are a potential commodity for developing functional derived products and meeting consumer demands.

Both carotenoids and phenolic compounds are usually enclosed by cell walls and organelle structures that hinder their release during digestion. Bioaccessibility refers to the percentage of a compound released from the food matrix and absorbed during digestion [3], which is more important than the actual content in a food matrix. The chemical structure, concentration, matrix structure, and processing are the most important factors that determine bioactive compound bioaccessibility [4]. Therefore, mechanical and thermal processes could disrupt the natural matrix, thus, modifying their further bioaccessibility [4,5].

A decrease in the particle size and depolymerization of pectin has been shown to improve β-carotene bioaccessibility in carrot purees [6–8]. Furthermore, carotenoid micellarization is conditioned by different factors, such as oil addition [9] and the application of thermal treatments [6]. Nonetheless, some studies have reported that cell wall fragments formed after thermal treatments may entrap carotenoids and compromise their bioaccessibility [10]. The bioaccessibility of phenolic compounds has been reported to increase in thermally treated grape and orange juices, whereas it was shown to decrease in pomelo (80 °C for 30 min) and fruit juice-based beverages (90 °C for 1 min) [11–13]. Information about the effect of the presence of oils and fats on phenolic bioaccessibility is limited since they are hydrophilic compounds that do not require micellarization prior to intestinal uptake. However, some literature works reported a positive effect when whole milk was added to juices [11–13].

Pulsed electric fields (PEF) is a non-thermal processing technology that delivers short pulses (ms or µs) of electric energy to a food product that is located between two electrodes. Electropermeabilization causes reversible or irreversible structural changes in the matrix depending on the applied intensity. Low and moderate intensities (0.1–5 kV cm^{-1}, 0.5–20 kJ kg^{-1}) have been reported to trigger a stress defense response in plant tissues, leading to the accumulation of bioactive compounds (e.g., carotenoids and phenolic compounds) in fruit and vegetables [14–16].

On the other hand, severe changes in structure may facilitate the extraction of bioactive compounds and, consequently, their bioaccessibility [17,18]. Results regarding the application of PEF to whole apples [19], tomatoes, and carrots to enhance the bioaccessibility of carotenoids and phenolic compounds are promising, suggesting that derivates, such as juices or purees, with higher and more bioaccessible antioxidant compounds, could be obtained from these commodities [20,21].

Therefore, the main aim of this study was to investigate the feasibility of applying PEF to whole carrots as a pre-treatment to enhance the bioaccessible fractions of carotenoids and phenolic compounds in different shelf-stabled derived products (juices, purees, and oil-added purees). Additionally, the influence of applying PEF and further processing strategies on the quality attributes, microstructure, and bioactive contents was investigated.

2. Materials and Methods

2.1. Chemicals and Reagents

HPLC grade methanol, acetone, and methyl tert-butyl ether were acquired from Fisher Scientific Scharlau Chemie (Loughborough, UK). Sodium chloride was purchased from POCH S.A. (Sowińskiego, Poland). Ultrapure water was obtained with a Milli-Q system (Millipore Ibérica, Madrid, Spain). Ammonium carbonate, acetonitrile, hexane, ethanol (HPLC grade), magnesium chloride hexahydrate, acetic acid, and ammonium acetate were acquired from Scharlab (Sentmenat, Spain). Butyl hydroxytoluene (BHT) was purchased from Scharlau Chemie S.A. (Barcelona, Spain). Calcium chloride dihydrate was obtained from Merck (Darmstadt, Germany). Sodium hydrogen carbonate and potassium dihydrogen phosphate were acquired from VWR (Llinars del Vallès, Spain). Potassium chloride was obtained from Panreac (Castellar del Vallès, Spain). Digestive enzymes (porcine pepsin, porcine bile extract, porcine pancreatin, and porcine lipase) were acquired from Sigma-Aldrich (Darmstadt, Germany).

Caffeic acid, ferulic acid, and p-coumaric acid commercial patterns were obtained from Sigma-Aldrich (St. Louis, MO, USA). Β-carotene standard was acquired from CaroteNature (Ostermundigen, Switzerland), and α-carotene was acquired from Supelco-Merck (Darmstadt, Germany).

2.2. Carrot Samples

Carrots (*Daucus carota* cv. Nantes) (17 ± 2 cm and 106 ± 7 g) were purchased in a local supermarket (Lleida, Spain) and stored at 4 °C within a week until processing. Carrots were washed with tap water, and the excess was removed with a paper cloth before PEF application.

2.3. Pulsed Electric Fields (PEF) Treatments

PEF treatments were conducted in a batch PEF system (Physics International, San Leandro, CA, USA) equipped with a TG-70 gas control unit and a PT55 pulse generator (Pacific Atlantic Electronics Inc., El Cerrito, CA, USA). The system delivers exponential pulses of 4 μs from a capacitor of 0.1 μF at a frequency of 0.1 Hz. The treatment chamber consists of a parallelepiped methacrylate container with two parallel stainless-steel electrodes (20 × 5 cm) separated by a gap of 5 cm. Whole carrots (0.1 kg) were immersed in an aqueous solution (conductivity of 10 μS cm^{-1}) and placed in parallel to the electrodes. Then, carrots were subjected to five pulses of 3.5 kV cm^{-1} (0.61 kJ kg^{-1}) and were stored at 4 °C for 24 h. The treatment conditions were selected based on previous results in which phenolic and carotenoid bioaccessibilities were enhanced in whole carrots [20].

The specific energy input was calculated based on Wiktor et al. [22], and the medium temperature was measured after treatment application to ensure that it remained constant after PEF application.

2.4. Preparation of Carrot Derived Products

Untreated and PEF-treated carrots were washed with tap water, and the excess was removed with a paper cloth before discarding their top and bottom ends. Two types of puree were obtained: one batch from untreated (U) and the other from PEF-treated carrots (PEF). Purees were prepared by mixing approximately 500 g of 1-cm thick carrot slices with water (1:1) (*w/w*) in a food processor (Taurus Mycook) operated with the crushing function at full power in two 10-s intervals.

To prepare oil-added carrot purees, extra virgin olive oil (Borges Branded Foods, S.L.U., Tàrrega, Lleida, Spain) was added (5% *w/w*), and the homogenates were stirred for 15 min at 8000 rpm with an Ultra-Turrax IKA equipped using a 3-blade stirring rod. The olive oil included 0.86 mg/100 g fresh weight (FW) of α-carotene, 2.59 mg/100 g FW of β-carotene, 0.18 mg kg^{-1} dry weight (DW) of coumaric acid, 0.02 mg kg^{-1} DW of caffeic acid, and 0.02 mg kg^{-1} DW of ferulic and isoferulic acids.

Two types of carrot juices were obtained from approximately 500 g of 1-cm thick carrot slices using a cold blender (Imetec Succovivo SJ1000 coupled to a filter of 0.4 mm). One batch was obtained from untreated carrots and another from PEF-treated carrots (0.61 kJ kg^{-1}). The resulting purees and juices were divided into two fractions, that were thermally treated (U/T or PEF/T) or remained unheated (U or PEF) as a reference for the former treatments.

Thermal treatment was applied in order to inactivate pectin methylesterase and peroxidase activities [23–25], thus, obtaining stable products. Carrot purees or juices (200 g) were packed in re-sealable polyethylene bags (20 × 15 cm) and heated in a water bath for 10 min at 70 °C. The product temperature was monitored during treatment to assure that the purees/juices did not exceed 70 °C. Thereafter, the purees were cooled under a constant flow of cold water for 3 min. Aliquots (20 mL) of the non-digested fractions were stored at −40 °C until the extraction and analysis of carotenoids were performed. Then, the samples were freeze-dried in order to extract the phenolic content. Additional aliquots (20 mL) were subjected to an in vitro digestion to determine their carotenoid and phenolic contents in digesta.

2.5. Evaluation of Quality Attributes

The color was evaluated by measuring the CIEL* a* and b* parameters with a colorimeter (Minolta CR-400, Konica Minolta Sensing, INC., Osaka, Japan). The total color difference (ΔE) was also calculated using Equation (1).

$$\Delta E = [[(L^* - L^*_0)^2 + (a^* - a^*_0)^2 + (b^* - b^*_0)^2]]^{0.5} \quad (1)$$

where L^*_0, a^*_0, and b^*_0 refer to untreated carrot products and L*, a*, and b* correspond to the data collected after treatments.

The pH was assessed in products using a pH meter (Crison Instruments S.A., Alella, Barcelona, Spain). The total soluble solids (TSS) was measured using a refractometer (Atago Company Ltd., Tokyo, Japan) and expressed as % of the total soluble solids.

2.6. Particle Size Distribution

A Mastersizer 3000™ (Malvern Instruments Ltd., Worcestershire, UK) was used to measure the particle size distribution of juices and purees. The results were expressed in terms of the volume and surface diameter, D [4, 3] and D [3, 2], respectively. The refractive index of water was 1.33, and particle calculation was set for irregular particles.

2.7. Microstructure

The microstructure was investigated using a light microscope (BX41, Olympus, Göttingen, Germany) equipped with UIS2 optical system. We mounted 10 µL drops on glass slides without staining and microscopically observed them. A general inspection of the samples was made, and representative photos were taken with the 10× lens. All images were processed using the instrument software Olympus CellSense (Barcelona, Spain).

2.8. In Vitro Digestion

The in vitro digestion procedure was performed according to the standardized COST Infogest protocol [26], in which electrolyte and enzymatic solutions to simulate the phases of human digestion are described. Digestions were performed in darkness, in the absence of oxygen (bottles were flushed with nitrogen gas) in an orbital incubator (Ovan, Badalona, Spain) at 37 °C and 120 rpm. Electrolyte concentrations and enzyme activities were prepared following the indications provided by Minekus et al. [26], and blank samples consisting of water instead of carrot products, were made in identical conditions.

The oral phase was omitted due to the very short residence times of purees and juices in the oral cavity [26]. Then, the gastric phase started by adding 20 mL of simulated gastric fluid (pH 3 and 37 °C) and pepsin to 20 g of juice/puree. This mixture was incubated at 37 °C for 2 h in agitation. The duodenal phase was initiated by inserting a cellulose-membrane dialysis bag (molecular weight cut-off 12,000 Da, Sigma-Aldrich, St. Louis, MO, USA), which contained simulated intestinal fluid (pH 7 and 37 °C). This dialysis bag simulates the intestinal epithelium, and it harbors phenolic compounds released from the matrix (bioaccessible fraction) [19,26].

After 30 min of incubation, the pH was adjusted to 7, and a solution containing simulated intestinal fluid (pH 7 and 37 °C), bile extract, pancreatin (and lipase in case of purees containing oil) was added and the mixture was further incubated for 2 h. At the end of digestion, the dialysis bags were rinsed with distilled water until clean, and its contents were collected. The remaining digesta, which contained carotenoid compounds, was centrifuged at 5000× g for 15 min at 4 °C [27–29], and the supernatant was also collected. Digested fractions were freeze-dried and stored at −40 °C until analysis.

2.9. Carotenoids Determination

2.9.1. Carotenoids Extraction

Carotenoids were extracted following the method described by Sadler et al. [30], with slight modifications. An extraction solution (50 mL) composed of hexane:acetone:ethanol (50:25:25) and 1 g·L^{-1} BHT was added to carrot puree (2 g) or juice (1 g) and was stirred for 20 min. Then, 15 mL of NaCl (10% (w/v)) solution was added, and the samples were stirred for 10 additional minutes. The samples were left to stand for ≥3 minutes, and the upper organic phase was microfiltered across a nylon filter (0.45 µm, ø 13 mm, Labbox Labware S.L., Barcelona, Spain) and analyzed using High-Performance Liquid Chromatography with Diode Array Detection (HPLC-DAD).

Recovery of the carotenes from the micellar digested fraction was performed by adding 5 mL of the extraction solution to 0.2 g of freeze-dried digesta. After that, the samples were vortexed for 20 s, and 1 mL of NaCl solution (10% (w/v)) was added. The samples were vortexed for another 20 s and centrifuged at 4000× g for 5 min [29]. An aliquot of the upper organic phase was microfiltered across a nylon filter (0.45 µm, ø 13 mm, Labbox Labware S.L., Barcelona, Spain) and analyzed by HPLC using the same method as for non-digested fractions.

All extractions were performed in duplicate, and the samples were protected from light throughout extraction and analysis to avoid carotenoid degradation and isomerization.

2.9.2. Identification and Quantification of Carotenoids by HPLC-DAD

The carotenoids were quantified by HPLC-DAD, following a procedure validated by Cortés et al. [31]. The HPLC system was equipped with a 600 Controller, a 486 Absorbance Detector, a thermostatic column compartment, and a 717 Plus Auto Sampler with a cooling system (Waters, Milford, MA, USA). An aliquot of 20 µL of the extracted samples was injected and the carotenoids were separated using a reverse-phase C18 Spherisorb ODS2 (5 µm) stainless steel column (4.6 × 250 mm).

The mobile phase consisted of: (A) methanol/ammonium acetate 0.1 M, (B) milli-Q water, (C) methyl tert-butyl ether, and (D) methanol. The flow rate was fixed at 1 mL min^{-1}, and the total run time was 60 min. The column was set at 30 °C, while sample amber vials on the autosampler were preserved at 4 °C. Identification was carried out by UV–vis spectral data and their retention times [31]. Carotenoids were quantified by using calibration curves and integrating peak areas. The results are expressed on a fresh weight basis.

2.10. Phenolic Compounds Determination

2.10.1. Phenolic Compounds Extraction

Phenolic compounds were extracted from freeze-dried non-digested or digested carrot puree/juice (0.2 g). For non-digested juices and digested fractions of purees and juices, 1 mL of methanol (80:20 *v*/*v*) was added, whereas 1.5 mL was needed for the non-digested purees. The samples were vortexed for 1 min and then centrifuged (16,209× *g*, 15 min, 4 °C). The clear supernatant was microfiltered using polyvinylidene difluoride (PVDF) filters (0.2 µm) (Scharlab, Barcelona, Spain) prior to injection into the chromatographic system.

2.10.2. Identification and Quantification of Phenolic Compounds by Ultra-Performance™ Liquid Chromatography (UPLC-MS/MS)

Phenolic compounds and their generated metabolites were determined in methanolic extracts obtained from freeze-dried non-digested or digested fractions. AcQuity Ultra-Performance™ liquid chromatography (UPLC) coupled to a triple quadrupole detector (TQD) mass spectrometer (Waters, Milford, MA, USA) was used. The analytical column was an AcQuity BEH C18 column (100 × 2.1 mm i.d., 1.7 µm), equipped with a VanGuard™ Pre-Column AcQuity BEH C18 (2.1 × 5 mm, 1.7 µm). During the analysis, the column was kept at 30 °C, and the flow rate was 0.3 mL min^{-1}.

Mobile phases were acetic acid (0.2%) and acetonitrile (Table 1). Tandem MS analyses were carried out on a triple quadrupole detector (TQD) mass spectrometer (Waters, Milford, MA, USA) equipped with a Z-spray electrospray interface (ESI). Ionization was achieved using the electrospray interface operating in the negative mode $[M-H]^-$, and the data were acquired through selected reaction monitoring (SRM). The dwell time established for each transition was 30 ms. Data acquisition was carried out with the MassLynx 4.1 software (Milford, MA, USA).

2.11. Bioaccessibility Calculation

The bioaccessibility of individual compounds was determined using Equation (2), and results are expressed as a percentage.

$$\text{Bioaccessibility (\%)} = CC_{\text{digested}}/CC_{\text{non-digested}} \times 100 \quad (2)$$

where CC_{digested} corresponds to the overall concentration of each compound in the absorbable fraction and $CC_{\text{non-digested}}$ is the concentration in non-digested samples.

The carotenoid bioaccessibility was calculated referring to the concentration found in the digested micellar fraction, whereas the phenolic compound bioaccessibility was calculated in reference to their concentration in the dialyzed digested fraction.

Table 1. Chromatographic conditions for phenolic compounds identification by UPLC.

Time (min)	Acetic Acid (0.2% v/v) (%)	Acetonitrile (%)
0	95	5
5	90	10
10	87.6	12.4
18	72	28
21	15	85
23	0	100
25.5	0	100
27	95	5
30	95	5

2.12. Statistical Analysis

Statistical analyses were carried out using the SigmaPlot 11.0 software (Systat Software Inc., Chicago, IL, USA). Three different replicates were submitted to each assayed treatment condition, and each analysis was conducted thrice, excepting the extraction and determination of phenolics and carotenoids, which were conducted twice. The results are reported as the mean ± standard deviation and were subjected to an analysis of variance (ANOVA) followed by the Tukey post hoc test to establish statistical differences among treatments. A two-way ANOVA was carried out for establishing differences between derived products. In the case where the results showed no homogeneity in their variance, they were subjected to ANOVA on ranks by the Kruskal–Wallis test. The statistical significance level was set up at $p < 0.05$.

3. Results

3.1. Quality Attributes

The quality attributes from differently treated carrot juices and purees are shown in Table 2. The TSS and pH were not generally influenced either by PEF treatment or by further processing conditions, although the color was differently affected. PEF application to carrot matrix did not cause significant changes in the L* or a* values of any carrot-derived product, although b* was lower in juices. Hence, the ΔE value of PEF purees was 1.6, whereas that of PEF juices reached 3.8. In the latter case, color changes would be visually apparent, as ΔE > 2 have been reported to be noticeable by consumers [32]. Similar results were reported by Aguiló-Aguayo et al. [33] and Xiang et al. [34] in PEF-treated carrot purees and juices, respectively.

Oil addition and thermal treatment also affected the color of different products. On the one hand, oil addition increased the values of the CIEL*a*b* coordinates compared to purees without added oil. These differences would be detectable by consumers, given that ΔE values were between 3.3 and 4.7. On the other hand, the temperature did not affect the L* nor b* in any derived product. However, thermally treated juices (U/T and PEF/T) exhibited higher a* values than those untreated, and the ΔE values were higher than 2. Moreover, U/T and PEF/T oil-added purees had lower a* compared with untreated ones.

Color differences could be associated with the disruption of cells and the breakage of the chromoplast carotenoid–protein complexes, leading to the release of carotenoids [33,35]. Likewise, non-enzymatic Maillard reactions could also be responsible for color changes after thermal treatments, since carrot juice had greater sugar contents, and high temperatures could favor this reaction [36]. In addition, Mutsokoti et al. [37] demonstrated that carotenoid transfer to oil was enhanced by thermal treatments, although carotenoid degradation was also more pronounced in the presence of oil [38], which could explain the decrease in a* values.

Table 2. Quality attributes and particle size of untreated carrot purees and juices (U), those obtained from PEF-treated (0.61 kJ kg^{-1}) carrots (PEF), and those thermally treated (10 min at 70 °C) (U/T and PEF/T).

Product	Treatment	L*	a*	b*	ΔE	pH	TSS (%)	D [4, 3] (μm)	D [3, 2] (μm)
Puree	U	$41.3 \pm 0.3^{a,A}$	$11.0 \pm 0.3^{a,A}$	$24.4 \pm 0.2^{a,A}$	-	$6.4 \pm 0.1^{a,b,c,A}$	$3.6 \pm 0.3^{a,A}$	$596 \pm 12^{a,A}$	$207 \pm 48^{a,A}$
	PEF	$40.5 \pm 0.1^{a,A}$	$10.3 \pm 0.1^{a,A}$	$23.1 \pm 0.2^{a,A}$	$1.6 \pm 0.2^{a,A}$	$6.5 \pm 0.1^{b,A}$	$3.4 \pm 0.2^{a,A}$	$589 \pm 16^{a,A}$	$183 \pm 34^{a,A}$
	U/T	$42.0 \pm 0.2^{a,A}$	$10.4 \pm 0.5^{a,A}$	$24.3 \pm 0.5^{a,A}$	$1.3 \pm 0.1^{a,A}$	$6.1 \pm 0.1^{d,A}$	$3.7 \pm 0.0^{a,A}$	$608 \pm 11^{a,A}$	$472 \pm 20^{b,A}$
	PEF/T	$41.0 \pm 0.4^{a,A}$	$9.9 \pm 0.5^{a,A}$	$23.1 \pm 0.8^{a,A}$	$1.8 \pm 0.7^{a,A}$	$6.3 \pm 0.0^{c,A}$	$3.5 \pm 0.3^{a,A}$	$601 \pm 12^{a,A}$	$431 \pm 53^{b,A}$
Oil-added puree	U	$56.0 \pm 1.7^{a',B}$	$15.2 \pm 0.7^{a',B}$	$44.8 \pm 1.9^{a',B}$	-	$6.3 \pm 0.0^{a',A}$	$4.0 \pm 0.6^{a',A}$	$449 \pm 41^{a',B}$	$15 \pm 3^{a',B}$
	PEF	$55.7 \pm 0.3^{a',B}$	$14.5 \pm 0.5^{a',b',B}$	$45.6 \pm 3.9^{a',B}$	$3.3 \pm 2.4^{a'A}$	$6.4 \pm 0.1^{a',A}$	$3.8 \pm 0.4^{a',A}$	$422 \pm 15^{a',B}$	$13.8 \pm 2.0^{a',B}$
	U/T	$55.3 \pm 0.7^{a',B}$	$13.8 \pm 0.4^{b',B}$	$45.6 \pm 3.9^{a',B}$	$3.6 \pm 1.9^{a',B}$	$6.0 \pm 0.0^{b',A}$	$4.2 \pm 0.4^{a',A}$	$460 \pm 30^{a',B}$	$20 \pm 4^{a',B}$
	PEF/T	$55.2 \pm 0.7^{a',B}$	$12.4 \pm 0.9^{d',B}$	$47 \pm 3^{a',B}$	$4.7 \pm 2.1^{a',B}$	$6.1 \pm 0.0^{c',B}$	$3.9 \pm 0.1^{a',A}$	$408 \pm 39^{a',B}$	$16 \pm 4^{a',B}$
Juice	U	$43.0 \pm 0.5^{a''}$	$15.8 \pm 0.7^{a'',b''}$	$29.2 \pm 0.7^{a''}$	-	$6.2 \pm 0.1^{a''}$	$8.2 \pm 0.5^{a''}$	$487 \pm 43^{a''}$	$66 \pm 17^{a'',b''}$
	PEF	$41.3 \pm 0.3^{a''}$	$13.4 \pm 0.7^{a'',b''}$	$26.9 \pm 0.7^{b''}$	$3.8 \pm 1.9^{a''}$	$6.2 \pm 0.1^{a''}$	$8.2 \pm 06^{a''}$	$499 \pm 4^{a''}$	$74.3 \pm 0.9^{a''}$
	U/T	$40.7 \pm 2.3^{a''}$	$18.6 \pm 1.6^{c''}$	$28.1 \pm 1.3^{a'',b''}$	$4.8 \pm 1.7^{a''}$	$6.1 \pm 0.0^{a''}$	$7.7 \pm 0.4^{a''}$	$359 \pm 48^{b''}$	$32 \pm 2^{b''}$
	PEF/T	$41.7 \pm 0.3^{a''}$	$17.9 \pm 0.6^{c''}$	$27.7 \pm 0.7^{a'',b''}$	$2.86 \pm 0.2^{a''}$	$6.3 \pm 0.2^{a''}$	$7.5 \pm 0.2^{a''}$	$408 \pm 61^{b''}$	$46 \pm 3^{b''}$

Values are means ± standard deviation. Different letters in the same column within the same product indicate significant ($p < 0.05$) differences among treatments (puree: lowercase letters, puree with oil: lowercase letters′, and juice: lowercase″). Different uppercase letters in the same row indicate significant differences between both purees. L*: lightness, a*: redness, b*: yellowness, ΔE: total color difference, TSS: total soluble solids, D [4, 3]: mean volume diameter, and D [3, 2]: mean surface diameter.

3.2. Particle Size Distribution

The mean particle size of differently treated carrot-based products is displayed in Table 2. Purees were constituted by 3–5% of particles smaller than 100 μm, which may suggest that they were mainly formed by cell fragments or single cells since carrot cell size has been reported to average ca. 125 μm [7]. On the other hand, juices had between 13–17% of particles below 100 μm, which is likely due to the removal of cell clusters during juicing. Hence, juices mainly contained small fragments of disrupted tissue and cells.

PEF treatment before further processing did not cause significant changes in the mean particle size of any derived product. Untreated and PEF juices were characterized by D [3, 2] of 66 ± 17 μm and 74.3 ± 0.9 μm, respectively. Otherwise, untreated and PEF purees showed D [3, 2] = 207 ± 48 μm and 183 ± 34 μm, respectively. Some authors reported that PEF treatment may facilitate juice-pressing due to structural rearrangements. Thus, derived products with smaller particle sizes than those untreated could be obtained [39]. However, the results from this work suggest that PEF treatment was not intense enough to cause a lower resistance to mechanical load and promote changes in particle size. Similarities in particle size distribution between PEF and untreated products could indicate that they will have similar stability during storage.

Thermal treatments (U/T and PEF/T) caused changes in the area-based mean diameter of juices and purees without oil, whereas those with oil added were not affected. Purees showed higher D [3, 2] values (472 ± 20 and 431 ± 53 μm) compared to those of untreated purees (207 ± 48 μm). This increase in the size of small particles after thermal treatments has been reported in thermally treated tomato suspensions [40], which was attributed to the swelling of cells or the formation of aggregates from cellular components [10].

However, it could also mean that the largest cell clusters were disintegrated in smaller cell aggregates, which increased D [3, 2]. On the other hand, juices (U/T and PEF/T) had between 29–34% of particles below 100 μm and D [3, 2] in the range of 32–46 μm. These results indicate that thermal treatments increased the number of small particles, reducing the mean particle size (D [3, 2]). This was also observed in pasteurized orange juices [41,42]. Thermal treatments between 65 and 95 °C have been reported to cause disruptions between chromoplast membrane structures and/or carotenoid protein interactions, which could cause a decrease in the particle size of juices [10,43].

On the other hand, the particle sizes (D [4, 3] and D [3, 2]) of oil-added purees were significantly lower than that of purees without oil. Oil-added purees had between 29–37% of particles below 100 μm. These values indicate that purees would be mainly formed by cell fragments or single cells together with oil droplets (20 μm) [7]. Any treatment affected the oil-added purees particle size. These results were likely highly influenced by the presence of oil droplets, which likely has a major influence on the particle size distribution and masked changes in the particle size of purees.

3.3. Microstructure

The microstructure of juices, purees, and oil-added purees is shown in Figure 1. Both purees showed clusters of whole cells with carotenoids inside. On the other hand, whole cells were rarely detected in juices (Figure 1) given that cells and chromoplasts are likely comminuted during processing. Juices are mainly composed of cloud particles of different densities formed by chromoplast fragments and carotenoids, which was verified through isopycnic gradient centrifugation by Marx et al. [44].

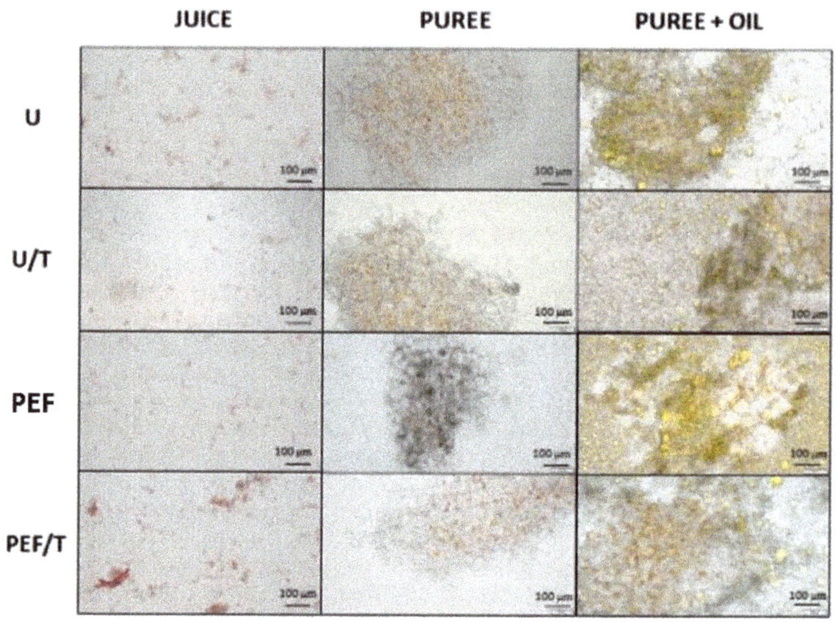

Figure 1. Representative light micrographs of juices and purees obtained from untreated carrots (U), obtained from PEF-treated (0.61 kJ kg^{-1}) carrots (PEF) and thermally treated (10 min at 70 °C) (U/T and PEF/T).

The application of PEF to whole carrots did not cause significant changes in the microstructure of the obtained juices. However, PEF purees showed some starch grains inside cells that were absent in untreated purees. Gómez Galindo et al. [45] demonstrated that PEF can trigger changes in the hexose pool due to stress induction. Specifically, it could have affected the AGPase activity, which is involved in starch biosynthesis, or starch-degrading enzymes [46]. To the best of our knowledge, microstructural studies of derived products obtained from PEF-treated matrices are limited. However, some authors have observed irregular cell walls, such as folds and loss of smoothness in PEF-treated whole tomatoes and carrots [18,47].

Thermal treatment of the derived products was shown to lead to the presence of thin cell walls in purees, probably as a consequence of depolymerization and pectin degradation [48]. Additionally, temperature likely favors interactions between dissolved particles [49], which would explain the increases in particle size, D [3, 2] (Table 2). On the other hand, thermally treated juices had a greater content of small particles than those untreated or PEF. Temperatures between 65–95 °C are able to break carotenoid-chromoplasts complexes [43], which could explain the decrease in particle size (Table 2) and increased carotenoid release (Figure 1).

The oil-added purees showed lipid droplets surrounding carrot clusters (Figure 1). Oil was especially internalized in thermally treated and PEF purees, given that the disruption of cell membranes may facilitate the efflux and influx of intra- and extracellular content. These results have been previously reported in excipient emulsions of olive oil and tomato by Li et al. [50], who attributed this lipid entry to capillary forces generated by pores in tissues [51].

3.4. Carotenoid Content

The carotenoid contents from differently treated products are shown in Figure 2. PEF application to whole carrots did not affect the content nor carotenoid profile, although further processing caused a significant effect in carotenoid concentration. Hence, different carotenoid contents were obtained depending on the type of derived product; however, neither thermal treatment nor oil addition caused significant changes. Carrot juices had higher carotenoid contents (43.4 ± 7–48 ± 8 mg/100 g FW) compared with both types of puree (20.7 ± 2.5–23.6 ± 1.8 mg/100 g FW). However, these differences were mainly related to dilution during the puree preparation. The obtained results are in accordance with Hedrén et al. [8], given that mechanical processing was more determinant to the release of carotenoids than thermal treatment or oil addition.

The main carotenoids present in carrot-derived products were α-carotene and β-carotene. These results are in agreement with those previously reported [52,53], although some authors have also detected lutein presence in carrots. Panozzo et al. [54] hypothesized that crystalloid chromoplasts were more easily disrupted by mechanical processing compared with globular chromoplasts, in which lutein is usually dissolved. In addition, lutein is more susceptible to degradation due to the presence of oxygen in its chemical structure [55]. Hence, it could be degraded during mechanical processing or be entrapped in chromoplasts.

The carotenoid contents of derived products obtained from PEF-treated whole matrices have been scarcely studied. However, increases in the carotenoid content of purees and juices obtained from PEF-treated tomatoes have been reported [18,21]. Such increases were attributed to two main causes: (1) their accumulation in the whole product resulting from the induction of a stress defense response and (2) their better extractability due to electropermeabilization. On the other hand, Rybak et al. [39] observed a decreased content in juices obtained from PEF-treated peppers (3 kJ kg^{-1}), which was attributed to promoted oxidation or isomerization reactions. Therefore, our results suggest that selected PEF conditions did not induce carotenoid biosynthesis during 24 h of storage nor improve their extractability in carrot-based products.

Thermal treatment did not enhance the carotenoid extractability nor induce their degradation in any derived product. Similar results were obtained in thermally treated oil-added carrot purees [6]. However, decreases in carrot juices [56] and increases in carrot purees [57] have been reported. Differences between our results and those previously mentioned are likely due to variations in the processing parameters (time or temperature) or derived product preparation procedures. These processing conditions likely also caused differences in particle size, structural properties, or enzyme activities, which are closely related to carotenoid extractability or degradation.

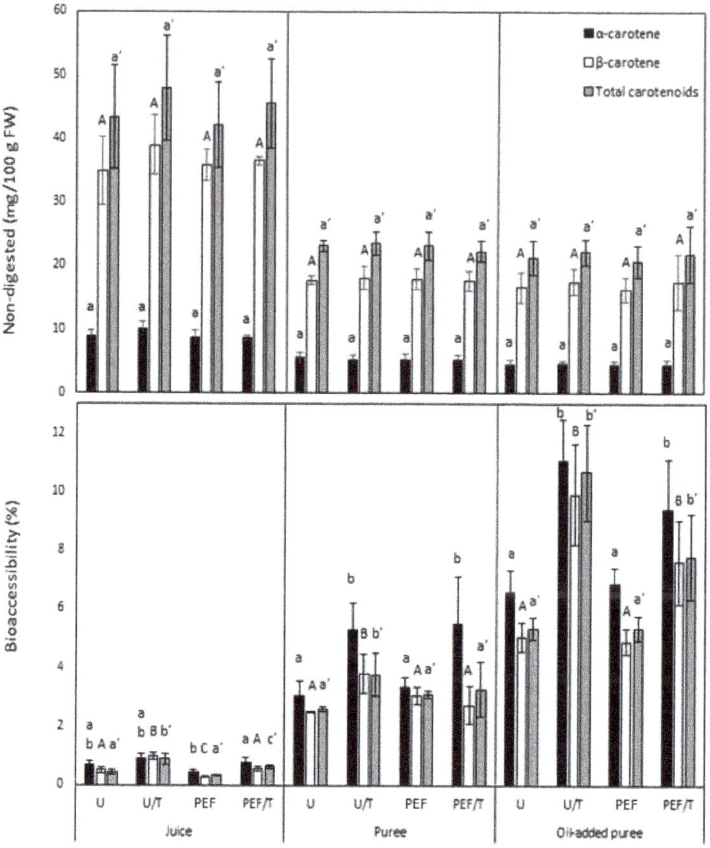

Figure 2. The carotenoid content and bioaccessibility of untreated purees and juices (U), those obtained from PEF-treated (0.61 kJ kg^{-1}) carrots (PEF), and those thermally treated (10 min at 70 °C) (U/T and PEF/T). Values are means ± standard deviation. Different letters within the same product and compound indicate significant ($p < 0.05$) differences among treatments (α-carotene: lowercase letters; β-carotene: uppercase letters; and total carotenoids: lowercase′).

3.5. Carotenoid Bioaccessibility

The carotenoid bioaccessibility was affected by PEF treatment and further processing. Mechanical processing was the main factor that influenced bioaccessibility, followed by oil addition and thermal treatment (Figure 2). The highest total bioaccessibility was obtained in oil-added purees (5.3%), whereas purees (2.6%) and juices (0.4%) had lower bioaccessibility.

The application of PEF to carrots before obtaining derived products did not affect the total bioaccessibility. Generally, thermal treatments were those that led to further enhancement of the total bioaccessibility: oil-added purees (10.7%) > purees (3.8%) > juices (0.9%). Regarding the bioaccessibility of individual carotenoids, α-carotene and β-carotene similarly increased in any of the studied products as a consequence of the application of thermal treatments, whereas PEF only caused a decrease in β-carotene bioaccessibility when it was applied before juicing (Figure 2).

Regarding the carotenoid bioaccessibility in derived products obtained from PEF-treated carrots, they contrast with those reported by González-Casado et al. [21]. The authors presented increases in the total carotenoids and β-carotene bioaccessibility of purees obtained from PEF-treated tomatoes, which were attributed to electropermeabilization and better carotenoids release. The initial content of such tomato purees was considerably higher than in the untreated samples, which could make a difference regarding our study since the content in non-digested products was similar regardless of the applied treatment.

In addition, an observed decrease in the β-carotene bioaccessibility from juices was also reported by Bot et al. [17] in PEF-treated tomato chromoplast fractions. This was attributed to induced modifications in carotenoid-protein complexes, which limit their bioaccessibility. Finally, some authors have also suggested that released carotenoids can be broken down during digestion into non-detected metabolites (e.g., oxidation products) [58].

Carotenoids are generally located inside chromoplasts or bound to the membrane [59]. Hence, chromoplasts are probably comminuted during juicing or blending, which makes carotenoids more available to be absorbed when the particle size decreases as a consequence of processing (Table 2). Apart from a particle size decrease, carotenoids bioaccessibility also depends on their chemical structure, interactions with other macromolecules, micellarization, content, and the characteristics of pectin [60]. The increase in bioaccessibility after thermal treatments has been related to carotenoid isomerization [44], given that cis isomers are better assimilated than trans [61]. On the other hand, thermal processing can degrade pectin and may improve carotenoid bioaccessibility [48,62], given that a high pectin content would entrap carotenoids or act as a barrier for lipase [63], hindering their micellarization.

The solubilization of carotenoids into micelles is another critical step in carotenoid absorption. The low amount of lipids in purees and juices could hinder their micellarization, given that carotenoids are not water-soluble. Hence, the presence of lipids facilitates carotenoid transference to micelles [64]. The obtained results indicate that the application of thermal treatment to oil-added purees caused the highest bioaccessibility. Previous studies have shown that thermal-treated carrot cell clusters (95–110 °C) improved carotenoid transfer to oil [65].

3.6. Phenolic Content

The phenolic content was affected by both PEF treatment and further processing (Table 3). Carrot juices showed the highest phenolic content (322 ± 56 mg kg^{-1} DW), whereas oil-added purees (81 ± 36 mg kg^{-1} DW) and purees (62 ± 23 mg kg^{-1} DW) exhibited similar contents. PEF treatment applied to whole carrots generally did not affect the phenolic content of both types of puree, although it decreased the content in juices (38.5%). Despite this reduction, the phenolic content in juices remained higher than in purees. On the other hand, the total phenolic content in purees without added oil was doubled when a thermal treatment was applied, whereas the contents in juices and oil-added purees were not significantly affected.

The main phenolic compounds found in carrot-derived products were hydroxycinnamic acids, namely 5-caffeoylquinic acid, coumaroylquinic acid or 5-feruloylquinic acid (Table 3). These results are in accordance with previous studies performed in whole carrots [20,66] or juices [67,68]. It has been reported that a low dietary fiber content in juices is beneficial for releasing phenolic compounds [69]. Furthermore, the extractability of phenolic compounds may be enhanced in juices due to their lower particle size (Table 2). Likewise, differences in phenolic composition between purees and juices could also be related to mechanical processing. The procedure applied to obtain juices may have enhanced the extraction of compounds tightly linked to cell walls, whereas carrot cells were not totally disrupted in purees.

Table 3. Phenolic content in untreated purees and juices (U), those obtained from PEF-treated (0.61 kJ kg^{-1}) carrots (PEF), and those thermally treated (10 min at 70 °C) (U/T and PEF/T).

Phenolic Compounds	Puree				Oil-Added Puree				Juice			
	U	PEF	U/T	PEF/T	U	PEF	U/T	PEF/T	U	PEF	U/T	PEF/T
Coumaric acid	5.05 ± 0.21 ^{a}	7.17 ± 1.82 ^{b}	2.91 ± 0.50 ^{c}	1.03 ± 0.24 ^{d}	4.33 ± 0.58 ^{A}	2.84 ± 0.30 ^{B,C}	1.67 ± 0.04 ^{C}	2.16 ± 0.71 ^{C}	0.70 ± 0.07 $^{a'}$	0.79 ± 0.04 $^{a'}$	0.71 ± 0.13 $^{a'}$	0.75 ± 0.05 $^{a'}$
Coumaroylquinic acid	17 ± 8 ^{a}	8 ± 3 ^{a,b}	9.0 ± 2.5 ^{a,b}	5.9 ± 1.23 ^{b}	3.22 ± 0.60 ^{A}	3.6 ± 0.74 ^{A}	6.2 ± 2.3 ^{A}	10 ± 4 ^{A}	15 ± 7 $^{a'}$	6.01 ± 0.37 $^{b'}$	19 ± 3 $^{a'}$	6.41 ± 0.48 $^{b'}$
Coumaric acid and its derivates	22 ± 8 ^{a}	15 ± 5 ^{a,b}	12 ± 3 ^{a,b}	6.9 ± 1.5 ^{b}	7.6 ± 0.8 ^{A}	6.4 ± 1.0 ^{A}	7.9 ± 2.2 ^{A}	12 ± 4 ^{A}	15 ± 7 $^{a'}$	6.80 ± 0.36 $^{b'}$	19 ± 4 $^{a'}$	7.2 ± 0.5 $^{b'}$
Caffeic acid	1.49 ± 0.92 ^{a}	0.83 ± 0.13 ^{a}	0.92 ± 0.30 ^{a}	0.42 ± 0.07 ^{b}	1.72 ± 0.64 ^{A}	0.61 ± 0.12 ^{B}	0.51 ± 0.24 ^{B}	0.39 ± 0.07 ^{B}	1.3 ± 0.2 $^{a'}$	0.90 ± 0.11 $^{a'}$	2.47 ± 0.34 $^{b'}$	1.13 ± 0.10 $^{a'}$
Caffeic acid arab/xiloside	nd ^{1,a}	nd ^{1,a}	nd ^{a}	nd ^{a}	nd ^{1,A}	nd ^{1,A}	nd ^{1,A}	nd ^{1,A}	0.12 ± 0.03 $^{a'}$	0.06 ± 0.01 $^{a'}$	0.29 ± 0.06 $^{b'}$	0.14 ± 0.02 $^{a'}$
Caffeoylshikimic acid	nd ^{1,a}	nd ^{1,a}	nd ^{1,a}	nd ^{1,a}	nd ^{1,A}	nd ^{1,A}	nd ^{1,A}	nd ^{1,A}	0.06 ± 0.02 $^{a'}$	0.02 ± 0.01 $^{b'}$	0.05 ± 0.01 $^{a'}$	0.02 ± 0.003 $^{b'}$
3-caffeoylquinic acid	nd ^{1,a}	nd ^{1,a}	0.18 ± 0.08 ^{b}	0.23 ± 0.02 ^{b}	nd ^{1,A}	nd ^{1,A}	nd ^{1,A}	0.08 ± 0.01 ^{A}	0.21 ± 0.02 $^{a'}$	0.14 ± 0.02 $^{a'}$	0.97 ± 0.11 $^{b'}$	0.57 ± 0.09 $^{c'}$
5-caffeoylquinic acid	32 ± 17 ^{a}	15.1 ± 1.0 ^{a}	106 ± 23 ^{b}	85 ± 13 ^{b}	67 ± 35 ^{A}	65 ± 20 ^{A}	79 ± 6 ^{A}	59 ± 19 ^{A}	264 ± 47 $^{a'}$	83 ± 7 $^{b'}$	264 ± 32 $^{a'}$	92 ± 7 $^{c'}$
4-caffeoylquinic acid	nd ^{1,a}	nd ^{1,a}	1.73 ± 0.66 ^{b}	1.62 ± 0.09 ^{b}	0.17 ± 0.10 ^{A}	0.30 ± 0.12 ^{A}	0.8 ± 0.1 ^{B}	0.65 ± 0.09 ^{B}	0.76 ± 0.14 $^{a'}$	0.34 ± 0.07 $^{a'}$	7.66 ± 1.15 $^{b'}$	3.4 ± 0.3 $^{c'}$
Dicaffeoylferuoylquinic acid	nd ^{1,a}	nd ^{1,a}	0.25 ± 0.10 ^{b}	0.21 ± 0.03 ^{b}	nd ^{1,A}	nd ^{1,A}	nd ^{1,A}	nd ^{1,A}	0.31 ± 0.06 $^{a'}$	0.06 ± 0.01 $^{b'}$	0.55 ± 0.07 $^{c'}$	0.07 ± 0.02 $^{b'}$
Caffeoylferuoylquinic acid	0.35 ± 0.10 ^{a}	0.38 ± 0.07 ^{a}	0.29 ± 0.11 ^{a}	0.35 ± 0.08 ^{a}	0.24 ± 0.05 ^{A}	0.14 ± 0.02 ^{A}	0.23 ± 0.09 ^{A}	0.18 ± 0.05 ^{A}	0.48 ± 0.02 $^{a'}$	0.56 ± 0.05 $^{b'}$	0.43 ± 0.06 $^{a'}$	0.62 ± 0.06 $^{b'}$
Caffeic acid arabinoside glucoside	0.06 ± 0.02 ^{a}	0.1 ± 0.02 ^{a}	0.07 ± 0.01 ^{a}	0.11 ± 0.02 ^{a}	0.05 ± 0.002 ^{A}	0.07 ± 0.02 ^{A}	0.05 ± 0.02 ^{A}	0.07 ± 0.01 ^{A}	0.12 ± 0.06 $^{a'}$	0.18 ± 0.01 $^{b'}$	0.09 ± 0.02 $^{a'}$	0.22 ± 0.03 $^{b'}$
Caffeic acid Glu Acetyl glucoside	0.84 ± 0.46 ^{a}	0.7 ± 0.19 ^{a}	0.8 ± 0.5 ^{a}	0.78 ± 0.07 ^{a}	0.5 ± 0.04 ^{A}	1.0 ± 0.35 ^{A}	0.8 ± 0.14 ^{A}	1.04 ± 0.13 ^{A}	3.8 ± 0.2 $^{a'}$	4.32 ± 0.24 $^{a',b'}$	3.34 ± 0.57 $^{a'}$	4.7 ± 0.4 $^{b'}$
Caffeic acid and its derivates	35 ± 17 ^{a}	17.2 ± 1.3 ^{a}	111 ± 23 ^{b}	89 ± 13 ^{b}	69 ± 35 ^{A}	67 ± 20 ^{A}	81 ± 5 ^{A}	62 ± 19 ^{A}	271 ± 48 $^{a'}$	90 ± 7 $^{b'}$	280 ± 34 $^{a'}$	103 ± 7 $^{c'}$
Ferulic acid	0.32 ± 0.07 ^{a}	0.40 ± 0.15 ^{a}	0.35 ± 0.14 ^{a}	0.40 ± 0.02 ^{a}	0.32 ± 0.03 ^{A}	0.2 ± 0.02 ^{A}	0.54 ± 0.18 ^{A}	0.49 ± 0.05 ^{A}	1.21 ± 0.12 $^{a'}$	1.69 ± 0.14 $^{a'}$	2.90 ± 0.55 $^{b'}$	2.64 ± 0.19 $^{b'}$
Isoferulic acid	0.08 ± 0.04 ^{a}	0.10 ± 0.03 ^{a}	0.07 ± 0.03 ^{a}	0.08 ± 0.02 ^{a}	0.12 ± 0.04 ^{A}	0.06 ± 0.02 ^{A}	0.09 ± 0.03 ^{A}	0.10 ± 0.02 ^{A}	0.20 ± 0.03 $^{a'}$	0.09 ± 0.02 $^{b'}$	0.18 ± 0.05 $^{a',b'}$	0.11 ± 0.02 $^{b'}$
3-feruloylquinic acid	nd ^{1,a}	nd ^{1,a}	0.17 ± 0.03 ^{b}	0.13 ± 0.02 ^{b}	0.11 ± 0.06 ^{A}	0.13 ± 0.02 ^{A}	0.16 ± 0.03 ^{A}	0.12 ± 0.03 ^{A}	0.94 ± 0.12 $^{a'}$	0.45 ± 0.10 $^{b'}$	0.9 ± 0.12 $^{a'}$	0.67 ± 0.05 $^{b'}$
5-feruloylquinic acid	1.34 ± 0.41 ^{a}	1.26 ± 0.05 ^{a}	2.49 ± 0.46 ^{a}	2.43 ± 0.19 ^{a}	1.07 ± 0.16 ^{A}	1.18 ± 0.1 ^{A}	1.35 ± 0.27 ^{A}	1.48 ± 0.47 ^{A}	13.15 ± 1.92 $^{a'}$	8.90 ± 0.37 $^{b'}$	9.73 ± 1.20 $^{b'}$	9.2 ± 0.8 $^{b'}$
4-feruloylquinic acid	0.11 ± 0.01 ^{a}	0.14 ± 0.01 ^{a}	0.33 ± 0.06 ^{a}	0.3 ± 0.04 ^{a}	0.13 ± 0.0 ^{A}	0.1 ± 0.09 ^{A}	0.18 ± 0.04 ^{A}	0.17 ± 0.02 ^{A}	1.74 ± 0.25 $^{a'}$	1.25 ± 0.04 $^{b'}$	1.62 ± 0.28 $^{a'}$	1.56 ± 0.11 $^{a'}$
Ferulic acid glucoside	nd ^{1,a}	nd ^{1,a}	0.05 ± 0.03 ^{a}	0.05 ± 0.004 ^{a}	nd ^{1,A}	nd ^{1,A}	nd ^{1,A}	0.49 ± 0.25 ^{B}	0.32 ± 0.03 $^{a'}$	0.25 ± 0.02 $^{a'}$	0.29 ± 0.06 $^{a'}$	0.19 ± 0.01 $^{a'}$
Ferulic acid coumaroyl glucoside	2.02 ± 0.50 ^{a}	1.46 ± 0.21 ^{a}	1.72 ± 0.43 ^{a}	1.41 ± 0.19 ^{a}	1.52 ± 0.24 ^{A}	1.44 ± 0.84 ^{A}	2.15 ± 0.48 ^{A}	1.32 ± 0.02 ^{A}	3.72 ± 0.35 $^{a'}$	3.44 ± 0.21 $^{a'}$	2.48 ± 0.42 $^{b'}$	1.65 ± 0.08 $^{c'}$
Ferulic acid caffeoyl glucoside	0.09 ± 0.03 ^{a}	0.07 ± 0.03 ^{a}	0.18 ± 0.12 ^{a}	0.14 ± 0.08 ^{a}	0.06 ± 0.03 ^{A}	nd ^{1,A}	0.08 ± 0.03 ^{A}	0.07 ± 0.03 ^{A}	8.56 ± 0.61 $^{a'}$	4.31 ± 0.31 $^{b'}$	7.3 ± 1.0 $^{a',c'}$	6.95 ± 0.53 $^{c'}$
Feruloylquinic acid derivative	0.85 ± 0.18 ^{a}	0.8 ± 0.1 ^{a}	0.68 ± 0.33 ^{a}	0.83 ± 0.10 ^{a}	0.43 ± 0.15 ^{A}	0.4 ± 0.02 ^{A}	0.43 ± 0.01 ^{A}	0.42 ± 0.05 ^{A}	2.92 ± 0.25 $^{a'}$	4.25 ± 0.10 $^{b'}$	2.52 ± 0.42 $^{a'}$	2.42 ± 0.23 $^{a'}$
Feruloylquinic acid derivative (2)	0.2 ± 0.08 ^{a}	0.1 ± 0.02 ^{a}	0.23 ± 0.10 ^{a}	0.11 ± 0.01 ^{a}	0.09 ± 0.03 ^{A}	0.11 ± 0.01 ^{A}	0.12 ± 0.01 ^{A}	0.11 ± 0.02 ^{A}	1.20 ± 0.11 $^{a'}$	1.34 ± 0.05 $^{a'}$	0.9 ± 0.12 $^{b'}$	1.34 ± 0.1 $^{a'}$
Ferulic acid and its derivatives	5.09 ± 0.12 ^{a}	4.4 ± 0.3 ^{a}	6.3 ± 1.3 ^{a}	5.9 ± 0.3 ^{a}	3.9 ± 0.4 ^{A}	3.7 ± 0.7 ^{A}	5.1 ± 0.8 ^{A}	4.8 ± 0.8 ^{A}	36 ± 4 $^{a'}$	26.8 ± 1.0 $^{b'}$	30 ± 4 $^{b'}$	27.8 ± 1.9 $^{b'}$
Total phenolic compounds	62 ± 23 ^{a}	37 ± 5 ^{a}	129 ± 25 ^{b}	101 ± 14 ^{b}	81 ± 36 ^{A}	77 ± 20 ^{A}	94 ± 6 ^{A}	79 ± 24 ^{A}	322 ± 56 $^{a'}$	124 ± 8 $^{b'}$	329 ± 40 $^{a'}$	138 ± 9 $^{b'}$

Values are means ± standard deviation. Different letters within the same product and compound indicate significant ($p < 0.05$) differences among treatments (puree: lowercase letters; puree with oil: uppercase letters; and juice: lowercase'). nd [1]: not detected.

PEF treatment caused different effects depending on the type of carrot-derived product and phenolic chemical structure. Hence, the application of PEF to whole carrots did not affect most of the compounds from both purees obtained from such carrots, although the juices had lower contents of coumaric, ferulic, and caffeic acid derivatives (e.g., 5-caffeoylquinic (68.5%)) (Table 3). The selected PEF treatment was based on previous results obtained by López-Gámez et al. [70] in which the total phenolic content of whole carrots was enhanced. Nevertheless, such an increase was not observed in derived products obtained from PEF-treated whole carrots. This may suggest that mechanical processing could affect the phenolic stability or favor their interactions with cell wall debris, which would hinder their extractability.

The results indicate that individual compounds were also differently affected by thermal treatment, depending on their structure and carrot-derived product. Hence, caffeic acid derivatives were strongly enhanced in purees (e.g., 5-caffeoylquinic acid (231%)), although caffeic acid decreased in those with oil added (Table 3). Juices showed increases in some caffeic acids derivates (e.g., 3-caffeoylquinic acid (400%)); however, a lower content was observed in some ferulic acid derivates (e.g., 5-feruloylquinic acid). Increases in phenolic content have been previously reported in thermally treated fruit juices (e.g., caffeoyl glucoside or caffeic acid) [11–13]. This was attributed to their better release due to cell wall disruptions, which is in agreement with the lower particle sizes observed in the juices (Table 2). The higher content in caffeic acid derivatives could result from the partial inactivation of enzymes responsible for phenolic degradation (e.g., polyphenol oxidase, PPO) [13].

3.7. Phenolic Bioaccessibility

The phenolic compound bioaccessibility was affected by PEF treatment application to whole carrots and further processing. The bioaccessibility differed depending on the evaluated carrot-derived product. Carrot purees had the highest total bioaccessibility (52%) followed by oil-added purees (31%) and juices (16.1%) (Table 4). PEF application to carrots before mechanical processing caused a large enhancement of the total bioaccessibility, hence, reaching 100% in purees, whereas that of juices and oil-added purees was not affected. On the other hand, thermal treatment did not significantly influence the total bioaccessibility in any carrot-based product (Table 4).

Individual compounds were differently affected depending on their chemical structure and processing. PEF treatments increased the bioaccessibility of most ferulic (e.g., ferulic acid glucoside) and caffeic acid derivatives (e.g., 5-caffeoylquinic acid) in purees, although some decreases were also observed (e.g., coumaric acid). Oil-added purees were similarly affected to those without lipids; however, the bioaccessibility of certain compounds decreased (e.g., 5-caffeoylquinic acid or feruloylquinic acid derivate (2)).

On the other hand, the individual phenol bioaccessibility in juices obtained from PEF-treated carrots was not affected, excepting that of isoferulic acid, which reached 100%. Limited information about the phenolic bioaccessibility in juices and purees obtained from PEF-treated matrices is available in the literature. Generally, phenolic compounds should be easily released from juices, due to their low content in dietary fiber [69]. However, the obtained results are controversial since the phenolic bioaccessibility in carrot juice was the lowest (Table 4).

Table 4. Phenolic bioaccessibility of untreated purees and juices (U), those obtained from PEF-treated (0.61 kJ kg^{-1}) carrots (PEF), and those thermally treated (10 min at 70 °C) (U/T and PEF/T).

Phenolic Compounds	Puree				Oil-Added Puree				Juice			
	U	PEF	U/T	PEF/T	U	PEF	U/T	PEF/T	U	PEF	U/T	PEF/T
Coumaric acid	91 ± 11^{a}	56 ± 12^{b}	53 ± 10^{b}	100 ± 0^{a}	63 ± 4^{A}	99.4 ± 1.1^{B}	79 ± 7^{A}	77 ± 28^{A}	$100 \pm 0^{a'}$	$100 \pm 0^{a'}$	$47 \pm 7^{b'}$	$45.1 \pm 1.1^{b'}$
Coumaroylquinic acid	80 ± 20^{a}	100 ± 0^{b}	100 ± 0^{b}	100 ± 0^{b}	84 ± 14^{A}	97.4 ± 2.2^{B}	42.7 ± 0.4^{C}	26.3 ± 2.2^{D}	$42 \pm 4^{a'}$	$35.9 \pm 2.1^{a'}$	$25 \pm 4^{b'}$	$50 \pm 4^{c'}$
Coumaric acid and its derivatives	84 ± 18^{a}	96.2 ± 6.5^{a}	100 ± 0^{a}	100 ± 0^{a}	74 ± 8^{A}	100 ± 0^{B}	48.8 ± 0.1^{C}	31.2 ± 0.2^{D}	$56 \pm 6^{a'}$	$60 \pm 3^{a'}$	$25 \pm 4^{b'}$	$50 \pm 4^{a'}$
Caffeic acid	0^{a}	0^{a}	52 ± 32^{b}	24 ± 4^{b}	0^{A}	0^{A}	43 ± 13^{B}	24 ± 5^{C}	$0^{a'}$	$0^{a'}$	$47 \pm 7^{b'}$	$34 \pm 5^{b'}$
Caffeic acid arab/xiloside	0^{*a}	0^{*a}	0^{a}	0^{a}	$0^{*,A}$	$0^{*,A}$	$0^{*,A}$	84.4 ± 18.6^{B}	$0^{a'}$	$0^{a'}$	$23 \pm 7^{b'}$	$55 \pm 9^{c'}$
Caffeoylshikimic acid	0^{a}	0^{a}	0^{a}	0^{a}	0^{A}	0^{A}	0^{A}	0^{A}	$0^{a'}$	$0^{a'}$	$0^{a'}$	$0^{a'}$
3-caffeoylquinic acid	$0^{*,a}$	$0^{*,A}$	100 ± 0^{b}	100 ± 0^{b}	$0^{*,A}$	$0^{*,A}$	$0^{*,A}$	$0^{*,A}$	$0^{a'}$	$0^{a'}$	$0^{a'}$	$0^{a'}$
5-caffeoylquinic acid	15 ± 9^{a}	56 ± 10^{b}	22 ± 4^{a}	12 ± 4^{a}	16 ± 11^{A}	8.4 ± 1.6^{B}	29 ± 3^{A}	28 ± 11^{A}	$0^{a'}$	$0^{a'}$	$26.1 \pm 1.7^{b'}$	$23 \pm 5^{b'}$
4-caffeoylquinic acid	100 ± 0^{a}	100 ± 0^{a}	100 ± 0^{a}	66 ± 6^{b}	100 ± 0^{A}	73 ± 35^{B}	100 ± 0^{A}	100 ± 0^{A}	$10.6 \pm 1.8^{a'}$	$19 \pm 5^{a'}$	$93 \pm 6^{b'}$	$84 \pm 14^{b'}$
Dicaffeoylferuoylquinic acid	0^{a}	0^{a}	0^{a}	0^{a}	0^{A}	0^{A}	0^{A}	0^{A}	$0^{a'}$	$0^{a'}$	$18 \pm 5^{b'}$	$0^{a'}$
Caffeoylferuoylquinic acid	39 ± 5^{a}	69 ± 16^{b}	46 ± 14^{a}	$66 \pm 18^{a,b}$	48 ± 17^{A}	48 ± 12^{A}	36 ± 12^{A}	25.9 ± 1.7^{A}	67 ± 5 a′	$47 \pm 1.7^{a',b'}$	$43 \pm 13^{a',b'}$	$28.7 \pm 1.8^{b'}$
Caffeic acid arabinoside glucoside	73 ± 27^{a}	55.2 ± 2.4^{a}	62 ± 6^{a}	48.0 ± 2.4^{a}	84 ± 27^{A}	42 ± 28^{B}	100 ± 0^{A}	71 ± 22^{A}	$74 \pm 29^{a'}$	$57 \pm 3^{a'}$	$86 \pm 8^{a'}$	$35.8 \pm 1.9^{a'}$
Caffeic acid Glu acetyl glucoside	89 ± 19^{a}	100 ± 0^{a}	99.1 ± 1.5^{a}	100 ± 0^{a}	100 ± 0^{A}	100 ± 0^{A}	100 ± 0^{A}	100 ± 0^{A}	$55 \pm 3^{a'}$	$45 \pm 7^{a'}$	$35 \pm 9^{a'}$	$31.9 \pm 0.9^{a'}$
Caffeic acid and its derivatives	19 ± 9^{a}	66 ± 9^{b}	26 ± 5^{a}	16 ± 4^{a}	$20 \pm 13^{A,B}$	11.7 ± 2.3^{B}	33 ± 3^{A}	33 ± 12^{A}	$1.3 \pm 0.2^{a'}$	$3.4 \pm 0.4^{a'}$	$29.1 \pm 1.9^{b'}$	$27 \pm 5^{b'}$
Ferulic acid	99 ± 0.7^{a}	100 ± 0^{a}	76 ± 24^{b}	99.2 ± 1.3^{a}	100 ± 0^{A}	100 ± 0^{A}	100 ± 0^{A}	95.1 ± 8.5^{A}	$100 \pm 0^{a'}$	$100 \pm 0^{a'}$	$68 \pm 15^{b'}$	$45 \pm 6^{c'}$
Isoferulic acid	100 ± 0^{a}	100 ± 0^{a}	98 ± 4^{a}	100 ± 0^{a}	100 ± 0^{A}	100 ± 0^{A}	100 ± 0^{A}	100 ± 0^{A}	$70 \pm 16^{a'}$	$100 \pm 0^{b'}$	$91.3 \pm 15^{b'}$	$99.2 \pm 1.3^{b'}$
3-feruloylquinic acid	100 ± 0^{a}	100 ± 0^{a}	100 ± 0^{a}	100 ± 0^{a}	$93.8 \pm 11^{A,B}$	81 ± 19^{B}	100 ± 0^{A}	100 ± 0^{A}	$79 \pm 11^{a'}$	$100 \pm 0^{a'}$	$98 \pm 4^{a'}$	$98.0 \pm 2.0^{a'}$
5-feruloylquinic acid	81 ± 17^{a}	99.6 ± 0.7^{b}	$95 \pm 5^{a,b}$	$96 \pm 4^{a,b}$	100 ± 0^{A}	91 ± 9^{A}	100 ± 0^{A}	100 ± 0^{A}	$40 \pm 6^{a'}$	$43 \pm 3^{a'}$	$56 \pm 9^{b'}$	$35 \pm 5^{a'}$
4-feruloylquinic acid	100 ± 0^{a}	100 ± 0^{a}	100 ± 0^{a}	100 ± 0^{a}	100 ± 0^{A}	100 ± 0^{A}	100 ± 0^{A}	99.5 ± 0.8^{A}	$86 \pm 12^{a',b'}$	$83 \pm 7^{a',b'}$	$94 \pm 5^{b'}$	$66 \pm 6^{c'}$
Ferulic acid glucoside	56 ± 4^{a}	99.1 ± 1.5^{b}	100 ± 0^{b}	91 ± 15^{b}	0^{A}	0^{A}	60.5 ± 35^{B}	8 ± 3^{C}	$34.7 \pm 0.9^{a'}$	$38 \pm 6^{a'}$	$36 \pm 8^{a'}$	$37 \pm 7^{a'}$
Ferulic acid coumaroyl glucoside	71 ± 19^{a}	100 ± 0^{a}	76 ± 20^{a}	94 ± 12^{a}	81 ± 24^{A}	57 ± 37^{A}	78 ± 22^{A}	82 ± 24^{A}	$44 \pm 8^{a'}$	$41.2 \pm 1.7^{a'}$	$99 \pm 3^{b'}$	$57.9 \pm 1.9^{a'}$
Ferulic acid caffeoyl glucoside	38 ± 15^{a}	65 ± 3^{b}	100 ± 0^{c}	92 ± 14^{c}	71 ± 31^{A}	92 ± 14^{A}	61 ± 14^{A}	78 ± 19^{A}	$36 \pm 3^{a'}$	$43.3 \pm 0.9^{a'}$	$37 \pm 7^{a'}$	$24 \pm 7^{a'}$
Feruloylquinic acid derivative	46 ± 11^{a}	89 ± 9^{b}	89 ± 19^{b}	93 ± 12^{b}	73 ± 27^{A}	78 ± 10^{A}	68 ± 14^{A}	78 ± 9^{A}	$46 \pm 7^{a'}$	$33 \pm 3^{a'}$	$41 \pm 14^{a'}$	$33.5 \pm 0.3^{a'}$
Feruloylquinic acid derivative (2)	18 ± 7^{a}	57 ± 12^{b}	53 ± 24^{b}	89 ± 19^{c}	77 ± 21^{A}	42 ± 17^{B}	39 ± 17^{B}	38 ± 10^{B}	$38.3 \pm 2.4^{a'}$	$33.6 \pm 1.2^{a'}$	$31 \pm 5^{a'}$	$19 \pm 4^{b'}$
Ferulic acid and its derivatives	77 ± 4^{a}	100 ± 0^{b}	97 ± 5^{b}	100 ± 0^{b}	98 ± 3^{A}	89 ± 12^{A}	100 ± 0^{A}	100 ± 0^{A}	$45 \pm 4^{a'}$	$50.0 \pm 1.8^{a',b'}$	$60 \pm 10^{b'}$	$37 \pm 5^{a'}$
Total phenolic compounds	52 ± 14^{a}	100 ± 0^{b}	49 ± 8^{a}	48 ± 6^{a}	31 ± 15^{A}	24 ± 5^{A}	40 ± 4^{A}	40 ± 17^{A}	$16.1 \pm 2.5^{a'}$	$27.9 \pm 2.0^{a'}$	$34 \pm 3^{a'}$	$33 \pm 2^{a'}$

Values are means ± standard deviation. Different letters within the same product and compound indicate significant ($p < 0.05$) differences among treatments (puree: lowercase letters; puree with oil: uppercase letters; and juice: lowercase′). The asterisk (*) indicates that the compound was detected in the dialyzed fraction, but it was not present in non-digested fractions.

The juices had a lower particle size (D [4, 3] = 487 μm) compared with the purees (D [4, 3] = 596 μm), which favored phenolic release from the matrix. Phenolic compounds are likely more exposed to degradation or entrapment by other macromolecules during digestion in juices than in purees, hence, limiting their bioaccessibility [71]. PEF application to carrots strongly enhanced the bioaccessibility in purees without oil, whereas treatment was not effective for juices or oil-added purees. Despite no correlation being found between the particle size and bioaccessibility values, cell permeability changes could have caused a better release and dialysis of phenols in purees. In addition, variations caused by PEF in the initial phenolic content of carrots would directly affect the bioaccessibility in carrot-derived products.

4. Conclusions

The content and bioaccessibility of carotenoids and phenolic compounds were affected by both PEF application to whole carrots and further processing conditions. Carrot juices had the highest phenolic and carotenoid content. However, only the phenolic content in purees further increased after a thermal treatment, whereas it decreased in juices obtained from PEF-treated carrots. On the other hand, PEF pre-treatment of whole carrots stands as a potential method for enhancing the phenolic bioaccessibility in purees since most of them were completely dialyzed after treatment.

Nevertheless, PEF was not helpful to improve the total phenolic bioaccessibility in purees with added oil and juices. Regarding carotenoids, PEF pre-treatment did not substantially enhance their total bioaccessibility, whereas thermal treatment and oil addition were more effective to improve the carotenoid bioaccessibility in carrot-based purees. Therefore, these results demonstrate that applying PEF as a pre-treatment is a feasible strategy for developing products with an enhanced nutritive value. Further studies focused on the matrix structure and composition (e.g., pectin characteristics) are necessary to understand the mechanisms governing changes in the bioaccessibility of health-related compounds in the different studied products.

Author Contributions: Conceptualization, G.L.-G.; Formal analysis, G.L.-G.; Investigation, G.L.-G.; Methodology, G.L.-G.; Supervision, P.E.-M., O.M.-B. and R.S.-F.; Writing—original draft, G.L.-G.; Writing—review and editing, P.E.-M., O.M.-B. and R.S.-F. All authors have read and agreed to the published version of the manuscript.

Funding: This research was funded by the Ministerio de Economía y Competitividad (Spain) grant number AGL2013-44851-R and Ministerio de Ciencia, Innovación y Universidades (Spain) grant number RTI2018-095560-B-I00. López-Gámez, G. thanks the Agència de Gestió d'Ajuts Universitaris i de Recerca (AGAUR) for the predoctoral grant (2019FI_B2 00129).

Institutional Review Board Statement: Not applicable.

Informed Consent Statement: Not applicable.

Acknowledgments: To Alba Macià for her technical support in analyzing phenolic compounds extracts and Joana Martínez for her assistance in performing in vitro digestions and carotenoid extractions.

Conflicts of Interest: The authors declare no conflict of interest.

References

1. Rao, A.V.; Rao, L.G. Carotenoids and human health. *Pharmacol. Res.* **2007**, *55*, 207–216. [CrossRef]
2. Cvejić, J.H.; Krstonošić, M.A.; Bursać, M.; Miljić, U. *Polyphenols*; Academic Press: Cambridge, MA, USA, 2017; ISBN 9780128052570.
3. Granado-Lorencio, F.; Olmedilla-Alonso, B.; Herrero-Barbudo, C.; Blanco-Navarro, I.; Pérez-Sacristán, B.; Blázquez-García, S. In vitro bioaccessibility of carotenoids and tocopherols from fruits and vegetables. *Food Chem.* **2007**, *102*, 641–648. [CrossRef]
4. Van Buggenhout, S.; Alminger, M.; Lemmens, L.; Colle, I.; Knockaert, G.; Moelants, K.; Van Loey, A.; Hendrickx, M. In vitro approaches to estimate the effect of food processing on carotenoid bioavailability need thorough understanding of process induced microstructural changes. *Trends Food Sci. Technol.* **2010**, *21*, 607–618. [CrossRef]
5. Parada, J.; Aguilera, J.M. Food microstructure affects the bioavailability of several nutrients. *J. Food Sci.* **2007**, *72*, 21–32. [CrossRef] [PubMed]

6. Knockaert, G.; Lemmens, L.; Van Buggenhout, S.; Hendrickx, M.; Van Loey, A. Changes in β-carotene bioaccessibility and concentration during processing of carrot puree. *Food Chem.* **2012**, *133*, 60–67. [CrossRef]
7. Lemmens, L.; Van Buggenhout, S.; Van Loey, A.M.; Hendrickx, M.E. Particle size reduction leading to cell wall rupture is more important for the β-carotene bioaccessibility of raw compared to thermally processed carrots. *J. Agric. Food Chem.* **2010**, *58*, 12769–12776. [CrossRef] [PubMed]
8. Hedrén, E.; Diaz, V.; Svanberg, U. Estimation of carotenoid accessibility from carrots determined by an in vitro digestion method. *Eur. J. Clin. Nutr.* **2002**, *56*, 425–430. [CrossRef] [PubMed]
9. Hornero-Méndez, D.; Mínguez-Mosquera, M.I. Bioaccessibility of carotenes from carrots: Effect of cooking and addition of oil. *Innov. Food Sci. Emerg. Technol.* **2007**, *8*, 407–412. [CrossRef]
10. Palmero, P.; Lemmens, L.; Hendrickx, M.; Van Loey, A. Role of carotenoid type on the effect of thermal processing on bioaccessibility. *Food Chem.* **2014**, *157*, 275–282. [CrossRef] [PubMed]
11. He, Z.; Tao, Y.; Zeng, M.; Zhang, S.; Tao, G.; Qin, F.; Chen, J. High pressure homogenization processing, thermal treatment and milk matrix affect in vitro bioaccessibility of phenolics in apple, grape and orange juice to different extents. *Food Chem.* **2016**, *200*, 107–116. [CrossRef] [PubMed]
12. Quan, W.; Tao, Y.; Qie, X.; Zeng, M.; Qin, F.; Chen, J.; He, Z. Effects of high-pressure homogenization, thermal processing, and milk matrix on the in vitro bioaccessibility of phenolic compounds in pomelo and kiwi juices. *J. Funct. Foods* **2020**, *64*, 103633. [CrossRef]
13. Rodríguez-Roque, M.J.; Ancos, B.; Sánchez-Moreno, C.; Cano, M.P.; Elez-Martínez, P.; Martín-Belloso, O. Impact of food matrix and processing on the in vitro bioaccessibility of vitamin C, phenolic compounds, and hydrophilic antioxidant activity from fruit juice-based beverages. *J. Funct. Foods* **2015**, *14*, 33–43. [CrossRef]
14. Jacobo-Velázquez, D.A.; Cuéllar-Villarreal, M.d.R.; Welti-Chanes, J.; Cisneros-Zevallos, L.; Ramos-Parra, P.A.; Hernández-Brenes, C. Nonthermal processing technologies as elicitors to induce the biosynthesis and accumulation of nutraceuticals in plant foods. *Trends Food Sci. Technol.* **2017**, *60*, 80–87. [CrossRef]
15. Ribas-Agustí, A.; Martín-Belloso, O.; Soliva-Fortuny, R.; Elez-Martínez, P. Enhancing hydroxycinnamic acids and flavan-3-ol contents by pulsed electric fields without affecting quality attributes of apple. *Food Res. Int.* **2019**, *121*, 433–440. [CrossRef]
16. López-Gámez, G.; Elez-Martínez, P.; Martín-Belloso, O.; Soliva-Fortuny, R. Enhancing phenolic content in carrots by pulsed electric fields during post-treatment time: Effects on cell viability and quality attributes. *Innov. Food Sci. Emerg. Technol.* **2020**, *59*, 102252. [CrossRef]
17. Bot, F.; Verkerk, R.; Mastwijk, H.; Anese, M.; Fogliano, V.; Capuano, E. The effect of pulsed electric fields on carotenoids bioaccessibility: The role of tomato matrix. *Food Chem.* **2018**, *240*, 415–421. [CrossRef]
18. Jayathunge, K.G.L.R.; Stratakos, A.C.; Cregenzán-Albertia, O.; Grant, I.R.; Lyng, J.; Koidis, A. Enhancing the lycopene in vitro bioaccessibility of tomato juice synergistically applying thermal and non-thermal processing technologies. *Food Chem.* **2017**, *221*, 698–705. [CrossRef]
19. Ribas-Agustí, A.; Martín-Belloso, O.; Soliva-Fortuny, R.; Elez-Martínez, P. Influence of pulsed electric fields processing on the bioaccessible and non-bioaccessible fractions of apple phenolic compounds. *J. Funct. Foods* **2019**, *59*, 206–214. [CrossRef]
20. López-Gámez, G.; Elez-Martínez, P.; Quiles-Chuliá, A.; Martín-Belloso, O.; Hernando-Hernando, I.; Soliva-Fortuny, R. Effect of pulsed electric fields on carotenoid and phenolic bioaccessibility and their relationship with carrot structure. *Food Funct.* **2021**, *12*, 2772–2783. [CrossRef]
21. González-Casado, S.; Martín-Belloso, O.; Elez-Martínez, P.; Soliva-Fortuny, R. Application of pulsed electric fields to tomato fruit for enhancing the bioaccessibility of carotenoids in derived products. *Food Funct.* **2018**, *9*, 2282–2289. [CrossRef]
22. Wiktor, A.; Sledz, M.; Nowacka, M.; Rybak, K.; Chudoba, T.; Lojkowski, W.; Witrowa-Rajchert, D. The impact of pulsed electric field treatment on selected bioactive compound content and color of plant tissue. *Innov. Food Sci. Emerg. Technol.* **2015**, *30*, 69–78. [CrossRef]
23. Balogh, T.; Smout, C.; Nguyen, B.L.; Van Loey, A.M.; Hendrickx, M.E. Thermal and high-pressure inactivation kinetics of carrot pectinmethylesterase: From model system to real foods. *Innov. Food Sci. Emerg. Technol.* **2004**, *5*, 429–436. [CrossRef]
24. Soysal, Ç.; Söylemez, Z. Kinetics and inactivation of carrot peroxidase by heat treatment. *J. Food Eng.* **2005**, *68*, 349–356. [CrossRef]
25. Houben, K.; Kermani, Z.J.; Van Buggenhout, S.; Van Loey, A.M.; Hendrickx, M.E. Thermal and high-pressure stability of pectin-converting enzymes in broccoli and carrot purée: Towards the creation of specific endogenous enzyme populations through processing. *Food Bioprocess Technol.* **2014**, *7*, 1713–1724. [CrossRef]
26. Minekus, M.; Alminger, M.; Alvito, P.; Ballance, S.; Bohn, T.; Bourlieu, C.; Carrì, F.; Boutrou, R.; Corredig, F.M.; Dupont, D.; et al. A standardised static in vitro digestion method suitable for food—An international consensus. *Food Funct.* **2014**, *5*, 1113–1124. [CrossRef]
27. Brodkorb, A.; Egger, L.; Alminger, M.; Alvito, P.; Assunção, R.; Ballance, S.; Bohn, T.; Bourlieu-Lacanal, C.; Boutrou, R.; Carrière, F.; et al. INFOGEST static in vitro simulation of gastrointestinal food digestion. *Nat. Protoc.* **2019**, *14*, 991–1014. [CrossRef]
28. Eriksen, J.N.; Luu, A.Y.; Dragsted, L.O.; Arrigoni, E. Adaption of an in vitro digestion method to screen carotenoid liberation and in vitro accessibility from differently processed spinach preparations. *Food Chem.* **2017**, *224*, 407–413. [CrossRef] [PubMed]
29. Svelander, C. *In Vitro Bioaccessibility of Carotenes. Influence of Microstructure in Tomato and Carrot as Modified by Processing*; Chalmers University of Technology: Göteborg, Sweden, 2011.

30. Sadler, G.; Davis, J.; Dezman, D. Rapid extraction of lycopene and β-Carotene from reconstituted tomato paste and pink grapefruit homogenates. *J. Food Sci.* **1990**, *55*, 1460–1461. [CrossRef]
31. Cortés, C.; Esteve, M.J.; Frígola, A.; Torregrosa, F. Identification and quantification of carotenoids including geometrical isomers in fruit and vegetable juices by liquid chromatography with ultraviolet-diode array detection. *J. Agric. Food Chem.* **2004**, *52*, 2203–2212. [CrossRef] [PubMed]
32. Barba, F.J.; Esteve, M.J.; Frígola, A. High pressure treatment effect on physicochemical and nutritional properties of fluid foods during storage: A review. *Compr. Rev. Food Sci. Food Saf.* **2012**, *11*, 307–322. [CrossRef]
33. Aguiló-Aguayo, I.; Hossain, M.B.; Brunton, N.; Lyng, J.; Valverde, J.; Rai, D.K. Pulsed electric fields pre-treatment of carrot purees to enhance their polyacetylene and sugar contents. *Innov. Food Sci. Emerg. Technol.* **2014**, *23*, 79–86. [CrossRef]
34. Xiang, B.; Sundararajan, S.; Solval, K.M.; Espinoza-Rodezno, L.; Aryana, K.; Sathivel, S. Effects of pulsed electric fields on physicochemical properties and microbial inactivation of carrot juice. *J. Food Process. Preserv.* **2013**, *38*, 1556–1564. [CrossRef]
35. Yan, B.; Martínez-Monteagudo, S.I.; Cooperstone, J.L.; Riedl, K.M.; Schwartz, S.J.; Balasubramaniam, V.M. Impact of Thermal and Pressure-Based Technologies on Carotenoid Retention and Quality Attributes in Tomato Juice. *Food Bioprocess Technol.* **2017**, *10*, 808–818. [CrossRef]
36. Kebede, B.T.; Grauwet, T.; Palmers, S.; Vervoort, L.; Carle, R.; Hendrickx, M.; Van Loey, A. Effect of high pressure high temperature processing on the volatile fraction of differently coloured carrots. *Food Chem.* **2014**, *153*, 340–352. [CrossRef] [PubMed]
37. Mutsokoti, L.; Panozzo, A.; Van Loey, A.; Hendrickx, M. Carotenoid transfer to oil during thermal processing of low fat carrot and tomato particle based suspensions. *Food Res. Int.* **2016**, *86*, 64–73. [CrossRef]
38. Knockaert, G.; Pulissery, S.K.; Lemmens, L.; Van Buggenhout, S.; Hendrickx, M.; Van Loey, A. Carrot β-carotene degradation and isomerization kinetics during thermal processing in the presence of oil. *J. Agric. Food Chem.* **2012**, *60*, 10312–10319. [CrossRef] [PubMed]
39. Rybak, K.; Samborska, K.; Jedlinska, A.; Parniakov, O.; Nowacka, M.; Witrowa-Rajchert, D.; Wiktor, A. The impact of pulsed electric field pretreatment of bell pepper on the selected properties of spray dried juice. *Innov. Food Sci. Emerg. Technol.* **2020**, *65*, 102446. [CrossRef]
40. Li, N.; Feng, Z.; Niu, Y.; Yu, L.L. Structural, rheological and functional properties of modified soluble dietary fiber from tomato peels. *Food Hydrocoll.* **2018**, *77*, 557–565. [CrossRef]
41. Velázquez-Estrada, R.M.; Hernández-Herrero, M.M.; Guamis-López, B.; Roig-Saguès, A.X. Influence of ultra-high pressure homogenisation on physicochemical and sensorial properties of orange juice in comparison with conventional thermal processing. *Int. J. Food Sci. Technol.* **2019**, *54*, 1858–1864. [CrossRef]
42. Stinco, C.M.; Sentandreu, E.; Mapelli-Brahm, P.; Navarro, J.L.; Vicario, I.M.; Meléndez-Martínez, A.J. Influence of high pressure homogenization and pasteurization on the in vitro bioaccessibility of carotenoids and flavonoids in orange juice. *Food Chem.* **2020**, *331*, 127259. [CrossRef]
43. Greve, L.C.; McArdle, R.N.; Gohlke, J.R.; Labavitch, J.M. Impact of Heating on Carrot Firmness: Changes in Cell Wall Components. *J. Agric. Food Chem.* **1994**, *42*, 2900–2906. [CrossRef]
44. Marx, M.; Stuparic, M.; Schieber, A.; Carle, R. Effects of thermal processing on trans-cis-isomerization of β-carotene in carrot juices and carotene-containing preparations. *Food Chem.* **2003**, *83*, 609–617. [CrossRef]
45. Galindo, F.G.; Dejmek, P.; Lundgren, K.; Rasmusson, A.G.; Vicente, A.; Moritz, T. Metabolomic evaluation of pulsed electric field-induced stress on potato tissue. *Planta* **2009**, *230*, 469–479. [CrossRef] [PubMed]
46. Blenkinsop, R.W.; Yada, R.Y.; Marangoni, A.G. Metabolic Control of Low-Temperature Sweetening in Potato Tubers during Postharvest Storage. In *Horticultural Reviews*; John Wiley & Sons: Hoboken, NJ, USA, 2004; Volume 30, ISBN 9780471354208.
47. Wiktor, A.; Gondek, E.; Jakubczyk, E.; Dadan, M.; Nowacka, M.; Rybak, K.; Witrowa-Rajchert, D. Acoustic and mechanical properties of carrot tissue treated by pulsed electric field, ultrasound and combination of both. *J. Food Eng.* **2018**, *238*, 12–21. [CrossRef]
48. Ribas-Agustí, A.; Van Buggenhout, S.; Palmero, P.; Hendrickx, M.; Loey, A. Van Investigating the role of pectin in carrot cell wall changes during thermal processing: A microscopic approach. *Innov. Food Sci. Emerg. Technol.* **2014**, *24*, 113–120. [CrossRef]
49. Augusto, P.E.D.; Ibarz, A.; Cristianini, M. Effect of high pressure homogenization (HPH) on the rheological properties of tomato juice: Time-dependent and steady-state shear. *J. Food Eng.* **2012**, *111*, 570–579. [CrossRef]
50. Li, Q.; Li, T.; Liu, C.; Chen, J.; Zhang, R.; Zhang, Z.; Dai, T.; Julian, D. Potential physicochemical basis of Mediterranean diet effect: Ability of emulsified olive oil to increase carotenoid bioaccessibility in raw and cooked tomatoes. *Food Res. Int.* **2016**, *89*, 320–329. [CrossRef]
51. Zhang, R.; Zhang, Z.; Zou, L.; Xiao, H.; Zhang, G.; Decker, E.A.; McClements, D.J. Enhancing Nutraceutical Bioavailability from Raw and Cooked Vegetables Using Excipient Emulsions: Influence of Lipid Type on Carotenoid Bioaccessibility from Carrots. *J. Agric. Food Chem.* **2015**, *63*, 10508–10517. [CrossRef]
52. Schweiggert, R.M.; Mezger, D.; Schimpf, F.; Steingass, C.B.; Carle, R. Influence of chromoplast morphology on carotenoid bioaccessibility of carrot, mango, papaya, and tomato. *Food Chem.* **2012**, *135*, 2736–2742. [CrossRef] [PubMed]
53. Knockaert, G.; De Roeck, A.; Lemmens, L.; Van Buggenhout, S.; Hendrickx, M.; Van Loey, A. Effect of thermal and high pressure processes on structural and health-related properties of carrots (*Daucus carota*). *Food Chem.* **2011**, *125*, 903–912. [CrossRef]

54. Panozzo, A.; Lemmens, L.; Van Loey, A.; Manzocco, L.; Nicoli, M.C.; Hendrickx, M. Microstructure and bioaccessibility of different carotenoid species as affected by high pressure homogenisation: A case study on differently coloured tomatoes. *Food Chem.* **2013**, *141*, 4094–4100. [CrossRef] [PubMed]
55. Morales-De La Peña, M.; Salvia-Trujillo, L.; Rojas-Graü, M.A.; Martín-Belloso, O. Changes on phenolic and carotenoid composition of high intensity pulsed electric field and thermally treated fruit juice-soymilk beverages during refrigerated storage. *Food Chem.* **2011**, *129*, 982–990. [CrossRef] [PubMed]
56. Kim, Y.S.; Park, S.J.; Cho, Y.H.; Park, J. Effects of combined treatment of high hydrostatic pressure and mild heat on the quality of carrot juice. *J. Food Sci.* **2001**, *66*, 1355–1360. [CrossRef]
57. Patras, A.; Brunton, N.; Da Pieve, S.; Butler, F.; Downey, G. Effect of thermal and high pressure processing on antioxidant activity and instrumental colour of tomato and carrot purées. *Innov. Food Sci. Emerg. Technol.* **2009**, *10*, 16–22. [CrossRef]
58. Blanquet-Diot, S.; Soufi, M.; Rambeau, M.; Rock, E.; Alric, M. Digestive stability of xanthophylls exceeds that of carotenes as studied in a dynamic in vitro gastrointestinal system. *J. Nutr.* **2009**, *139*, 876–883. [CrossRef]
59. Faulks, R.M.; Southon, S. Challenges to understanding and measuring carotenoid bioavailability. *Biochim. Biophys. Acta BBA Mol. Basis Dis.* **2005**, *1740*, 95–100. [CrossRef]
60. Gence, L.; Servent, A.; Poucheret, P.; Hiol, A.; Dhuique-Mayer, C. Pectin structure and particle size modify carotenoid bioaccessibility and uptake by Caco-2 cells in citrus juices: Vs. concentrates. *Food Funct.* **2018**, *9*, 3523–3531. [CrossRef]
61. Meléndez-Martínez, A.J.; Paulino, M.; Stinco, C.M.; Mapelli-Brahm, P.; Wang, X.D. Study of the time-course of cis/trans (Z/E) isomerization of lycopene, phytoene, and phytofluene from tomato. *J. Agric. Food Chem.* **2014**, *62*, 12399–12406. [CrossRef]
62. Lemmens, L.; Van Buggenhout, S.; Oey, I.; Van Loey, A.; Hendrickx, M. Towards a better understanding of the relationship between the β-carotene in vitro bio-accessibility and pectin structural changes: A case study on carrots. *Food Res. Int.* **2009**, *42*, 1323–1330. [CrossRef]
63. Palmero, P.; Panozzo, A.; Colle, I.; Chigwedere, C.; Hendrickx, M.; Van Loey, A. Role of structural barriers for carotenoid bioaccessibility upon high pressure homogenization. *Food Chem.* **2016**, *199*, 423–432. [CrossRef]
64. Huo, T.; Ferruzzi, M.G.; Schwartz, S.J.; Failla, M.L. Impact of fatty acyl composition and quantity of triglycerides on bioaccessibility of dietary carotenoids. *J. Agric. Food Chem.* **2007**, *55*, 8950–8957. [CrossRef]
65. Palmero, P.; Panozzo, A.; Simatupang, D.; Hendrickx, M.; Van Loey, A. Lycopene and β-carotene transfer to oil and micellar phases during in vitro digestion of tomato and red carrot based-fractions. *Food Res. Int.* **2014**, *64*, 831–838. [CrossRef]
66. Becerra-Moreno, A.; Redondo-Gil, M.; Benavides, J.; Nair, V.; Cisneros-Zevallos, L.; Jacobo-Velázquez, D.A. Combined effect of water loss and wounding stress on gene activation of metabolic pathways associated with phenolic biosynthesis in carrot. *Front. Plant Sci.* **2015**, *6*, 837. [CrossRef] [PubMed]
67. Quitão-Teixeira, L.J.; Odriozola-Serrano, I.; Soliva-Fortuny, R.; Mota-Ramos, A.; Martín-Belloso, O. Comparative study on antioxidant properties of carrot juice stabilised by high-intensity pulsed electric fields or heat treatments. *J. Sci. Food Agric.* **2009**, *89*, 2636–2642. [CrossRef]
68. Szczepańska, J.; Barba, F.J.; Skąpska, S.; Marszałek, K. High pressure processing of carrot juice: Effect of static and multi-pulsed pressure on the polyphenolic profile, oxidoreductases activity and colour. *Food Chem.* **2020**, *307*, 125549. [CrossRef]
69. Palafox-Carlos, H.; Ayala-Zavala, J.F.; González-Aguilar, G.A. The role of dietary fiber in the bioaccessibility and bioavailability of fruit and vegetable antioxidants. *J. Food Sci.* **2011**, *76*, 6–15. [CrossRef] [PubMed]
70. López-Gámez, G.; Elez-Martínez, P.; Martín-Belloso, O.; Soliva-Fortuny, R. Pulsed electric fields affect endogenous enzyme activities, respiration and biosynthesis of phenolic compounds in carrots. *Postharvest Biol. Technol.* **2020**, *168*, 111284. [CrossRef]
71. Jakobek, L. Interactions of polyphenols with carbohydrates, lipids and proteins. *Food Chem.* **2015**, *175*, 556–567. [CrossRef] [PubMed]

foods

Article

How Do the Different Types of Carrier and Drying Techniques Affect the Changes in Physico-Chemical Properties of Powders from Chokeberry Pomace Extracts?

Anna Michalska-Ciechanowska [1,*], Aleksandra Hendrysiak [1], Jessica Brzezowska [1], Aneta Wojdyło [1] and Agnieszka Gajewicz-Skretna [2]

1 Department of Fruit, Vegetable and Plant Nutraceutical Technology, the Faculty of Biotechnology and Food Science, Wrocław University of Environmental and Life Sciences, Chełmońskiego 37, 51-630 Wrocław, Poland; 109211@student.upwr.edu.pl (A.H.); jessica.brzezowska@upwr.edu.pl (J.B.); aneta.wojdylo@upwr.edu.pl (A.W.)

2 Laboratory of Environmental Chemometrics, Faculty of Chemistry, University of Gdansk, Wita Stwosza 63, 80-308 Gdansk, Poland; agnieszka.gajewicz@ug.edu.pl

* Correspondence: anna.michalska@upwr.edu.pl

Abstract: Chokeberry fruit, one of the richest plant sources of bioactives, is processed into different foodstuffs, mainly juice, which generates a considerable amount of by-products. To follow the latest trends in the food industry considering waste management, the study aimed to produce chokeberry pomace extract powders and conduct experimental and chemometric assessment of the effect of different carriers and drying techniques on the physico-chemical properties of such products. The PCA analysis showed that the examined powders were classified into two groups: freeze-dried (variation in case of moisture content, water activity, colour, and browning index) and vacuum-dried (bulk density). No clear pattern was observed for the physical properties of carrier added products. The sum of polyphenolics (phenolic acids, anthocyanins and flavonols) ranged from 3.3–22.7 g/100 g dry matter. Drying techniques had a stronger effect on the polyphenols profile than the type of carrier. Hydroxymethyl-*L*-furfural formation was enhanced by inulin addition during high-temperature treatment. Overall, the addition of maltodextrin and trehalose mixture for freeze drying and vacuum drying at 90 °C caused the highest retention of polyphenolics and the lowest formation of hydroxymethyl-*L*-furfural; however, an individual and comprehensive approach is required when the obtainment of high-quality chokeberry powders is expected.

Keywords: *Aronia melanocarpa* L.; by-products; sustainability; inulin; trehalose; polyphenols; HMF; unsupervised chemometric analysis

Citation: Michalska-Ciechanowska, A.; Hendrysiak, A.; Brzezowska, J.; Wojdyło, A.; Gajewicz-Skretna, A. How Do the Different Types of Carrier and Drying Techniques Affect the Changes in Physico-Chemical Properties of Powders from Chokeberry Pomace Extracts? *Foods* **2021**, *10*, 1864. https://doi.org/10.3390/foods10081864

Academic Editor: Adamantini Paraskevopoulou

Received: 22 June 2021
Accepted: 28 July 2021
Published: 12 August 2021

1. Introduction

As of late, a new trend has become increasingly evident in the food industry, with consumers shifting their preferences from animal to plant-based products. This is mainly driven by consumers' growing awareness of a healthy lifestyle, of which a balanced diet is an indispensable part, but also by ecological and ethical issues [1]. For this reason, the production and processing of plant products, mainly fruit and vegetables, has been increasing significantly for some time. However, this leads to the generation of an enormous amount of by-products, including pomace, the management of which is currently one of the biggest challenges for the food industry sector [2]. As it was reported earlier in the literature, fruit and vegetable pomace is a valuable source of numerous bioactive compounds [3,4]. One of the most frequently processed raw materials is black chokeberry (*Aronia melanocarpa* L.), which, due to its characteristic astringent taste, is not usually consumed as a fresh fruit. The main direction of its use is the production of juices, jams, or fruit wines, which results in significant amounts of by-products [5]. Due to its unfavourable sensory properties, it often remains wasted. However, a body of evidence has demonstrated that chokeberry pomace is a rich source of polyphenolic compounds, which have strong antioxidant properties, but has also indicated the beneficial effects in, among others, obesity, glucose metabolic disorders, pro-inflammatory conditions, hypertension, dyslipidaemia, etc. [6]. It has to be stressed that chokeberry pomace has almost an eight times higher content of polyphenols than juice [7]. Due to the proven health-promoting properties of chokeberry pomace, it can be an excellent raw material when developing functional foods [7]. In this setting, the processing of chokeberry by-products is of great importance and the production of a powdered form is new and one of the most promising alternatives for its utilisation, while at the same time being an effective tool for introducing sustainable food management [8,9]. One of the interesting approaches is to obtain powders from chokeberry pomace extracts. This type of product, due to its easy-to-handle form, possible high solubility (in contrast to pomace), and a relatively high microbial stability, is an attractive additive to other foodstuffs as a sustainable natural colouring or functional enrichment agent [5,10]. However, in order to retain the satisfactory amount of selected polyphenolics, a correctly chosen extraction method is essential, as are the subsequent steps: solvent evaporation, purification of the extracts on a polymer-bed-type Amberlite XAD, and next solvent evaporation leading to a purified pomace polyphenolic extract obtainment [11]. Moreover, in order to acquire the powdered form, it is necessary to carry out drying, which may significantly affect the physical properties, but also can lead to alterations of the chemical composition, particularly in the profile of polyphenolic compounds. For this reason, the choice of an appropriate drying technique and its parameters is pivotal to maintain a relatively high possible content of polyphenolic compounds in the final product [12]. Freeze drying, as a low-temperature water removal process, is considered to be the least intrusive with the slightest impact on the transformation of the dried matrix [13]. On the other hand, recent studies by Michalska-Ciechanowska et al. [12] on black chokeberry juice demonstrated that the application of high temperatures during drying can positively influence the polyphenolic profile, leading to the release of significant amounts of selected bioactive constituents from more complex structures [12]. One technique that enables the use of high drying temperatures while excluding the effects of oxygen on the dried matrix is vacuum drying [14]. However, the thermolability of polyphenols, especially anthocyanins, to thermal processes, as well as the different yields for obtaining powders depending on the raw material composition, induce the necessity of using a carrier additive in the drying process [15]. Substances of carbohydrate origin, such as maltodextrin, are widely used for this purpose and have proven protective properties, allowing for an increased retention of anthocyanins at high drying temperatures compared with other carriers [14]. Further interesting additives include trehalose, which has been found to be inactive in the Maillard reaction due to its non-reducing properties, and inulin, which is well known for its functional properties [16]. Importantly, the selection of carrier is of high importance as it was previously demonstrated that the type and concentration of selected carriers may not only cause a decrease in the polyphenolics content, but may also accelerate the formation of undesirable process contaminants, especially in fruit-based matrices [12,14]. Taking the above into consideration, it is hypothesised that the drying techniques and type of carrier will simultaneously moderate the polyphenolic composition in chokeberry pomace extract powders. Thus, this study aimed to evaluate different drying techniques and parameters as well as carrier types on the alteration in the polyphenolics composition, antioxidant capacity, and formation of Maillard reaction and caramelisation products in chokeberry pomace extract powders.

2. Materials and Methods

2.1. Materials

The material used in the study was composed of chokeberry fruits (approximately 70 kg) obtained from Rolniczo-Sadownicze Gospodarstwo Doświadczalne 'Przybroda' (Rokietnica near Poznań, Poland). The fruits were ground in a Thermomix (Wuppertal,

Vorkwek, Germany) and pressed on a hydraulic press (SRSE, Warszawa, Poland). The pomace gained was frozen before the extraction process at −20 °C.

2.2. Methods

2.2.1. Extraction Procedure

The extraction of polyphenolic compounds from chokeberry pomace (initial moisture content of 44.45 ± 0.01%) was performed according to the patent Oszmiański and Krzywicki [17]. Thawed pomace (approximately 10 kg) was mixed with 30% acetone (1:4, *w*/*v*) and sonicated for 15 min. The solution was left for 24 h and sonicated again for 15 min. The acetone was evaporated (Unipan 350P, Warsaw, Poland) and the solution gained in a quantity of approximately 17 L (4.9 ± 0.07 Bx) was introduced into the Amberlite XAD-16 (Brenntag, Poland) according to the procedure of Kammerer et al. [11] in order to recover the selected polyphenolics in the extracts [18]. This resulted in approximately 2.5 L of final solution (6.9 ± 0.1 Bx), which was submitted to the formulation of the drying compositions (control, samples with carrier addition). The procedure was performed in duplicate (n = 2).

2.2.2. Preparation of Chokeberry Pomace Extract Powders

The pomace extracts were mixed with carriers, i.e., maltodextrin (M) (DE 9.3; PEPEES S.A, Poland), inulin (I) (Beneo-Orafti, Belgium), trehalose (T) (Hayashibara, Co., Okayama, Japan), and their mixtures maltodextrin-inulin (M-I), maltodextrin-trehalose (M-T), and inulin-trehalose (I-T) at the level of 10% (*w*/*w*) (chosen on the basis of experimental work). No carrier was added to the control sample. Prepared solutions were submitted to drying processes: freeze drying at −60 °C/+ 24 °C for 24 h (FreeZone freeze dryer, Labconco Corp., Kansas, MO, USA), and vacuum drying at 60 and 90 °C for, respectively, 22 and 16 h (Vacucell 111 Eco Line, MMM Medcenter Einrichtungen GmbH, Germany). The drying processes were performed in duplicate (n = 2). After the drying, the obtained powders were vacuum packed and stored at −20 °C until analysis.

2.2.3. Physical Properties

Moisture Content

The moisture content (*Mc*) was determined in duplicate (n = 2) [12] at 80 °C. The results were expressed as %.

Water Activity

The water activity (a_w) was done at 25 °C in duplicate (n = 2) using the water activity meter Dew Point Water Activity Meter 4TE (AQUA LAB, Pullman, WA, USA).

Bulk Density

The bulk density (ρ_b) of powders was performed in duplicate (n = 2) using a graduated cylinder (10 cm^3) and laboratory scale. This was calculated as follows:

$$\rho_b = \frac{m}{V_b} \quad (1)$$

m—mass of the powder, V_b—volume of the powder

Colour

The colour of the powders was measured in triplicate (n = 3) using a Minolta Chroma Meter CR-400 colorimeter (Minolta Co. Ltd., Osaka, Japan) according to the CIE *L*a*b** system. Based on the results, the browning index (BI) was calculated according to the equation described by Mexis and Kontominas [19]:

$$BI = \frac{[100(x - 0.31)]}{0.17} \quad (2)$$

where:

$$x = \frac{a^* + 1.75L^*}{5.64L^* + a^* - 3.012b^*} \quad (3)$$

2.2.4. Chemical Properties

Preparation of Extracts

The extraction of polyphenolics from the chokeberry pomace powders was performed according to the procedure described by Wojdyło et al. [20]. Samples for qualitative and quantitative determination of these compounds and hydroxymethyl-*L*-furfural (HMF) by Ultrahigh Performance Liquid Chromatography were done in duplicate (n = 2) and extracted with 1.7 mL of aqueous solution of MeOH (30%; v/v) with ascorbic acid (0.2%) and 0.1% CH_3COOH while those for antioxidant capacity analyses (n = 2) were prepared with 1.7 mL of MeOH (80%; v/v) with HCl (1 mL/L). All samples were sonicated for 15 min and refrigerated at 4 °C for 24 h. After this time, the samples were ultrasonicated again (15 min) and then centrifuged (19,515× g, 20 °C; MPW-251, MPW Med. Instruments, Poland). The obtained extracts were subjected to further analyses.

Qualitative and Quantitative Determination of Polyphenolics and Hydroxymethyl-*L*-furfural

The qualitative and quantitative analyses of polyphenolics were performed using an Acquity UPLC system (Waters, Milford, MA, USA) with a PDA detector, equipped with a binary pump system and a solvent manager. The separation of the individual compounds was done at a flow rate of 0.42 mL/min in an ACQUITY BEH C_{18} analytical column (100 mm × 1.7 µm, Waters, Milford, MA, USA). The column was conditioned with acetonitrile (100%) and an aqueous solution of acetonitrile (10%; v/v). The separation was performed by the gradient elution method using 4.5% formic acid (solvent A) and acetonitrile (solvent B): 0–10 min linear gradient, 1–15% solvent B; 10–11.5 min linear gradient, 25–100% solvent B. Anthocyanins, flavonols, phenolic acids, and HMF were detected at λ = 520 nm, 360 nm, 320 nm, and 280 nm, respectively. The determination was performed in duplicate (n = 2). The polyphenolics were identified by LC-MS QTof and assessed using the MassLynx 4.0 ChromaLynx Application Manager software [21]. Results were expressed as g/100 g dry basis (db) for individual polyphenols and as µg/100 g of db for HMF.

Antioxidant Capacity

The antioxidant capacity of polyphenolic extracts from chokeberry pomace extract powders was evaluated by $ABTS^+$ radical cation scavenging [22] and FRAP [23] by *in vitro* assays using a Synergy H1 spectrophotometer (BioTek Instruments Inc., Winooski, VT, USA). The determination was made in duplicate (n = 2) and the results were expressed as mmol Trolox equivalent (TE) per 100 g of db.

2.2.5. Statistical Analysis and Chemometrics

The results were statistically analysed using STATISTICA 13 software (StatSoft, Tulsa, OK, USA). One-way analysis of variance (ANOVA) was performed ($p < 0.05$) and the Tukey post hoc test (also referred to as the Tukey's honestly significant difference (Tukey HSD)) was applied to compare the pairwise differences between group means ($p \leq 0.05$). To investigate the dependence between the selected variables, Pearson's correlation coefficient was also calculated.

To identify the direction that maximises the variance of the projected data and to explore the trends and relations between observations and variables simultaneously, principal component analysis (PCA) [24] was performed using the 'factoextra' R package [25]. The radar plots were prepared using the 'fmsb' R package [26], allowing differences in sub-groups to be visualised.

3. Results and Discussion

3.1. Physical Properties

3.1.1. Moisture Content

The moisture content (*Mc*) of the analysed powders ranged from 0.47% to 7.59% for the products gained, respectively, after vacuum drying at 90 °C and after freeze drying with the addition of trehalose (Table 1). In general, the average moisture content of the control samples (no addition of carriers) was lower by approximately 29% when compared to the average moisture content of the products gained with the addition of carriers. The drying techniques and parameters applied for the powders' preparation had a significant effect on this parameter: the higher the drying temperature, the lower the *Mc* of the powders [27].

When the drying techniques and carrier type were concerned, it was noted that during the freeze drying process, the addition of maltodextrin resulted in the lowest *Mc*, whereas the addition of trehalose led to the highest moisture content of the powders (Table 1).

In the case of freeze and vacuum drying at 60 °C, the powders gained with the addition of inulin had a lower moisture content than the samples obtained with the addition of its mixes (M-I and I-T). However, when vacuum drying at 90 °C was considered, the powders obtained with the addition of inulin and its mixes had the highest *Mc* when compared to the rest of the applied carriers. It can be concluded that the type of carrier might influence the moisture content due to the different water-holding capacity, which could be additionally altered by the drying technique used for the powders' preparation [28]. However, taking into account the moisture content of the carriers (i.e., *Mc* of M: 10.07 ± 0.93; *Mc* of I: 3.05 ± 0.01; *Mc* of T: 1.04 ± 0.04; *Mc* of M-I: 5.52 ± 0.01; *Mc* of M-T: 3.68 ± 0.18; *Mc* of I-T: 2.66 ± 0.13), it can be observed that *Mc* of the powders was influenced more by possible interactions between the compounds present in the extracts and the constituents of the carriers (Table 1). This was also confirmed by the lack of a significant correlation between the moisture content of the carriers themselves and *Mc* of the powders gained with their addition.

3.1.2. Water Activity (a_w)

In all the analysed powders, the water activity values were below 0.45 (Table 1), indicating that these products can be considered as stable from a biochemical and microbiological point of view [29]. In the case of the controls and powders with the addition of carriers and their mixtures, the water activity values were higher after the freeze drying process than in the products gained after vacuum drying at 60 and 90 °C [27]. This might be due to the more porous structure of the products gained after freeze drying, when compared to the other drying technique used [30]. Similarly to da Silva Calvaho et al. [29], the type of carrier influenced a_w of the powders. In the products obtained with the application of maltodextrin and its mixes for freeze drying, I and I-T for vacuum drying at 60 °C and T for vacuum drying at 90 °C resulted in higher water activity in the products obtained. A positive correlation between water activity and moisture content was noted (r = 0.67) (Figure S1). A similar observation was made in case of apple juice powders [27]. In general, the application of different carriers and their mixture led to alterations in the water activity of the samples, which was also dependent on the drying technique used. There are reports suggesting that in relation to the type of carrier used for drying, the formation of a crust on the outer layer of samples was observed, which influences the water activity values in the products gained [31]. Thus, water activity may be connected with the physical changes that occur on the surface of the samples during drying that might differ in terms of their carrier properties.

Table 1. Moisture content, water activity, bulk density, colour (CIE $L^*a^*b^*$), and browning index of chokeberry powders obtained after freeze and vacuum drying (n = 2; average ± standard deviation).

Drying Technique	Process Conditions	Carrier	Mc (%)	a_w (−)	ρ_b (g/cm^3)	Colour L^*	a^*	b^*	BI (AU)
FD	−60 °C/24 °C	(−)	4.38 ± 0.54 ^{cdef}	0.34 ± 0.02 ^{h}	0.09 ± 0.01 ^{a}	24.20 ± 0.11 ^{i}	16.14 ± 0.23 ^{i}	4.65 ± 0.02 ^{m}	0.42 ± 0.01 ^{ij}
		Maltodextrin	4.20 ± 0.42 ^{cdef}	0.41 ± 0.02 ^{i}	0.12 ± 0.01 ^{a}	25.78 ± 0.25 ^{j}	22.31 ± 0.07 ^{j}	4.09 ± 0.01 ^{l}	0.43 ± 0.01 ^{k}
		Inulin	4.74 ± 1.04 ^{def}	0.15 ± 0.01 ^{ab}	0.32 ± 0.02 ^{bc}	28.59 ± 0.06 ^{k}	25.93 ± 0.04 ^{k}	2.26 ± 0.04 ^{i}	0.42 ± 0.02 ^{i}
		Trehalose	7.59 ± 1.05 ^{g}	0.29 ± 0.02 ^{gh}	0.61 ± 0.01 ^{defg}	13.57 ± 0.16 ^{a}	12.43 ± 0.19 ^{g}	1.53 ± 0.03 ^{h}	0.43 ± 0.01 ^{jk}
		Maltodextrin-Inulin	6.15 ± 1.25 ^{efg}	0.42 ± 0.02 ^{i}	0.17 ± 0.01 ^{ab}	23.44 ± 0.71 ^{i}	22.94 ± 0.07 ^{i}	4.35 ± 0.11 ^{l}	0.45 ± 0.01 ^{l}
		Maltodextrin-Trehalose	6.53 ± 0.01 ^{fg}	0.43 ± 0.01 ^{i}	0.29 ± 0.01 ^{bc}	17.28 ± 0.03 ^{bc}	12.69 ± 0.05 ^{g}	3.32 ± 0.02 ^{k}	0.43 ± 0.01 ^{jk}
		Inulin-Trehalose	5.42 ± 0.70 ^{efg}	0.31 ± 0.01 ^{h}	0.32 ± 0.03 ^{bc}	18.27 ± 0.18 ^{cde}	12.89 ± 0.33 ^{g}	3.37 ± 0.07 ^{k}	0.42 ± 0.01 ^{ij}
VD	60 °C	(−)	1.12 ± 0.39 ^{a}	0.24 ± 0.01 ^{def}	0.64 ± 0.05 ^{defgh}	18.48 ± 0.36 ^{cdef}	9.76 ± 0.87 ^{d}	2.46 ± 0.36 ^{ij}	0.39 ± 0.01 ^{g}
		Maltodextrin	2.75 ± 0.57 ^{abcd}	0.23 ± 0.01 ^{def}	0.70 ± 0.01 ^{efghi}	16.57 ± 0.11 ^{b}	7.38 ± 0.09 ^{c}	0.49 ± 0.02 ^{cde}	0.37 ± 0.02 ^{cd}
		Inulin	2.65 ± 0.04 ^{abcd}	0.32 ± 0.01 ^{h}	0.65 ± 0.06 ^{defgh}	18.55 ± 0.54 ^{def}	15.51 ± 0.24 ^{i}	1.05 ± 0.02 ^{g}	0.41 ± 0.01 ^{h}
		Trehalose	2.74 ± 1.16 ^{abcd}	0.19 ± 0.01 ^{bcde}	0.53 ± 0.05 ^{d}	20.55 ± 0.12 ^{gh}	10.71 ± 0.03 ^{ef}	0.81 ± 0.04 ^{fg}	0.38 ± 0.01 ^{ef}
		Maltodextrin-Inulin	2.89 ± 0.45 ^{abcd}	0.21 ± 0.01 ^{cdef}	0.75 ± 0.04 ^{fghi}	17.93 ± 0.18 ^{cd}	6.34 ± 0.15 ^{ab}	0.19 ± 0.03 ^{ab}	0.35 ± 0.01 ^{ab}
		Maltodextrin-Trehalose	2.02 ± 0.42 ^{abc}	0.25 ± 0.01 ^{fg}	0.60 ± 0.01 ^{def}	19.61 ± 0.19 ^{fg}	6.12 ± 0.07 ^{ab}	0.39 ± 0.04 ^{bcd}	0.35 ± 0.01 ^{a}
		Inulin-Trehalose	3.91 ± 0.54 ^{bcde}	0.32 ± 0.01 ^{h}	0.77 ± 0.08 ^{hi}	20.29 ± 0.36 ^{gh}	13.79 ± 0.18 ^{h}	1.41 ± 0.03 ^{h}	0.40 ± 0.01 ^{g}
	90 °C	(−)	0.98 ± 0.47 ^{a}	0.24 ± 0.01 ^{efg}	0.55 ± 0.03 ^{de}	17.92 ± 0.11 ^{cd}	6.07 ± 0.21 ^{ab}	2.71 ± 0.06 ^{j}	0.38 ± 0.01 ^{f}
		Maltodextrin	0.89 ± 0.38 ^{a}	0.14 ± 0.01 ^{a}	0.65 ± 0.08 ^{defgh}	17.97 ± 0.11 ^{cd}	5.66 ± 0.01 ^{a}	0.26 ± 0.01 ^{abc}	0.35 ± 0.01 ^{a}
		Inulin	1.37 ± 0.25 ^{a}	0.15 ± 0.02 ^{ab}	0.76 ± 0.02 ^{ghi}	19.31 ± 0.12 ^{efg}	10.49 ± 0.22 ^{de}	0.02 ± 0.04 ^{a}	0.37 ± 0.01 ^{de}
		Trehalose	0.47 ± 0.02 ^{a}	0.18 ± 0.01 ^{abcd}	0.34 ± 0.02 ^{c}	21.06 ± 0.18 ^{h}	11.44 ± 0.03 ^{f}	0.70 ± 0.03 ^{ef}	0.38 ± 0.01 ^{ef}
		Maltodextrin-Inulin	1.69 ± 0.28 ^{ab}	0.14 ± 0.02 ^{a}	0.80 ± 0.04 ^{hi}	17.55 ± 0.69 ^{bcd}	7.50 ± 0.12 ^{c}	0.07 ± 0.09 ^{a}	0.36 ± 0.02 ^{bc}
		Maltodextrin-Trehalose	0.88 ± 0.12 ^{a}	0.16 ± 0.02 ^{ab}	0.57 ± 0.06 ^{de}	18.15 ± 0.13 ^{cde}	6.80 ± 0.25 ^{bc}	0.52 ± 0.04 ^{cde}	0.36 ± 0.01 ^{b}
		Inulin-Trehalose	2.62 ± 0.09 ^{abcd}	0.17 ± 0.01 ^{abc}	0.83 ± 0.01 ^{i}	21.75 ± 0.77 ^{h}	15.8 ± 0.27 ^{i}	0.67 ± 0.04 ^{def}	0.39 ± 0.01 ^{g}

(−)—no carrier addition, FD—freeze drying, VD—vacuum drying, M_C—moisture content, a_w—water activity, ρ_b—bulk density, AU—arbitrary units; $^{a–m}$ the same letters within a column indicate no statistically significant differences (HSD Tukey test; $p \leq 0.05$).

3.1.3. Bulk Density

The bulk density of chokeberry pomace extracts powders ranged from 0.09 g/cm^3 to 0.83 g/cm^3 for, respectively, the control sample gained after freeze drying and powders obtained after vacuum drying at 90 °C with the addition of I-T (Table 1). When the control samples and products gained with different carriers were concerned, the average value of bulk density was the lowest in the case of controls, whereas the highest bulk density was noticed for powders gained with the addition of carriers, especially when inulin and its mixtures were considered (M-I, I-T). In general, as observed by Michalska and Lech [27], freeze drying resulted in lower values of the bulk density of powders gained followed by VD at 60 °C and VD at 90 °C, except powders produced with trehalose. The addition of trehalose to pomace extract formulations caused the lowest bulk density of powders gained after VD at 60 °C and VD at 90 °C. This might be connected with the higher soluble solids content in the formulation containing trehalose [32] during the freeze drying process and/or with its different behaviour when a relatively high temperature of drying was applied. It could also be related to the fluidizing properties of this carrier [33] as trehalose, due to its chemical composition, might react differently with the compounds present in the extracts additionally moderated by the vacuum drying. What is more, the influence of the carrier type on the bulk density of powders was also noticed in the case of cranberry juice [34]. Taking the above into consideration, the selection of a carrier for powder preparation should also consider the carrier physico-chemical properties as it may moderate the physical properties of the material subjected to drying to a different extent. For the analysed powders, a negative correlation ($r = -0.60$) between bulk density and water activity (Figure S1) was noted, indicating that the specific structure of the powders gained by different drying techniques might differently influence the ability of samples to trap the water molecules [35].

3.1.4. Colour

The colour parameters measured in terms of L^*, a^*, and b^* values as well as the browning index (BI) are indicated in Table 1. In general, powders gained after the freeze drying process were the lightest, with the exception of samples prepared with trehalose, whereas vacuum drying at 60 and 90 °C resulted in slightly lower values of coordinate L^*. In the latest samples, trehalose addition caused the highest values of coordinate L^* in the analysed products. The type of carrier had a significant impact on the lightness of the powders gained [36], which was also moderated by the drying technique used. In general, when the values of coordinate a^* were concerned, it was noted that the carrier addition altered the level of red pigments in chokeberry powders. Contrary to the juice powders [37], the 10% addition of carrier led to the obtainment of products that were visually similar to the control samples. Thus, the addition of carriers, besides an improvement of the efficiency of the sustainable powder production [38], may not visually alter the colour of the products gained. In this case, the composition of chokeberry pomace extracts played a key role in colour preservation as the addition of carriers into the chokeberry juice modified the lightness to a high extent [37].

The strongest retention of red pigment was noted after freeze drying when compared to samples gained after vacuum drying, except samples prepared with the addition of trehalose and its mixes (M-T, I-T). In general, the application of vacuum drying for the powder preparation caused a decrease in the coordinate a^* values, indicating the influence of the high-temperature processing on the thermolability of the red components [39]. The strongest pigment degradation was noted in the case of powders produced with maltodextrin and its mixes (I-M; I-T). The addition of I and I-T seems to prevent deterioration of the coordinate a^* values during vacuum drying. It was observed that the b^* parameter was the highest in the control samples (no addition of carrier), regardless of the drying technique used for their preparation (Table 1). The addition of carriers led to a decrease in the coordinate b^* values, which was the strongest in the case of trehalose application for the freeze drying process. This might be connected with the presence of natural colour compounds in the chokeberry pomace extracts (an average for FD: 3.37), as vacuum drying at 60 °C (an average: 0.97) and 90 °C (an average: 0.71) resulted in significantly lower values of parameter b^*. In the current study, the browning index was applied for determination of overall alterations in the browning colour [40]. In contrast to sea buckthorn powders [41], the highest values of BI were noted for freeze-dried samples, regardless of the type of carrier used for their preparation. It was assumed that the dominant reddish colour could mask the brown pigments present in the analysed powders [42]. Additionally, the BI values might result from the complexes formed from polyphenols [43]. The application of vacuum drying for chokeberry pomace extract powder production caused a decrease in BI when compared to freeze drying. The addition of maltodextrin and its mixes (M-I, M-T) led to lower values of BI. Additionally, a strong correlation between BI and a^* (r = 0.82) as well as b^* (r = 0.83) confirmed the masking effect of the reddish and bluish pigments (Figure S1) [44].

3.1.5. PCA Analysis

The PCA biplot (Figure 1a) shows that freeze-dried samples had a greater spread and more variance than vacuum-dried samples (at 60 and 90 °C).

The first principal component (PC1) clearly separates freeze-dried samples (positive scores) from vacuum-dried samples (negative scores). The explanatory variables (vectors) with the greatest influence on the separation of chokeberry pomace extract powders in PC1 were the colour parameters, including the browning index (BI), and coordinate a^*, b^* as well as Mc, a_w (positively correlated), and ρ_b (negatively correlated). PC2 loadings showed that negatively correlated L^* has the greatest influence on the sample distinctions (Figure 1b). Considering the locations of the samples in the space defined by the first two principal components (PCs), it can be stated that, due to low PC1 scores and positive loading values, vacuum-dried samples (at 60 and 90 °C) were characterised by a relatively low browning index (BI), and coordinate a^*, b^*, as well as Mc, a_w. At the same time, powders gained after vacuum drying had the highest values of bulk density. Interestingly, although the type of carrier was also concerned, no straightforward trends were observed for the powders gained (Figure 1c).

3.2. Chemical Properties

3.2.1. Polyphenols Content

In the powders obtained from chokeberry pomace extracts, three major groups of polyphenolic compounds were identified, i.e., phenolic acids (3), anthocyanins (4), and flavonols (8). The extraction of chokeberry pomace and usage of absorber technology led to modification of the polyphenolics composition [45] as proanthocyanidins were not identified in the powders obtained. This may be connected with the results reported by Sójka et al. [46] of major absorption of proanthocyanidins in the cell wall of chokeberry pomace. The extraction procedure applied and clarification of the polyphenols by Amberlite 16 of pomace extracts led to an absence of these constituents in the powders. As previously reported by Wang et al. [18], this could be linked to the different affinity of the particular groups of polyphenolics for the stationary phase, which might affect the elution time and thus the presence or absence of these constituents in the extract. Besides this, the presence of phenolic acids, anthocyanins, and flavonols was confirmed [46].

The sum of identified polyphenols was, on average, 3.9-fold higher for control samples (Table 2) (average 19.07 g/100 g db) when compared to powders gained with 10% (w/w) carrier addition to the extracts before drying, regardless of the drying technique applied (Figure 2; Table S1). A similar observation was previously noted in the case of cranberry juices and extracts [47]. The drying techniques influenced the sum of identified polyphenols to a high extent. In the case of control samples, the highest content of polyphenols was noted after freeze drying, followed by vacuum drying at 90 and 60 °C (Table 2).

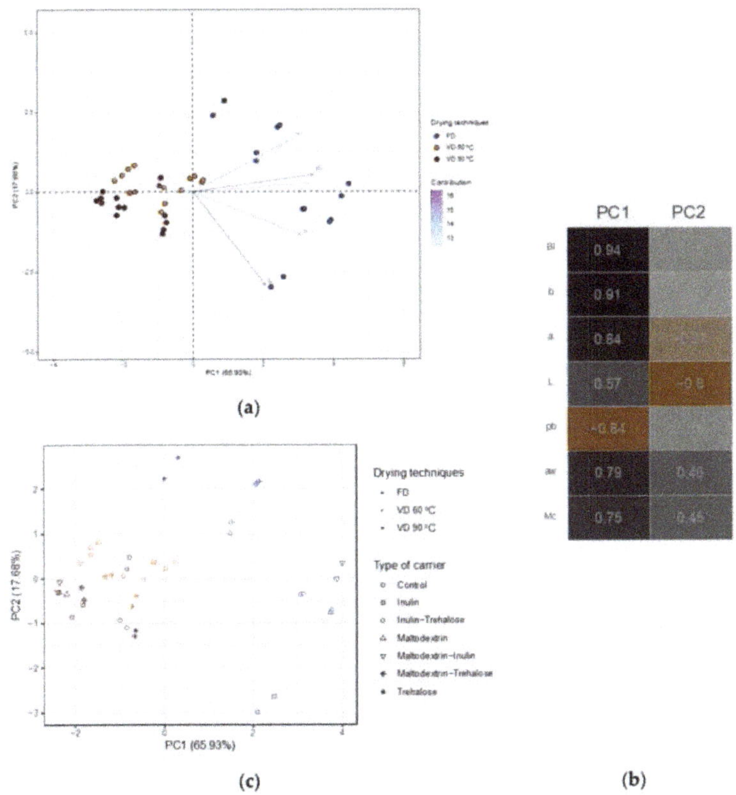

Figure 1. (**a**) The PCA biplot of the first two principal components that simultaneously shows PC scores of chokeberry pomace extract powders (points) and loadings of explanatory variables (vectors). The marker colour corresponds to the drying techniques (i.e., freeze and vacuum drying at 60 and 90 °C), while the length and the transparency of the arrows indicate the variance of the physical properties of powders from chokeberry pomace extracts and their contributions to the principal components, respectively. Together, the first two principal components explain 83.61% of the variability; (**b**) The plot of normalised factor loadings that quantify the extent to which the explanatory variable is related with a given principal component. Applying the so-called Malinowski rule (i.e., normalised factor loading cut-off of | 0.70 |) simplifies assigning physical meaning to each principal component; (**c**) Score plot in the space defined by the first two principal components illustrating relations and trends of chokeberry pomace extract powders gained after freeze and vacuum drying at 60 and 90 °C with the addition of maltodextrin (M), inulin (I), trehalose (T), maltodextrin—inulin (M—I), maltodextrin—trehalose (M—T), and inulin—trehalose (I—T).

In the case of powders produced with the addition of carriers and their mixes, no statistical differences were noted between the average content of polyphenols in the samples gained with the addition of carriers after freeze drying and vacuum drying at 60 °C (Table S1). The strongest influence was noted after the application of vacuum drying at 90 °C. Going into the details the application of the freeze drying process resulted in the highest retention of all identified polyphenols when maltodextrin and the mix of maltodextrin and trehalose (M-T) was used, whereas the usage of inulin resulted in the lowest retention of these constituents (Figure 2); however, the results were not statistically significantly different. One probable cause may be interactions between the carrier and polyphenolics as raised by Tomas et al. [48].

Table 2. The content of identified polyphenolics (g/100 g db), hydroxymethyl-*L*-furfural (μg/100 g db) and antioxidant capacity measured by TEAC ABTS and FRAP methods (mmol TE/100 g db) in chokeberry pomace extract powders (controls) obtained after freeze and vacuum drying (average ± standard deviation; n = 2).

	FD	VD	
	−60 °C/+24 °C	60 °C	90 °C
Total polyphenols	22.70 ± 0.12 [c]	15.88 ± 0.44 [a]	18.64 ± 0.47 [b]
Phenolic acids			
Neochlorogenic acid	4.86 ± 0.18 [b]	3.01 ± 0.09 [a]	4.47 ± 0.67 [ab]
Cryptochlorogenic acid	0.10 ± 0.01 [ab]	0.07 ± 0.01 [a]	0.14 ± 0.02 [b]
Chlorogenic acid	5.15 ± 0.15 [b]	3.43 ± 0.11 [a]	4.72 ± 0.37 [b]
Sum	10.10 ± 0.02 [b]	6.51 ± 0.20 [a]	9.33 ± 1.06 [b]
Anthocyanins			
Cyanidin-3-*O*-galactoside	5.90 ± 0.14 [b]	3.68 ± 0.09 [a]	3.93 ± 0.37 [a]
Cyanidin-3-*O*-glucoside	0.33 ± 0.05 [b]	0.18 ± 0.01 [a]	0.22 ± 0.02 [ab]
Cyanidin-3-*O*-arabinoside	2.56 ± 0.05 [b]	1.68 ± 0.05 [a]	1.61 ± 0.13 [a]
Cyanidin-3-*O*-xyloside	0.41 ± 0.01 [b]	0.28 ± 0.02 [a]	0.29 ± 0.03 [a]
Sum	9.19 ± 0.13 [b]	5.82 ± 0.12 [a]	6.06 ± 0.56 [a]
Flavonols			
Quercetin-dihexoside 1	0.26 ± 0.01 [b]	0.17 ± 0.01 [a]	0.30 ± 0.01 [c]
Quercetin-dihexoside 2	0.14 ± 0.01 [a]	0.21 ± 0.01 [a]	0.21 ± 0.04 [a]
Quercetin-3-*O*-vicianoside	0.14 ± 0.01 [b]	0.12 ± 0.01 [b]	0.06 ± 0.02 [a]
Kaempferol-3-*O*-robinobioside	0.45 ± 0.02 [b]	0.24 ± 0.01 [a]	0.43 ± 0.02 [b]
Kaempferol-3-*O*-rutinoside	0.53 ± 0.02 [b]	0.30 ± 0.01 [a]	0.48 ± 0.02 [b]
Kaempferol-3-*O*-galactoside	1.06 ± 0.02 [a]	1.72 ± 0.07 [b]	1.03 ± 0.01 [a]
Kaempferol-3-*O*-glucoside	0.78 ± 0.02 [b]	0.59 ± 0.02 [a]	0.72 ± 0.02 [b]
Derivative of quercetin	0.05 ± 0.01 [b]	0.20 ± 0.01 [c]	0.04 ± 0.01 [a]
Sum	3.40 ± 0.01 [a]	3.55 ± 0.12 [a]	3.26 ± 0.03 [a]
Hydroxymethyl-*L*-furfural	11.57 ± 0.23 [a]	2.62 ± 0.24 [b]	14.03 ± 1.59 [a]
TEAC ABTS	364.95 ± 11.98 [a]	357.33 ± 11.01 [a]	374.80 ± 12.92 [a]
FRAP	292.95 ± 7.61 [a]	285.14 ± 1.02 [a]	299.38 ± 1.01 [a]

FD—freeze drying, VD—vacuum drying; TE—Trolox equivalent; [a,b,c]—the same letters within a row indicate no statistically significant differences (HSD Tukey test, $p \leq 0.05$).

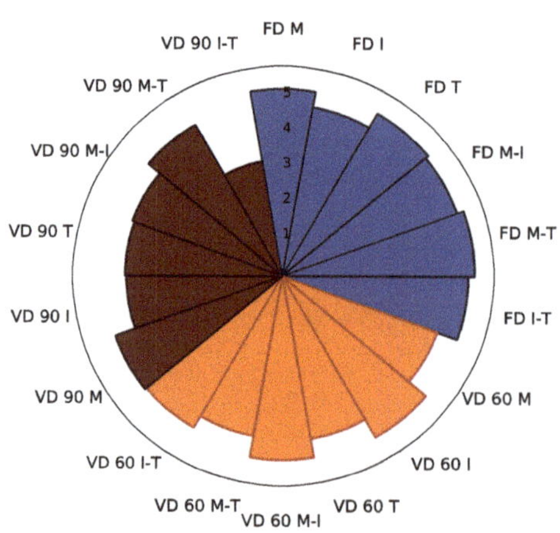

Figure 2. The radar plot of sum of identified polyphenolic compounds in chokeberry pomace extract powders gained with addition of maltodextrin (M), inulin (I), trehalose (T), maltodextrin—inulin (M—I), maltodextrin—trehalose (M—T) and inulin—trehalose (I—T) by freeze drying (FD) and vacuum drying at 60 °C (VD 60) and 90 °C (VD 90) (n = 2) (g/100 g db).

The reverse effect was noted in the case of powders gained after the application of vacuum drying at 60 °C: the addition of inulin resulted in the highest retention of these constituents. Among the powders gained after vacuum drying at 90 °C, the highest retention of the sum of polyphenolics was indicated in the samples obtained after the addition of M and M-T (Figure 2). A statistically significantly lower content of polyphenols was noted when I-T was applied for drying (Table S1). Previously, an influence on the retention of polyphenolics during application of different drying techniques and selected types of carriers was noted for blackcurrant [14] and chokeberry [12] juices.

Going into the details, the percentage share of phenolic acids, anthocyanins, and flavonols in the powders obtained by the selected drying techniques was as follows: 42.9%, 41.2%, and 15.9% for freeze-dried samples; 44%, 37.8%, and 18.2% for vacuum-dried samples at 60 °C; and 47.9%, 31.1%, and 21% for vacuum-dried samples at 90 °C. In order to follow these changes, each group of polyphenolics was examined (Table S2).

Similar to chokeberry juice powders [12], the dominant identified group of polyphenolics present in the controls and powders made with the addition of carriers consisted of phenolic acids [49], among which chlorogenic (50.9% of total phenolic acids), neochlorogenic (47.9% of total phenolic acids), and cryptochlorogenic (1.2% of total phenolic acids) acids were quantified (Table 2 and Table S2). In comparison, Sójka et al. [46] identified only chlorogenic and neochlorogenic acids in chokeberry pomace dried at 70 °C. In the current study, the content of phenolic acids was the highest in the control samples gained after freeze and vacuum drying at 90 °C, whereas the application of VD at 60 °C resulted, on average, in a 33% lower content of these constituents (Table 2). When the addition of selected carriers was concerned, the average content of the sum of phenolic acids was at a similar level, regardless of the drying technique and parameters applied (Figure 3a–d). There were no statistically significant differences noted between samples gained with the addition of M, I, T, M-I, M-T, and I-T after freeze and vacuum drying at 60 °C. Vacuum drying at 90 °C caused significant changes in the content of phenolic acids (Table S2). The lowest content of these constituents was noted when I and I-T were added for drying (Figure 3a). A similar observation was noted in the case of chokeberry juice drying, in which addition of inulin resulted in the lowest content of phenolic acids in powders [12]. When the single compounds were concerned (Figure 3b–d), the chlorogenic and neochlorogenic acids followed the comparable alterations caused by the carrier type and drying technique applied. The strongest changes were noted in the case of cryptochlorogenic acid, the content of which was the lowest. Maltodextrin and trehalose preserved the greatest content of this constituent after freeze drying and vacuum drying at 90 °C; however, the latest technique led to the strongest degradation of this compound in the powders gained. Regardless of the quantity of the selected phenolic acids present in the products gained, their thermolability, moderated by the carrier type due to their chemical structure, might be significantly different.

Overall, similar to Tkacz et al. [41], when the type of carrier was concerned, the highest retention of phenolic acids was noted in products gained with maltodextrin in this particular setting of drying techniques. It can be stated that the selection of an appropriate carrier type and drying technique used for the possible highest retention of phenolic acids in plant powders should be tested for specific products (including the initial chemical composition of raw materials) as it may differ due to the interactions between carriers and individual bioactive compounds present in plants [50].

The second group of polyphenolics identified in chokeberry pomace extract powders consisted of anthocyanins, among which the presence of cyanidin-3-*O*-galactoside, -glucoside, -arabinoside, and -xyloside was confirmed (Tables 2 and 3). Among the controls, freeze drying led to the highest retention of the sum of these constituents followed by VD at 90 and 60 °C (Table 2). A similar observation was made in the case of formulations with carriers, with some exceptions. When inulin was added to the chokeberry extracts, the application of FD and VD at 60 °C resulted in a similar content of identified anthocyanins, whereas the usage of maltodextrin and mix composed of maltodextrin and trehalose resulted in a similar retention of these compounds in the products gained after vacuum drying at 60 and 90 °C. The strongest degradation of these constituents was indicated after vacuum drying at 90 °C for the formulation containing the I-T mixture.

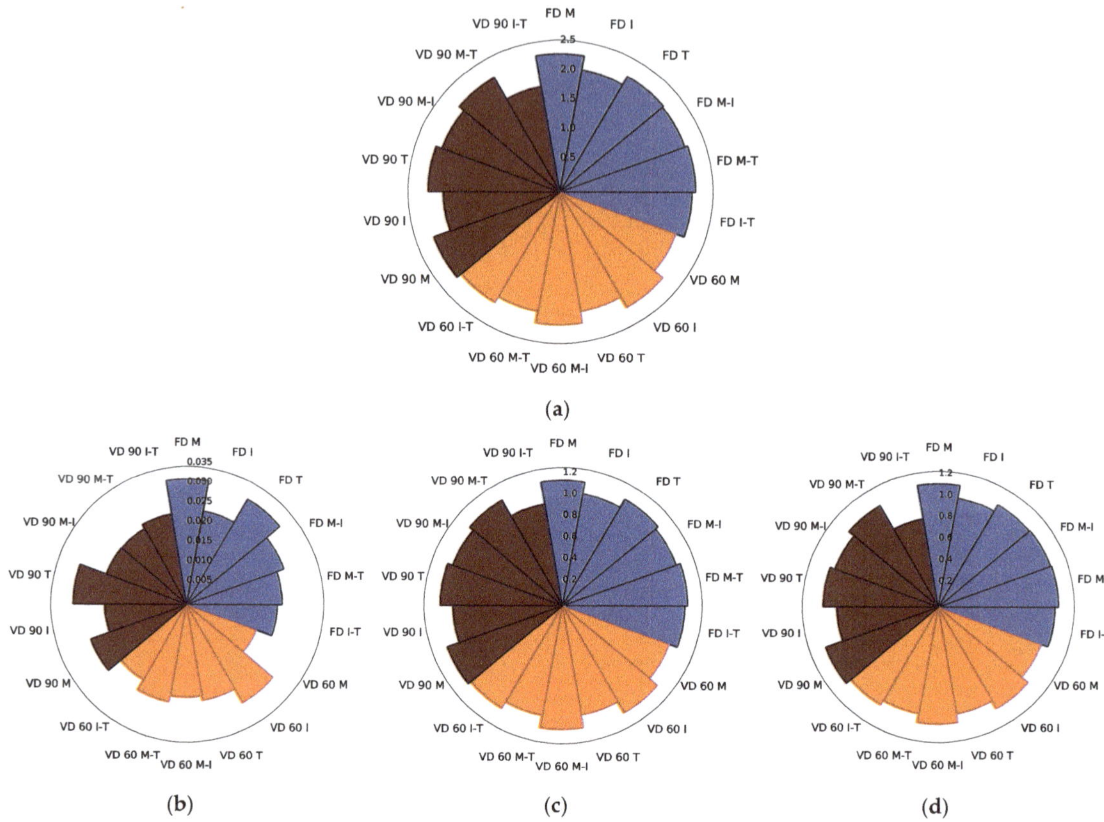

Figure 3. The radar plot of identified phenolic acids ((**a**)—sum of phenolic acids, (**b**)—chlorogenic acid, (**c**)—neochlorogenic acid, (**d**)—cryptochlorogenic acid) in chokeberry pomace extract powders gained with the addition of maltodextrin (M), inulin (I), trehalose (T), maltodextrin—inulin (M—I), maltodextrin—trehalose (M—T) and inulin—trehalose (I—T) by freeze drying (FD) and vacuum drying at 60 (VD 60) and 90 °C (VD 90) (n = 2) (g/100 g db).

Going into the details, the dominant anthocyanin present in chokeberry pomace extracts' powders was cyanidin-3-*O*-galactoside, which consisted, on average, of 64.5% of the sum of identified anthocyanins present in powders (controls, carrier added powders), followed by cyanidin-3-*O*-arabinoside, cyanidin-3-*O*-xyloside, and cyanidin-3-*O*-glucoside. A similar percentage share of anthocyanins was previously noted in the case of chokeberry juice powders [12]. What is more, the influence of the carriers used in the study on individual anthocyanins followed the path of the sum of anthocyanins (Table 3).

Table 3. The content of identified anthocyanins in chokeberry pomace extracts powders (g/100 g db) (n = 2; average ± standard deviation).

Drying Technique	Process Conditions	Carrier	Anthocyanins				
			Cyanidin-3-*O*-Galactoside	Cyanidin-3-*O*-Glucoside	Cyanidin-3-*O*-Arabinoside	Cyanidin-3-*O*-Xyloside	Sum of Anthocyanins
FD	−60 °C/24 °C	Maltodextrin	1.36 ± 0.04 ^{fg}	0.08 ± 0.01 ^{g}	0.59 ± 0.01 ^{fgh}	0.10 ± 0.01 ^{fg}	2.12 ± 0.06 ^{ghi}
		Inulin	1.28 ± 0.06 ^{efg}	0.07 ± 0.01 ^{defg}	0.56 ± 0.03 ^{efgh}	0.10 ± 0.01 ^{efg}	2.00 ± 0.10 ^{fghi}
		Trehalose	1.42 ± 0.08 ^{g}	0.07 ± 0.01 ^{efg}	0.62 ± 0.04 ^{gh}	0.11 ± 0.01 ^{g}	2.22 ± 0.14 ^{hi}
		Maltodextrin—Inulin	1.38 ± 0.05 ^{g}	0.06 ± 0.01 ^{cdefg}	0.59 ± 0.02 ^{fgh}	0.10 ± 0.01 ^{fg}	2.13 ± 0.07 ^{ghi}
		Maltodextrin—Trehalose	1.45 ± 0.02 ^{g}	0.07 ± 0.01 ^{fg}	0.64 ± 0.02 ^{h}	0.11 ± 0.01 ^{g}	2.27 ± 0.04 ^{i}
		Inulin—Trehalose	1.41 ± 0.04 ^{g}	0.07 ± 0.01 ^{defg}	0.62 ± 0.02 ^{gh}	0.09 ± 0.01 ^{defg}	2.18 ± 0.08 ^{hi}
VD	60 °C	Maltodextrin	1.13 ± 0.03 ^{cde}	0.05 ± 0.01 ^{bc}	0.49 ± 0.01 ^{de}	0.08 ± 0.01 ^{bcd}	1.75 ± 0.05 ^{cdef}
		Inulin	1.37 ± 0.06 ^{fg}	0.06 ± 0.01 ^{cdef}	0.60 ± 0.03 ^{fgh}	0.09 ± 0.01 ^{efg}	2.12 ± 0.09 ^{ghi}
		Trehalose	1.12 ± 0.04 ^{cde}	0.05 ± 0.01 ^{bc}	0.48 ± 0.02 ^{cde}	0.08 ± 0.01 ^{cde}	1.73 ± 0.07 ^{cdef}
		Maltodextrin—Inulin	1.27 ± 0.10 ^{efg}	0.06 ± 0.01 ^{cdef}	0.55 ± 0.05 ^{efg}	0.09 ± 0.01 ^{cdef}	1.97 ± 0.15 ^{efgh}
		Maltodextrin—Trehalose	1.10 ± 0.04 ^{cde}	0.05 ± 0.01 ^{bc}	0.47 ± 0.02 ^{cde}	0.07 ± 0.01 ^{bc}	1.70 ± 0.07 ^{cde}
		Inulin—Trehalose	1.19 ± 0.01 ^{def}	0.06 ± 0.01 ^{cde}	0.52 ± 0.01 ^{ef}	0.09 ± 0.01 ^{cdef}	1.85 ± 0.01 ^{defg}
	90 °C	Maltodextrin	1.16 ± 0.01 ^{de}	0.06 ± 0.01 ^{cdef}	0.49 ± 0.01 ^{de}	0.08 ± 0.01 ^{cdef}	1.80 ± 0.01 ^{def}
		Inulin	1.03 ± 0.02 ^{cd}	0.05 ± 0.01 ^{bc}	0.43 ± 0.01 ^{bcd}	0.07 ± 0.01 ^{bcd}	1.58 ± 0.04 ^{cd}
		Trehalose	0.82 ± 0.02 ^{b}	0.04 ± 0.01 ^{b}	0.35 ± 0.01 ^{b}	0.06 ± 0.01 ^{b}	1.27 ± 0.03 ^{b}
		Maltodextrin—Inulin	0.95 ± 0.04 ^{bc}	0.05 ± 0.01 ^{bc}	0.40 ± 0.02 ^{bc}	0.07 ± 0.01 ^{bc}	1.47 ± 0.07 ^{bc}
		Maltodextrin—Trehalose	1.15 ± 0.02 ^{de}	0.05 ± 0.01 ^{bcd}	0.48 ± 0.01 ^{cde}	0.08 ± 0.01 ^{cdef}	1.76 ± 0.02 ^{def}
		Inulin—Trehalose	0.39 ± 0.01 ^{a}	0.02 ± 0.01 ^{a}	0.16 ± 0.01 ^{a}	0.03 ± 0.01 ^{a}	0.60 ± 0.02 ^{a}

FD—freeze drying, VD—vacuum drying; $^{a–i}$—the same letters within a column indicate no statistically significant differences (HSD Tukey test, $p \leq 0.05$).

The last of the identified groups of polyphenols in all chokeberry pomace extract powders were flavonols [12] (about 18.3% of all determined compounds). Among them, eight constituents were detected: kaempferol-3-*O*-galactoside, kaempferol-3-*O*-glucoside, kaempferol-3-*O*-rutinoside, kaempferol-3-*O*-robinobioside, quercetin-dihexoside 1, quercetin-3-*O*-vicianoside, quercetin-dihexoside 2, and a derivative of quercetin (Table S3). In comparison with the other groups of polyphenols, the sum of flavonols had the smallest fluctuations in their content depending on the drying technique or carrier type (no statistically significant differences), which confirmed their high stability during the powdering process (Figure 4a; Table S3). However, going into detail, it should be noted that their content was lower in powders obtained by freeze drying (15.9% of all identified polyphenols), while the highest content was reported in products obtained by vacuum drying at 90 °C (21% of all identified polyphenols) (Table S3). As indicated by Hamrouni-Sellami et al. [51], this may be due to the release of these compounds from more polymerised structures during heating. A similar trend was observed for quercetin and its glucoside. In this case, heating up to 120 °C resulted in an increase in their content, while further processing up to 150 °C caused their degradation [52]. However, it is worth looking at the changes in individual flavonols depending on the drying technique used and the addition of the carrier (Figure 4a–i).

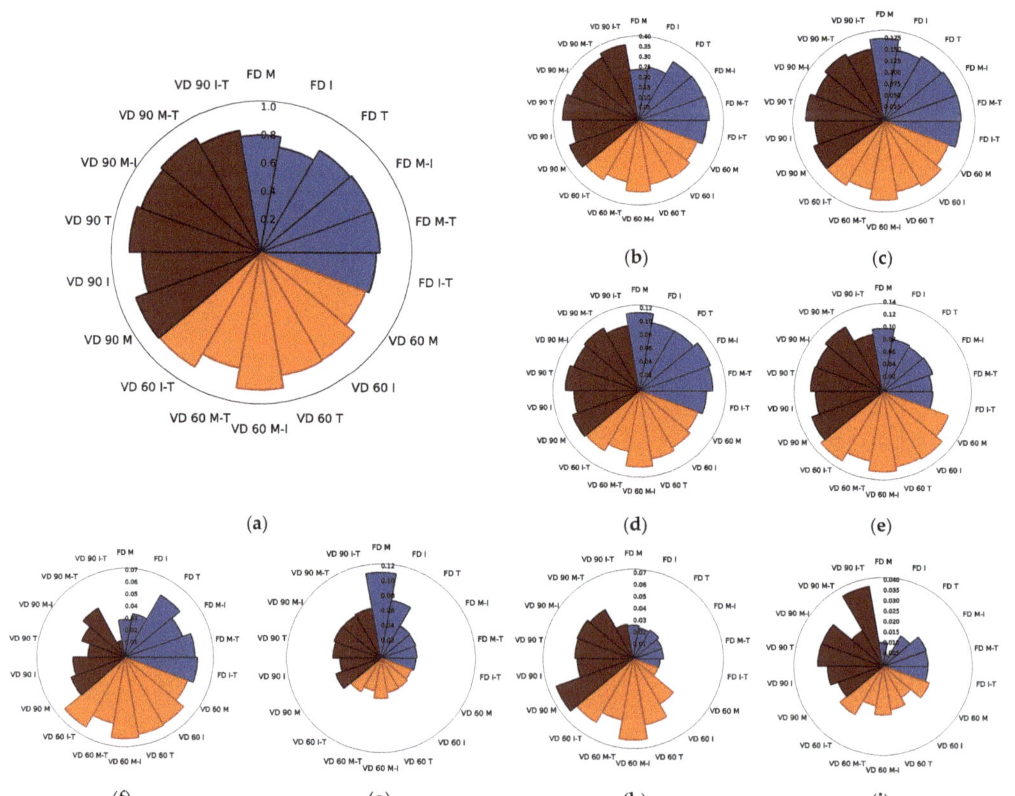

Figure 4. The radar plot of identified flavonols ((**a**)—sum of flavonols, (**b**)—kaempferol-3-*O*-galactoside, (**c**)—kaempferol-3-*O*-glucoside, (**d**)—kaempferol-3-*O*-rutinoside, (**e**)—kaempferol-3-*O*-robinobioside, (**f**)—quercetin-3-*O*-vicianoside, (**g**)—quercetin-dihexoside 1, (**h**)—quercetin-dihexoside 2, (**i**)—derivative of quercetin) in chokeberry pomace extract powders gained with addition of maltodextrin (M), inulin (I), trehalose (T), maltodextrin—inulin (M—I), maltodextrin—trehalose (M—T) and inulin—trehalose (I—T) by freeze drying (FD) and vacuum drying at 60 (VD 60) and 90 °C (VD 90) (n = 2) (g/100 g db).

The highest content of the dominant flavonol, i.e., kaempferol-3-O-galactoside, in the control powders was noted after VD at 60 °C, while the lowest when the FD and VD at 90 °C were applied (Table 2). Considering the powders with carrier addition (Figure 4b), vacuum drying at 90 °C and inulin-trehalose, trehalose, and maltodextrin application was the most suitable, while freeze-dried powders with M and I resulted in the lowest content of this compound, for which statistically significant differences were found (Table S3).

A very similar relationship was observed for the quercetin derivative in the control samples (Table 2); however, in the case of the carrier-added powders, the lowest content of this constituent was found in freeze-dried products with inulin, while an approximately 7-times higher content was determined in those obtained by the vacuum drying at 90 °C with the addition of I-T mixture (Figure 4i). Interestingly, the parameters and carriers mixture, on the one hand, allowed the highest quercetin derivative content, and on the other hand, resulted in the lowest quercetin-3-O-vicianoside concentration (Figure 4f). Comparing the drying techniques, VD at 60 °C was the most favourable in the context of the quercetin-3-O-vicianoside level in the powders analysed. Upon comparison of the content of this compound in powders without a carrier, FD and VD at 60 °C proved to be the best methods, while vacuum drying at 90 °C resulted in an approximately 2 times lower content of quercetin-3-O-vicianoside (Table 2). Noteworthy, kaempferol-3-O-glucoside and kaempferol-3-O-rutinoside proved to be the most stable flavonols as their content changed depending on the drying method and the type of added carrier were relatively small compared to the other compounds of the same group (Figure 4c,d). Additionally, fluctuations in the concentration of these compounds depending on the drying parameters and media type were comparable. For kaempferol-3-O-robinobioside in the control samples, the lowest concentration of this compound was found in vacuum-dried powders at 60 °C, while FD and VD at 90 °C yielded the highest levels (no statistically significant differences) of this flavonol (Table 2). A reverse effect was observed in the case of powders with carriers (Figure 4e), for which the lowest content of kaempferol-3-O-robinobioside was determined in freeze-dried products, while vacuum drying at 60 and 90 °C resulted in the highest and similar content of this constituent in the analysed powders. Taking into account the effect of the carrier agent, the highest content of this compound was found in vacuum-dried powders at 60 °C with I, M-I, and I-T, while the lowest content was found in freeze-dried products with M-T. An interesting relationship was also observed for the two quercetin-dihexosides. In case of quercetin-dihexoside 1 for no-carrier added samples, VD at 60 °C resulted in powders with the lowest concentration of this compound being obtained, while VD at 90 °C resulted in the highest concentration (Table 2). When analysing the powders with added carriers (Figure 4g), the quercetin-dihexoside 1 content was the highest in powders obtained by FD with the addition of M, while considering other drying methods, the same carrier did not have such a noticeable effect as in the case of freeze drying. Moreover, except for the samples with added maltodextrin obtained by FD, no statistically significant differences were found for all the rest of the powders (Table S3). Regarding quercetin-dihexoside 2 for the control powders, no statistically significant differences were found for the content of this compound (Table 2). Interestingly, in the case of powders with added carriers (Figure 4h), freeze drying, irrespective of the type of carrier, as well as vacuum drying at 90 °C and the addition of I-T mixture and VD at 60 °C with M resulted in the lowest content of this compound, while the other variants of powders were characterised by significantly higher quercetin-dihexoside 2 levels (Table S3). Moreover, while for quercetin-dihexoside 1, vacuum drying at 60 °C resulted in its lowest content in the powders obtained, for quercetin-dihexoside 2, the same drying allowed products with the highest level of this compound. Therefore, it can be concluded that the highest quercetin-dihexoside 1 content is possible to obtain if low-temperature drying is used, while the use of high temperatures during drying favours a high quercetin-dihexoside 2 content.

3.2.2. Hydroxymethyl-*L*-furfural

In the current study, the hydroxymethyl-*L*-furfural was identified in all analysed powders (Table 2, Figure 5). In the case of control samples, the highest content of HMF was noted in powders gained after VD at 90 °C and freeze drying (no statistically significant differences) while, interestingly, a lower content was noted after the application of vacuum drying at 60 °C (Table 2). Contrary to the expectations that freeze drying as a low-temperature treatment should result in the lowest content of HMF (or its absence), its presence in the analysed powders might be connected with the particular composition of chokeberry pomace extracts. The substrates for the HMF formation present in these extracts might additionally react with bioactives during the storage of fruit products [53] and even after freeze drying. As a confirmation of this, its occurrence was previously noted in freeze-dried products [54] and fruit juice-based foodstuffs [12,41]. Interestingly, the studies of Zhang et al. [55] showed that the formation of HMF was accelerated by the chlorogenic acid in model systems, which is in accordance with the present study as a strong correlation between HMF and phenolic acids ($r = 0.84$) was noted. What is more, Zhang and An [56] proved that its formation might be inhibited by the interactions with flavonols, which was in line with the present research where the sum of these identified compounds was concerned ($r = -0.85$) (Figure S2). However, this inhibition mechanism of single flavonols is ambiguous as between HMF and two dominant constituents, namely, kaempferol-3-*O*-galactoside and kaempferol-3-*O*-glucoside, strong negative ($r = -0.97$) and positive ($r = 0.86$) correlations, respectively, were observed at the same time (Figure S2).

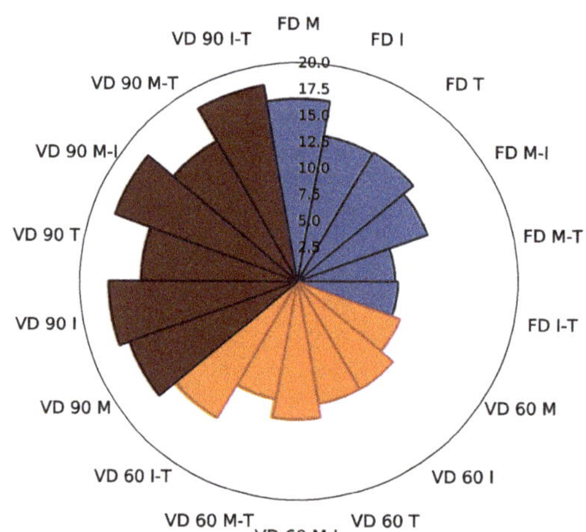

Figure 5. The radar plot of HMF in chokeberry pomace extract powders gained with the addition of maltodextrin (M), inulin (I), trehalose (T), maltodextrin—inulin (M-I), maltodextrin—trehalose (M-T) and inulin—trehalose (I-T) by freeze drying (FD) and vacuum drying at 60 (VD 60) and 90 °C (VD 90) ($n = 2$) (µg/100 g db).

Similar to the control samples, the strongest formation of HMF was also noted for powders obtained with carriers gained after VD at 90 °C, followed by freeze drying. Its content was the lowest when VD at 60 °C was used (Figure 5). Its slower formation at 60 °C was also reported by Olivares-Tenorio et al. [57] and Michalska et al. [58] during the drying of fruit-based products, whereas a further increase in drying temperature caused its rapid formation. As it was previously reported, such phenomena might be linked to the formation of intermediary compounds during the application of 60 °C [57]. When the type of carrier was considered, the highest content of HMF was noted when inulin and its mixes (I-T, I-M) were used for powder production at VD at 90 °C. Thus, the inulin might enhance the formation of this process contaminant during powder preparation [12].

As the different bioactives might influence the formation of HMF during drying, in the case of powders gained with the addition of carriers, no significant correlation between the sum of identified polyphenols and HMF was noted, with the exception of quercetin-3-*O*-vicianoside ($r = -0.65$) (Figure S3). Probably, as stated by Zhang and An [56], its formation might be inhibited by the interactions with some flavonols; however, until now, no particular compounds were indicated. To sum up, when the quality of chokeberry extract powders is considered, the formation of HMF could be controlled by the initial composition of material submitted for drying, process parameters, and type of carrier used for the powder preparation.

3.2.3. Antioxidant Capacity

The antioxidant capacity of chokeberry pomace extract powders determined by the TEAC ABTS and FRAP methods showed that the ability of control samples (Table 2) to scavenge free radicals was about 4.7 times (TEAC ABTS assay) and about 4.2 times higher (FRAP assay) when compared to the powders gained with the application of selected carriers. Going into detail, the phenolic acids significantly influenced the antioxidant capacity measured by TEAC ABTS ($r = 0.56$) and FRAP ($r = 0.64$) in samples produced without the addition of carriers. What is more, a negative correlation between FRAP and the sum of identified flavonols in these products was noticed ($r = -0.75$), which could be connected with the different actions of these compounds present in the analysed powders toward the reducing potential of ferric ion (Fe^{3+} to Fe^{2+}) [59]. In the case of the products with added carriers, it was observed that the antioxidant capacity varied slightly depending on the carrier used (Figure 6a,b).

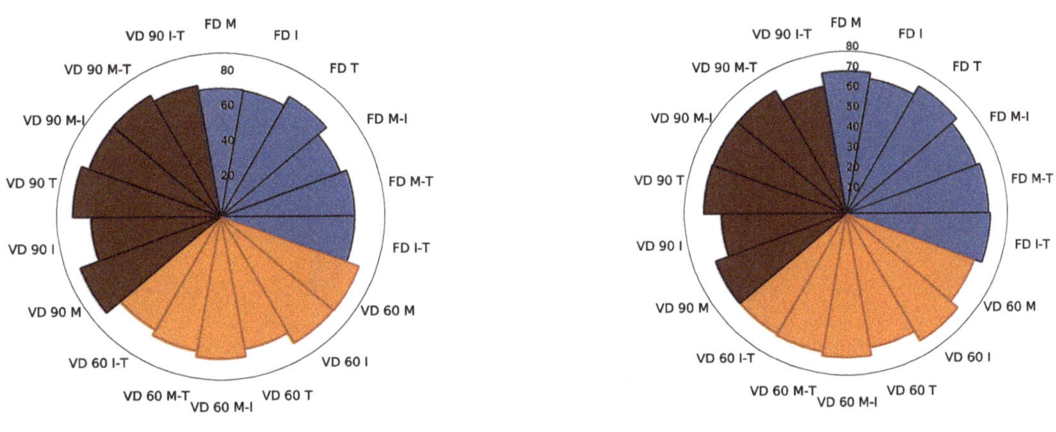

(a) (b)

Figure 6. The radar plot of TEAC ABTS (**a**) and FRAP (**b**) of chokeberry pomace extract powders gained with the addition of maltodextrin (M), inulin (I), trehalose (T), maltodextrin—inulin (M—I), maltodextrin—trehalose (M—T), and inulin—trehalose (I—T) by freeze drying (FD) and vacuum drying at 60 (VD 60) and 90 °C (VD 90) ($n = 2$) (mmol Trolox/100 g db).

Interestingly, the highest average antioxidant capacity values measured by the TEAC ABTS method were observed in the powders obtained by VD at 90 °C, followed by VD at 60 °C, and freeze drying (Table S1). The reverse effect was observed in the case of FRAP. The antioxidant capacity values may be related to a significant content of certain polyphenolic compounds in those powders, such as predominant phenolic acids [60]. This was confirmed by a high positive correlation ($r = 0.78$) between the content of these constituents in the analysed powders and their antioxidant capacity measured by the FRAP method (Figure S3). Moreover, there was a moderate positive correlation between the sum of anthocyanins and antioxidant capacity measured by FRAP ($r = 0.58$) and no linear relationship between the content of these compounds and TEAC ABTS analysis values ($r = -0.13$) (Figure S3). It might be connected with the lower content of anthocyanins than phenolic acids in the carrier-added powders; however, in the literature, there are also some reports indicating that the antioxidant capacity is related more to the total content of polyphenolic compounds than to the content of anthocyanins, which may be due to the lower free radical scavenging capacity of these compounds compared to other polyphenolics [61]. Furthermore, in the case of TEAC ABTS, and FRAP analysis, the powders obtained with the addition of trehalose and its mixes after freeze drying had the highest values. In general, during VD at 60 °C, the application of inulin resulted in higher TEAC ABTS and FRAP values of the powders analysed, whereas the addition of this carrier and its mix with trehalose lowered the antioxidant properties measured by these two methods when vacuum drying at 90 °C was applied. This proved the selectivity of the action of individual compounds and/or carrier substances [62] toward the free radical scavenging properties of products, which could be additionally moderated by the drying parameters [50].

3.2.4. PCA Analysis

The PCA biplot for chemical properties (Figure 7a) showed quite the opposite results from those shown for physical properties (Figure 1a).

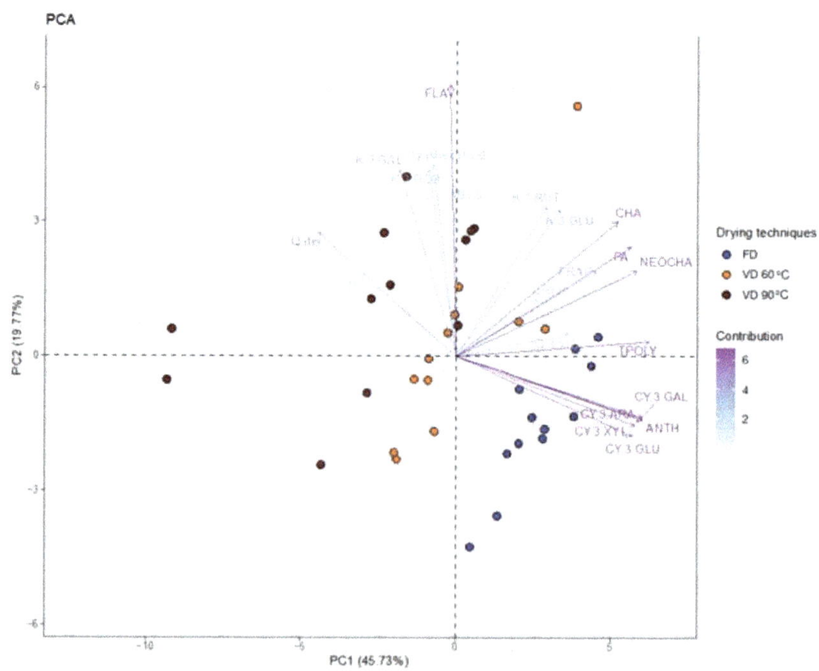

(a)

Figure 7. *Cont.*

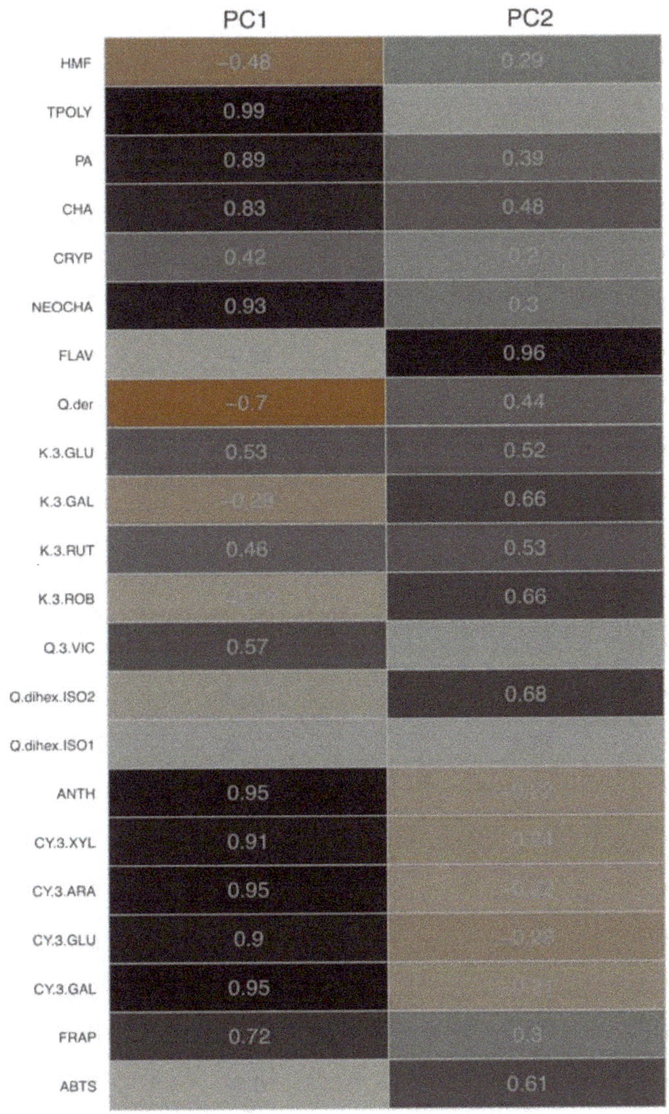

(b)

Figure 7. *Cont.*

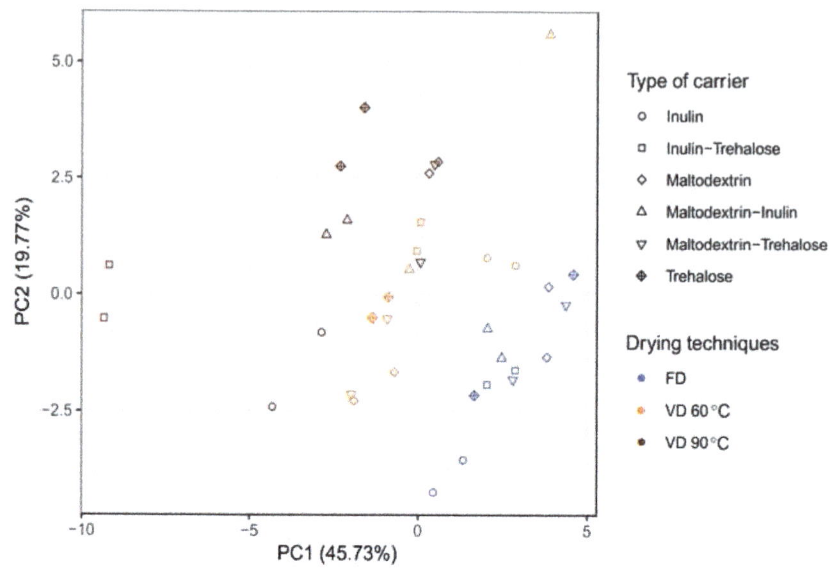

(c)

Figure 7. (**a**) The PCA biplot of the first two principal components. The marker colour corresponds to the drying techniques (i.e., freeze and vacuum drying at 60 and 90 °C), while the length and the transparency of the arrows indicate the variance of the chemical properties of powders from chokeberry pomace extracts and their contributions to the principal components, respectively. Together, the first two principal components explain 65.50% of the variability; (**b**) The plot of normalised factor loadings; (**c**) Score plot in the space defined by the first two principal components illustrating the relations and trends of the chokeberry pomace extract powders gained after freeze and vacuum drying at 60 and 90 °C with the addition of maltodextrin (M), inulin (I), trehalose (T), maltodextrin—inulin (M—I), maltodextrin—trehalose (M—T), and inulin—trehalose (I—T).

The vacuum-dried samples at 90 °C had much greater variance than the freeze-dried samples. Chokeberry pomace extract powders gained after the application of different drying techniques were split along PC1. PC1 was positively correlated with the sum of identified polyphenolics (TPOLY), phenolic acids (PA), including chlorogenic acid (CHA), neochlorogenic acid (NEOCHA), the sum of identified anthocyanins (ANTH), including cyanidin-3-*O*-xyloside (CY.3.XYL), -arabinoside (CY.3.ARA), -glucoside (CY.3.GLU), and -galactoside (CY.3.GAL) (Figure 7b). Therefore, it was straightforward to see that drying techniques changed from VD at 90 °C through freeze drying with an increasing content of anthocyanins and phenolic acids, when moving from left to right along the *X*-axis (PC1). PC1 was also negatively correlated with the derivative of quercetin (Q.der). Due to low PC1 scores and negative loading values, samples vacuum dried at 90 °C were characterised by a relatively higher value of Q.der compared to the freeze-dried samples. The most influential variable in PC2 was positively correlated with the sum of flavonols (FLAV). PC2 did not indicate variation, which clearly distinguished between the samples gained after the application of different drying techniques. Hence, only a general statement could be made that chokeberry pomace extract powders toward the top of the PCA biplot (Figure 7a) were described by the highest content of flavonols due to the positive correlations between PC2 and FLAV. A closer look at the score plot shown in Figure 7c revealed that no clear trends were observed for the powders produced with the addition of different carriers and their mixes. Nonetheless, chokeberry pomace extract powders gained after vacuum drying at 90 °C with addition of I-T differed significantly from the rest of the samples in their low content of anthocyanins and phenolic acids.

4. Conclusions

The current study evaluated the possibility of obtaining powders from chokeberry pomace extracts by drying techniques and different carrier types as one waste management practice in the food industry. The quality of such products should consider the priorities of their potential application as the moderation of powder properties is a multifactor issue. Taking the above into consideration, the PCA analysis indicated that freeze-dried samples exhibit more variation than those produced by vacuum drying at 60 and 90 °C, especially in terms of moisture content, water activity, colour, and browning index. The bulk density was higher for products obtained after vacuum drying. No straightforward trends in physical properties were observed for products that has selected carriers added.

In the analysed powders, three groups of polyphenols were identified and quantified, i.e., phenolic acids (3), anthocyanins (4), and flavonols (8). Drying techniques significantly influenced the polyphenolics in the powders gained with the addition of selected carriers. In general, the application of freeze drying resulted in a higher content of anthocyanins and phenolic acids, while vacuum drying at 90 °C allowed for the obtainment of products with high quantities of flavonols. Where the analysed carriers were concerned, the highest retention of the sum of identified polyphenolics was noted when maltodextrin and its mixture with trehalose were applied for powder production by freeze drying and vacuum drying at 90 °C, whereas during VD at 60 °C, it was inulin and its mixes. In the case of phenolic acids and anthocyanins, a similar observation was made for FD and VD 60 in that maltodextrin and trehalose protect most of the mentioned compounds; however, in case of VD 90, trehalose caused the lowest retention of anthocyanins. Regarding flavonols, this group was characterised by the highest stability during drying, regardless of the carrier type used. A detailed analysis showed very diverse behaviour of the individual compounds with respect to the applied processing parameters, thus making it impossible to identify any specific method of powder production that results in flavonols' highest retention.

As the content of hydroxymethyl-*L*-furfural is of high importance to monitor in processed foods' quality, the lowest concentration of this compound was determined in powders gained after vacuum drying at 60 °C, while its highest level was noted after VD at 90 °C. The current study confirmed [12] that the addition of inulin and its mixes during high-temperature treatment (vacuum drying at 90 °C) should be carefully considered as this carrier may influence the formation of HMF in fruit-based products.

To sum up, the retention of polyphenolics and formation of HMF in chokeberry pomace extracts' powders was affected simultaneously by the initial composition of raw material, carrier type, drying techniques, and parameters applied. Taking all these factors into account, including interactions between the matrix composition during drying, 10% addition of maltodextrin and trehalose mixture for freeze drying and vacuum drying at 90 °C allowed the production of powders with the highest retention of polyphenolic compounds and the lowest HMF level, at the same time. The outcome of the current study supported by the chemometric analyses can provide guidance for further research as well as give directions for work on designing functional foodstuff based on powders from chokeberry pomace extracts.

Supplementary Materials: The following are available online at https://www.mdpi.com/article/10.3390/foods10081864/s1, Figure S1: A correlation matrix showing the Pearson's correlation coefficients between each pair of variables (i.e., physical properties) in chokeberry pomace extract powders data set (controls and powders with carrier addition), Figure S2: A correlation matrix showing the Pearson's correlation coefficients between each pair of variables (i.e., chemical properties) in chokeberry pomace extract powders data set (control samples), Figure S3: A correlation matrix showing the Pearson's correlation coefficients between each pair of variables (i.e., chemical properties) in chokeberry pomace extract powders data set (powders with addition of carriers; no controls included), Table S1: The content of sum of polyphenols, hydroxymethyl-*L*-furfural and the antioxidant capacity measured by TEAC ABTS and FRAP methods of chokeberry pomace extracts powders made with the addition maltodextrin, inulin, trehalose and a mixture of them using different drying methods (average ± standard deviation; n = 2), Table S2: The content of identified phenolic acids in chokeberry pomace extracts powders made with the addition maltodextrin, inulin, trehalose and a mixture of them using different drying methods (g/100 g db) (n = 2; average ± standard deviation), Table S3: The content of identified flavonols in chokeberry pomace extracts powders made with the addition maltodextrin, inulin, trehalose and a mixture of them using different drying methods (g/100 g db) (n = 2; average ± standard deviation).

Author Contributions: Conceptualization, A.M.-C. and A.W.; methodology, A.W., A.M.-C., J.B., A.G.-S.; software, A.G.-S.; formal analysis, A.M.-C., J.B., A.G.-S.; investigation, A.H., A.M.-C., J.B; resources, A.M.-C., A.W., A.G.-S.; data curation, A.M.-C., J.B., A.H., A.W., A.G.-S.; writing—original draft preparation, A.M.-C., J.B., A.G.-S., A.H.; writing—review and editing, A.M.-C., J.B., A.W.; supervision, A.M.-C.; project administration, A.M.-C.; funding acquisition, A.H., A.M.-C. All authors have read and agreed to the published version of the manuscript.

Funding: The research is financed under the individual student research project 'Młode umysły—Young Minds Project' from the subsidy increased for the period 2020–2025 in the amount of 2% of the subsidy referred to Art. 387 (3) of the Law of 20 July 2018 on Higher Education and Science, obtained in 2019.

Institutional Review Board Statement: Not applicable.

Informed Consent Statement: Not applicable.

Acknowledgments: The publication is the result of the research group activity: 'Plants4FOOD' (Leading Research Group of the University of Environmental and Life Sciences).

Conflicts of Interest: The authors declare no conflict of interest. The funders had no role in the design of the study; in the collection, analyses, or interpretation of data; in the writing of the manuscript, or in the decision to publish the results.

References

1. Salehi, G.; Díaz, E.; Redondo, R. Consumers' switching to vegan, vegetarian, and plant-based (Veggan) diets: A systematic Review of literature. In Proceedings of the 19th International Congress on Public and Nonprofit Marketing Sustainability: New Challenges for Marketing and Socioeconomic Development, Madrid, Spain, 2–4 July 2020.
2. Campos, D.A.; Garcia, R.G.; Vilas-Boas, A.A.; Madureira, A.R.; Pintado, M.M. Management of fruit industrial by-products—A case study on circular economy approach. *Molecules* **2020**, *25*, 320. [CrossRef]
3. Mayer-Miebach, E.; Briviba, K.; Schiffer, C.; Geiger, L.; Behsnilian, D.; Greiner, R. Particle size of milled chokeberry pomace did not influence in vitro cellular absorption and transport efficiencies of anthocyanins, phenolic acids and flavonols. *Int. J. Food Sci. Nutr.* **2019**, *70*, 932–940. [CrossRef] [PubMed]
4. Majerska, J.; Michalska, A.; Figiel, A. A review of new directions in managing fruit and vegetable processing by-products. *Trends Food Sci. Technol.* **2019**, *88*, 207–219. [CrossRef]
5. Vagiri, M.; Jensen, M. Influence of juice processing factors on quality of black chokeberry pomace as a future resource for colour extraction. *Food Chem.* **2017**, *217*, 409–417. [CrossRef]
6. Sidor, A.; Drożdżyńska, A.; Gramza-Michałowska, A. Black chokeberry (*Aronia melanocarpa*) and its products as potential health-promoting factors—An overview. *Trends Food Sci. Technol.* **2019**, *89*, 45–60. [CrossRef]
7. Witczak, T.; Stępień, A.; Gumul, D.; Witczak, M.; Fiutak, G.; Zięba, T. The influence of the extrusion process on the nutritional composition, physical properties and storage stability of black chokeberry pomaces. *Food Chem.* **2021**, *334*, 127548. [CrossRef] [PubMed]
8. Comunian, T.A.; Silva, M.P.; Souza, C.J. The use of food by-products as a novel for functional foods: Their use as ingredients and for the encapsulation process. *Trends Food Sci. Technol.* **2021**, *108*, 269–280. [CrossRef]
9. Roda-Serrat, M.C.; Andrade, T.; Rindom, J.; Lund, P.B.; Norddahl, B.; Errico, M. Optimization of the recovery of anthocyanins from chokeberry juice pomace by homogenization in acidified water. *Waste Biomass Valorization* **2020**, *12*, 1815–1827. [CrossRef] [PubMed]
10. Bhandari, B.; Bansal, N.; Zhang, M.; Schuck, P. Woodhead Publishing Series in Food Science, Technology and Nutrition. In *Handbook of Food Powders*; Elsevier: Amsterdam, The Netherlands, 2013.
11. Kammerer, D.; Kljusuric, J.G.; Carle, R.; Schieber, A. Recovery of anthocyanins from grape pomace extracts (Vitis vinifera L. cv. Cabernet Mitos) using a polymeric adsorber resin. *Eur. Food Res. Technol.* **2005**, *220*, 431–437. [CrossRef]

12. Michalska-Ciechanowska, A.; Brzezowska, J.; Wojdyło, A.; Gajewicz-Skretna, A.; Ciska, E.; Majerska, J. Chemometric contribution for deeper understanding of thermally-induced changes of polyphenolics and the formation of hydroxymethyl-*L*-furfural in chokeberry powders. *Food Chem.* **2021**, *342*, 128335. [CrossRef]
13. Nowak, D.; Jakubczyk, E. The Freeze-Drying of Foods—The characteristic of the process course and the effect of its parameters on the physical properties of food materials. *Foods* **2020**, *9*, 1488. [CrossRef]
14. Michalska, A.; Wojdyło, A.; Brzezowska, J.; Majerska, J.; Ciska, E. The influence of inulin on the retention of polyphenolic compounds during the drying of blackcurrant juice. *Molecules* **2019**, *24*, 4167. [CrossRef]
15. Sobulska, M.; Zbicinski, I. Advances in spray drying of sugar-rich products. *Dry. Technol.* **2020**, 1–26. [CrossRef]
16. Wan, X.; Guo, H.; Liang, Y.; Zhou, C.; Liu, Z.; Li, K.; Niu, F.; Zhai, X.; Wang, L. The physiological functions and pharmaceutical applications of inulin: A review. *Carbohydr. Polym.* **2020**, *246*, 116589. [CrossRef] [PubMed]
17. Oszmiański, J. Sposób otrzymywania barwników antocyjanowych. PL-158707. 30 July 1993. Available online: https://grab.uprp.pl/sites/WynalazkiWzoryUzytkowe/Opisy/Patenty%20i%20Wzory%20uytkowe/158707_B1.pdf (accessed on 26 July 2021). (In Polish)
18. Wang, W.; Yagiz, Y.; Buran, T.J.; Nunes, C.D.N.; Gu, L. Phytochemicals from berries and grapes inhibited the formation of advanced glycation end-products by scavenging reactive carbonyls. *Food Res. Int.* **2011**, *44*, 2666–2673. [CrossRef]
19. Mexis, S.; Kontominas, M. Effect of oxygen absorber, nitrogen flushing, packaging material oxygen transmission rate and storage conditions on quality retention of raw whole unpeeled almond kernels (*Prunus dulcis*). *LWT* **2010**, *43*, 1–11. [CrossRef]
20. Wojdyło, A.; Oszmiański, J.; Bielicki, P. Polyphenolic composition, antioxidant activity, and polyphenol oxidase (PPO) activity of quince (*Cydonia oblonga* Miller) varieties. *J. Agric. Food Chem.* **2013**, *61*, 2762–2772. [CrossRef] [PubMed]
21. Michalska, A.; Wojdyło, A.; Łysiak, G.P.; Figiel, A. Chemical composition and antioxidant properties of powders obtained from different plum juice formulations. *Int. J. Mol. Sci.* **2017**, *18*, 176. [CrossRef]
22. Re, R.; Pellegrini, N.; Proteggente, A.; Pannala, A.; Yang, M.; Rice-Evans, C. Antioxidant activity applying an improved ABTS radical cation decolorization assay. *Free. Radic. Biol. Med.* **1999**, *26*, 1231–1237. [CrossRef]
23. Benzie, I.; Strain, J. The ferric reducing ability of plasma (FRAP) as a measure of "antioxidant power": The FRAP assay. *Anal. Biochem.* **1996**, *239*, 70–76. [CrossRef]
24. Jolliffe, I.T.; Cadima, J. Principal component analysis: A review and recent developments. *Philos. Trans. R. Soc. A Math. Phys. Eng. Sci.* **2016**, *374*, 20150202. [CrossRef]
25. Kassambara, A.; Mundt, F. Factoextra: Extract and Visualize the Results of Multivariate Data, R Package Version 1.0.7; CRAN: 2020. Available online: https://cran.r-project.org/web/packages/factoextra/index.html (accessed on 26 July 2021).
26. Nakazawa, M. Fmsb: Functions for Medical Statistics Book with Some Demographic, Data Version 0.7.1; CRAN: 2019. Available online: https://cran.r-project.org/web/packages/fmsb/index.html (accessed on 26 July 2021).
27. Michalska, A.; Lech, K. The Effect of carrier quantity and drying method on the physical properties of apple juice powders. *Beverages* **2018**, *4*, 2. [CrossRef]
28. Tontul, I.; Topuz, A. Spray-drying of fruit and vegetable juices: Effect of drying conditions on the product yield and physical properties. *Trends Food Sci. Technol.* **2017**, *63*, 91–102. [CrossRef]
29. Carvalho, A.G.D.S.; Machado, M.T.D.C.; da Silva, V.M.; Sartoratto, A.; Rodrigues, R.A.F.; Hubinger, M. Physical properties and morphology of spray dried microparticles containing anthocyanins of Jussara (*Euterpe edulis Martius*) extract. *Powder Technol.* **2016**, *294*, 421–428. [CrossRef]
30. Shafiur Rahman, M. *Handbook of Food Preservation*, 2nd ed.; Shafiur Rahman, M., Ed.; CRC Press: Boca Raton, FL, USA, 2007.
31. Adetoro, A.O.; Opara, U.L.; Fawole, O.A. Effect of carrier agents on the physicochemical and technofunctional properties and antioxidant capacity of freeze-dried pomegranate juice (*Punica granatum*) powder. *Foods* **2020**, *9*, 1388. [CrossRef]
32. Muhoza, B.; Xia, S.; Wang, X.; Zhang, X. The protection effect of trehalose on the multinuclear microcapsules based on gelatin and high methyl pectin coacervate during freeze-drying. *Food Hydrocoll.* **2020**, *105*, 105807. [CrossRef]
33. Kilburn, D.; Townrow, S.; Meunier, V.; Richardson, R.; Alam, A.; Ubbink, J. Organization and mobility of water in amorphous and crystalline trehalose. *Nat. Mater.* **2006**, *5*, 632–635. [CrossRef]
34. Michalska-Ciechanowska, A.; Majerska, J.; Brzezowska, J.; Wojdyło, A.; Figiel, A. The Influence of maltodextrin and inulin on the physico-chemical properties of cranberry juice powders. *ChemEngineering* **2020**, *4*, 12. [CrossRef]
35. Chang, K.S.; Kim, D.W.; Kim, S.S.; Jung, M.Y. Bulk flow properties of model food powder at different water activity. *Int. J. Food Prop.* **1998**, *1*, 45–55. [CrossRef]
36. Khalifa, I.; Li, M.; Mamet, T.; Li, C. Maltodextrin or gum Arabic with whey proteins as wall-material blends increased the stability and physiochemical characteristics of mulberry microparticles. *Food Biosci.* **2019**, *31*, 10445. [CrossRef]
37. Bednarska, M.A.; Janiszewska-Turak, E. The influence of spray drying parameters and carrier material on the physico-chemical properties and quality of chokeberry juice powder. *J. Food Sci. Technol.* **2019**, *57*, 564–577. [CrossRef]
38. Sarabandi, K.; Peighambardoust, S.H.; Mahoonak, A.R.S.; Samaei, S. Effect of different carriers on microstructure and physical characteristics of spray dried apple juice concentrate. *J. Food Sci. Technol.* **2018**, *55*, 3098–3109. [CrossRef] [PubMed]
39. Wojdyło, A.; Lech, K.; Nowicka, P. Effects of different drying methods on the retention of bioactive compounds, on-line antioxidant capacity and color of the novel snack from red-fleshed apples. *Molecules* **2020**, *25*, 5521. [CrossRef]
40. Pathare, P.; Opara, U.L.; Al-Said, F.A.-J. Colour measurement and analysis in fresh and processed foods: A review. *Food Bioprocess Technol.* **2012**, *6*, 36–60. [CrossRef]

41. Tkacz, K.; Wojdyło, A.; Michalska-Ciechanowska, A.; Turkiewicz, I.P.; Lech, K.; Nowicka, P. Influence carrier agents, drying methods, storage time on physico-chemical properties and bioactive potential of encapsulated sea buckthorn juice powders. *Molecules* **2020**, *25*, 3801. [CrossRef]
42. Michalska, A.; Wojdyło, A.; Lech, K.; Łysiak, G.P.; Figiel, A. Physicochemical properties of whole fruit plum powders obtained using different drying technologies. *Food Chem.* **2016**, *207*, 223–232. [CrossRef] [PubMed]
43. Liu, S.-C.; Chang, H.-M.; Wu, J.S.-B. A study on the mechanism of browning in mei liqueur using model solutions. *Food Res. Int.* **2003**, *36*, 579–585. [CrossRef]
44. Dorris, M.R.; Voss, D.M.; Bollom, M.; Krawiec-Thayer, M.P.; Bolling, B.W. Browning index of anthocyanin-rich fruit juice depends on pH and anthocyanin loss more than the gain of soluble polymeric pigments. *J. Food Sci.* **2018**, *83*, 911–921. [CrossRef]
45. Sójka, M.; Kołodziejczyk, K.; Milala, J.; Abadias, M.; Viñas, I.; Guyot, S.; Baron, A. Composition and properties of the polyphenolic extracts obtained from industrial plum pomaces. *J. Funct. Foods* **2015**, *12*, 168–178. [CrossRef]
46. Sójka, M.; Kołodziejczyk, K.; Milala, J. Polyphenolic and basic chemical composition of black chokeberry industrial by-products. *Ind. Crop. Prod.* **2013**, *51*, 77–86. [CrossRef]
47. Michalska, A.; Wojdyło, A.; Honke, J.; Ciska, E.; Andlauer, W. Drying-induced physico-chemical changes in cranberry products. *Food Chem.* **2018**, *240*, 448–455. [CrossRef] [PubMed]
48. Tomas, M.; Beekwilder, J.; Hall, R.; Simon, C.D.; Sagdic, O.; Capanoglu, E. Effect of dietary fiber (inulin) addition on phenolics and in vitro bioaccessibility of tomato sauce. *Food Res. Int.* **2018**, *106*, 129–135. [CrossRef]
49. Szopa, A.; Kokotkiewicz, A.; Kubica, P.; Banaszczak, P.; Wojtanowska-Krośniak, A.; Krosniak, M.; Marzec-Wróblewska, U.; Badura, A.; Zagrodzki, P.; Bucinski, A.; et al. Comparative analysis of different groups of phenolic compounds in fruit and leaf extracts of *Aronia* sp.: *A. melanocarpa*, *A. arbutifolia*, and *A.* $\times$*prunifolia* and their antioxidant activities. *Eur. Food Res. Technol.* **2017**, *243*, 1645–1657. [CrossRef]
50. Turkiewicz, I.; Wojdyło, A.; Tkacz, K.; Lech, K.; Michalska-Ciechanowska, A.; Nowicka, P. The influence of different carrier agents and drying techniques on physical and chemical characterization of Japanese quince (*Chaenomeles japonica*) microencapsulation powder. *Food Chem.* **2020**, *323*, 126830. [CrossRef]
51. Hamrouni-Sellami, I.; Rahali, F.Z.; Rebey, I.B.; Bourgou, S.; Limam, F.; Marzouk, B. Total phenolics, flavonoids, and antioxidant activity of sage (*Salvia officinalis* L.) plants as affected by different drying methods. *Food Bioprocess Technol.* **2012**, *6*, 806–817. [CrossRef]
52. Sharma, K.; Ko, E.Y.; Assefa, A.; Ha, S.; Nile, S.; Lee, E.T.; Park, S.W. Temperature-dependent studies on the total phenolics, flavonoids, antioxidant activities, and sugar content in six onion varieties. *J. Food Drug Anal.* **2015**, *23*, 243–252. [CrossRef]
53. Aktağ, I.G.; Gökmen, V. Multiresponse kinetic modelling of α-dicarbonyl compounds formation in fruit juices during storage. *Food Chem.* **2020**, *320*, 126620. [CrossRef] [PubMed]
54. Fitzpatrick, K.; Kendrick, B.; Santos, C.; Green, P.; Zhang, B.; Hunt, D.; Ronk, M.; Luo, Y. Freeze-dry mediated formation of 5-(Hydroxylmethyl)furfural. In *Developments in Biotechnology and Bioprocessing*; ACS Symposium Series; American Chemical Society: Washington, DC, USA, 2013; Volume 1125, pp. 129–145, ISBN 978-0-8412-2910-5.
55. Zhang, Z.; Zou, Y.; Wu, T.; Huang, C.; Pei, K.; Zhang, G.; Lin, X.; Bai, W.; Ou, S. Chlorogenic acid increased 5-hydroxymethylfurfural formation when heating fructose alone or with aspartic acid at two pH levels. *Food Chem.* **2016**, *190*, 832–835. [CrossRef]
56. Zhang, Y.; An, X. Inhibitory mechanism of quercetin against the formation of 5-(hydroxymethyl)-2-furaldehyde in buckwheat flour bread by ultra-performance liquid chromatography coupled with high-resolution tandem mass spectrometry. *Food Res. Int.* **2017**, *95*, 68–81. [CrossRef] [PubMed]
57. Olivares-Tenorio, M.-L.; Verkerk, R.; van Boekel, M.A.; Dekker, M. Thermal stability of phytochemicals, HMF and antioxidant activity in cape gooseberry (*Physalis peruviana* L.). *J. Funct. Foods* **2017**, *32*, 46–57. [CrossRef]
58. Michalska-Ciechanowska, A.; Wojdyło, A.; Łysiak, G.P.; Lech, K.; Figiel, A. Functional relationships between phytochemicals and drying conditions during the processing of blackcurrant pomace into powders. *Adv. Powder Technol.* **2017**, *28*, 1340–1348. [CrossRef]
59. Firuzi, O.; Lacanna, A.; Petrucci, R.; Marrosu, G.; Saso, L. Evaluation of the antioxidant activity of flavonoids by "ferric reducing antioxidant power" assay and cyclic voltammetry. *Subj. Biochim. Biophys. Acta Gen. Subj.* **2005**, *1721*, 174–184. [CrossRef]
60. Chen, J.; Yang, J.; Ma, L.; Li, J.; Shahzad, N.; Kim, C.K. Structure-antioxidant activity relationship of methoxy, phenolic hydroxyl, and carboxylic acid groups of phenolic acids. *Sci. Rep.* **2020**, *10*, 1–9. [CrossRef]
61. Wilkowska, A.; Ambroziak, W.; Czyżowska, A.; Adamiec, J. Effect of microencapsulation by spray drying and freeze drying technique on the antioxidant properties of blueberry (*Vaccinium myrtillus*) juice polyphenolic compounds. *Pol. J. Food Nutr. Sci.* **2016**, *66*, 11–16. [CrossRef]
62. Silva-Espinoza, M.A.; García-Martínez, E.; Martínez-Navarrete, N. Protective capacity of gum Arabic, maltodextrin, different starches, and fibers on the bioactive compounds and antioxidant activity of an orange puree (*Citrus sinensis* (L.) Osbeck) against freeze-drying and in vitro digestion. *Food Chem.* **2021**, *357*, 129724. [CrossRef]

foods

Review

Ozone and Bioactive Compounds in Grapes and Wine

Margherita Modesti [1,*], Monica Macaluso [2], Isabella Taglieri [2,*], Andrea Bellincontro [1] and Chiara Sanmartin [2,3]

[1] Department for Innovation in Biological, Agro-Food and Forest Systems, University of Tuscia, Via S. Camillo de Lellis, 01100 Viterbo, Italy; bellin@unitus.it

[2] Department of Agriculture, Food and Environment, University of Pisa, Via del Borghetto 80, 56124 Pisa, Italy; monica.macaluso@phd.unipi.it (M.M.); chiara.sanmartin@unipi.it (C.S.)

[3] Interdepartmental Research Center "Nutraceuticals and Food for Health", University of Pisa, Via del Borghetto 80, 56124 Pisa, Italy

* Correspondence: margherita.modesti@unitus.it (M.M.); isabella.taglieri@for.unipi.it (I.T.)

Abstract: Ozone is widely used in the agri-food and food processing industries mainly as a sanitizing agent. However, it has recently become clear that ozone exposition leads to another important benefit: in living tissues, the induced-oxidative stress triggers the antioxidant response, and, therefore, it enhances the production of antioxidant and stress-related secondary metabolites. As such, ozone can be considered an abiotic elicitor. The goal of the present review was to critically summarize knowledge about the possibility of improving bioactive compounds and, consequently, the health-related properties of grapes and wine, by using ozone. The greatest interest has been given not only to the pre- and post-harvest treatment of table and wine grapes, but also to the explanation of the mechanisms involved in the ozone-related response and the main secondary metabolites biosynthetic pathways. From the literature available, it is clear that the effect of ozone treatment on health-related properties and secondary metabolites accumulation depends on many factors, such as the cultivar, but also the form (water or gaseous), doses, and application method of ozone. Most of the published papers report an increase in antioxidant compounds (e.g., polyphenols) and stress-related volatiles, confirming the hypothesis that ozone could be used to improve berry and wine compositional and sensory quality.

Keywords: ozonisation; antioxidants; elicitation; table grape; wine grape; wine

Citation: Modesti, M.; Macaluso, M.; Taglieri, I.; Bellincontro, A.; Sanmartin, C. Ozone and Bioactive Compounds in Grapes and Wine. *Foods* **2021**, *10*, 2934. https://doi.org/10.3390/foods10122934

Academic Editors: Marina Cano Lamadrid and Francisco Artés-Hernández

Received: 29 October 2021
Accepted: 24 November 2021
Published: 28 November 2021

1. Introduction

Ozone (O_3) is the gaseous triatomic molecule of oxygen, naturally part of the stratosphere [1,2]. O_3 has a strong oxidative potential, which makes it highly instable [2,3], and leads to a rapid oxidation of any organic matter (such as fungi, bacteria, yeasts, and viruses) which it is in contact with, and a rapid reconversion to O_2 without the production of any harmful by-products [3]. In 2001, O_3 was identified and generally recognized as a safe (GRAS) substance by the U.S. Food and Drug Administration (FDA), and it has, therefore, been widely used in the food industry [4]. Its use in the agri-food sector represents an environmental and human health friendly approach to sanitize and to preserve fresh food. Its application on fresh products has revealed many advantages, such as the reduction of post-harvest disease development, reduction of spore production, oxidation of ethylene, and the slowing down of fruit respiration and, in general, the ripening process, thus, overall increasing the shelf-life [1,5,6].

Many patents have been developed for O_3 application on plants, fruits, and vegetables. The first method was patented in 1988 by Cantelli [7], who developed a method based on the storage of fresh products in closed containers with a constant O_3 concentration of 0.05 ppm. In 1990, Karg [8] set up the sterilization of minimally processed fruits and vegetables from different contaminants using O_3 mixed with other gases (CO_2 and N_2) within the processing and packaging rooms [9]. Finally, in 2012, the first application in the wine industry was proposed by Mencarelli and Catelli [10]. The patented method, called Purovino®, involves the use of O_3 with the goal to decontaminate grapes and winery facilities during winemaking, by reducing the employ of sulphur dioxide.

In addition, O_3 exposure of plants and harvested fruits and vegetables has been demonstrated to induce important metabolic shifts. Hence, it can promote the biosynthesis of secondary metabolites, such as polyphenols and volatile organic compounds (VOCs), in plants and in plants products. The resulted metabolic shifts are strictly related to different factors, such as concentration and length of exposure, and are cultivar-dependent [11–14].

Therefore, in the recent years, many studies have focused on the possibility to use O_3 as an abiotic stressor to trigger the content of bioactive compounds in plants and plant products [2,15–18]. The following paragraphs will focus on the main secondary metabolic responses induced in vines and grapes after O_3 exposition, focusing on those metabolites important for grapes and wine quality, as well as on their health-promoting properties. The most relevant papers selected and discussed in the present review are summarized in Table 1.

Table 1. Effect of ozone treatment on bioactive compounds accumulation in table and wine grapes, and in wine made starting from ozonated grapes.

Product	Cultivar	Ozone Form	Ozone Treatment, Dose and Duration	[1] Effect	Reference
Table grapes	Napoleón	Gas	38 days of storage at 0 °C under 0.1 mg/L of O_3, and stored in 2.5 L glass jar with 8 mg/L of O_3 for 30 min every 2.5 h	+ total stilbenes	Artés-Hernández et al., 2003 [19]
	Autumn Seedless	Gas	Continuous flow with 0.1 µL/L, and discontinuous with 8 µL/L for 30 min every 2.5 h	+ total flavanols + total hydroxycinnamic acid derivatives + total phenolics	Artés-Hernández et al., 2007 [20]
	Superior Seedless Regina Victoria Cardinal CL80	Gas	Continuous and discontinuous (12 h/days) O_3 flows (2 mg/L) during 72 days of storage	+ resveratrol (with discontinuous flow)	Cayuela et al., 2009 [21]
	Scarlotta	Gas	Pre-storage treatment with 5, 10, and 20 µL for 30 min at 0 °C	+ total polyphenols + antioxidant activity + total anthocyanin	Admane et al., 2018 [22]
	Seedless Black grapes	Water	Immersion in pre-storage with ozonated water (2, 4, 6, or 8 mg/L) for 4 min at 5 °C	+ total polyphenols + antioxidant activity	Silveira et al., 2018 [23]
	Perlette Thompson Zeiny Alphonse Lavallee Barlinka	Gas	Pre-storage treatment with 16 mg/L for 5 to 10 min	+ phytoalexins (resveratrol, pterostilbene)	Sarig et al., 1996 [24]
	Superior	Gas	Pre-storage treatment with 1.67 and 3.88 g/h 1, 3, and 5 h	+ total stilbenes	González-Barrio et al., 2006 [25]
Wine grapes	Barbera Nebbiolo	Gas	Post-harvest treatment pre-vinification for 24 and 72 h with 30 µL/L	+ total proanthocyanidins extraction + flavanols extraction + total anthocyanin extraction	Paissoni et al., 2017 [26]
	Petit Verdot	Gas	Post-harvest treatment pre-vinification for 12 h with 20 g/h	+ total anthocyanin	Bellincontro et al., 2017 [15]
	Maturano	Gas	Pre-harvest treatment	+ chlorogenic acid	Valletta et al., 2016 [27]
	Grechetto	Gas	Post-harvest treatment pre-vinification for 12 h with 1.5 g/h	+ catechins	Carbone and Mencarelli 2015 [28]

Table 1. *Cont.*

Product	Cultivar	Ozone Form	Ozone Treatment, Dose and Duration	[1] Effect	Reference
	Pignola	Gas	Post-harvest treatment pre-vinification for 18 h with 1.5 g/h	+ polyphenols ± carotenoids ± anthocyanin + anthocyanin extractability + flavonol extractability	Botondi et al., 2015 [11]
	Romanesco	Gas	Post-harvest treatment pre-dehydration process with 20 g/h for 12 h	+ polyphenols	Modesti et al., 2018 [29]
	Nebbiolo	Gas	Constant flow during dehydration process with 30 μL/L	± total anthocyanin − anthocyanins extraction	Rìo Segade et al., 2020 [30]
	Barbera	Gas	Constant flow during dehydration process with 30 μL/L	− total anthocyanin + anthocyanins extraction	Rìo Segade et al., 2020 [30]
	Moscato bianco	Gas	Pre-dehydration treatment with 60 mL/L for 24 or 48 h	+ glycosylated VOCs + free linalool + cis-furan linalool oxide + terpineol	Rìo Segade et al., 2018 [14]
	Moscato bianco	Gas	Constant flow during dehydration process with 30 μL/L	+ glycosylated VOCs + linalool + geraniol + nerol	Rìo Segade et al., 2017 [31]
	Merlot	Gas	Post-harvest treatment with 1 and 3 mg/L for 12 and 24 h	+ total polyphenols	Modesti et al., 2021 [32]
Wine	Bobal	Water	Pre-harvest singles treatment	+ total polyphenols + phenolic acids + flavanols + flavonols + anthocyanins + free terpenoids	Campayo et al., 2020 [33]
	Bobal	Water	Pre-harvest treatments (three treatments performed between fruit set and harvest)	+ phenolic acids + flavanols + stilbenes + farnesol + nerodiol	Campayo et al., 2020 [33]
	Petit Verdot	Gas	Post-harvest treatment pre-vinification for 12 h with 20 g/h	+ anthocyanins + skin tannins	Bellincontro et al., 2017 [15]
Patent	PCT/IB2012/000036 "Process for the Treatment and the Winemaking of Grapes"	Gas	Post-harvest treatment with 20 g/h for 12 h	+ gallic acids + catechins + epicatechins	Mencarelli and Catelli, 2012 [10]

[1] Notes: + = increase; − = decrease; ± = both increase or decrease.

2. Review Methodology

The bibliographic identification was conducted between December 2020 and August 2021 using relevant electronic bibliographic databases (e.g., Web of Science, Science Direct, and PubMed) to ensure the highest coverage for significant papers. The primary keywords (ozone, table and wine grapes, wine) were combined using the set operator AND with secondary keywords (i.e., pre- and post-harvest treatment, bioactive compounds, volatile organic compounds, polyphenols, and antioxidants). To find and select documents, recently published reviews were firstly analysed. Then, starting from pre-selected papers, literature older than the mentioned time period was included if considered helpful to improve topic description. Authors independently evaluated the available literature using predefined eligibility criteria, resolving disagreement by discussion. To restrict and focus the aim of the research, only papers sections dealing with polyphenols and volatiles following ozone treatments were selected (excluding sections dealing with other grapes and wine traits). A total of 88 papers were selected in the end.

3. Ozone and Plant Interaction

3.1. Reactive Oxygen Species Production in Plant Tissue

Once O_3 penetrates the cells, it is immediately converted in reactive oxygen species (ROS), and, therefore, an endogenous ROS production, known as *oxidative burst*, contributes to the overall oxidizing potential of O_3 [34]. ROS production has been correlated with cellular stress responses which, consequentially, leads to important shifts in secondary metabolism [35]. The oxidative stress is characterized by an imbalance of the cellular redox status, and shifts of cell metabolism are needed to rebalance the basal levels of ROS [36]. The scavenging antioxidant system is characterized by the activity of specific enzymes having an antioxidant role, and by the antioxidant compound biosynthesis (stress-related volatiles, such as C6 (aldehydes, alcohols, and esters) and terpenoids and polyphenols, stilbenes, isoprene, and ascorbic acid) [37,38].

Secondary metabolites are biologically active compounds produced under specific condition in plants tissues, and are generally involved in plants adaptation as a response to changes in external condition. Several secondary metabolites (such as polyphenols, flavonols and tannins, essential oils, sterols, phytoalexins, and monoterpenes) have been demonstrated to have important functional effects on human health, and, therefore, techniques aimed at increasing their content in plants and plant products are becoming popular [39,40]. Elicitation can be defined as a controlled stress induced by elicitors which leads to the production of secondary metabolites [41,42], and can, therefore, improve the biological activity of plant products. Given the well-established efficacy of O_3 for the agri-food preservation and in inducing secondary metabolism shifts, recently, O_3 has been suggested as a pre-and post-harvest elicitor [16,19,43,44].

3.2. Ozone and Polyphenols

Grape and wine polyphenols have considerable importance not only for their contribution to wine quality parameters, such as color, flavor, astringency, bitterness, and ageing behavior [45], but also for their antioxidant capacity. In particular, phenolic acids, flavonoids, anthocyanins, and tannins are well-known to have many health benefits [5,20,46] due to their activity as radical scavengers.

Given the antioxidant role that polyphenols play in the cells, most of them are biosynthesized in both vines and grapes as a response to biotic and abiotic stresses thanks to the activation of the phenylpropanoid pathway [47]. Researchers have indeed observed that a higher polyphenol accumulation is due to the increase of activity of enzymes, such as phenylalanine ammonia lyase (PAL), stilbene synthases (STS), and flavonol synthases (FLS) [5,48,49]. The key point is that O_3, playing as stressor, induces the defence mechanism to protect plants and fruits tissue against oxidative stress-related damages.

However, the effect on polyphenol content after ozonisation is not always clear [5,40]: some papers describe a positive effect of O_3 in terms of polyphenols accumulation [11,15,29], whereas others talk about negative ones, especially their oxidation with consequent reduction [18,50,51].

The effect of O_3 in inducing the biosynthesis of polyphenols mainly depends on the concentration and method, as well as the length of exposure. Generally, high ozone doses lead to an over oxidation which will induce a phenol decrease [18,51]. Nevertheless, a more controlled and limited amount of O_3 results in a controlled oxidative stress which may stimulate the biosynthesis of these compounds [52].

Apart from the doses utilized, the method of application is another crucial aspect which can influence the effects and the consequences of the ozonisation process. O_3 in gaseous form is much more stable and effective in potentially inducing the oxidation. On the contrary, ozonated water is less stable and less oxidative, and, therefore, a negative impact on phenol accumulation is unlikely to occur [53]. Lastly, the duration and the type of exposition (single, continuous, or intermittent) also matters. For example, it has been observed that an overnight treatment of wine grapes with O_3 at 1.5 g/L increases phenol and anthocyanin content, whereas longer and continues exposition decreases phenolics [11].

3.3. Ozone and Volatile Organic Compounds

Volatiles are important secondary metabolites which are often involved in defence mechanisms. VOCs are indeed produced in, and released from, leaves, flowers, and fruits with the main functions of: (i) attracting pollinators and beneficial insects; (ii) protecting plants against pathogen infection and herbivore attack; (iii) creating molecules signal for plant–plant communication [54]. Because of these roles, VOCs are often synthesized under stressed condition. Specifically, in the case of the oxidative stress, the mechanism is still linked with ROS production, which determines the membrane lipid peroxidation [55] and VOCs associated with lipid peroxidation. Compounds, such as C6 (mainly aldehydes and alcohols), methanol, and methyl salicylate, are monitored as signalling volatiles [56], and have antibacterial and fungicidal properties [56].

In living tissue, such as vine leaves and grapes, stress responses occur when O_3 is applied, especially if accurately managed, and stress-related VOCs biosynthesis generally is induced [57,58]. Remarkably, C6 volatiles (especially aldehydes and alcohols) formed in grapes are generally associated with herbaceous aroma and flavour. Additionally, when C6 volatiles are converted in their acetate esters, the result is pleasant fruity nuances [34].

Another class of VOCs known to play a key role in the antioxidant response, and, therefore, also produced after O_3 exposition, belongs to the terpenoid family, i.e., isoprene, monoterpenes, and sesquiterpenes [59]. The stimulation of terpenoid biosynthesis in fruits and vegetables is extremely significant, considering the crucial role they play in the floral and fruity aroma and taste. Two different pathways are involved in terpenoid biosynthesis: the methylerythritol phosphate (MEP) and the mevalonate (MVE) pathway [60], which are known to be strongly influenced by biotic and abiotic stresses [57,61–63]. Several fruits, vegetables, and plant tissues show an increase of terpenoid biosynthesis as a response to ozonisation [14,38,57,64]. It has been hypothesized that terpenes biosynthesis is induced by the oxidative stress because they might act as ROS scavengers, reducing their reactivity [65]. In addition, in O_3 stressed plants tissues, the inhibition of terpenes biosynthesis results in higher oxidative damage [38], supporting the hypothesis of an antioxidant role, as suggested by Calogirou, 1999 [66].

Lastly, another defence mechanism induced by oxidative stress is the stimulation of the activity of different enzymes, including uridine5′-diphospho-gluconosyltransferases (UGTs), which play an indirect role in ROS-detoxification [67]. Glycosylation allows compartmentalization of small and toxic/reactive molecules, such as ROS, but also VOCs, by reducing their volatility through derivatization [68]. As such, the UGTs plays a key role not only in plant defence mechanisms [69], but also in the formation of glycosylated volatiles. Therefore, the induced oxidative stress stimulating UGTs activity can, in turn, increase glycosylated volatiles [70], also known as aromatic precursors.

4. Ozone and Bioactive Compounds in Table and Wine Grape

4.1. Table Grape

Table grapes (*Vitis vinifera* L.) are affected in postharvest life by evident quality depletion mainly caused by loss of water, berry softening, browning, and microbiological contamination, the latter mainly due to grey mold action [45]. As stated by many researches, O_3 treatment ranging from 0.1 mg/L/day or higher allows to prolong table grape shelf-life, inhibiting the growth of grey mold [71,72]. Furthermore, O_3 was reported to boost the phenolic and aromatic compound biosynthesis, since O_3 induces in living tissues different defense mechanisms at the genetic, transcriptional, and biochemical level [73].

Researchers have evaluated the effect of O_3 atmosphere on Napoleón grapes both packed in a macro-perforated polypropylene basket during 38 days of storage at 0 °C under

0.1 mg/L of O_3, and stored in a 2.5 L glass jar with 8 mg/L of O_3 flushed for 30 min every 2.5 h. They observed an increase of total stilbenoids for both O_3 treatments, even if the O_3 concentrations were not enough to avoid *Botrytis cinerea* spread [19,48,74].

A similar study [20] tested different O_3 treatments (continuous: 0.1 μL/L; discontinuous. 8 μL/L for 30 min every 2.5 h) on the quality of Autumn Seedless grapes both after protracted storage at low temperature (60 days; 0 °C), and after one week at retail conditions (15 °C). Either continuous or discontinuous treatments, at low temperature, determined an increase of total flavan-3-ol. Furthermore, continuous treatment also retained the initial content of hydroxycinnamates. After the retail period, a significant increase of total polyphenols was observed for both treatments [20].

Cayuela et al., [74] treated two white table grapes (Superior Seedless and Regina Victoria) varieties and one red grape (Cardinal CL80) variety with continuous and discontinuous (12 h/days) O_3 flows (2 mg/L) during storage (5 °C; 72 days). The shelf-life of O_3-treated grapes, regardless of the method of application, was significantly prolonged in comparison with the control grapes stored in air [74]. Additionally, discontinuous treatment led to the highest resveratrol content. On the other hand, continuous application induced a reduction of these compounds. In this study, the authors suggested that the continuous presence of O_3 blocked the resveratrol biosynthesis, whereas a discontinuous action could trigger its biosynthesis [74]. This was also observed by Artés-Hernández et al. and Tomás-Barberán et al. [19,48], where the discontinuous treatment (8 mg/L O_3 for 30 min, every 2.5 h) boosted the resveratrol content in Napoléon grapes. However, some discrepancies in the results have been evidenced in the literature, probably as a consequence of a cultivar [20,48] or dose-dependent effect of O_3 treatments: continuous highly concentrated (2 mg/L) treatment could deplete antioxidant compounds as a defensive mechanism toward oxidative stress [75], whereas discontinuous applications could improve their accumulation [21].

Admane et al. [22] evaluated the effects of O_3 pre-treatments at three different concentrations (5, 10, and 20 μL/L on harvested organic Scarlotta table grapes packed under modified atmosphere packaging (MAP) (2% O_2 and 5% CO_2), monitoring the quality decay trend, sensory traits, and antioxidant compounds profile during 45 days at 0 ± 0.5 °C, followed by 7 days at higher temperature (15 ± 1.0 °C). The results showed a higher level of polyphenols and antioxidant capacity for all samples O_3-treated, confirming the O_3 activity as an elicitor of phenolic compounds biosynthesis. Moreover, the content of anthocyanins in the berry skins of O_3-treated grapes was significantly higher than in the control ones [22]. The increase of anthocyanins content was also observed in Cardinal grapes stored for 12 days at low temperature [76].

Silveira et al. [23] observed similar results in Thompson Seedless and Black Seedless grapes, previously immersed in ozonated water at different concentrations (2, 4, 6, or 8 mg/L) for 4 min at 5 °C, and then stored for 21 days at 5 °C. Namely, a 23–50% and 18.5–28% improvement of total polyphenols was observed in Thompson Seedless and Black Seedless grapes, respectively. Furthermore, all the O_3 treatments determined a doubling of the antioxidant capacity in Thompson samples, whereas only the treatment with 6 and 8 mg/L increased antioxidant activity in Black Seedless grapes [23].

4.2. Wine Grape

As previously described, O_3 exposure can cause modifications in grape secondary metabolism, improving the synthesis of phenolics such as stilbenes and anthocyanins [26,77].

Paissoni et al. [26] observed that O_3 treatments for 24 and 72 h affected the initial phases of skin maceration in red vinification for both Barbera and Nebbiolo grapes, favoring the extraction of di-substituted anthocyanins in Nebbiolo grapes. Namely, O_3 treatment did not affect the final individual anthocyanin extractability, thus, the varietal anthocyanin fingerprint was maintained. O_3 also influenced the flavanol extraction, which was slowed down in both varieties. In another study, Bellincontro et al. [15] reported a faster fermentation rate, as well as a higher anthocyanin concentration by an overnight O_3 treatment of Petit Verdot grapes.

Valetta et al. [27] observed that O_3 treatment did not activate the stilbene synthesis in Maturano white grapes, even if the leafy chlorogenic acid content was increased; thus, they proposed chlorogenic acid as a biochemical marker of O_3-induced stress in the *V. vinifera* plant.

Short-term treatments with O_3 on fresh grapes have proven effective in determining changes in flavanol fraction, with a significant increase in catechins, and a slightly decreased epicatechin [28]. As suggested by Carbone and Mencarelli [28], the triggering of low-molecular-weight antioxidant biosynthesis could be considered as a defense mechanism against the O_3-induced oxidative stress.

At the same time, O_3 could play a protective role against the oxidation of flavanols since, during post-harvest partial grape dehydration, O_3 exposure both promotes antioxidant enzymes, and inhibits the oxidant activity of polyphenol oxidase and lipoxygenase enzymes [29]. Botondi et al. [11] showed that O_3 fumigation at 1.5 g/h for 18 h in continuous flow (shock treatment) could preserve the polyphenol and anthocyanin content. On the other hand, a long-term O_3 treatment (1.5 g/h, continuous flow, subsequent dehydration with 0.5 g/h of O_3, 4 h per day) determined a significant oxidation of the polyphenol content.

Both dehydration and O_3 effects are cultivar-dependent. As reported by Rìo Segade et al. [77], in Barbera skins, the combination of the two post-harvest stresses (e.g., oxidative and water stress) determined a limited proanthocyanidin loss, together with increased prodelphinidin and limited galloylation percentages. Besides, in Nebbiolo skin, richer in proanthocyanidin, an increased galloylation was observed during dehydration when associated with O_3 treatment [30].

Another study [77] carried out on Moscato Bianco wine grapes (*Vitis vinifera* L.) showed that short-term O_3 exposure (60 μL/L for 48 h) on fresh grapes did not determine an immediate resveratrol accumulation, but it induced an elicitor effect on total stilbenes (+36%) in dehydrated grapes (20% of weight loss), with a considerable overproduction of *trans*-resveratrol and *trans*-piceatannol.

Furthermore, O_3 treatments during grapes post-harvest seem to stimulate the berry skin cell wall degradation, affecting the extractability of oligomeric flavanols and proanthocyanidins [26]. As observed by Botondi et al. [11], O_3 shock treatment on Pignola grapes did not affect the pectin methylesterase and polygalacturonase activities, which, in turn, affect cell wall composition and porosity, and are responsible for different anthocyanin and flavonol extractability [32]. Moreover, polyphenols are retained by the cell wall according to their structure. In this regard, the analysis of the texture has been proven to be an effective tool to correlate phenolic compounds extractability to skin mechanical properties. In particular, a significant correlation has been found between skin hardness and the extraction of anthocyanins and flavanols with low and high molecular mass [78]. Recently, Laureano et al. [79] observed a hardening of table (Italia and Muscat Hamburg) and wine (Merlot and Barbera) grapes' berry skin as a consequence of post-harvest gaseous O_3 exposure (30 μL/L) for 24 h, evidencing a role of O_3 treatment on the berry skin mechanical features. The use of O_3 for the treatment of wine grapes in post-harvest is currently being explored to improve polyphenol extractability, which is mainly affected by the cultivar and, to a lesser extent, by the time of the O_3 exposure [26,30,77].

Lastly, the effect of O_3 exposition (30 mg/L continuously supplied) on the aromatic composition of Moscato Bianco grapes during the dehydration process was investigated by Segade et al. [31]. O_3 significantly increased not only the amount of total VOCs, but also the amount of terpenoids, which are the major aromatic markers of the Moscato scent. Accordingly, the biosynthetic pathways involved in terpenoids and C6 biosynthesis (i.e., MEP and LOX-HPL) were up-regulated following the O_3 exposition. On the contrary, higher doses (60 mg/L) supplied for a shorter time significantly reduce total VOCs, due to terpenoids oxidation [14]. These findings highlight again that the effect of O_3 on VOCs profile strongly depend on the concentration and exposure time.

5. Ozone Strategy to Increase Bioactive Compounds in Wine

With respect to wine production, the use of O_3 has become popular, and its employment is increasing in a significant number of wineries. As already discussed, O_3 has been widely used not only as sanitizing agent and for increasing the shelf-life of harvested grapes, but also as an exogenous elicitor to enrich grapes of secondary metabolites, significant factors for grape quality and human health. Hence, at berry level, the oxidative stress modifies the accumulation of different compounds to defend the cells from possible oxidative damages [73]. Most of the studies available in the literature focused on the elicitor effect on grape metabolism, confirming the aptitude of O_3 in increasing the bioactive compound presence. However, when dealing with wine grapes, it is decisive to understand if the changes induced in grapes are transferred in the resulting wine, considering that the accumulation of those metabolites can increase wine health-promoting attributes thanks to their antioxidant activity [42].

As is obvious, the accumulation of secondary metabolites in grapes have a great influence on wine quality. For instance, modifications of the polyphenol profile in grapes will result in changes in color, astringency, bitterness, and body of the resulting wines. Furthermore, the accumulation of antioxidant volatiles, such as terpenoids and C6 compounds, induced by the oxidative stress [29,31,61] can strongly affect the aromatic profile of the wine. Nevertheless, very few studies on the accumulation of bioactive compounds in wine are currently available.

5.1. Effect of Pre-Harvest O_3 Application on Wine Features

Recently, O_3 has been proposed for pest management in viticulture as a possible alternative to traditional pesticides, considering its environmental and human health friendliness. Studies related to the in-field O_3 application on grapes, and associated to the wine quality and composition, are very limited. It is well established that viticulture practices (such as clusters thinning, defoliation, elicitor application, etc.) can strongly influence grape development and metabolism [58], and, consequently, wine composition. The main problem when dealing with pre-harvest applications is that O_3 has always been considered as an environmental pollutant, associated with yield reduction, and the development of physiological disorders in plants (such as chlorophyll degradation, and a decrease of carbon and nitrogen assimilation) and fruits [80]. However, it is also clear that the possibility to cause damages is strongly affected by the cultivar, phenological stage, and environmental conditions. As such, there is a small risk of toxicity if O_3 is applied under controlled conditions. Additionally, the controlled oxidative stress induces antioxidant responses which could enrich fruit and wine of secondary metabolites, exactly as they occur in post-harvest applications [29,72,81].

Recently, ozonated water was applied to control grapevine diseases, and the effect on grapes and wine quality (*Vitis vinifera* L. cv. Bobal) was studied by Campayo et al. [33,73,81]. These authors tested two different applications (i.e., a single treatment with ozonated water during the ripening period, and three treatments with ozonated water performed between fruit set-up to harvest), evaluating their effect on grapes, and during winemaking. The treatments induced a different response even within the same family of compounds. The first ozonated water treatment increased total polyphenol by about 130%. Concerning the effect of ozonated water on different classes of polyphenols, it is possible to observe how the single treatment led to a wine with higher (more than double of the control wine) phenolic acids, flavanols, flavonols, and anthocyanins. On the other hand, the trial performed using three ozonated water treatments increased stilbenes and flavanols, while reduced the amount of anthocyanins. Specifically, phenolic acids such as gallic, vanillic, syringic, *trans*-caftaric, and *trans-p*-coutaric acids were in higher amount after vines ozonation, regardless of the dose and type of application. Furthermore, among the different stilbenes identified by the authors, *trans*-resveratrol and glucoside piceid-*trans*-resveratrol increased in wine made starting from ozonated vines [33]. These compounds belong to the class of phytoalexins, and they are produced as defensive molecules in plants as a response to different stresses, as well as the oxidative stress [82]. Concerning flavanols, higher concentrations of catechin were found in wines made starting from ozonated vines. Flavanols are known as the most powerful ROS scavengers in grapes and wines, presenting an extremely high antioxidant activity [83], and, therefore, it is not surprising their increase in wines derived from ozonated vines.

The amount of non-acylated anthocyanins (i.e., delphinidin, petunidin, peonidin, and malvidin 3-*O*-glucosides) and the peonidin and malvidin 3-*O*-glucosides acetylated increased in the wine made starting from ozonated vines. However, when the ozonated water treatments were repeated three times, the amounts of all the non-acylated anthocyanins decreased. The authors reported that these antioxidant compounds produced in grapes against oxidative stress would be oxidized and depleted under long exposure to ozone.

Regarding wine aroma, when plants are subjected to abiotic stresses, they increase the biosynthesis and emission of VOCs. Specifically, it is well known that plants produce isoprene, terpenes, and C6 in response to the oxidative stress after O_3 exposition [37,38]. This was again confirmed by Campayo et al. [73], who observed an increase of the total free terpenoids in wine made starting from ozonated vine, mainly due to the accumulation of farnesol and nerolidol. When treatments were repeated during the growing season, citronellol and nerolidol increased. Based on recent studies, terpenoids have been demonstrated to induce health-beneficial effects, mainly for their anti-inflammatory and antimicrobial properties [84].

Based on the only one study available, it seems clear that the application of O_3 on vines, by modifying grapes metabolism, increases the content of important health-related bioactive and antioxidant compounds in the resulting wine (i.e., polyphenols and terpenoids).

5.2. Effect of Post-Harvest O_3 Application on Wine Features

Though the pre-harvest use of O_3 is still very debated, its use for post-harvest management of wine grapes is widely used. Hence, in the context of new technologies for grapes preservation and wine making, O_3 is currently a common tool not only to control spoilage microorganisms, but also to increase the nutritional value of grapes and wine. As already discussed in Section 4, there are many papers discussing the effect of post-harvest O_3 treatment on grape composition and metabolic responses, whereas the effect of ozonation on wine quality is less studied. Many winemakers considered the use of O_3 as a potential risk of oxidation of important compounds for the sensory and quality attributes of wine (i.e., polyphenols and volatiles). However, it is currently known that, if used under controlled conditions, it can represent a good option to increase the amount of these compounds. Hence, growing attention has been paid to O_3, related to its stress action, which enhances the biosynthesis of bioactive compounds (such as phenolic substances) in table and wine grapes [20,28,85].

Mencarelli and Catelli [10] showed, for the first time, an enrichment of polyphenols and anthocyanins extraction in red wines treated with O_3 according to the Purovino® method (12 h, at 4 °C and 70% RH with max 20 g/h with 6% w/w of O_3 and a flow rate at maximum 150 normal liter/h). The authors explain that the increase of these compounds (namely gallic acids, catechins, and epicatechins) is obtained thanks to two different, but linked, mechanisms [10]. On one hand, grapes exposition at a right dose of O_3 for the right amount of time promotes biosynthesis of metabolites [10,14,86], and on the other hand, it modifies skin permeability, determining a consequent greater extraction of polyphenols. The treatment also promotes biosynthesis and extraction of antioxidant-related volatiles. Furthermore, the wines produced through the mentioned method were characterized by a polyphenols fraction which remained unaltered during malolactic fermentation and also in bottled wines (up to two years), indicating a greater stability of wines produced employing

O_3-treated grapes. Bellincontro et al. [15] observed an increase of anthocyanins and skin tannins in wine derived from O_3-treated harvested grapes. The increase in anthocyanins is probably due to the berries' reaction to a moderate stress, which induces polyphenols biosynthesis, which are then transferred in the resulting wine.

Furthermore, after ozonation, the extractability index increased in O_3-treated grapes, and, therefore, O_3 treatment not only induces the biosynthesis of phenolic substances, but also affects the cell wall structure and cell membrane composition [87], facilitating the extraction of these compounds during winemaking. Ref. [11] thus represents a great advantage for the vinification process in term of time, cost, and health-related properties as well.

The increase of polyphenols in wine grapes after post-harvest O_3 exposition was also pointed out by Carbone and Mencarelli [28], Paissoni et al. [26], and Segade et al. [77], who reported an increase of catechin [28], and higher flavanol and anthocyanins content, thanks to a greater extraction [26,77]. The same increase of polyphenols, specifically of flavonoids, has been found even when O_3 has been used at the beginning of controlled dehydration of Pignola and Romanesco grapes for "passito" wine production [11,29]. Lastly, post-harvest O_3 treatments used to reduce the *smoke taint* in wine, which means wine made from grapes exposed to bushfires with undesirable sensory characters (smoky, burnt, and ashy), lead to higher polyphenol content as well [32,88]. Unfortunately, all these studies do not report any information about the resulting wines.

6. Conclusions

Overall, the O_3 treatments operated on table and wine grapes reveal many important advantages, especially related to the phenolic and aromatic fractions. However, the effect of O_3 treatment strongly depends on the treated cultivar, O_3 form (gas, water), and method used for the treatment (i.e., dose, duration, intermittence or constant exposition). Both gaseous and water ozone treatment exert an elicitor effect on grape bioactive compounds (i.e., polyphenols, anthocyanins, flavanols, etc.), although most of the studies refer to gaseous treatments. Moreover, it is often reported that high doses and long exposition could result in an excessive oxidative stress which potentially decreases grape quality parameters, as bioactive compounds may be oxidized. On the other hand, when O_3 is applied under studied and controlled conditions, an increase of bioactive compounds, such as polyphenols and antioxidant volatiles, is often reported. The literature suggests that in most of the cases, the best results are obtained with a low concentration and short treatment. Moreover, continuous ozone treatment during post-harvest dehydration increases the total VOCs. However, it is also suggested that the internal composition of the berries (cultivar-dependent) strongly influences the final result of O_3 exposition. For example, in cultivars with higher flavanol and anthocyanins content, the result of O_3 treatment is a greater extraction of polyphenols. Furthermore, in cultivars characterized by the prevalence of anthocyanins di-substituted, the result is a lower anthocyanin extractability, whereas in cultivars with tri-substituted anthocyanins, the extractability after O_3 exposition is higher.

In the light of all the above considerations, the factors affecting the bioactive compounds content in grapes, and, as a consequence, in wine, are many, and a unique strategy is, therefore, difficult to identify. Nevertheless, all considered, the potential role of O_3 to stimulate the biosynthesis of bioactive compounds is clear, and, considering that O_3 treatment is very practical, it can be easily incorporated into the wine production chain not only as sanitizing agent, but also to promote the health-related compounds of wines, inducing an improvement of their general quality. Moreover, post-harvest grape exposure could significantly reduce the use of sulphur dioxide in winemaking due to its bactericidal and fungicidal properties. However, in a large-scale application, especially in the case of winemaking goals, an adaptation of O_3 treatment depending on the cultivar and on the target wine would also be highly desirable.

Author Contributions: Conceptualization, M.M. (Margherita Modesti), C.S., I.T., M.M. (Monica Macaluso), A.B.; Writing—original draft writing M.M. (Margherita Modesti), C.S., I.T., M.M. (Monica

Macaluso); review and editing M.M. (Margherita Modesti), C.S., I.T., M.M. (Monica Macaluso), A.B.; supervision A.B., C.S. All authors have read and agreed to the published version of the manuscript.

Funding: This research was funded in the frame of VIOLoC project (FISR2019_03020).

Institutional Review Board Statement: Not applicable.

Informed Consent Statement: Not applicable.

Acknowledgments: The authors gratefully acknowledge Fabio Mencarelli of DAFE, University of Pisa, who contributed to the conceptualization and the structure of the review.

Conflicts of Interest: The authors declare no conflict of interest.

References

1. Prabha, V.; Barma, R.D.; Singh, R.; Madan, A. Ozone Technology in Food Processing: A Review. *Trends Biosci.* **2015**, *8*, 4031–4047.
2. Forney, C. Postharvest Response of Horticultural Products to Ozone. In *Postharvest Oxidative Stress in Horticultural Crops*; Hodges, D., Ed.; Food Products Press: New York, NY, USA, 2003; pp. 13–54.
3. Gabler, F.M.; Smilanick, J.L.; Mansour, M.F.; Karaca, H. Influence of fumigation with high concentrations of ozone gas on postharvest gray mold and fungicide residues on table grapes. *Postharvest Biol. Technol.* **2010**, *55*, 85–90. [CrossRef]
4. Glowacz, M.; Rees, D. Exposure to ozone reduces postharvest quality loss in red and green chilli peppers. *Food Chem.* **2016**, *210*, 305–310. [CrossRef]
5. Shezi, S.; Magwaza, L.S.; Mditshwa, A.; Tesfay, S.Z. Changes in biochemistry of fresh produce in response to ozone postharvest treatment. *Sci. Hortic.* **2020**, *269*, 109397. [CrossRef]
6. Tzortzakis, N.; Chrysargyris, A. Food Reviews International Postharvest ozone application for the preservation of fruits and vegetables. *Food Rev. Int.* **2016**, *33*, 270–315. [CrossRef]
7. Cesar, C. Process and Device for Improving the Preservation of Fruits and Vegetables. France Patent No. 8612739, 1988. Available online: https://patentscope.wipo.int/search/en/detail.jsf?docId=FR185876064&docAn=8612739 (accessed on 18 November 2021).
8. Karg, J.E. Ozone Treatment for Sterilization of Highly Contaminated Natural Products, Especially Herbs, Spices, Fruits, Vegetables and Cereals. German Federal Republic Patent No. DE3917250A1, 1990. Available online: https://scholar.google.com.hk/scholar?hl=zh-CN&as_sdt=0%2C5&q=Ozone+Treatment+for+Sterilization+of+Highly+Contaminated+Natural+Products%2C+Especially+Herbs%2C+Spices%2C+Fruits%2C+Vegetables+and+Cereals&btnG= (accessed on 18 November 2021).
9. Mitsuda, H.; Ohminami, H.; Nagasawa, H. Process for Sterilizing Food Stuffs. U.S. Patent No. US 5011699, 7 September 1989.
10. Mencarelli, F.; Catelli, C. Process for the Treatment and the Winemaking of Grapes. Italy Patent No. PCT/IB2012/000036, 8 November 2013.
11. Botondi, R.; De Sanctis, F.; Moscatelli, N.; Vettraino, A.M.; Catelli, C.; Mencarelli, F. Ozone fumigation for safety and quality of wine grapes in postharvest dehydration. *Food Chem.* **2015**, *188*, 641–647. [CrossRef] [PubMed]
12. Desanctis, F.; Ceccantoni, B.; Bellincontro, A.; Botondi, R.; Mencarelli, F.; D'onofrio, C.; Ducci, E.; Catelli, C. Ozone Fumigation Postharvest Treatment for the Quality of Wine Grape. *Acta Hortic.* **2015**, *1071*, 795–800. [CrossRef]
13. Guzzon, R.; Nardin, T.; Micheletti, O.; Nicolini, G.; Larcher, R. Antimicrobial activity of ozone. Effectiveness against the main wine spoilage microorganisms and evaluation of impact on simple phenols in wine. *Aust. J. Grape Wine Res.* **2013**, *19*, 180–188. [CrossRef]
14. Río Segade, S.; Vilanova, M.; Pollon, M.; Giacosa, S.; Torchio, F.; Rolle, L. Grape VOCs response to postharvest short-term ozone treatments. *Front. Plant Sci.* **2018**, *9*, 1826. [CrossRef] [PubMed]
15. Bellincontro, A.; Catelli, C.; Cotarella, R.; Mencarelli, F. Postharvest ozone fumigation of Petit Verdot grapes to prevent the use of sulfites and to increase anthocyanin in wine. *Aust. J. Grape Wine Res.* **2017**, *23*, 200–206. [CrossRef]
16. Lv, Y.; Tahir, I.I.; Olsson, M.E. Effect of ozone application on bioactive compounds of apple fruit during short-term cold storage. *Sci. Hortic.* **2019**, *253*, 49–60. [CrossRef]
17. Karaca, H.; Velioglu, Y.S. Effects of ozone treatments on microbial quality and some chemical properties of lettuce, spinach, and parsley. *Postharvest Biol. Technol.* **2014**, *88*, 46–53. [CrossRef]
18. Allende, A.; Marín, A.; Buendía, B.; Tomás-Barberán, F.; Gil, M.I. Impact of combined postharvest treatments (UV-C light, gaseous O_3, superatmospheric O_2 and high CO_2) on health promoting compounds and shelf-life of strawberries. *Postharvest Biol. Technol.* **2007**, *46*, 201–211. [CrossRef]
19. Artés-Hernández, F.; Artés, F.; Tomás-Barberán, F.A. Quality and enhancement of bioactive phenolics in cv. Napoleon table grapes exposed to different postharvest gaseous treatments. *J. Agric. Food Chem.* **2003**, *51*, 5290–5295. [CrossRef] [PubMed]
20. Artés-Hernández, F.; Aguayo, E.; Artés, F.; Tomás-Barberán, F.A. Enriched ozone atmosphere enhances bioactive phenolics in seedless table grapes after prolonged shelf life. *J. Sci. Food Agric.* **2007**, *87*, 824–831. [CrossRef]
21. Cayuela, J.A.; Vázquez, A.; Pérez, A.G.; García, J.M. Control of table grapes postharvest decay by ozone treatment and resveratrol induction. *Food Sci. Technol. Int.* **2009**, *15*, 495–502. [CrossRef]

22. Admane, N.; Genovese, F.; Altieri, G.; Tauriello, A.; Trani, A.; Gambacorta, G.; Verrastro, V.; Di Renzo, G.C. Effect of ozone or carbon dioxide pre-treatment during long-term storage of organic table grapes with modified atmosphere packaging. *LWT* **2018**, *98*, 170–178. [CrossRef]
23. Silveira, A.C.; Oyarzún, D.; Escalona, V. Oxidative enzymes and functional quality of minimally processed grape berries sanitised with ozonated water. *Int. J. Food Sci. Technol.* **2018**, *53*, 1371–1380. [CrossRef]
24. Sarig, P.; Zahavi, T.; Zutkhi, Y.; Yannai, S. Ozone for control of post-harvest decay of table grapes caused by *Rhizopus stolonifer*. *Physiol. Mol. Plant Pathol.* **1996**, *48*, 403–415. [CrossRef]
25. González-Barrio, R.; Beltrán, D.; Cantos, E.; Gil, M.I.; Espín, J.C.; Tomás-Barberán, F.A. Comparison of ozone and UV-C treatments on the postharvest stilbenoid monomer, dimer, and trimer induction in var. "Superior" white table grapes. *J. Agric. Food Chem.* **2006**, *54*, 4222–4228. [CrossRef]
26. Paissoni, M.A.; Río Segade, S.; Giacosa, S.; Torchio, F.; Cravero, F.; Englezos, V.; Rantsiou, K.; Carboni, C.; Gerbi, V.; Teissedre, P.L.; et al. Impact of post-harvest ozone treatments on the skin phenolic extractability of red winegrapes cv Barbera and Nebbiolo (*Vitis vinifera* L.). *Food Res. Int.* **2017**, *98*, 68–78. [CrossRef]
27. Valletta, A.; Salvatori, E.; Rita Santamaria, A.; Nicoletti, M.; Toniolo, C.; Caboni, E.; Bernardini, A.; Pasqua, G.; Manes, F. Ecophysiological and phytochemical response to ozone of wine grape cultivars of *Vitis vinifera* L. *Nat. Prod. Res.* **2016**, *30*, 2514–2522. [CrossRef] [PubMed]
28. Carbone, K.; Mencarelli, F. Influence of Short-Term Postharvest Ozone Treatments in Nitrogen or Air Atmosphere on the Metabolic Response of White Wine Grapes. *Food Bioprocess Technol.* **2015**, *8*, 1739–1749. [CrossRef]
29. Modesti, M.; Petriccione, M.; Forniti, R.; Zampella, L.; Scortichini, M.; Mencarelli, F. Methyl jasmonate and ozone affect the antioxidant system and the quality of wine grape during postharvest partial dehydration. *Food Res. Int.* **2018**, *112*, 369–377. [CrossRef] [PubMed]
30. Segade, S.R.; Belén, A.; Bautista-Ortín, B.; Paissoni, M.A.; Giacosa, S.; Gerbi, V.; Rolle, L.; Gómezgómez-Plaza, E. Changes in Skin Flavanol Composition as a Response to Ozone-Induced Stress during Postharvest Dehydration of Red Wine Grapes with Different Phenolic Profiles. *J. Agric. Food Chem.* **2020**, *68*, 13439–13449. [CrossRef] [PubMed]
31. Río Segade, S.; Vilanova, M.; Giacosa, S.; Perrone, I.; Chitarra, W.; Pollon, M.; Torchio, F.; Boccacci, P.; Gambino, G.; Gerbi, V.; et al. Ozone Improves the Aromatic Fingerprint of White Grapes. *Sci. Rep.* **2017**, *7*, 16301. [CrossRef]
32. Modesti, M.; Szeto, C.; Ristic, R.; Jiang, W.; Culbert, J.; Bindon, K.; Catelli, C.; Mencarelli, F.; Tonutti, P.; Wilkinson, K. Potential Mitigation of Smoke Taint in Wines by Post-Harvest Ozone Treatment of Grapes. *Molecules* **2021**, *26*, 1798. [CrossRef] [PubMed]
33. Campayo, A.; de la Hoz, K.S.; García-Martínez, M.M.; Salinas, M.R.; Alonso, G.L. Spraying ozonated water on bobal grapevines: Effect on wine quality. *Biomolecules* **2020**, *10*, 213. [CrossRef]
34. Kim, J.-G.; Yousef, A.E.; Dave, S. Application of Ozone for Enhancing the Microbiological Safety and Quality of Foods: A Review. *J. Food Prot.* **1999**, *62*, 1071–1087. [CrossRef]
35. Conrath, U.; Beckers, G.J.M.; Flors, V.; García-Agustín, P.; Jakab, G.; Mauch, F.; Newman, M.-A.; Pieterse, C.M.J.; Poinssot, B.; Pozo, M.J.; et al. Priming: Getting Ready for Battle Prime-A-Plant Group. *Mol. Plant-Microbe Interact. MPMI* **2006**, *19*, 1062–1071. [CrossRef] [PubMed]
36. Mittler, R.; Vanderauwera, S.; Gollery, M.; Breusegem, F. Van Reactive oxygen gene network of plants. *Trends Plant Sci.* **2004**, *9*, 490–498. [CrossRef] [PubMed]
37. Loreto, F.; Schnitzler, J.P. Abiotic stresses and induced BVOCs. *Trends Plant Sci.* **2010**, *15*, 154–166. [CrossRef] [PubMed]
38. Loreto, F.; Mannozzi, M.; Maris, C.; Nascetti, P.; Ferranti, F.; Pasqualini, S. Ozone Quenching Properties of Isoprene and Its Antioxidant Role in Leaves. *Plant Physiol.* **2001**, *126*, 993–1000. [CrossRef]
39. Gawlik-Dziki, U.; Dziki, D.; Nowak, R.; Świeca, M.; Olech, M.; Pietrzak, W. Influence of sprouting and elicitation on phenolic acids profile and antioxidant activity of wheat seedlings. *J. Cereal Sci.* **2016**, *70*, 221–228. [CrossRef]
40. Sachadyn-Król, M.; Agriopoulou, S. Ozonation as a method of abiotic elicitation improving the health-promoting properties of plant products-A review. *Molecules* **2020**, *25*, 2416. [CrossRef]
41. Baenas, N.; García-Viguera, C.; Moreno, D. Elicitation: A Tool for Enriching the Bioactive Composition of Foods. *Molecules* **2014**, *19*, 13541–13563. [CrossRef]
42. Thakur, M.; Bhattacharya, S.; Khosla, P.K.; Puri, S. Improving production of plant secondary metabolites through biotic and abiotic elicitation. *J. Appl. Res. Med. Aromat. Plants* **2019**, *12*, 1–12. [CrossRef]
43. Sandermann, H.; Ernst, D.; Heller, W.; Langebartels, C. Ozone: An abiotic elicitor of plant defence reactions. *Trends Plant Sci.* **1998**, *3*, 47–50. [CrossRef]
44. Keutgen, A.J.; Pawelzik, E. Influence of pre-harvest ozone exposure on quality of strawberry fruit under simulated retail conditions. *Postharvest Biol. Technol.* **2008**, *49*, 10–18. [CrossRef]
45. Ribéreau-Gayon, P.; Dubourdieu, D.; Doneche, B.; Lonvaud, A. Handbook of Enology. Volume 1. The Microbiology of Wine and Vinifications, 2nd ed.John Wiley & Sons, Ltd.: West Sussex, UK, 2006; ISBN 0-470-01034-7.
46. Li, L.; Sun, B. Critical Reviews in Food Science and Nutrition Grape and wine polymeric polyphenols: Their importance in enology. *Crit. Rev. Food Sci. Nutr.* **2017**, *59*, 563–579. [CrossRef]
47. Dixon, A.; Lahoucine, A.; Parvathi, K.; Chang-Jun, L.; Srinivasa, R.; Liangjiang, W. The phenylpropanoid pathway and plant defence-a genomics perspective. *Mol. Plant Pathol.* **2002**, *1*, 371–390. [CrossRef]

48. Tomás-Barberán, F.A.; Espín, J.C. Phenolic compounds and related enzymes as determinants of quality in fruits and vegetables. *J. Sci. Food Agric.* **2001**, *81*, 853–876. [CrossRef]
49. Chen, C.; Zhang, H.; Dong, C.; Ji, H.; Zhang, X.; Li, L.; Ban, Z.; Zhang, N.; Xue, W. Effect of ozone treatment on the phenylpropanoid biosynthesis of postharvest strawberries. *RSC Adv.* **2019**, *9*, 25429–25438. [CrossRef]
50. Jia, X.; Li, J.; Du, M.; Zhao, Z.; Song, J.; Yang, W.; Zheng, Y.; Chen, L.; Li, X. Combination of Low Fluctuation of Temperature with TiO_2 Photocatalytic/Ozone for the Quality Maintenance of Postharvest Peach. *Foods* **2020**, *9*, 234. [CrossRef] [PubMed]
51. Goffi, V.; Modesti, M.; Forniti, R.; Botondi, R. Quality of green (*Actinidia chinensis* var. *deliciosa* 'Hayward') and yellow (*A. Chinensis* var. *chinensis* 'Soreli') kiwifruit during cold storage at 0°C in normal atmosphere and with gaseous ozone. *Acta Hortic.* **2018**, *1218*, 473–480. [CrossRef]
52. Shah, N.; Sulaiman, N.; Sidek, A.; Supian, N.S.M. Quality assessment of ozone-treated citrus fruit juices. *Int. Food Res. J.* **2019**, *26*, 1405–1415.
53. Augpole, I.; Rakejeva, T.; Dukalska, L.; Skudra, L. Providing Quality of Shredded Carrots during Storage by Treatment with Ozonated Water. *Mater. Sci. Appl. Chem.* **2014**, *30*, 10–17. [CrossRef]
54. Dudareva, N.; Pichersky, E. Metabolic engineering of plant volatiles. *Curr. Opin. Biotechnol.* **2008**, *19*, 181–189. [CrossRef]
55. Zhang, X.; Zhang, Z.; Wang, L.; Zhang, Z.; Li, J.; Zhao, C. Impact of ozone on quality of strawberry during cold storage. *Front. Agric. China* **2011**, *5*, 356–360. [CrossRef]
56. Sivasankar, S.; Sheldrick, B.; Rothstein, S.J. Expression of allene oxide synthase determines defense gene activation in tomato. *Plant Physiol.* **2000**, *122*, 1335–1342. [CrossRef]
57. Modesti, M.; Baccelloni, S.; Brizzolara, S.; Aleandri, M.P.; Bellincontro, A.; Mencarelli, F.; Tonutti, P. Effects of treatments with ozonated water in the vineyard (cv Vermentino) on microbial population and fruit quality parameters. *BIO Web Conf.* **2019**, *13*, 04011. [CrossRef]
58. Mencarelli, F.; Bellincontro, A. Recent advances in postharvest technology of the wine grape to improve the wine aroma. *J. Sci. Food Agric.* **2020**, *100*, 5046–5055. [CrossRef] [PubMed]
59. Velikova, V.B. Isoprene as a tool for plant protection against abiotic stresse. *J. Plant Interact.* **2008**, *3*, 1–15. [CrossRef]
60. Luan, S. Signalling drought in guard cells. *Plant Cell Environ.* **2002**, *25*, 229–237. [CrossRef] [PubMed]
61. Loreto, F.; Dicke, M.; Schnitzler, J.P.; Turlings, T.C.J. Plant volatiles and the environment. *Plant Cell Environ.* **2014**, *37*, 1905–1908. [CrossRef]
62. Hjelmeland, A.K.; Ebeler, S.E. Glycosidically bound volatile aroma compounds in grapes and wine: A review. *Am. J. Enol. Vitic.* **2015**, *66*, 1–11. [CrossRef]
63. Kalua, C.M.; Boss, P.K. Evolution of Volatile Compounds during the Development of Cabernet Sauvignon Grapes (*Vitis vinifera* L.). *J. Agric. Food Chem.* **2009**, *57*, 3818–3830. [CrossRef]
64. Pinto, D.M.; Blande, J.D.; Souza, S.R.; Nerg, A.M.; Holopainen, J.K. Plant volatile organic compounds (vocs) in ozone (o3) polluted atmospheres: The ecological effects. *J. Chem. Ecol.* **2010**, *36*, 22–34. [CrossRef]
65. Sharkey, T.D.; Loreto, F. Water stress, temperature, and light effects on the capacity for isoprene emission and photosynthesis of kudzu leaves. *Oecologia* **1993**, *95*, 328–333. [CrossRef]
66. Calogirou, A.; Larsen, B.R.; Kotzias, D. Gas-phase terpene oxidation products: A review. *Atmos. Environ.* **1999**, *33*, 1423–1439. [CrossRef]
67. Farooq, M.A.; Niazi, A.K.; Akhtar, J.; Saifullah; Farooq, M.; Souri, Z.; Karimi, N.; Rengel, Z. Acquiring control: The evolution of ROS-Induced oxidative stress and redox signaling pathways in plant stress responses. *Plant Physiol. Biochem. PPB* **2019**, *141*, 353–369. [CrossRef]
68. Härtl, K.; Huang, F.C.; Giri, A.P.; Franz-Oberdorf, K.; Frotscher, J.; Shao, Y.; Hoffmann, T.; Schwab, W. Glucosylation of Smoke-Derived Volatiles in Grapevine (*Vitis vinifera*) is Catalyzed by a Promiscuous Resveratrol/Guaiacol Glucosyltransferase. *J. Agric. Food Chem.* **2017**, *65*, 5681–5689. [CrossRef] [PubMed]
69. Verma, V.; Ravindran, P.; Kumar, P.P. Plant hormone-mediated regulation of stress responses. *BMC Plant Biol.* **2016**, *16*, 86. [CrossRef] [PubMed]
70. Behr, M.; Neutelings, G.; El Jaziri, M.; Baucher, M. You Want it Sweeter: How Glycosylation Affects Plant Response to Oxidative Stress. *Front. Plant Sci.* **2020**, *11*, 1443. [CrossRef] [PubMed]
71. Feliziani, E.; Romanazzi, G.; Smilanick, J.L. Application of low concentrations of ozone during the cold storage of table grapes. *Postharvest Biol. Technol.* **2014**, *93*, 38–48. [CrossRef]
72. Duarte-Sierra, A.; Aispuro-Hernández, E.; Vargas-Arispuro, I.; Islas-Osuna, M.A.; González-Aguilar, G.A.; Martínez-Téllez, M.Á. Quality and PR gene expression of table grapes treated with ozone and sulfur dioxide to control fungal decay. *J. Sci. Food Agric.* **2016**, *96*, 2018–2024. [CrossRef]
73. Campayo, A.; Savoi, S.; Romieu, C.; López-Jiménez, A.J.; de la Hoz, K.S.; Salinas, M.R.; Torregrosa, L.; Alonso, G.L. The application of ozonated water rearranges the *Vitis vinifera* L. leaf and berry transcriptomes eliciting defence and antioxidant responses. *Sci. Rep.* **2021**, *11*, 8114. [CrossRef]
74. Cayuela-Sánchez, J.A.; Caballero-Guerrero, B. Fresh extra virgin olive oil, with or without veil. *Trends Food Sci. Technol.* **2019**, *83*, 78–85. [CrossRef]
75. Schraudner, M.; Moeder, W.; Wiese, C.; Van Camp, W.; Inzé, D.; Langebartels, C.; Sandermann, H. Ozone-induced oxidative burst in the ozone biomonitor plant, tobacco Bel W3. *Plant J.* **1998**, *16*, 235–245. [CrossRef]

76. de Pascual-Teresa, S.; Sanchez-Ballesta, M.T. Anthocyanins: From plant to health. *Phytochem. Rev.* **2008**, *7*, 281–299. [CrossRef]
77. Río Segade, S.; Paissoni, M.A.; Giacosa, S.; Bautista-Ortín, A.B.; Gómez-Plaza, E.; Gerbi, V.; Rolle, L. Winegrapes dehydration under ozone-enriched atmosphere: Influence on berry skin phenols release, cell wall composition and mechanical properties. *Food Chem.* **2019**, *271*, 673–684. [CrossRef]
78. Rolle, L.; Torchio, F.; Zeppa, G.; Gerbi, V. Anthocyanin extractability assesment of grape skins by texture analysis. *OENO One* **2008**, *42*, 157–162. [CrossRef]
79. Laureano, J.; Giacosa, S.; Río Segade, S.; Torchio, F.; Cravero, F.; Gerbi, V.; Englezos, V.; Carboni, C.; Cocolin, L.; Rantsiou, K.; et al. Effects of Continuous Exposure to Ozone Gas and Electrolyzed Water on the Skin Hardness of Table and Wine Grape Varieties. *J. Texture Stud.* **2016**, *47*, 40–48. [CrossRef]
80. Fumagalli, I.; Cieslik, S.; De Marco, A.; Proietti, C.; Paoletti, E. Grapevine and Ozone: Uptake and Effects. *Climate* **2019**, *7*, 140. [CrossRef]
81. Campayo, A.; de la Hoz, K.S.; García-Martínez, M.M.; Sánchez-Martínez, J.F.; Salinas, M.R.; Alonso, G.L. Spraying ozonated water on Bobal grapevines: Effect on grape quality. *Food Res. Int.* **2019**, *125*, 108540. [CrossRef] [PubMed]
82. Grimmig, B.; Schubert, R.; Fischer, R.; Hain, R.; Schreier, P.H.; Betz, C.; Langebartels, C.; Ernst, D.; Sandermann, H. Ozone- and ethylene-induced regulation of a grapevine resveratrol synthase promoter in transgenic tobacco. *Acta Physiol. Plant.* **1997**, *19*, 467–474. [CrossRef]
83. Castillo-Muñoz, N.; Gómez-Alonso, S.; García-Romero, E.; Hermosín-Gutiérrez, I. Flavonol profiles of *Vitis vinifera* red grapes and their single-cultivar wines. *J. Agric. Food Chem.* **2007**, *55*, 992–1002. [CrossRef] [PubMed]
84. Perestrelo, R.; Silva, C.L.; Câmara, J.S. Quantification of furanic derivatives in fortified wines by a highly sensitive and ultrafast analytical strategy based on digitally controlled microextraction by packed sorbent combined with ultrahigh pressure liquid chromatography. *J. Chromatogr. A* **2015**, *1381*, 54–63. [CrossRef] [PubMed]
85. Yaseen, T.; Ricelli, A.; Albanese, P.; Nicoletti, I.; Essakhi, S.; D'Onghia, A.M. Influence of postharvest ozone treatment on decay, catalase, lipoxygenase activities, and anthocyanin content of 'Redglobe' table grapes. In Proceedings of the International Conference Ozone and Related Oxidants: Advances in Science and Technology, Dublin, Ireland, 3–5 June 2014; pp. 2–9.
86. Pott, D.M.; Vallarino, J.G.; Osorio, S. Metabolite Changes during Postharvest Storage: Effects on Fruit Quality Traits. *Metabolites* **2020**, *10*, 187. [CrossRef]
87. Pellegrini, E.; Campanella, A.; Paolocci, M.; Trivellini, A.; Gennai, C.; Muganu, M.; Nali, C.; Lorenzini, G. Functional leaf traits and diurnal dynamics of photosynthetic parameters predict the behavior of grapevine varieties towards ozone. *PLoS ONE* **2015**, *10*, e0135056. [CrossRef]
88. Antolini, A.; Forniti, R.; Modesti, M.; Bellincontro, A.; Catelli, C.; Mencarelli, F. First Application of Ozone Postharvest Fumigation to Remove Smoke Taint from Grapes. *Ozone Sci. Eng.* **2021**, *43*, 254–262. [CrossRef]

foods

Article

High Hydrostatic Pressure vs. Thermal Pasteurization: The Effect on the Bioactive Compound Profile of a Citrus Maqui Beverage

Francisco J. Salar [1], Paula M. Periago [2,3], Vicente Agulló [1], Cristina García-Viguera [1,3,*] and Pablo S. Fernández [2,3]

1 Phytochemistry and Healthy Foods Lab (LabFAS), Department of Food Science and Technology (CEBAS-CSIC), University Campus of Espinardo, Edif. 25, 30100 Murcia, Spain; fjsalar@cebas.csic.es (F.J.S.); vagullo@cebas.csic.es (V.A.)
2 Agronomic Engineering Department, Universidad Politécnica de Cartagena (UPCT), Paseo Alfonso XIII 48, 30203 Cartagena, Spain; paula.periago@upct.es (P.M.P.); pablo.fernandez@upct.es (P.S.F.)
3 Associated Unit of Food Quality and Risk Assessment CEBAS-CSIC/UPCT, 30100 Murcia, Spain
* Correspondence: cgviguera@cebas.csic.es; Tel.: +34-968-396-200

Abstract: The effects of high hydrostatic pressure (HHP) compared to thermal pasteurization (TP) were studied in healthy citrus-maqui beverages. The impact of the processing technologies on the microbiological and phytochemical profile was assessed by applying two HHP treatments at 450 and 600 MPa for 180 s and TP at 85 °C for 15 s. The shelf life under refrigeration (4 °C) and room temperature (20 °C) was monitored for 90 days. All treatments ensured microbiological stability at both storage temperatures. Aside from that, the physicochemical parameters were not significantly different after processing or throughout the storage period. Regarding color parameters, an increase in the reddish coloration was observed during storage for those beverages treated by HHP. In general, phenolic compounds were little affected by the processing technique, even when treatment under HHP was more stable than by TP during storage. On the other hand, vitamin C showed great degradation after processing under any condition. It can be concluded that HHP is an effective alternative to thermal treatments, achieving effective microbial inactivation and extending the shelf life of the juices by contributing to a better preservation of color and bioactive compounds.

Keywords: high hydrostatic pressure; thermal pasteurization; anthocyanin; vitamin C; flavanones

Citation: Salar, F.J.; Periago, P.M.; Agulló, V.; García-Viguera, C.; Fernández, P.S. High Hydrostatic Pressure vs. Thermal Pasteurization: The Effect on the Bioactive Compound Profile of a Citrus Maqui Beverage. *Foods* **2021**, *10*, 2416. https://doi.org/10.3390/foods10102416

Academic Editors: Marina Cano Lamadrid and Francisco Artés-Hernández

Received: 13 September 2021
Accepted: 9 October 2021
Published: 12 October 2021

Publisher's Note: MDPI stays neutral with regard to jurisdictional claims in published maps and institutional affiliations.

1. Introduction

Nowadays, there is a growing trend focused on the implementation of healthy and balanced diets by consumers, minimizing the consumption of processed foods. In this respect, recent studies revealed that over 70% of consumers associate emerging non-thermal technologies with better nutrient or sensory quality [1], since these novel techniques are capable of maintaining a very similar freshness for unprocessed foods, minimizing losses in the physicochemical and organoleptic profiles and ensuring microbial safety during subsequent storage, as well as being environmentally friendly [2,3]. Moreover, consumers are increasingly aware of including fruit- and vegetable-based products in their diets due to their nutritional importance and potential health-promoting profiles, which can be useful to support the immune system, prevent chronic diseases and micronutrient deficiencies, or counteract or inhibit the progression of degenerative diseases caused by pro-oxidant agents [4]. Consequently, there is a marked increase in the demand for high-quality foods, with attractive organoleptic and nutritional characteristics similar to fresh equivalents, ensuring maximum food safety in addition to being minimally processed and without using preservatives, additives, or colorings [5,6].

This study follows previous studies focused on the design of healthy plant-based soft drinks, which showed that thermally treated beverages based on maqui and citrus fruits are an excellent source of bioavailable bioactive compounds [7–9], mainly vitamins and phenolic compounds with wide biological activity [10,11]. This is based on the health-protective properties of berries, namely maqui berry (*Aristotelia chilensis* (Mol.) Stunz), a native plant from Chile and Argentina particularly studied as a potential healthy food ingredient [12,13]. The functional capacity of this berry has been related to its high anthocyanin concentration, which is partially responsible for its antioxidant and anti-inflammatory capacity [14], in addition to cardio or neuroprotective effects [15] as well as preventing chronic degenerative diseases such as obesity [16] or cancer [17,18]. That aside, its contribution to postprandial hyperglycemia decreasing has been pointed out, with inhibitory effects regarding enzymatic activity involved in metabolic syndromes [19] and those effects related to aging and antioxidant activity [20].

On the other hand, citrus fruits are a source of nutraceutical and bioactive phytochemicals [21] such as flavonoids, particularly flavanones, and phenolic acids, in addition to vitamins, carotenoids, minerals, dietary fiber, and essential fatty acids that provide antimicrobial [22], anti-inflamatory [23,24], and antitumor [25] functions, as well as biological activity against obesity [26], diabetes [27], and neuroprotective [28] and cardioprotective effects [29].

Concerning the different beverage processing methods, conventional thermal methods have been capable of ensuring food safety and achieving a high degree of enzymatic inactivation in the fruit- and vegetable-based beverage industry. However, thermal processing (TP) tends to degrade the overall quality in juices, such as the sensory and phytochemical profiles [30], affecting the functional capacity of the drinks. This general loss of quality is directly related to consumer acceptance, resulting in recent years in a substantial stagnation and decrease of processed beverages with thermal treatments in comparison with minimally or non-thermally processed beverages, which are rapidly expanding worldwide [31,32]. Currently, the primary new non-thermal technologies that are on the rise are high hydrostatic pressure (HHP), pulsed electric fields (PEF), ultrasound (US), ultraviolet light (UV), and cold plasma, among others [33]. In this sense, HHP has achieved the greatest success [34], with a continuously expanding market due to its capacity to reduce the microbial load of both pathogens and disruptive microorganisms in multiple juice matrices, consequently resulting in an extension of the storage period [35] while preserving the organoleptic, nutritional, and phytochemical characteristics of the juices [36,37]. Due to the intrinsic characteristic of this technology, at moderate temperatures, high pressures do not affect the covalent bonds during the pressurization process [38,39], with food biomolecules minimally affected in vegetable- and fruit-based drinks [40,41]. What is more, an increase in the levels of nutrients and bioactive compounds has been reported in HHP-treated juices, as this technology would enhance the extractability of these substances [42]. In addition, HHP at pressures usually >400 MPa would also assist to inhibit, totally or in most part, the endogenous enzyme activities of cases only partially, which are responsible for undesired changes in the global quality of the juices in storage [43]. On this basis, fruit or vegetable HHP beverages generally give rise to a better overall quality than samples treated with conventional heat treatments [44,45].

Given these antecedents, the purpose of the present study was the optimization of the processing design for healthy maqui-citrus soft drinks using HHP in order to minimize the loss of bioactive compounds and, consequently, increase the healthy properties previously reported. Thereby, a comparison between HHP (at 450 and 600 MPa) and TP (under standard industrial processing conditions) for their quality parameters (physicochemical and microbial profiles) and bioactive compound contents (flavonoids and vitamin C) over 90 days of storage at 4 °C and 20 °C was set. Moreover, to our knowledge, this is the first study comparing the effect of high hydrostatic pressure relative to TP on the levels of functional compounds and shelf life in juices containing a mixture of bioactive compounds from berry and citrus fruits.

2. Materials and Methods

2.1. Chemicals and Reagents

Hesperidin was obtained from Merck (Darmstadt, Germany), and cyanidin 3-*O*-glucoside was purchased from TransMIT (Geiben, Germany). Acetonitrile, formic acid, methanol, and ethylenediaminetetraacetic acid disodium salt 2-hydrate (EDTA), were obtained from Panreac (Barcelona, Spain). Buffered peptone water, Plate Count Agar (PCA), Rose Bengal Agar (RBA), Brilliant Green Bile 2% Broth (BGBB) and Man Rogosa Sharpe Agar (MRSA) were purchased from Scharlab (Barcelona, Spain). L-ascorbic acid (AA) and dehydroascorbic acid (DHAA) were acquired from Acros Oganics (Thermo Fisher Scientific Inc., Madrid, Spain) and Sigma-Aldrich (St. Louis, MO, USA), respectively. All solutions were prepared with ultrapure water from a Milli-Q Advantage A10 ultrapure water purification system (Millipore, Burlington, MA, USA).

2.2. Ingredients

Fresh dry maqui powder organic was provided by Maqui New Life S.A. (Santiago, Chile), citrus juices by Cítricos de Murcia S.L. (Ceutí, Murcia, Spain) and AMC Grupo Alimentación S.A. (Espinardo, Murcia, Spain) and sucrose by AB Azucarera Iberia S.L. (Madrid, Spain).

2.3. Beverage Preparation

Juice preparation was conducted according to previous studies of the group [7]. Briefly, maqui powder was mixed with citrus juices and sugar to obtain the base beverage.

2.4. HHP Processing and Thermal Pasteurization

The juices were poured into PET clear bottles (80 mm, 27.5 mm Ø; vol. 30 mL) with plastic screw caps (Sunbox, Barcelona, Spain) prior to HHP processing and subsequently in the case of TP. The beverages were pressurized using a commercial Hiperbaric 135 for HHP (Hiperbaric, Burgos, Spain). This equipment has a capacity of 135 L, which applies a maximum treatment of 6000 bares (600 MPa). The samples underwent two HHP treatments at 450 and 600 MPa for 3 min at 20 °C. The times to reach 450 MPa or 600 MPa were 120 and 165 s, respectively. The initial water temperature in the vessel ranged from 8–10 °C, and the rate of temperature rise during compression was 3 °C/100 MPa, while the decompression of the beverages was almost instantaneous (less than 5 s). Selection of processing parameters was mainly based on HHP common industrial processing conditions for ensuring microbial safety in juices with low pH levels.

The thermal treatment was carried out with a Mastia thermoresistometer [46]. The initial temperature was 20 °C. The thermoresistometer was programmed with a ramp at a heating rate of 40 °C/min to reach the target temperature of 85 °C. The temperature was held at 85 °C for 15 s, according to common industrial standards for similar beverage compositions and pH levels. Then, the product was cooled down at a speed of 40 °C/min, reaching a temperature of 20 °C, and stored in the same PET clear bottles as mentioned before. It is important to note that equivalent HHP processing and thermal conditions were selected in terms of industrial microbial safety for an equal comparison of both preservation technologies.

2.5. Sampling

All treated beverages were stored immediately after processing at 4 and 20 °C in darkness for 90 days. The samples were labeled according to the codification specified in Table 1, where all the test conditions are included. All juices and experimental conditions tested were prepared in triplicate ($n = 3$), and all analytical determinations were performed in triplicate ($n = 3$). The samples were analyzed at 0, 7, 15, 30, 45, 60, and 90 days. Moreover, untreated samples were stored and analyzed as controls against the treated ones for day 0.

Table 1. Codification of samples included in the experimental design.

Code	Beverage and Storage Conditions
Control	Untreated sample
HHP—450 MPa 4	Beverage subjected to high hydrostatic pressure (450 MPa) stored at 4 °C under darkness conditions
HHP— 450 MPa 20	Beverage subjected to high hydrostatic pressure (450 MPa) stored at 20 °C under darkness conditions
HHP— 600 MPa 4	Beverage subjected to high hydrostatic pressure (600 MPa) stored at 4 °C under darkness conditions
HHP— 600 MPa 20	Beverage subjected to high hydrostatic pressure (600 MPa) stored at 20 °C under darkness conditions
TP—85 °C 4	Beverage subjected to thermal pasteurization stored at 4 °C under darkness conditions
TP—85 °C 20	Beverage subjected to thermal pasteurization stored at 20 °C under darkness conditions

2.6. pH, Titratable Acidity, and Total Soluble Solids

The pH values were measured using a GLP 21 pH meter (Crison Ltd., Barcelona, Spain). The TA was determined using a 794 Basic Titrino (Metrohm) by titrating 2 g of juice (up to 30 g of Milli-Q water) with 0.1 mol/L NaOH to an end point of pH 8.1. The results were expressed as grams of citric acid per 100 mL of the sample (g CA/100 mL). The TSS contents of the samples were recorded in a refractometer (Abbe WYA-S, Optic Ivymen® System; Biotech SL, Barcelona, Spain) at 20 °C, with values being expressed as °Brix [7].

2.7. Microbiology Analysis

The samples were aseptically diluted in buffered peptone water and then analyzed for aerobic mesophilic bacteria, aerobic psychrophilic bacteria, molds and yeasts, Enterobacteriaceae, and Lactic Acid Bacteria (LAB). They were analyzed at day 0 and after 7, 15, 30, 45, 60, and 90 days of storage at 4 °C and 20 °C. Microbial analyses were carried out in order to measure the short- and long-term effectiveness of HHP processing and TP at day 0 just after processing and throughout 90 days of shelf life. The quantifications of the molds and yeasts were determined by plating the samples in Rose Bengal Agar (RBA), followed by incubation for 5 days at 25 °C. The counts for aerobic mesophilic and psychrophilic bacteria were performed using the Plate Count Agar (PCA) medium, incubated for 48 h at 30 °C and for 10 days at 5 °C, respectively. Enterobacteriaceae were determined using the more probable number method with Brilliant Green Bile 2% Broth (BGBB) after incubation for 24–48 h at 37 °C. Lactic Acid Bacteria (LAB) were determined using Man Rogosa Sharpe Agar (MRSA) after incubation in anaerobic conditions for 5 days at 37 °C. Microbial counts were expressed as colony-forming units per milliliter (CFU/mL).

2.8. Qualitative and Quantitative Analysis of Phenolic Compounds by RP-HPLC-DAD

Juice samples were processed following the method previously described [7]. The identification and quantification of phenolic compounds was carried out by applying the method previously reported [7,8]. The diverse phenolic compounds in the samples were identified by comparison with authentic standard compounds of analytical grade. Flavanones were quantified as hesperidin at 280 nm and anthocyanins as cyanidin 3-*O*-glucoside at 520 nm. The concentration of phenolic compounds was expressed as mg per 100 mL of juice.

2.9. Extraction and Analysis of Vitamin C

The content of vitamin C was found by applying the UHPLC-ESI-QqQ-MS/MS-based method recently developed [47] and calculated by comparison with freshly prepared ascorbic acid (AA) and dehydroascorbic acid (DHAA) authentic standard curves. The results were expressed as mg per 100 mL of juice.

2.10. Color Measurements

The color was determined using a Konica Minolta CM-5 Chroma Meter (Osaka, Japan). The results were expressed in accordance with the CIEL*a*b* system with reference to a visual angle of 10° and a light source set on D65. Three measurements of each sample were performed, and the values were averaged. The color parameters determined were the luminosity (CIEL*), redness (a*), and yellowness (b*) using the CM-5 spectrophotometer in reflection mode. The *Hue* angle (H), *Chroma* (C) and total color differences (ΔE) were calculated from $\tan^{-1}(b^*/a^*)$, $(a^{*2} + b^{*2})^{1/2}$, and $(da^{*2} + db^{*2} + dL^{*2})^{1/2}$, respectively [7].

2.11. Statistical Analyses

The results are presented as the means ± SD ($n = 3$). A paired *t*-test was developed to compare two parameters, and analysis of variance (ANOVA) and Tukey's multiple range tests were carried out to compare three or more conditions. All statistical analyses were performed using SPSS 19.0 software (LEAD Technologies, Inc., Chicago, IL, USA). The level of statistical significance was established at $p < 0.05$.

3. Results and Discussion

3.1. Initial Impact of Proccesing on the Overall Quality Parameters in Juices

In general, neither HHP nor TP presented differences in their pH, acidity, or °Brix values, which remained stable immediately after processing ($p > 0.05$) in all treated drinks (Table 2). Hence, the preservation treatment was not a relevant parameter concerning physicochemical characteristics. These results agree with those previously obtained by Chung et al. and Nayak et al. in fruit juices [48,49].

Table 2. Physicochemical parameters and antioxidant biomolecules of non-processed, pressurized, and pasteurized citrus-maqui beverages at zero time of storage.

Condition Z	Physicochemical Parameters				Bioactive Compounds (mg/100 mL)		
	Color (ΔE)	TSS (°Brix)	pH	TA (g CA/100 mL)	Anthocyanins	Flavanones	Vitamin C
Non-processed	0.00c Y	13.60 a	3.41 b	2.95 a	16.54 b	14.99 b	10.90 a
HHP—450 MPa	1.39bc	13.60 a	3.45 a	2.91 a	15.98 c	15.11 b	6.03 b
HHP—600 MPa	1.82b	13.60 a	3.40 b	2.95 a	16.1 c	15.14 b	7.49 b
TP—85 °C	3.59a	13.60 a	3.41 b	2.95 a	18.37 a	23.33 a	7.79 b
LSD ($p < 0.05$)	1.02	<0.01	0.03	<0.01	0.31	0.48	1.70
p-value	*** X	N.s.	*	N.s.	***	***	**

Z HHP— 450 MPa, processing at 450 MPa/3 min; HHP— 600 MPa, processing at 600 MPa/3 min; TP— 85 °C, thermal pasteurization at 85 °C/15 s. Y Data (means) within each column, with values followed by different letters for each processing condition, are significantly different at $p < 0.05$ according to the analysis of variance (ANOVA) and Tukey's multiple range test. X Significant at * ($p < 0.05$), ** ($p < 0.01$), and *** ($p < 0.001$) according to a paired *t*-test. N.s. = not significant differences.

On the other hand, the samples treated under both conditions exhibited significant color differences (ΔE) in comparison with the untreated beverage ($p < 0.05$) in the aftermath of processing (Table 2). These changes were more prone to thermal treatment, mainly due to a decrease in the L value (lightness) when high temperatures were applied, in concordance with some authors [50], while HHP processing caused minor variations in the drinks' colors, as previously reported by others [51], although according to Orellana-Palma et al., these differences were not detectable with the naked eye for values of ΔE lower than four units [52].

Regarding flavonoids (anthocyanins and flavanones), slight significant changes ($p < 0.05$) were noticed after all treatments (Table 2). These results indicated that the initial total content of anthocyanins slightly decreased after both HHP conditions for processing, while this content increased up to 11% after TP. These results differ from other authors [53,54], who have reported higher recovery of anthocyanins for processing treatments under pressure in comparison with heat treatments. Moreover, concerning the total content of flavanones, there were significant differences ($p < 0.05$) in the pasteurized juices just after processing related to HHP and the unprocessed beverages. In this sense, TP managed to enhance the content and, consequently, bioaccessibility of these bioactive molecules up to 55% more, while both HHP treatments were capable of maintaining the initial flavanone content at a rather stable value compared with the untreated juice. This increase in flavonoids after thermal treatment may have been associated with the increased extractability of them as they were released when applying moderate temperatures during thermal treatment due to vegetable cell wall disruption, as previously reported [55,56]. It should be noted that the results of this study were in concordance with those found by He et al., who after thermal processing increased the total phenolic content up to 39% in orange juice, mainly due to the individual contribution of hesperetin-rutinoside, which increased its initial value by 27% post-pasteurization [57].

Meanwhile, changes regarding other bioactive compounds such as vitamin C were noticeable between the untreated and treated samples (Table 2). Either pressurized or pasteurized beverages showed lower total vitamin C contents ($p < 0.05$) compared with the unprocessed citrus-maqui drink by 35% on average after all the processes (Table 2). These findings agree with Spira et al., who also reported similar results for HHP and pasteurization treatment in orange juice [58]. In the same context, Andrés et al. also indicated a significant decrease in ascorbic acid immediately after processing for all preservation treatments, with the specimen under HHP treatment (7% loss on average) being lower than that under TP treatment (12% loss on average) in a fruit-based beverage [59].

On the other hand, concerning the microbiological aspects, the untreated samples were characterized by an initial concentration of aerobic mesophilic bacteria of 80 ± 5 CFU/mL (Table 3), lower than the detection limits for molds and yeasts (<100 CFU/mL), and for aerobic psychrophilic bacteria, Enterobacteriaceae and lactic acid bacteria concentrations were less than 10 CFU/mL. Nevertheless, either HHP or TP processing managed to reduce these values below the detection limits immediately after applying treatments for aerobic mesophilic bacteria. Similar results have been previously reported by other authors, who achieved, just after processing by HHP or TP, microbial counts below the detection limits in plant-based beverages [60–62].

Table 3. Microbiological quality of non-processed, pressurized, and pasteurized citrus-maqui beverages at zero time of storage.

Condition [Z]	Microbiological Count (CFU/mL)				
	Aerobic Mesophilic Bacteria	**Aerobic Psycrophilic Bacteria**	**Molds and Yeast**	**Enterobacteriae**	**Lactic Acid Bacteria**
Non-processed	80 ± 5	<10 [Y]	<100 [Y]	<10 [Y]	<10 [Y]
HHP—450 MPa	<10 [Y]	<10 [Y]	<100 [Y]	<10 [Y]	<10 [Y]
HHP—600 MPa	<10 [Y]	<10 [Y]	<100 [Y]	<10 [Y]	<10 [Y]
TP—85°C	<10 [Y]	<10 [Y]	<100 [Y]	<10 [Y]	<10 [Y]

[Z] HHP— 450 MPa, high hydrostatic pressure at 450 MPa/3 min; HHP— 600 MPa, high hydrostatic pressure at 600 MPa/3 min; TP— 85°C, thermal pasteurization at 85 °C/15 s. [Y] Values below the detection limit for aerobic mesophilic and psychrophilic bacteria, Enterobacteriaceae and lactic acid bacteria (<10 CFU/mL), and molds and yeasts (<100 CFU/mL).

3.2. *Effect of HHP and TP Treatments on pH, Tritrable Acidity (TA), and Total Soluble Solids (°Brix) during Storage*

According to the data related to the physicochemical parameters obtained in the present work, the total soluble solids (°Brix) values were similar in all treated beverages ($p > 0.05$) (Table 4), with only a slight increase ranging from 13.60 to 14.20 at the end of the storage time. Concerning the pH and total titratable acidity, there were no significant differences between both HHP treatments and TP, remaining rather stable during the monitored storage period. These results are consistent with other authors that have also reported no differences regarding these physicochemical parameters over the storage period in similar juice matrices after processing by HHP or thermal treatment [63,64].

Table 4. pH, titratable acidity (TA), and total soluble solids (TSS) measured at day 0 (initial) and after 90 days of storage (final) for beverages subjected to HHP and TP and stored under two different conditions.

Condition [Z]	TSS (°Brix)			pH			TA (g CA/100 mL)		
	Initial	Final	*p*-Value	Initial	Final	*p*-Value	Initial	Final	*p*-Value
HHP—450 MPa 4	13.60 a [Y]	14.20 a	*** [X]	3.45 a	3.41 a	*	2.91 a	3.00 a	*
HHP—600 MPa 4	13.60 a	14.20 a	***	3.40 b	3.41 a	N.s.	2.95 a	3.00 a	N.s.
TP—85°C 4	13.60 a	14.20 a	***	3.41 b	3.41 a	N.s.	2.95 a	2.98 a	N.s.
LSD (p <0.05)	<0.01	<0.01		0.03	<0.01		<0.01	<0.01	
p-value	N.s.	N.s.		*	N.s.		N.s.	N.s.	
HHP—450 MPa 20	13.60 a	14.20 a	***	3.45 a	3.40 a	*	2.91 a	2.98 a	*
HHP —450 MPa 20	13.60 a	14.20 a	***	3.40 b	3.40 a	N.s.	2.95 a	3.00 a	*
TP— 85°C 20	13.60 a	14.20 a	***	3.41 b	3.41 a	N.s.	2.95 a	2.99 a	*
LSD (p <0.05)	<0.01	<0.01		0.03	<0.01		<0.01	<0.01	
p-value	N.s.	N.s.		*	N.s.		N.s.	N.s.	

[Z] HHP—450MPa 4, high hydrostatic pressure at 450 MPa/4 °C; HHP—600MPa 4, high hydrostatic pressure at 600 Pa/4°C; TP—85°C 4, thermal pasteurization at 85 °C/4 °C; HHP—450 MPa 20, high hydrostatic pressure at 450 MPa/20°C; HHP—600 MPa 20, high hydrostatic pressure at 600 MPa/20 °C; TP—85°C 20, thermal pasteurization at 85 °C/20 °C,. Initial and final values were significantly different according to a paired t-test at * ($p < 0.05$), ** ($p < 0.01$), and *** ($p < 0.001$). [Y] For the data (means) within each column, values followed by different letters for each processing condition are significantly different at $p < 0.05$ according to the analysis of variance (ANOVA) and Tukey's multiple range test. [X] Significant at * ($p < 0.05$) and *** ($p < 0.001$) according to a paired t-test. N.s. = not significant differences.

3.3. Changes in the Microbiological Profiles during Storage

The microbiological profiles of the citrus-maqui juice samples treated by HHP (450 and 600 MPa/3 min) and TP (85 °C/15 s) were analyzed during storage for 90 days at both 4 and 20 °C. In the current study for all the processed juices, no signs of microbial growth were observed throughout the 90 days of storage, with microbial counts remaining below the detection limit for all microorganisms analyzed, indicating that the citrus-maqui blends were microbiologically safe and stable at both refrigerated and room temperatures. Previous studies of similar technologies have also reported efficacy in preserving microbiological safety over the storage period in vegetable- and fruit-based juices. In this respect, Chen et al. reported that both HHP and thermal treatment ensured microbiological stability during 90 days of storage at 4 °C in pomegranate juice [65]. Moreover, Hsu et al. also pointed out that HHP and TP managed to achieve a microbiologically stable product during 28 days of refrigerated storage in tomato juice [66]. Finally, Bull et al. and Parish indicated that orange juice subjected to HHP or TP was able to maintain its microbiological quality for a shelf life of up to 4 months in chilled storage [67,68].

3.4. Effect of HHP and TP Treatments on Vitamin C during Storage

The content of vitamin C was mainly due to citrus plants, which are natural sources of this antioxidant, as maqui powder is not a relevant source. It was quantified by UHPLC-ESI-QqQ-MS/MS and reported by Salar et al. [7]. The initial concentration of vitamin C (calculated as the sum of AA and DHAA) of all the treated drinks was not significantly different (7 mg per 100 mL on average). Nevertheless, when monitoring the concentration of vitamin C over 90 days of shelf life, independent from the treatment, a significant rapid decrease ($p < 0.05$) of this compound was found (Figure 1A,B). However, the degradation rate was noticeably higher for the HHP-treated samples than for the thermally treated beverages. Moreover, the higher losses occurred mainly during the first 7 days of storage, exhibiting losses of 45% and 85% on average at 4 and 20 °C, respectively, for both HHP treatments. On the other hand, 30% loss and 60% loss were found at 4 and 20 °C, respectively, for TP during the same period. Finally, the beverages presented a total degradation of vitamin C after 30 and 15 days for the HHP-treated drinks and after 45 and 60 days of storage for thermal treatment at 4 and 20 °C, respectively. Therefore, the storage temperature was identified as a critical factor affecting vitamin C breakdown, which is in agreement with previous authors [69,70]. In addition, another factor that may contribute to the loss of vitamin C is the mutual degradation between vitamin C and anthocyanins [71,72], as discussed below.

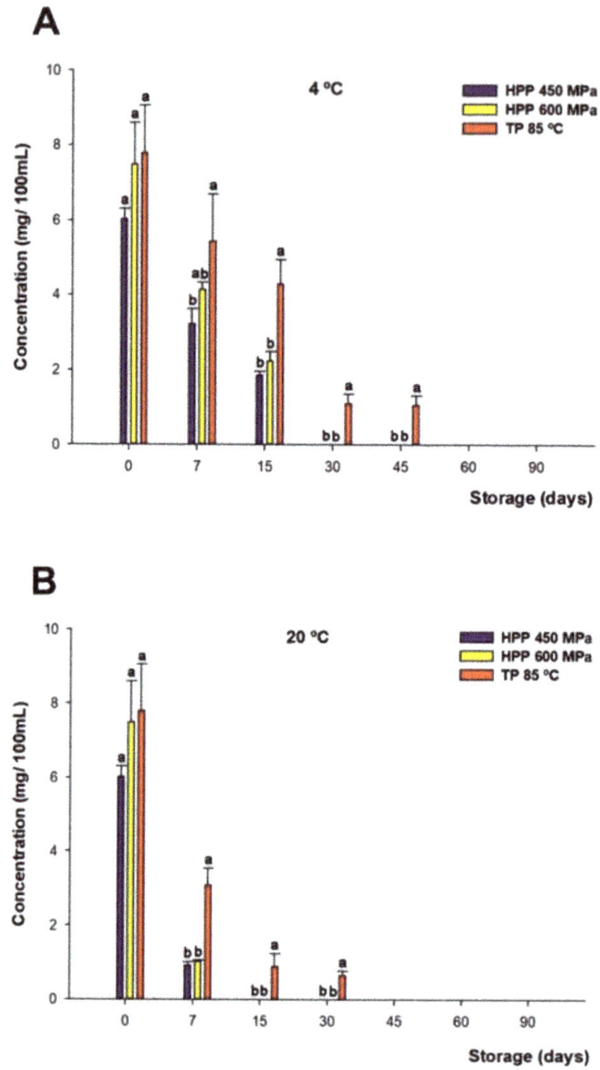

Figure 1. Changes in content of vitamin C (mg/100 mL) for juices subjected to high hydrostatic pressure (HHP—450 MPa and HHP—600 MPa) and thermal pasteurization (TP—85 °C), measured during storage for 90 days under refrigerated conditions at 4 °C (**A**) and 20 °C (**B**) under darkness conditions. Bars with different lowercase letters within each time point were different statistically at $p < 0.05$ according to the analysis of variance (ANOVA) and Tukey's multiple range test.

Regarding individual treatments, there were significant differences ($p < 0.05$) between the HHP-treated and thermally treated juices, since beverage processing by TP managed to extend a portion of the content of vitamin C up to 30 days more than juices subjected to HHP for both storage conditions during their shelf lives. Among other factors, these results may be due to the degradation of the active forms of vitamin C (AA and DHA) by enzymes with oxidase activity [73,74], such as peroxidase (POD), polyphenol oxidase

(PPO), and particularly ascorbate oxidase (AO), which catalyze ascorbic acid oxidation, playing a major role in oxidizing ascorbic acid to dehydroascorbic acid (which is likewise rapidly oxidized to diketogulonic acid, the inactive form of vitamin C) in the early stages of storage in processed foods [75,76]. In this regard, HHP would not degrade some of these enzymes in citrus juices [77], unlike TP which, for temperatures over 80 °C, is capable of achieving a greater degree of inactivation of the oxidative enzymes [78,79]. Nevertheless, these results differ from other authors in the literature [80,81], who previously reported better conservation of vitamin C in HHP-treated juices than in thermally treated ones in fruit- and vegetable-based juices.

It is noteworthy that no differences were found between both HPP treatments.

3.5. *Effect of HHP and TP Treatments on Phenolic Composition during Storage*

3.5.1. Flavanones

Regarding the flavanones of the citrus-maqui beverage, they were provided by citrus juices, being the most abundant eriocitrin (eriodyctiol 7-*O*-rutinoside), narirutin (naringenin 7-*O*-rutinoside), and hesperidin (hesperetin 7-*O*-rutinoside), characterized in preliminary studies by Salar et al. [7]. The total content of flavanones at the beginning of the beverage's shelf life was 15.05 mg/100 mL on average for both HHP treatments and 23.33 mg/100 mL for TP.

Related to the variations in the content of total flavanones over the storage period, significant differences ($p < 0.05$) were found between both treatments submitted to HHP compared with TP (Figure 2A,B). In this frame, the HHP-processed juices managed to remain completely constant in their initial concentrations of flavanones over 90 days of storage at 4 °C, with a negligible final loss by 3% on average. Similar protective effects on the profile of flavanones in citrus juices processed by HHP in cold storage have been reported by Sanchez-Moreno et al. [82]. On the contrary, Plaza et al. pointed out important losses of 50% in the flavanone content just after 20 days of refrigerated storage in orange juice [83]. Aside from that, drinks processed by TP underwent losses of 14.3% just after 7 days, with a progressive degradation of the original content of flavanones during their shelf lives, reaching losses up to 30% at the end of refrigerated storage according previous descriptions in the literature [84].

On the other hand, the samples stored at room temperature followed a similar trend under HHP heat treatment with respect to cold storage, even if a final 6% loss was found for both HHP treatments on average, and for TP (35%) at the end of storage. This indicates that even if a higher concentration of these compounds is reached after treatment, phenolic compounds are more stable after both HHP treatments. Again, there were no significant differences between both HPP conditions.

Moreover, considering the contribution of individual flavanones to the total concentration of the phenolic compounds, neither *O*-tryglycosil-naringenin nor eriocitrin and narirutin displayed significant losses over 90 days of storage (data not shown), whereas hesperidin was quite affected under all processing treatments, as previously reported [8,85]. Overall, the results obtained in this study showed clearly that the total content of flavanones in the citrus-maqui juices was closely dependent on the processing treatment and subsequent storage conditions.

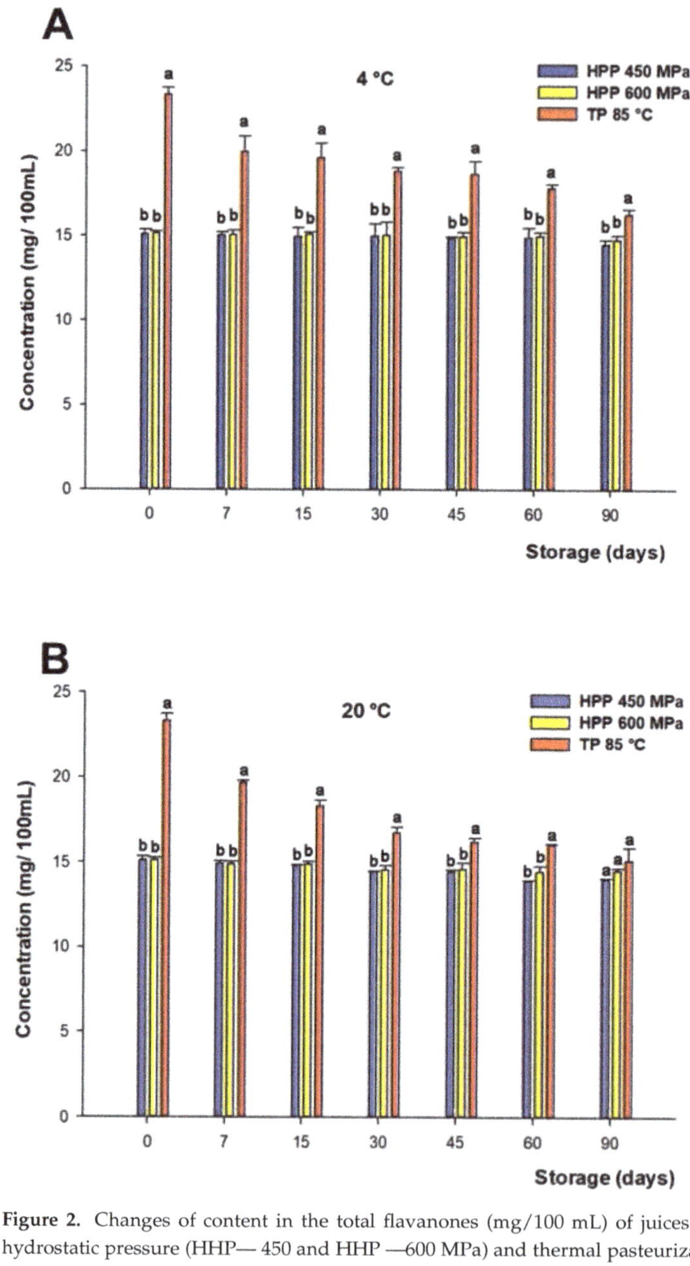

Figure 2. Changes of content in the total flavanones (mg/100 mL) of juices subjected to high hydrostatic pressure (HHP— 450 and HHP —600 MPa) and thermal pasteurization (TP— 85 °C), measured during storage for 90 days in refrigerated conditions at 4 °C (**A**) and 20 °C (**B**) under darkness conditions. Bars with different lowercase letters within each time point were different statistically at $p < 0.05$ according with the analysis of variance (ANOVA) and Tukey's multiple range test.

3.5.2. Anthocyanins

The range of anthocyanins present in the citrus-maqui beverages was due to *A. Chilensis* [8], widely described and characterized in previous studies by Salar et al. [7]. In this connection, the initial content of the total anthocyanins of the juices recorded at the beginning of the storage period was 16.04 mg/100 mL on average for both HHP treatments and 18.37 mg/100 mL for TP. In the present study, the concentration of total anthocyanins remained constant during the first 15 days of storage at 4 °C for all processing treatments. These results are in contrast to some studies on berry juices, which reported higher losses for anthocyanins (over 30–60%) only 9 days after processing for samples submitted to HHP and storage at refrigerated temperatures [86,87].

After the first fortnight of the storage period, the rate of anthocyanin degradation was gradual for all treatments, decreasing with time and without significant differences between both HHP processing treatments, reaching 30% loss on average at the end of the beverage's shelf life regardless of the treatment under refrigerated storage at 4 °C (Figure 3A). These findings differ from other previous studies with juices containing anthocyanins, which have reported a higher final concentration of these phenolic compounds in HHP-treated drinks versus thermally treated ones during refrigerated storage [88]. On the other hand, the loss of anthocyanins in those samples stored at 20 °C (Figure 3B) displayed a higher degradation in anthocyanin content against those stored at 4 °C. This degradation was gradual, even if slower for TP during the first 15 days, reaching a 71% loss on average after 90 days for both the HHP and TP samples. Therefore, the degradation rate of the total anthocyanins was significantly accelerated with increasing storage temperatures. In this regard, some authors have described similar results in previous studies that reported a negative relationship between the storage temperature and the degradation of these colored flavonoids [89–92].

Moreover, anthocyanin stability in plant-based products depends on the interaction of various factors such as temperature, light, pH, presence of oxygen, metal ions, and solvents, among others [93,94]. In this sense, the mechanism involved in condensation reactions between ascorbic acid and anthocyanins [95] or the reaction of anthocyanins with free radicals generated by the degradation of ascorbic acid [71,72,96] could contribute to anthocyanin breakdown. It has also been reported by Castañeda-Ovando et al. that anthocyanin degradation takes place through the reactions of oxidation and condensation with other phenolic compounds, generating colorless compounds [97], even though the residual enzymatic activity of endogenous enzymes such as polyphenol oxidase (PPO), β-glucosidase(β-GLC), or peroxidase (POD) present in vegetable- and fruit-based beverages during storage could give rise to anthocyanins pigment degradation as well [98–100].

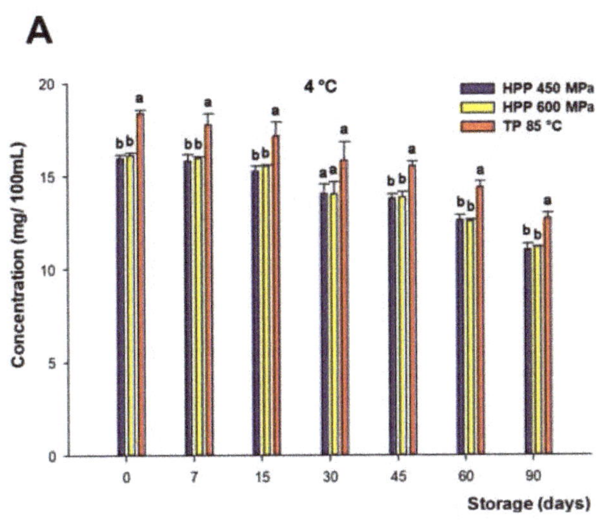

Figure 3. *Cont.*

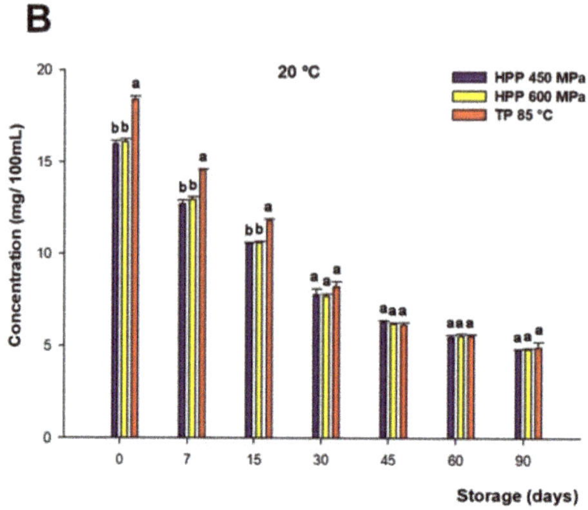

Figure 3. Changes in content of total anthocyanins (mg/100 mL) of juices subjected to high hydrostatic pressure (HHP—450 and HHP—600 MPa) and thermal pasteurization (TP—85 °C), measured during storage for 90 days in refrigerated conditions at 4 °C (**A**) and 20 °C (**B**) under darkness conditions. Bars with different lowercase letters within each time point were different statistically at $p < 0.05$ according to the analysis of variance (ANOVA) and Tukey's multiple range test.

3.6. Color Changes of Juices during Storage

The color changes during 90 days of shelf life at 4 and 20 °C were determined by measuring the CIE$L^*a^*b^*$ color parameters. The reddish coloration of the new citrus-maqui blend was due to the total content of anthocyanins, which contribute to providing a more attractive appearance in the beverages for consumers.

Regarding the lightness (CIEL^*) value, statistically significant differences were found between individual treatments ($p < 0.05$) throughout the storage period (Tables 5 and 6). Overall, in the present study, the CIEL^* value tended to increase independently of the preservation treatments and the storage temperatures. However, this increase was more evident for those beverages processed with HHP technology, being more pronounced for those processed at 600 MPa than at 450 MPa and stored at 20 °C. This is in accordance with previous studies but in contrast to other studies of juices rich in anthocyanins, which rose to create darker beverages during storage after processing by HHP [101–103] and which could most likely be attributed to distinct processing parameters and varied food arrays. On the other hand, the unequal rise of the CIEL^* values among individual treatments could be due to the varying and usually low rate of enzymatic inactivation achieved by this emerging preservation technology in the processing of fruit- and vegetable-based beverages [44,104]. In this regard, the remaining residual enzyme activity of enzymes such as pectinases (pectin methyl esterase, polygalacturonase, and pectate lyase) in citrus-based juices can lead to destabilization of the cloud as a consequence of the sediment of cloud particles in the form of calcium pectate complexes [105,106], which may involve higher CIEL^* values and a higher degree of clarification of citrus-based beverage processing by HHP during storage [67]. In this frame, some authors have described HHP processing at room temperature in vegetable- and fruit-based beverages as having limited effectiveness toward the inactivation of the main plant-based endogenous enzymes [77,107].

Table 5. Stability of CIE$L^*a^*b^*$ values in beverages stored at 4 °C.

Parameter	Storage (Days)	HHP-450 MPa	HHP-600 MPa	TP-85 °C	LSD ($p < 0.001$)
[W] CIEL^*	0	23.39 aB [Z]	22.55 aAB	21.38 deA	1.04
	7	23.53 aB	24.74 bC	19.57 abA	0.80
	15	24.64 bB	33.54 cC	19.72 aA	0.51
	30	30.96 cB	38.13 dC	20.54 bcA	0.49
	45	33.58 dB	39.08 eC	20.31 abA	0.13
	60	33.49 dB	40.31 fC	21.23 cA	0.41
	90	34.11 dB	41.12 gC	21.97 eA	0.63
	LSD ($p < 0.001$)	0.69	0.41	0.46	
CIEa^*	0	47.87 aB	47.48 aAB	46.78 bA	0.86
	7	47.86 aB	48.93 bC	44.80 aA	0.79
	15	48.40 aB	56.53 cC	45.06 aA	0.50
	30	53.48 cB	59.01 eC	45.26 aA	0.49
	45	54.83 dB	58.50 eC	44.60 aA	0.40
	60	53.35 cB	57.49 dC	45.03 aA	0.21
	90	52.39 bB	56.42 cC	44.89 aA	0.59
	LSD ($p < 0.001$)	0.57	0.41	0.43	
CIEb^*	0	37.15 aB	36.30 cAB	34.82 cdA	1.41
	7	37.39 aB	38.54 eC	32.41 aA	0.80
	15	38.52 bB	37.85 dB	32.63 aA	0.59
	30	40.95 deC	32.68 aA	33.85 abcB	0.28
	45	39.73 cC	32.73 aA	33.59 abB	0.53
	60	40.29 cdC	32.81 aA	34.71 bcdB	0.60
	90	41.70 eC	33.95 bA	35.79 dB	0.49
	LSD ($p < 0.001$)	0.68	0.41	0.76	
Chroma	0	60.60 aB	59.76 aAB	58.31 cA	1.52
	7	60.73 abB	62.28 bC	55.30 aA	1.10
	15	61.86 bB	68.03 dfC	55.63 aA	0.73
	30	67.36 cB	67.46 efB	56.52 abA	0.25
	45	67.71 cC	67.04 deB	55.84 aA	0.14
	60	66.85 cC	66.20 cdB	56.86 abA	0.52
	90	66.96 cA	65.85 cB	57.41 bcA	0.32
	LSD ($p < 0.001$)	0.77	0.54	0.78	
Hue angle	0	37.81 bcB	37.40 fAB	36.66 bcA	0.62
	7	38.00 cdB	38.22 eB	35.88 abA	0.19
	15	38.51 bdC	33.80 dA	35.91 aB	0.27
	30	37.44 bcC	28.98 aA	36.79 bcB	0.02
	45	35.92 aB	29.23 aA	36.95 cC	0.14
	60	37.06 bB	29.71 bA	37.62 dB	0.66
	90	38.52 dB	31.04 cA	38.56 eB	0.59
	LSD ($p < 0.001$)	0.51	0.18	0.39	
ΔE	0	0.00 a	0.00 a	0.00 a	<0.01
	7	0.47 aA	3.45 bB	2.86 cdB	1.26
	15	1.93 bA	14.32 cC	3.24 dB	0.90
	30	10.16 cB	19.72 dC	1.99 bcA	0.28
	45	12.61 dB	20.18 dC	2.72 cdA	0.13
	60	12.34 dB	20.68 dC	1.76 bA	0.31
	90	12.49 dB	20.74 dC	2.22 bcA	0.64
	LSD ($p < 0.001$)	0.48	0.66	0.57	

[Z] Means ($n = 3$) within a column followed by a different letter (storage time point comparison, lowercase letter) or within a row (treatment comparison, capital letter) are significantly different at $p < 0.001$. [W] CIEL^* = lightness; CIEa^* = redness; CIEb^* = yellowness; ΔE = difference or distance between two colors.

Table 6. Stability of CIE$L^*a^*b^*$ values in beverages stored at 20 °C.

Parameter	Storage (Days)	HHP-450 MPa	HHP-600 MPa	TP-85 °C	LSD ($p < 0.001$)
CIEL^*	0	23.39 aB [Z]	22.55 aAB	21.38 bA	1.04
	7	24.84 bB	31.80 bC	20.51 aA	0.74
	15	28.84 cB	37.44 cC	22.23 bA	0.51
	30	32.55 dB	40.13 dC	22.31 bA	0.64
	45	37.06 eB	43.61 eC	23.86 cA	0.43
	60	40.07 gB	45.71 fC	23.66 cA	1.59
	90	38.52 fB	47.81 gC	24.13 cA	0.11
	LSD ($p < 0.001$)	0.69	0.92	0.66	
CIEa^*	0	47.87 dB	47.48 dAB	46.78 eA	0.86
	7	47.08 cdB	51.89 fC	44.36 dA	0.68
	15	47.75 dB	53.61 gC	44.15 dA	0.38
	30	46.34 cB	49.76 eC	41.51 cA	0.24
	45	45.19 bB	46.88 cC	40.19 bA	0.30
	60	44.60 bB	45.12 bC	39.41 aA	0.18
	90	42.19 aB	43.92 aC	38.92 aA	0.32
	LSD ($p < 0.001$)	0.51	0.19	0.49	
CIEb^*	0	37.17 aB	36.30 aAB	34.82 abA	1.41
	7	39.14 bB	41.84 cC	33.76 aA	0.86
	15	43.15 cB	40.34 bC	36.19 bcA	0.64
	30	46.56 dC	44.07 dB	36.57 cA	0.73
	45	48.31 eC	44.76 dB	38.78 dA	0.55
	60	48.68 eC	45.63 eB	38.64 dA	0.91
	90	50.36 fC	46.23 eB	39.24 dA	0.14
	LSD ($p < 0.001$)	0.67	0.54	0.92	
Chroma	0	60.60 aB	59.76 aAB	58.31 cA	1.52
	7	61.22 aB	66.66 deC	55.75 abA	1.05
	15	64.35 bB	67.10 eC	57.09 bcA	0.69
	30	65.69 cB	66.48 dB	55.32 aA	0.63
	45	66.15 cA	64.81 cB	55.85 abA	0.58
	60	66.03 cC	64.17 bcB	55.19 aA	0.60
	90	65.70 cC	63.77 bB	55.26 aA	0.32
	LSD ($p < 0.001$)	0.81	0.39	0.93	
Hue angle	0	37.81 aB	37.40 aAB	36.66 aA	0.62
	7	39.74 bC	38.88 bB	37.27 aA	0.28
	15	42.10 cC	36.96 aA	39.45 bB	0.66
	30	45.14 dB	41.53 cA	41.38 cA	0.46
	45	46.92 eB	43.67 dA	43.98 dA	0.25
	60	47.50 fC	45.32 eB	44.43 dA	0.63
	90	50.04 gC	46.46 fB	45.24 eA	0.17
	LSD ($p < 0.001$)	0.21	0.37	0.47	
ΔE	0	0.00 a	0.00 a	0.00 a	<0.01
	7	2.63 bA	11.64 bB	2.80 bA	0.97
	15	8.10 cB	16.60 cC	3.14 bA	0.33
	30	13.23 dB	19.35 dC	5.67 cA	0.40
	45	17.85 eB	22.70 eC	8.09 dA	0.19
	60	20.53 fB	25.10 fC	8.61 dA	1.10
	90	20.88 fB	27.37 gC	9.46 eA	0.19
	LSD ($p < 0.001$)	0.47	0.64	0.43	

[Z] Means ($n = 3$) within a column followed by a different letter (storage time point comparison, lowercase letter) or within a row (treatment comparison, capital letter) are significantly different at $p < 0.001$.

Furthermore, an overall trend to decrease redness (CIEa^*) was observed for most of the samples as the shelf life increased (Tables 5 and 6), which was associated with the degradation in total anthocyanins, in agreement with previous studies [94,108]. However, it is important to point out that this evolution in the CIEa^* value was moved significantly toward the positive direction during storage in both HHP treatments when stored at 4 °C, being slightly higher for HHP-treated samples at 600 MPa, compared with thermal treatment. The increase in the CIEa* value observed in HHP-treated beverages stored at 4 °C could be associated with the breakdown of the citrus beverage cloud, owing to the formation and sedimentation of suspended particles (pectate complexes) in citrus-based juices [109], as previously mentioned. This breakdown may be due to the residual enzyme activity and supplementarily enhanced by a low storage temperature, which could contribute to faster particle precipitation [110], providing a lighter bright reddish color in the citrus-maqui drinks for those treated with HHP methods. It is also important to notice that the attractive reddish coloration of the beverages remained quite stable for 90 days for all treatments. Therefore, slight variations regarding the reddish color of the drinks over time could be associated to the formation of other newly colored polymers by co-pigmentation between anthocyanins and different phenolic compounds (flavonols, ferrulic acid, flavones, etc.) that could mask detrimental color losses or variations [111–113].

Moreover, slight changes were observed in the yellowness (CIE*b** value), as this parameter increased along with the shelf life time for most of the drinks, mainly in HHP-treated beverages at 450 MPa relative to the HHP-treated ones at 600 MPa or under thermal treatment (Tables 5 and 6). With respect to the Chroma and Hue angle parameters, an overall increase was noted for all treatments and temperatures, indicating a numerical browning inclination, which could barely be detected by the naked eye.

On the other hand, the total color difference (ΔE) increased along the 90 days of storage (Tables 5 and 6), indicating significant color changes between individual treatments in the citrus-maqui juices. In the current study, both HHP treatments led to a higher increase in ΔE, significantly more emphasized in those beverages treated at 600 MPa compared with those under thermal treatment and being more accentuated in those samples stored at 20 °C. With regard to the foregoing, the drinks maintained a stable color up to 15 days of storage without visual alterations, and visual differences were only appreciated by the human eye when the ΔE values surpassed 12 units. These results are in contrast with those found by other authors in red fruit juices [114,115], which exposed that noticeable visual differences for ΔE could be appreciated by the naked eye for value differences of just three units. In the present study, the differences among treatments in the ΔE values were mainly due to outstandingly higher CIE*L** values in those HHP-treated samples over time compared with the TP-treated samples, probably due to the cloud loss, as previously discussed.

On the whole, the shelf life had a significant impact on the color parameters monitored for beverages processed under either HHP or TP, although those samples processed by TP retained their original colors better than HHP, in accordance with previous reports [65,116]. Finally, it should be noticed that despite most consumers preferring fresh, cloudy juices without signs of sedimentation [117], in the present study, cloud loss in the citrus-maqui beverages provided a more intense and bright reddish color over storage time, as shown in Figure 4a,b, which could result in a more visually attractive commercial beverage for potential consumers [118].

Figure 4. Visual appearance of citrus-maqui based beverages processed by HHP at different pressure levels (HHP—450 MPa and HHP—600MPa) and TP—85°C over 90 days of storage at 4°C (**a**) and 20°C (**b**).

4. Conclusions

In this study, the application of HHP showed interesting results related to the microbial safety and stability of all parameters analyzed in the citrus-maqui beverages. Neither the processing nor storage conditions had an impact on the physicochemical parameters. Moreover, both HHP and TP guaranteed the microbiological safety throughout the shelf life. Regarding bioactive compounds, although TP promoted an increase in the content of flavonoids immediately after processing, HHP displayed a higher protective effect on bioactive flavonoids (flavanones and anthocyanins), giving rise to a lower rate of degradation during storage and preserving the healthy properties for the beverages. On the other hand, vitamin C underwent a rapid degradation for all processing conditions. Aside from that, the color parameters remained rather stable for all treated samples, keeping an attractive reddish coloration, although the use of HHP and refrigeration better preserved the intensity of the color during storage. It is noteworthy that color was the only studied parameter slightly affected by the HPP conditions (450 or 600 MPa). Finally, the storage temperature was shown to be the most critical parameter for the degradation of all the studied bioactive compounds, with 4 °C being the most suitable temperature for storage. Due to these results, HHP could be considered an effective alternative to conventional TP in the food industry for the production of high functional quality fruit-based beverages.

Author Contributions: Experimental design, sampling, analytical determination, data processing, and drafting, F.J.S.; experimental design, sampling, analytical determination, data processing, and drafting, P.M.P.; analytical determination, data processing, and drafting, V.A.; experimental design, drafting, and supervision of data analysis and final version of the manuscript, P.S.F.; project supervision, experimental design, beverage formula design, and contribution to the final version of the manuscript upon critical revision of the texts, C.G.-V. All authors have read and agreed to the published version of the manuscript.

Funding: This work was funded by the Spanish MICINN through Research Project PID2019-104212RB-I00/AEI/10.13039/501100011033.

Data Availability Statement: Not applicable.

Acknowledgments: F.J.S. was supported by an FPU (FPU18/00332) grant of the fellowship program from the Spanish Ministry of Science, Innovation, and Universities (MICIU). We also thank Hiperbaric S.A. for their support applying high pressure treatments to the beverages.

Conflicts of Interest: The authors declare no conflict of interest.

References

1. Khouryieh, H.A. Novel and emerging technologies used by the U.S. food processing industry. *Innov. Food Sci. Emerg. Technol.* **2020**, *67*, 102559. [CrossRef]
2. Sunil, N.C.J.S.; Chandra, S.; Chaudhary, V.; Kumar, V. "Non-thermal techniques: Application in food industries" A review. *J. Pharmacogn. Phytochem.* **2018**, *7*, 1507–1518.
3. Khan, M.K.; Ahmad, K.; Hassan, S.; Imran, M.; Ahmad, N.; Xu, C. Effect of novel technologies on polyphenols during food processing. *Innov. Food Sci. Emerg. Technol.* **2018**, *45*, 361–381. [CrossRef]
4. Meier, T.; Gräfe, K.; Senn, F.; Sur, P.; Stangl, G.I.; Dawczynski, C.; März, W.; Kleber, M.E.; Lorkowski, S. Cardiovascular mortality attributable to dietary risk factors in 51 countries in the WHO European Region from 1990 to 2016: A systematic analysis of the Global Burden of Disease Study. *Eur. J. Epidemiol.* **2018**, *34*, 37–55. [CrossRef]
5. Rastogi, N.K.; Raghavarao, K.S.M.S.; Balasubramaniam, V.; Niranjan, K.; Knorr, D. Opportunities and Challenges in High Pressure Processing of Foods. *Crit. Rev. Food Sci. Nutr.* **2007**, *47*, 69–112. [CrossRef]
6. Hooshyar, L.; Hesari, J.; Azadmard-Damirchi, S. Investigation of selected thermal and non-thermal preservative techniques to produce high quality and safe to drink sour cherry, red grape and pomegranate juices. *J. Food Sci. Technol.* **2020**, *57*, 1689–1697. [CrossRef]
7. Salar, F.J.; Agulló, V.; García-Viguera, C.; Domínguez-Perles, R. Stevia vs. Sucrose: Influence on the Phytochemical Content of a Citrus–Maqui Beverage—A Shelf Life Study. *Foods* **2020**, *9*, 219. [CrossRef]
8. Gironés-Vilaplana, A.; Mena, P.; Garcia-Viguera, C.; Moreno-Fernández, D. A novel beverage rich in antioxidant phenolics: Maqui berry (*Aristotelia chilensis*) and lemon juice. *LWT* **2012**, *47*, 279–286. [CrossRef]
9. Agulló, V.; García-Viguera, C.; Domínguez-Perles, R. Beverages Based on Second Quality Citrus Fruits and Maqui Berry, a Source of Bioactive (Poly)phenols: Sorting Out Urine Metabolites upon a Longitudinal Study. *Nutrients* **2021**, *13*, 312. [CrossRef]
10. Villaño, D.; Masoodi, H.; Marhuenda, J.; García-Viguera, C.; Zafrilla, P. Stevia, sucralose and sucrose added to a maqui-Citrus beverage and their effects on glycemic response in overweight subjects: A randomized clinical trial. *LWT* **2021**, *144*, 111173. [CrossRef]
11. Agulló, V.; González-Trujano, M.E.; Hernandez-Leon, A.; Estrada-Camarena, E.; Pellicer, F.; García-Viguera, C. Antinociceptive effects of maqui-berry (*Aristotelia chilensis* (Mol.) Stuntz). *Int. J. Food Sci. Nutr.* **2021**, 1–9. [CrossRef] [PubMed]
12. Araos, J.P. *Aristotelia chilensis*: A Possible Nutraceutical or Functional Food. *Med. Chem.* **2015**, *5*, 378–382. [CrossRef]

13. Ortiz, T.; Argüelles-Arias, F.; Begines, B.; García-Montes, J.-M.; Pereira, A.; Victoriano, M.; Vázquez-Román, V.; Bernal, J.P.; Callejón, R.; De-Miguel, M.; et al. Native Chilean Berries Preservation and In Vitro Studies of a Polyphenol Highly Antioxidant Extract from Maqui as a Potential Agent against Inflammatory Diseases. *Antioxidants* **2021**, *10*, 843. [CrossRef] [PubMed]
14. Scarano, A.; Butelli, E.; De Santis, S.; Cavalcanti, E.; Hill, L.; De Angelis, M.; Giovinazzo, G.; Chieppa, M.; Martin, C.; Santino, A. Combined Dietary Anthocyanins, Flavonols, and Stilbenoids Alleviate Inflammatory Bowel Disease Symptoms in Mice. *Front. Nutr.* **2018**, *4*, 75. [CrossRef] [PubMed]
15. Vega-Galvez, A.; Rodríguez, A.; Stucken, K. Antioxidant, functional properties and health-promoting potential of native South American berries: A review. *J. Sci. Food Agric.* **2020**, *101*, 364–378. [CrossRef]
16. Bastías-Montes, J.M.; Monterrosa, K.; Muñoz-Fariña, O.; García, O.; Acuña-Nelson, S.M.; Martín, C.V.-S.; Quevedo, R.; Kubo, I.; Avila-Acevedo, J.G.; Domiguez-Lopez, M.; et al. Chemoprotective and antiobesity effects of tocols from seed oil of Maqui-berry: Their antioxidative and digestive enzyme inhibition potential. *Food Chem. Toxicol.* **2019**, *136*, 111036. [CrossRef]
17. Mena, J.; Elgueta, E.; Espinola-Gonzales, F.; Cardenas, H.; Orihuela, P.A. Hydroethanolic Extracts of the Aristotelia Chilensis (Maqui) Berry Reduces Cellular Viability and Invasiveness in the Endometrial Cancer Cell Line Ishikawa. *Integr. Cancer Ther.* **2021**, *20*, 15347354211007560. [CrossRef]
18. Céspedes-Acuña, C.L.; Xiao, J.; Wei, Z.-J.; Chen, L.; Bastias, J.M.; Avila, J.G.; Alarcon-Enos, J.; Werner-Navarrete, E.; Kubo, I. Antioxidant and anti-inflammatory effects of extracts from Maqui berry *Aristotelia chilensis* in human colon cancer cells. *J. Berry Res.* **2018**, *8*, 275–296. [CrossRef]
19. Ávila, F.; Jiménez-Aspee, F.; Cruz, N.; Gómez, C.; González, M.A.; Ravello, N. Additive effect of maqui (*Aristotelia chilensis*) and lemon (*Citrus x limon*) juice in the postprandial glycemic responses after the intake of high glycemic index meals in healthy men. *NFS J.* **2019**, *17*, 8–16. [CrossRef]
20. Ma, H.; Johnson, S.L.; Liu, W.; DaSilva, N.A.; Meschwitz, S.; Dain, J.A.; Seeram, N.P. Evaluation of Polyphenol Anthocyanin-Enriched Extracts of Blackberry, Black Raspberry, Blueberry, Cranberry, Red Raspberry, and Strawberry for Free Radical Scavenging, Reactive Carbonyl Species Trapping, Anti-Glycation, Anti-β-Amyloid Aggregation, and Microglial Neuroprotective Effects. *Int. J. Mol. Sci.* **2018**, *19*, 461. [CrossRef]
21. Cebadera-Miranda, L.; Morales, P.; Cámara, M. Bioactive compounds in oranges from the Mediterranean climate area. In *The Mediterranean Diet: An Evidence-Based Approach*; Elsevier: Amsterdam, The Netherlands, 2020; pp. 293–309, ISBN 9780128186497.
22. Ballistreri, G.; Fabroni, S.; Romeo, F.V.; Timpanaro, N.; Amenta, M.; Rapisarda, P. Anthocyanins and Other Polyphenols in Citrus Genus: Biosynthesis, Chemical Profile, and Biological Activity. In *Polyphenols in Plants*; Academic Press: Cambridge, MA, USA, 2018; pp. 191–215, ISBN 9780128137680.
23. Miles, E.A.; Calder, P.C. Effects of Citrus Fruit Juices and Their Bioactive Components on Inflammation and Immunity: A Narrative Review. *Front. Immunol.* **2021**, *12*, 712608. [CrossRef]
24. Martinez, P.F.; de Carvalho, M.R.; Mendonça, M.L.M.; Okoshi, M.P.; de Oliveira-Junior, S.A. Efeito Antioxidante e Anti-inflamatório do Suco de Laranja. *Arq. Bras. De Cardiol.* **2021**, *116*, 1137–1138. [CrossRef]
25. ElSawy, H.; Algefare, A.I.; Alfwuaires, M.; Khalil, M.; Elmenshawy, O.M.; Sedky, A.; Abdel-Moneim, A.M. Naringin alleviates methotrexate-induced liver injury in male albino rats and enhances its antitumor efficacy in HepG2 cells. *Biosci. Rep.* **2020**, *40*, BSR20193686. [CrossRef]
26. Xiong, H.; Wang, J.; Ran, Q.; Lou, G.; Peng, C.; Gan, Q.-X.; Hu, J.; Sun, J.; Yao, R.; Huang, Q. Hesperidin: A Therapeutic Agent for Obesity. *Drug Des. Dev. Ther.* **2019**, *13*, 3855–3866. [CrossRef]
27. Zhang, J.; Sun, C.; Yan, Y.; Chen, Q.; Luo, F.; Zhu, X.; Li, X.; Chen, K. Purification of naringin and neohesperidin from Huyou (*Citrus changshanensis*) fruit and their effects on glucose consumption in human HepG2 cells. *Food Chem.* **2012**, *135*, 1471–1478. [CrossRef]
28. Mizrahi, A.; Knekt, P.; Montonen, J.; Laaksonen, M.A.; Heliövaara, M.; Järvinen, R. Plant foods and the risk of cerebrovascular diseases: A potential protection of fruit consumption. *Br. J. Nutr.* **2009**, *102*, 1075–1083. [CrossRef]
29. Testai, L.; Calderone, V. Nutraceutical Value of Citrus Flavanones and Their Implications in Cardiovascular Disease. *Nutrients* **2017**, *9*, 502. [CrossRef] [PubMed]
30. Roobab, U.; Aadil, R.M.; Madni, G.M.; Bekhit, A.E.-D. The Impact of Nonthermal Technologies on the Microbiological Quality of Juices: A Review. *Compr. Rev. Food Sci. Food Saf.* **2018**, *17*, 437–457. [CrossRef]
31. Daher, D.; Pérez-Lamela, C.; Le Gourrierec, S. Effect of High Pressure Processing on the Microbial Inactivation in Fruit Preparations and Other Vegetable Based Beverages. *Agriculture* **2017**, *7*, 72. [CrossRef]
32. Patrignani, F.; Siroli, L.; Serrazanetti, D.I.; Gardini, F.; Lanciotti, R. Innovative strategies based on the use of essential oils and their components to improve safety, shelf-life and quality of minimally processed fruits and vegetables. *Trends Food Sci. Technol.* **2015**, *46*, 311–319. [CrossRef]
33. Zhang, Z.-H.; Wang, L.-H.; Zeng, X.-A.; Han, Z.; Brennan, C.S. Non-thermal technologies and its current and future application in the food industry: A review. *Int. J. Food Sci. Technol.* **2018**, *54*, 1–13. [CrossRef]
34. Rojo, M.C.; Cristiani, M.; Szerman, N.; Gonzalez, M.L.; Lerena, M.C.; Mercado, L.A.; Combina, M. Reduction of Zygosaccharomyces rouxii Population in Concentrated Grape Juices by Thermal Pasteurization and Hydrostatic High Pressure Processing. *Food Bioprocess Technol.* **2019**, *12*, 781–788. [CrossRef]
35. Abera, G. Review on high-pressure processing of foods. *Cogent Food Agric.* **2019**, *5*, 1568725. [CrossRef]

36. Narjabadi Fam, S.; Khosravi-Darani, K.; Massoud, R.; Massoud, A. High-Pressure Processing in Food. *Biointerface Res. Appl. Chem.* **2020**, *11*, 11553–11561. [CrossRef]
37. Zhang, Y.; Liu, X.; Wang, Y.; Zhao, F.; Sun, Z.; Liao, X. Quality comparison of carrot juices processed by high-pressure processing and high-temperature short-time processing. *Innov. Food Sci. Emerg. Technol.* **2015**, *33*, 135–144. [CrossRef]
38. Muntean, M.-V.; Marian, O.; Barbieru, V.; Cătunescu, G.M.; Ranta, O.; Drocas, I.; Terhes, S. High Pressure Processing in Food Industry—Characteristics and Applications. *Agric. Agric. Sci. Procedia* **2016**, *10*, 377–383. [CrossRef]
39. Balda, F.P.; Aparicio, B.V.; Samson, C.T. Industrial High Pressure Processing of Foods: Review of Evolution and Emerging Trends. *J. Food Sci. Eng.* **2012**, *2*, 543. [CrossRef]
40. De Oliveira, P.M.; Júnior, B.R.D.C.L.; Martins, E.M.F.; Cristianini, M.; Martins, M.L.; Vieira, E.N.R.; Binoti, M.L.; Paula, D.D.A.; Ramos, A.M. Impact of high pressure and thermal processing on probiotic mixed mango and carrot juices. *J. Food Process. Preserv.* **2020**, *44*, e14530. [CrossRef]
41. Wongfhun, P.; Gordon, M.H.; Apichartsrangkoon, A. Flavour characterisation of fresh and processed pennywort (*Centella asiatica* L.) juices. *Food Chem.* **2010**, *119*, 69–74. [CrossRef]
42. Huang, H.-W.; Hsu, C.-P.; Wang, C.-Y. Healthy expectations of high hydrostatic pressure treatment in food processing industry. *J. Food Drug Anal.* **2019**, *28*, 1–13. [CrossRef]
43. Kaşıkcı, M.B. High Hydrostatic Pressure Treatment of Fruit, Fruit Products and Fruit Juices: A Review on Phenolic Compounds. *J. Food Health Sci.* **2015**, *2*, 27–39. [CrossRef]
44. Chang, Y.-H.; Wu, S.-J.; Chen, B.-Y.; Huang, H.-W.; Wang, C.-Y. Effect of high-pressure processing and thermal pasteurization on overall quality parameters of white grape juice. *J. Sci. Food Agric.* **2016**, *97*, 3166–3172. [CrossRef]
45. Yi, J.; Kebede, B.T.; Dang, D.N.H.; Buvé, C.; Grauwet, T.; Van Loey, A.; Hu, X.; Hendrickx, M.E. Quality change during high pressure processing and thermal processing of cloudy apple juice. *LWT* **2017**, *75*, 85–92. [CrossRef]
46. Conesa, R.; Andreu, S.; Fernandez, P.S.; Esnoz, A.; Palop, A. Nonisothermal heat resistance determinations with the thermoresistometer Mastia. *J. Appl. Microbiol.* **2009**, *107*, 506–513. [CrossRef]
47. Baenas, N.; Salar, F.J.; Domínguez-Perles, R.; García-Viguera, C. New UHPLC-QqQ-MS/MS Method for the Rapid and Sensitive Analysis of Ascorbic and Dehydroascorbic Acids in Plant Foods. *Molecules* **2019**, *24*, 1632. [CrossRef]
48. Wang, C.-Y.; Wang, Y.-T.; Wu, S.-J.; Shyu, Y.-T. Quality changes in high hydrostatic pressure and thermal pasteurized grapefruit juice during cold storage. *J. Food Sci. Technol.* **2018**, *55*, 5115–5122. [CrossRef]
49. Nayak, P.K.; Rayaguru, K.; Krishnan, K.R. Quality comparison of elephant apple juices after high-pressure processing and thermal treatment. *J. Sci. Food Agric.* **2016**, *97*, 1404–1411. [CrossRef]
50. Ganje, M.; Jafari, S.M.; Farzaneh, V.; Malekjani, N. Kinetics modelling of color deterioration during thermal processing of tomato paste with the use of response surface methodology. *Heat Mass Transf.* **2018**, *54*, 3663–3671. [CrossRef]
51. Kruszewski, B.; Zawada, K.; Karpiński, P. Impact of High-Pressure Homogenization Parameters on Physicochemical Characteristics, Bioactive Compounds Content, and Antioxidant Capacity of Blackcurrant Juice. *Molecules* **2021**, *26*, 1802. [CrossRef] [PubMed]
52. Orellana-Palma, P.; Tobar-Bolaños, G.; Casas-Forero, N.; Zúñiga, R.N.; Petzold, G. Quality Attributes of Cryoconcentrated Calafate (*Berberis microphylla*) Juice during Refrigerated Storage. *Foods* **2020**, *9*, 1314. [CrossRef] [PubMed]
53. Subasi, B.; Alpas, H. Effect of high hydrostatic pressure processing and squeezing pressure on some quality properties of pomegranate juice against thermal treatment. *High Press. Res.* **2016**, *37*, 78–92. [CrossRef]
54. Da Silveira, T.F.F.; Cristianini, M.; Kuhnle, G.G.; Ribeiro, A.B.; Filho, J.T.; Godoy, H.T. Anthocyanins, non-anthocyanin phenolics, tocopherols and antioxidant capacity of açaí juice (*Euterpe oleracea*) as affected by high pressure processing and thermal pasteurization. *Innov. Food Sci. Emerg. Technol.* **2019**, *55*, 88–96. [CrossRef]
55. Ahmed, M.; Eun, J.-B. Flavonoids in fruits and vegetables after thermal and nonthermal processing: A review. *Crit. Rev. Food Sci. Nutr.* **2017**, *58*, 3159–3188. [CrossRef]
56. Wang, Z.; Wu, G.; Shu, B.; Huang, F.; Dong, L.; Zhang, R.; Su, D. Comparison of the phenolic profiles and physicochemical properties of different varieties of thermally processed canned lychee pulp. *RSC Adv.* **2020**, *10*, 6743–6751. [CrossRef]
57. He, Z.; Tao, Y.; Zeng, M.; Zhang, S.; Tao, G.; Qin, F.; Chen, J. High pressure homogenization processing, thermal treatment and milk matrix affect in vitro bioaccessibility of phenolics in apple, grape and orange juice to different extents. *Food Chem.* **2016**, *200*, 107–116. [CrossRef]
58. Spira, P.; Bisconsin-Junior, A.; Rosenthal, A.; Monteiro, M. Effects of high hydrostatic pressure on the overall quality of Pêra-Rio orange juice during shelf life. *Food Sci. Technol. Int.* **2018**, *24*, 507–518. [CrossRef]
59. Andrés, V.; Villanueva, M.J.; Tenorio, M.D. The effect of high-pressure processing on colour, bioactive compounds, and antioxidant activity in smoothies during refrigerated storage. *Food Chem.* **2016**, *192*, 328–335. [CrossRef] [PubMed]
60. Wang, F.; Du, B.-L.; Cui, Z.-W.; Xu, L.-P.; Li, C. Effects of high hydrostatic pressure and thermal processing on bioactive compounds, antioxidant activity, and volatile profile of mulberry juice. *Food Sci. Technol. Int.* **2016**, *23*, 119–127. [CrossRef]
61. Zhao, L.; Wang, S.; Liu, F.; Dong, P.; Huang, W.; Xiong, L.; Liao, X. Comparing the effects of high hydrostatic pressure and thermal pasteurization combined with nisin on the quality of cucumber juice drinks. *Innov. Food Sci. Emerg. Technol.* **2013**, *17*, 27–36. [CrossRef]
62. Liu, F.; Wang, Y.; Li, R.; Bi, X.; Liao, X. Effects of high hydrostatic pressure and high temperature short time on antioxidant activity, antioxidant compounds and color of mango nectars. *Innov. Food Sci. Emerg. Technol.* **2014**, *21*, 35–43. [CrossRef]

63. Queirós, R.; Rainho, D.; Santos, M.D.; Fidalgo, L.; Delgadillo, I.; Saraiva, J.A. High pressure and thermal pasteurization effects on sweet cherry juice microbiological stability and physicochemical properties. *High Press. Res.* **2014**, *35*, 69–77. [CrossRef]
64. Elez-Martínez, P.; Soliva-Fortuny, R.C.; Martín-Belloso, O. Comparative study on shelf life of orange juice processed by high intensity pulsed electric fields or heat treatment. *Eur. Food Res. Technol.* **2005**, *222*, 321–329. [CrossRef]
65. Chen, D.; Xi, H.; Guo, X.; Qin, Z.; Pang, X.; Hu, X.; Liao, X.; Wu, J. Comparative study of quality of cloudy pomegranate juice treated by high hydrostatic pressure and high temperature short time. *Innov. Food Sci. Emerg. Technol.* **2013**, *19*, 85–94. [CrossRef]
66. Hsu, K.-C.; Tan, F.-J.; Chi, H.-Y. Evaluation of microbial inactivation and physicochemical properties of pressurized tomato juice during refrigerated storage. *LWT* **2008**, *41*, 367–375. [CrossRef]
67. Bull, M.K.; Zerdin, K.; Howe, E.; Goicoechea, D.; Paramanandhan, P.; Stockman, R.; Sellahewa, J.; Szabo, E.A.; Johnson, R.L.; Stewart, C.M. The effect of high pressure processing on the microbial, physical and chemical properties of Valencia and Navel orange juice. *Innov. Food Sci. Emerg. Technol.* **2004**, *5*, 135–149. [CrossRef]
68. Parish, M. Orange Juice Quality After Treatment by Thermal Pasteurization or Isostatic High Pressure. *LWT* **1998**, *31*, 439–442. [CrossRef]
69. Martí, N.; Mena, P.; Cánovas, J.A.; Micol, V.; Saura, D. Vitamin C and the Role of Citrus Juices as Functional Food. *Nat. Prod. Commun.* **2009**, *4*, 677–700. [CrossRef]
70. Njoku, P.; Ayuk, A.; Okoye, C. Temperature Effects on Vitamin C Content in Citrus Fruits. *Pak. J. Nutr.* **2011**, *10*, 1168–1169. [CrossRef]
71. Garcia-Viguera, C.; Bridle, P. Influence of structure on colour stability of anthocyanins and flavylium salts with ascorbic acid. *Food Chem.* **1999**, *64*, 21–26. [CrossRef]
72. Chung, C.; Rojanasasithara, T.; Mutilangi, W.; McClements, D.J. Stabilization of natural colors and nutraceuticals: Inhibition of anthocyanin degradation in model beverages using polyphenols. *Food Chem.* **2016**, *212*, 596–603. [CrossRef]
73. Monteiro, A.B.M. Effect of High Hydrostatic Pressure on Ascorbic Acid, Phenolic Compounds and Antioxidant Activity of Pera Rio Orange Juice. *J. Food Process. Technol.* **2015**, 1–7. [CrossRef]
74. El-Ishaq, A. Effect of Temperature and Storage on Vitamin C Content in Fruits Juice. *Int. J. Chem. Biomol. Sci.* **2015**, *1*, 17–21.
75. Munyaka, A.W.; Makule, E.E.; Oey, I.; Van Loey, A.; Hendrickx, M. Thermal Stability of l-Ascorbic Acid and Ascorbic Acid Oxidase in Broccoli (*Brassica oleracea* var. italica). *J. Food Sci.* **2010**, *75*, C336–C340. [CrossRef] [PubMed]
76. Zaritzky, N.E. Chemical and physical deterioration of frozen foods. In *Chemical Deterioration and Physical Instability of Food and Beverages*; Woodhead Publishing Limited: Sawston, UK, 2010; pp. 561–607, ISBN 9781845694951.
77. Terefe, N.S.; Buckow, R. High-Pressure Processing Effects on Endogenous Enzymes in Fruits and Vegetables. In *High Pressure Processing of Fruit and Vegetable Products*; Houska, M., Marques da Silva, F.V., Eds.; CRC Press: Boca Raton, FL, USA, 2017; pp. 39–62, ISBN 9781315121123.
78. Bai, J.-W.; Gao, Z.-J.; Xiao, H.-W.; Wang, X.-T.; Zhang, Q. Polyphenol oxidase inactivation and vitamin C degradation kinetics of Fuji apple quarters by high humidity air impingement blanching. *Int. J. Food Sci. Technol.* **2013**, *48*, 1135–1141. [CrossRef]
79. Vishwasrao, C.; Ananthanarayan, L. Kinetics of inactivation of quality-deteriorating enzymes and degradation of selective phytoconstituents in pink guava pulp during thermal processing. *J. Food Sci. Technol.* **2018**, *55*, 3273–3280. [CrossRef]
80. Polydera, A.; Stoforos, N.; Taoukis, P. Quality degradation kinetics of pasteurised and high pressure processed fresh Navel orange juice: Nutritional parameters and shelf life. *Innov. Food Sci. Emerg. Technol.* **2005**, *6*, 1–9. [CrossRef]
81. Barba, F.J.; Esteve, M.J.; Frigola, A. Ascorbic Acid Is the Only Bioactive That Is Better Preserved by High Hydrostatic Pressure than by Thermal Treatment of a Vegetable Beverage. *J. Agric. Food Chem.* **2010**, *58*, 10070–10075. [CrossRef]
82. Sánchez-Moreno, C.; Plaza, L.; de Ancos, B.; Cano, P.M. Effect of high-pressure processing on health-promoting attributes of freshly squeezed orange juice (*Citrus sinensis* L.) during chilled storage. *Eur. Food Res. Technol.* **2003**, *216*, 18–22. [CrossRef]
83. Plaza, L.; Sánchez-Moreno, C.; de Ancos, B.; Elez-Martínez, P.; Martín-Belloso, O.; Cano, M.P. Carotenoid and flavanone content during refrigerated storage of orange juice processed by high-pressure, pulsed electric fields and low pasteurization. *LWT* **2011**, *44*, 834–839. [CrossRef]
84. González-Molina, E.; Moreno-Fernández, D.; García-Viguera, C. A new drink rich in healthy bioactives combining lemon and pomegranate juices. *Food Chem.* **2009**, *115*, 1364–1372. [CrossRef]
85. Zhang, L.; Ling, W.; Yan, Z.; Liang, Y.; Guo, C.; Ouyang, Z.; Wang, X.; Kumaravel, K.; Ye, Q.; Zhong, B.; et al. Effects of storage conditions and heat treatment on the hesperidin concentration in Newhall navel orange (*Citrus sinensis* Osbeck cv. Newhall) juice. *J. Food Compos. Anal.* **2019**, *85*, 103338. [CrossRef]
86. Kouniaki, S.; Kajda, P.; Zabetakis, I. The effect of high hydrostatic pressure on anthocyanins and ascorbic acid in blackcurrants(*Ribes nigrum*). *Flavour Fragr. J.* **2004**, *19*, 281–286. [CrossRef]
87. Suthanthangjai, W.; Kajda, P.; Zabetakis, I. The effect of high hydrostatic pressure on the anthocyanins of raspberry (*Rubus idaeus*). *Food Chem.* **2005**, *90*, 193–197. [CrossRef]
88. Chaikham, P. Comparison of high hydrostatic pressure and thermal processing on physicochemical and antioxidant properties of Maoberry (*Antidesma thwaitesianum* Müell. Arg.) juice. *Int. Food Res. J.* **2015**, *22*, 1993–2001.
89. Roobha, J.J.; Saravanakumar, M.; Aravindhan, K.M.; Suganya, P. The effect of light, temperature, pH on stability of anthocyanin pigments in Musa acuminata bract. *Res. Plant Biol.* **2011**, *1*, 5–12.
90. Hellström, J.; Mattila, P.; Karjalainen, R. Stability of anthocyanins in berry juices stored at different temperatures. *J. Food Compos. Anal.* **2013**, *31*, 12–19. [CrossRef]

91. Ertan, K.; Türkyılmaz, M.; Ozkan, M. Effect of sweeteners on anthocyanin stability and colour properties of sour cherry and strawberry nectars during storage. *J. Food Sci. Technol.* **2018**, *55*, 4346–4355. [CrossRef] [PubMed]
92. Yu, Y.; Lin, Y.; Zhan, Y.; He, J.; Zhu, S. Effect of high pressure processing on the stability of anthocyanin, ascorbic acid and color of Chinese bayberry juice during storage. *J. Food Eng.* **2013**, *119*, 701–706. [CrossRef]
93. Sipahli, S.; Mohanlall, V.; Mellem, J. Stability and degradation kinetics of crude anthocyanin extracts from *H. sabdariffa*. *Food Sci. Technol.* **2017**, *37*, 209–215. [CrossRef]
94. Song, H.-N.; Ji, S.-A.; Park, H.-R.; Kim, H.-H.; Hogstrand, C. Impact of Various Factors on Color Stability of Fresh Blueberry Juice during Storage. *Prev. Nutr. Food Sci.* **2018**, *23*, 46–51. [CrossRef]
95. Poei-Langston, M.S.; Wrolstad, R.E. Color Degradation in an Ascorbic Acid-Anthocyanin-Flavanol Model System. *J. Food Sci.* **1981**, *46*, 1218–1236. [CrossRef]
96. Torres, B.; Tiwari, B.; Patras, A.; Cullen, P.; Brunton, N.; O'Donnell, C. Stability of anthocyanins and ascorbic acid of high pressure processed blood orange juice during storage. *Innov. Food Sci. Emerg. Technol.* **2011**, *12*, 93–97. [CrossRef]
97. Castañeda-Ovando, A.; Hernandez, M.D.L.P.; Páez-Hernández, M.E.; Rodríguez, J.A.; Galán-Vidal, C.A. Chemical studies of anthocyanins: A review. *Food Chem.* **2009**, *113*, 859–871. [CrossRef]
98. Martynenko, A.; Chen, Y. Degradation kinetics of total anthocyanins and formation of polymeric color in blueberry hydrothermodynamic (HTD) processing. *J. Food Eng.* **2016**, *171*, 44–51. [CrossRef]
99. Martín, C.V.-S.; Bastías-Montes, J.; Villagra-Jorquera, C.; Salinas-Huenchulao, G.; Flores-Ríos, A.; Gonzáles-Díaz, N.; Tamarit-Pino, Y.; Muñoz-Fariña, O.; Quevedo-León, R. Effect of Cryoconcentration Assisted by Centrifugation-Filtration on Bioactive Compounds and Microbiological Quality of Aqueous Maqui (*Aristotelia chilensis* (Mol.) Stuntz) and Calafate (*Berberis microphylla* G. Forst) Extracts Pretreated with High-Pressure Homogenization. *Processes* **2021**, *9*, 692. [CrossRef]
100. Ferrari, G.; Maresca, P.; Ciccarone, R. The Effects of High Hydrostatic Pressure on the Polyphenols and Anthocyanins in Red Fruit Products. *Procedia Food Sci.* **2011**, *1*, 847–853. [CrossRef]
101. Benjamin, O.; Gamrasni, D. Microbial, nutritional, and organoleptic quality of pomegranate juice following high-pressure homogenization and low-temperature pasteurization. *J. Food Sci.* **2020**, *85*, 592–599. [CrossRef]
102. Feng, X.; Zhou, Z.; Wang, X.; Bi, X.; Ma, Y.; Xing, Y. Comparison of High Hydrostatic Pressure, Ultrasound, and Heat Treatments on the Quality of Strawberry–Apple–Lemon Juice Blend. *Foods* **2020**, *9*, 218. [CrossRef]
103. Cao, X.; Bi, X.; Huang, W.; Wu, J.; Hu, X.; Liao, X. Changes of quality of high hydrostatic pressure processed cloudy and clear strawberry juices during storage. *Innov. Food Sci. Emerg. Technol.* **2012**, *16*, 181–190. [CrossRef]
104. Huang, W.; Bi, X.; Zhang, X.; Liao, X.; Hu, X.; Wu, J. Comparative study of enzymes, phenolics, carotenoids and color of apricot nectars treated by high hydrostatic pressure and high temperature short time. *Innov. Food Sci. Emerg. Technol.* **2013**, *18*, 74–82. [CrossRef]
105. Aghajanzadeh, S.; Kashaninejad, M.; Ziaiifar, A.M. Cloud stability of sour orange juice as affected by pectin methylesterase during come up time: Approached through fractal dimension. *Int. J. Food Prop.* **2017**, *20*, S2508–S2519. [CrossRef]
106. Croak, S.; Corredig, M. The role of pectin in orange juice stabilization: Effect of pectin methylesterase and pectinase activity on the size of cloud particles. *Food Hydrocoll.* **2006**, *20*, 961–965. [CrossRef]
107. Chakraborty, S.; Kaushik, N.; Rao, P.S.; Mishra, H.N. High-Pressure Inactivation of Enzymes: A Review on Its Recent Applications on Fruit Purees and Juices. *Compr. Rev. Food Sci. Food Saf.* **2014**, *13*, 578–596. [CrossRef]
108. Kopjar, M.; Piližota, V.; Tiban, N.; Subaric, D.; Babić, J.; Ačkar, Đ.; Sajdl, M. Strawberry jams: Influence of different pectins on colour and textural properties. *Czech. J. Food Sci.* **2009**, *27*, 20–28. [CrossRef]
109. Timmermans, R.; Mastwijk, H.; Knol, J.; Quataert, M.; Vervoort, L.; Van der Plancken, I.; Hendrickx, M.E.; Matser, A. Comparing equivalent thermal, high pressure and pulsed electric field processes for mild pasteurization of orange juice. Part I: Impact on overall quality attributes. *Innov. Food Sci. Emerg. Technol.* **2011**, *12*, 235–243. [CrossRef]
110. Baker, R.A.; Cameron, R.G. Clouds of citrus juices and juice drinks. *Food Technol.* **1999**, *53*, 64–69.
111. Weber, F.; Boch, K.; Schieber, A. Influence of copigmentation on the stability of spray dried anthocyanins from blackberry. *LWT* **2017**, *75*, 72–77. [CrossRef]
112. Khoo, H.E.; Azlan, A.; Tang, S.T.; Lim, S.M. Anthocyanidins and anthocyanins: Colored pigments as food, pharmaceutical ingredients, and the potential health benefits. *Food Nutr. Res.* **2017**, *61*, 1361779. [CrossRef]
113. Chatham, L.A.; Howard, J.E.; Juvik, J.A. A natural colorant system from corn: Flavone-anthocyanin copigmentation for altered hues and improved shelf life. *Food Chem.* **2019**, *310*, 125734. [CrossRef]
114. Zheng, X.; Yu, Y.; Xiao, G.; Xu, Y.; Wu, J.; Tang, D.; Zhang, Y. Comparing product stability of probiotic beverages using litchi juice treated by high hydrostatic pressure and heat as substrates. *Innov. Food Sci. Emerg. Technol.* **2014**, *23*, 61–67. [CrossRef]
115. Barba, F.J.; Jäger, H.; Meneses, N.; Esteve, M.; Frígola, A.; Knorr, D. Evaluation of quality changes of blueberry juice during refrigerated storage after high-pressure and pulsed electric fields processing. *Innov. Food Sci. Emerg. Technol.* **2012**, *14*, 18–24. [CrossRef]
116. Aaby, K.; Grimsbo, I.H.; Hovda, M.B.; Rode, T.M. Effect of high pressure and thermal processing on shelf life and quality of strawberry purée and juice. *Food Chem.* **2018**, *260*, 115–123. [CrossRef] [PubMed]

117. Stinco, C.M.; Szczepańska, J.; Marszałek, K.; Pinto, C.; Inácio, R.S.; Mapelli-Brahm, P.; Barba, F.J.; Lorenzo, J.M.; Saraiva, J.; Melendez-Martinez, A.J. Effect of high-pressure processing on carotenoids profile, colour, microbial and enzymatic stability of cloudy carrot juice. *Food Chem.* **2019**, *299*, 125112. [CrossRef]
118. Gao, G.; Zhao, L.; Ma, Y.; Wang, Y.; Sun, Z.; Liao, X. Microorganisms and Some Quality of Red Grapefruit Juice Affected by High Pressure Processing and High Temperature Short Time. *Food Bioprocess. Technol.* **2015**, *8*, 2096–2108. [CrossRef]

foods

MDPI

Review

High–Pressure Processing vs. Thermal Treatment: Effect on the Stability of Polyphenols in Strawberry and Apple Products

Gabriela Lorena Salazar-Orbea [1], Rocío García-Villalba [1], Francisco A. Tomás-Barberán [1,*] and Luis Manuel Sánchez-Siles [2,3]

[1] Quality, Safety and Bioactivity of Plant-Derived Foods, Centro de Edafología y Biología Aplicada del Segura-Consejo Superior de Investigaciones Científicas (CEBAS-CSIC), 30100 Murcia, Spain; glsalazar@cebas.csic.es (G.L.S.-O.); rgvillalba@cebas.csic.es (R.G.-V.)
[2] Research and Nutrition Department, Hero Group, 30820 Alcantarilla, Spain; luisma.sanchez@hero.es
[3] Institute for Research and Nutrition, Hero Group, 5600 Lenzburg, Switzerland
* Correspondence: fatomas@cebas.csic.es

Abstract: Polyphenols are important bioactive compounds that are affected by processing. The consumer's demand for minimally processed products contributes to the increase in non-thermal technologies such as high-pressure processing (HPP) in the food industry. This review is aimed at critically discussing the positive and negative effects of thermal treatment (TT) and HPP on the stability of different polyphenol families in agro-food products obtained from strawberry and apple, two of the most used fruits in food processing. Our findings show that the phenolic content was affected by processing, fruit type, polyphenol family, and storage conditions (time and temperature) of the final product. To increase shelf life, manufacturers aiming to preserve the natural content of polyphenols need to find the sweet spot between polyphenol stability and product shelf-life since the residual enzyme activity from HPP can affect polyphenols negatively.

Keywords: polyphenols; apple; strawberry; bioactive compounds; food processing; high-pressure processing; thermal processing; stability; storage

Citation: Salazar-Orbea, G.L.; García-Villalba, R.; Tomás-Barberán, F.A.; Sánchez-Siles, L.M. High–Pressure Processing vs. Thermal Treatment: Effect on the Stability of Polyphenols in Strawberry and Apple Products. *Foods* **2021**, *10*, 2919. https://doi.org/10.3390/foods10122919

Academic Editors: Francisco Artés-Hernández and Marina Cano Lamadrid

Received: 18 October 2021
Accepted: 22 November 2021
Published: 25 November 2021

1. Introduction

Strawberries and apples are fruits consumed worldwide, both fresh and processed (juices, jams, purees, smoothies, dried), due to their desirable sensory characteristics, nutritional value, and bioactive compounds. The main bioactive compounds found are polyphenols, the most significant dietary antioxidants present in fruit and vegetables, with a wide variety of biological activities linked to the associated health benefits [1]. These include flavonoid (flavonols, flavan-3-ols and proanthocyanidins, anthocyanins, dihydrochalcones, among others) and non-flavonoid (ellagitannins, ellagic acid, hydroxycinnamic acid derivatives, gallotannins) compounds. They are responsible for some of their organoleptic properties (color, aroma, astringency, bitterness) and also have relevant effects on nutrition and human health (antioxidants, anti-inflammatory, modulators of gut microbiota, etc.) [2]. There is considerable variability among the phenolic compounds found in strawberry and apple products (Figure 1), which depends on the inherent fruit characteristics (e.g., origin, variety, ripeness degree) as well as the type of food processing, and the part of the fruit (e.g., flesh, peel, achenes) [3,4].

In particular, in strawberries, flesh and achenes have a different polyphenol profile. In the flesh, the major phenols are flavan-3-ols/proanthocyanidins (F3OLs/PACs) and anthocyanins (ATs), while in achenes, the highest concentrations are for ellagitannins (ETs) and ellagic acid (EA) [3,5,6]. Similarly, in apples, there are differences in the phenolic compounds derived from flesh, peel, and seeds. In apple flesh, the major phenols are F3OLs/PACs and hydroxycinnamic acids (HCAs), being the latter mainly represented by chlorogenic acid. The principal polyphenols in apple peel are F3OLs/PACs having even higher concentrations than flesh, followed by flavonols (FOLs). In contrast, apple seeds are a more significant source of dihydrochalcones (DHCs) [4,7,8].

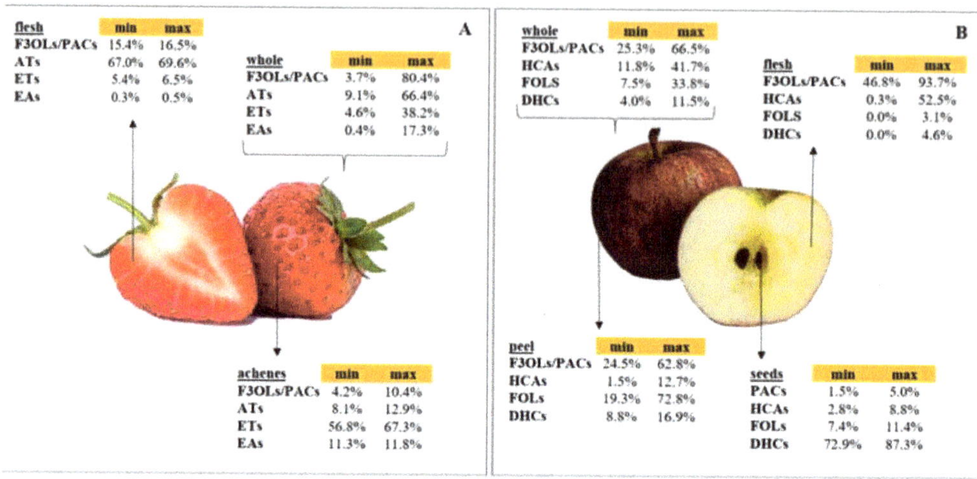

Figure 1. Percentage contribution of the main phenol families to the total polyphenols in different tissues of strawberry (**A**) and apple (**B**). (F3OLs/PACs) flavan-3-ols/proanthocyanidins, (ATs) anthocyanins, (ETs) ellagitannins, (EAs) ellagic acid and conjugates, (HCAs) hydroxycinnamic acids, (FOLs) flavonols, (DHCs) dihydrochalcones, (PACs) proanthocyanidins. Data retrieved from [3–6,8–13].

There is growing consumer awareness of the importance of consuming food products that preserve nutritional and functional properties [14]. Consumers' concern for food processing is increasing in many dimensions: from its link to healthiness and naturalness to its impact on the environment [15]. Nevertheless, preservation techniques are required to retain the nutritional value, ensure food safety during shelf life, and maintain its organoleptic characteristics. A wide variety of industrial thermal and non-thermal treatment technologies are available for food preservation. Thermal treatments include conventional techniques such as pasteurization and sterilization and new thermal technologies such as ohmic heating and dielectric heating (radiofrequency and microwave).

Meanwhile, non-thermal treatments include high-pressure processing, dehydration, freezing, pulsed electric fields, cold plasma, ultrasound processing, magnetic field, and ozone [16,17]. Thermal treatment is the most widely used preservation technique. According to the intensity of heat treatment, thermal treatments may be differentiated into pasteurization (70–80 °C), sterilization (110–120 °C), and ultra-high-temperature treatment (140–160 °C) [18]. Although thermal treatments produce some adverse effects, such as loss of some nutrients (vitamin C), formation of undesirable compounds (acrylamide, heterocyclic amines, sulfur compounds), and in some cases the degradation of organoleptic characteristics, it also exerts many beneficial effects such as inactivation of foodborne pathogens, inactivation of toxins and enzymes polyphenoloxidase (PPO), peroxidase (POD), pectin-methyl esterase, improvement of digestibility and bioavailability of nutrients, improvement of some sensory characteristics (texture and flavor), enhancement of extraction of health beneficial compounds and extension of shelf life [19,20]. Strawberry and apple polyphenol oxidases and peroxidases are responsible for the oxidation of phenolic compounds when fruit tissues are disrupted in the presence of oxygen, leading to the formation of brown polymers and other discolorations that affect the organoleptic properties of the fruit and derived products negatively.

Furthermore, one of the most studied non-thermal treatments is HPP. It works with high hydrostatic pressure and is commonly used at the industrial level between 300 and 600 MPa for 3–5 min [21,22]. As a consequence of pressure, there may be some changes in physical properties (solubility, density, viscosity), kinetic reactions (acceleration or delay of reactions rate), as well as, some equilibrium processes (dissociation of weak acids, acid-base equilibria, and ionization) [23]. The effect of pressure on food components depends on the type of bond between their molecules and the interatomic distance. In general, the high-pressure treatment minimally affects compounds with covalent bonds (vitamins, minerals, folates, antioxidants, anthocyanins, and flavor compounds) [17,24]. A disadvantage of HPP is that it is not entirely effective in polyphenol oxidase, peroxidase, and pectin methylesterase inactivation at the industrial processing conditions and subsequent storage [25]. Therefore, it triggers oxidation reactions that degrade phenolic compounds and shorten the shelf life [26,27].

Briefly, given the above facts and considerations summarized as follows: (1) polyphenols are important bioactive compounds that are affected by processing, (2) consumers are increasingly demanding "minimally processed" products, and (3) several non-thermal processing technologies such as HPP are now prevalent on the market, here, the objective of this article is to conduct a review of the literature about the differences between TT and HPP on the impact in polyphenol degradation in strawberry and apple products, two of the most used fruits in food processing.

2. Overview of the Studies Included in the Review

We systematically searched, evaluated, and synthesized research articles on the effects of traditional TT and HPP and their subsequent storage on the main polyphenols found in strawberries and apples. In strawberries, we analyzed F3OLs/PACs, ATs, ETs, and EA. Whereas for apple, F3OLs/PACs, FOLs, DHCs, and HCAs were reviewed. The major databases for the topic were used (Web of Science, ScienceDirect, SCOPUS, PubMed, and Google Scholar). The search terms used were: polyphenols or proanthocyanidins or anthocyanins or ellagic acid or flavonols or dihydrochalcones or hydroxycinnamic acids and thermal treatment or high-pressure processing or high hydrostatic pressure and strawberry or apple.

The search was focused on studies published over the last 10 years. Articles were included: (i) if the phenols were quantified by high-performance liquid chromatography (HPLC); (ii) when at least three studies were found for strawberry and apple identifying the effect of the related processing technology; (iii) when there was a control of fresh strawberry and apple or their unprocessed products. Articles were excluded: (i) when the analyzed sample was a mix of more than one fruit; (ii) when the HPP processing was combined with any thermal treatment higher than 50 °C; (iii) when the TT led to a concentration or reduction in the final product, e.g., jams.

For comparison purposes, the quantification of the different strawberry and apple polyphenols in the literature were converted to mg/100 g FW (fresh weight) to provide values closer to the fruit servings used in nutritional studies. The conversion of values based on DW (dry weight) to FW assumed that a fruit retains 10% of the original fresh weight after drying.

Overall, 26 articles were reviewed to address the study question: what are the effects of TT and HPP on the stability of polyphenols?

- Strawberry was the most studied fruit with 15 articles, whereas 11 were examined for apple products;
- A total of 19 articles reported the effects of thermal treatment and 12 for high-pressure processing. The reviewed studies examined 199 different trials at different processing and storage conditions (Figure 2).

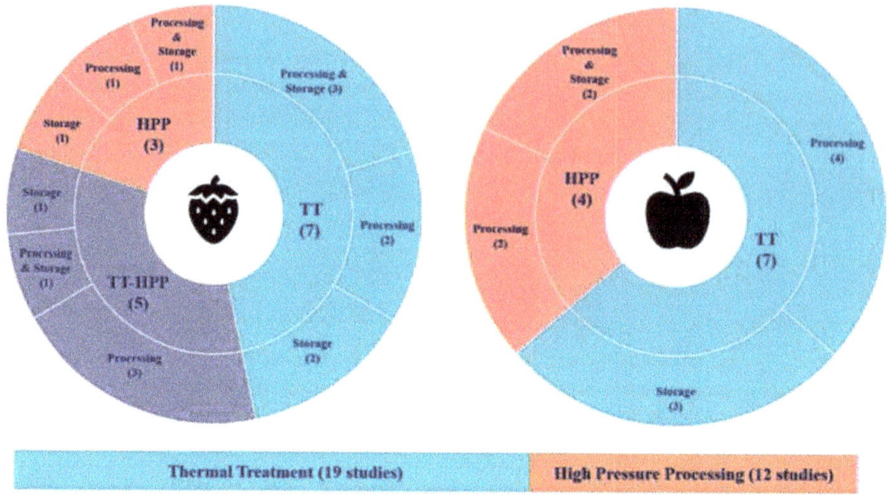

Figure 2. Overview of the studies evaluating the thermal treatment (TT) and high-pressure processing (HPP) impact and their subsequent storage on the primary polyphenols in strawberry and apple products.

The concentrations reported in these studies for phenols of the same family were added. Each group's total polyphenols were compared before and after processing, obtaining a percentage of increase or decrease attributed to the processing effect. A total of 4 of the 26 articles reviewed did not provide quantitative information to calculate these percentages, and therefore they were not included in the figures. The phenolic concentrations recorded after processing were considered control to determine the storage effect, and the results obtained during storage were compared against the control.

3. Effects of Processing and Storage Conditions on the Stability of Polyphenols

This section will discuss the positive and negative effects of TT and HPP treatments and their subsequent storage on the stability of the main polyphenols in strawberries (Table 1) and apples (Table 2). A summary is presented in Table 3.

3.1. Effects on the Stability of Proanthocyanidins in Strawberry and Apple Products

Proanthocyanidins, or condensed tannins, are plant secondary metabolites that could be found as oligomers or polymers of flavan-3-ol monomeric units. The most common monomeric building blocks are catechin, (epi)-catechin, (epi)-afzelechin, and (epi)-gallocatechin [28]. F3OLs/PACs are found in a wide variety of fruits, nuts, legumes, and cereals. The ranges reported highly depend on whether the study measured flavan-3-ol monomers (F3OLs), oligomers, and polymers or flavanol-3-ol units after hydrolysis (PACs). For most of the studies published, F3OLs and PACs were not differentiated, and therefore here we consider them together (F3OLs/PACs) unless differentiation is made.

In the case of strawberry, Nowicka et al. [3] recorded ranges of PACs polymers between 3.4 and 37.5 mg/100 g FW, while Buendia et al. [6] registered levels of total flavan-3-ol monomers from 53.9 to 168.1 mg/100 g FW after phloroglucinolysis. For apple, studies reported PACs polymers concentration from 12.6 to 203.3 mg/100 g FW in the whole fruit [12,29]. Nevertheless, higher levels were quantified for F3OLs units in peel and flesh. The flesh has from 1.4 to 548.2 mg/100 g FW, whereas in the peel, levels varied between 41.4 and 171.2 mg/100 g FW [7]. Numerous health benefits have been attributed to F3OLs/PACs, such as free radical scavenging, antioxidant, antiviral, antibacterial, anti-carcinogenic, anti-inflammatory, and anti-allergic activities [28,30].

Table 1. Effects of TT and HPP treatments and storage on the main phenolic compounds in strawberry products.

Material	Treatment: Conditions	Storage	Impact of Processing Conditions on Polyphenols [a]	Impact of Storage Conditions on Polyphenols [b]	Mechanisms	Ref.
Strawberry fresh	TT: 90 °C/5 min	360 days −20 °C	F3OLs/PACs: ↑30% CA, ↑73% EGC; ↑45% EC ATs: ↓16% pel-3-glu; ↓5% pel-3-rut; ≈cya-3-glu EA: ↑143%	F3OLs/PACs: ↓19% CA, ↓39% EGC ATs: ↓pel-3-glu; ↓pel-3-rut; ↑cya-3-glu EA: ↓65%	↑F3OLs/PACs: cleavage → release of dimers and monomers; release from cellular tissue ↓ATs: cleavage of covalent bonds, polymerization, and derivatization; ↑pH in the food matrix; enzymatic oxidation after pasteurization and during storage ↑EA: ETs hydrolysis	[31] [d]
Strawberry fresh	TT: 90 °C/5 min	90 days 23 °C	F3OLs/PACs: ↑34% CA; ↑134% EC; ↑119% EGCG; ↑30% *T.F3OLs/PACs* ATs: ↓30% cya-3-glu; ↓35% pel-3-glu; ≈ pel-3-rut; ↓30% *T.ATs* EA: ↑66%	F3OLs/PACs: ↓42% CA; ↓62% EGCG; ↓67% EC ATs: ↓87% cya-3-glu, ↓97% pel-3-glu; ↓92%pel-3-rut EA: ↓32%	TT ↑F3OLs/PACs: PACs cleavage → release of dimers and monomers; release from cellular tissue ↓ATs: cleavage of covalent bonds, polymerization, and derivatization ↑EA: release from cell walls; hydrolysis from ETs to EA Storage ↓F3OLs/PACs, ↓ATs and ↓EA: oxidation; non-ezymatic and enzymatic oxidation	[32]
Strawberry puree	TT1 (SB): 85 °C/3 min TT2 (P): 85 °C/3 min	No	SB: F3OLs/PACs: from ↑*42% to*↓*16%* ATs: ↓*34–43%* P: F3OLs/PACs: from ↑*53% to* ↓*35%* ATs: ↑*2–18%*	-	↓F3OLs/PACs: TT conditions degraded heat-labile flavan-3-ols ↓ATs: SB may influence the ATs stability	[33]
Strawberry puree (2 yearsharvest)	TT: 90 °C/2 min	No	Puree with seeds EA: ↓8% (2011 harvest), ↓13% (2012 harvest) Puree without seeds EA: ↓35% (2011 harvest) ↓19% EA (2012 harvest)	-	↓EA: oxidation by membrane breakage	[34]

Table 1. *Cont.*

Material	Treatment: Conditions	Storage	Impact of Processing Conditions on Polyphenols [a]	Impact of Storage Conditions on Polyphenols [b]	Mechanisms	Ref.
Strawberry puree	TT: 100 °C/10 min	8 weeks 25 °C	-	ATs: ↓80–88% pel-3-glu (8 weeks); ↓53–74% pel-3-rut (2 weeks); ↓63–78% pel-3-mal-glu (2 weeks); ↓70–86% pel-3-ace-glu (2 weeks); *↓90–93% T.ATs (8 weeks)*	↓ATs: oxidation by PPO; formation of dark condensation products	[35]
Strawberry puree (2 varieties)	TT1 (SB): 100 °C/3 min TT2 (P): 60, 75, 90 °C/3 min	28 days 20 °C	Ats—Elsanta var.: from ≈ *to ↓5% (P60; P75); ↓10–12% (SB; P90)* ATs—Everest var.: ≈ *(P90); ↓8% (SB, P60, P75)*	Elsanta var.: *↓62–65% T.ATs* (all treatments) Everest var.: *↓64–73% T.ATs* (all treatments)	SB and P ↓ATs: cleavage of covalent bonds, polymerization, and derivatization Storage ↓ATs: PPO partial reactivation during storage → oxidation	[36]
Strawberry juice (7 varieties)	TT: 90 °C/2 min	6 months at 4 and 20 °C	-	4 °C: F3OLs/PACs: from*↓0.3% to ↑27% polymeric PACs* ATs: ↓69% pel-3-glu; ↓73% pel-3-mal-lglu; ↓68% cya-3-glu; ↓56% cya-3-mal-glu; *↓59–89% T.ATs* 20 °C: F3OLs/PACs: from *↓10% to ↑11% polymeric PACs* ATs: ↓97% pel-3-glu; ↓99% pel-3-malonylglu; ↓98% cya-3-glu; ↓85% cya-3-mal-glu; *↓94–99% T.ATs*	↑F3OLs/PACs: protective effect of colloidal suspensions ↓F3OLs/PACs and ↓ATs: cleavage of covalent bonds, polymerization, and derivatization; enzymatic oxidation after pasteurization and during storage	[37]
Strawberry puree	HPP: 100–400 MPa /15 min at 20 and 50 °C	No	ATs: ≈ pel-3-glu; ≈cya-3-glu; ≈pel-3-rut (all HPP); from ≈ *to ↑15% T.ATs*	-	≈ ATs: Sufficient enzyme inactivation	[38]

Table 1. *Cont.*

Material	Treatment: Conditions	Storage	Impact of Processing Conditions on Polyphenols [a]	Impact of Storage Conditions on Polyphenols [b]	Mechanisms	Ref.
Strawberry puree (2 years harvest)	HPP: 300 and 600 MPa/ 15 min at 50 °C	28 weeks 6 °C	ATs - 300 MPa harvest 2011: ↓12% cy-3-glu; ↓13% pel-3-glu; ↓36% pel-3-rut; ↓*15% T.ATs* ATs - 600 MPa harvest 2012: ↓22% cy-3-glu; ↓21% pel-3-glu; ↓10% pel-3-rut; ↓*21% T.ATs*	ATs–300 MPa harvest 2011: 86 days half-life ATs–600 MPa harvest 2012: 62 days half-life	↓ATs: residual PPO activity → oxidation	[39]
Strawberry pure	TT: 72 °C/1 min HPP: 600 MPa/1 min	No	TT F3OLs/PACs: ↑122% CA; ↑33% proanthocyanidin B1; ↑*78% T.F3OLs/PACs* ATs: ↑40% cya-3-O-glu; ↑26% pel-3-O-glu; ↑22% pel-3-O-rut; ↑34% pel-3-O-mal-glu; ↑39% pel-3-O-acetylglu; ↑*32% T.ATs* EA: ↑8% HPP F3OLs/PACs: ↑68% CA; ↑19% proanthocyanidin B1; ↑*43% T.F3OLs/PACs* ATs: ↑12% cya-3-O-glu; ↑8% pel-3-O-glu; ↑10% pel-3-O-rut; ↑13% pel-3-O-mal-glu; ↑12% pel-3-O-ace-glu; ↑*11% T.ATs* EA: ≈	-	TT ↑F3OLs/PACs: cleavage → release of dimers and monomers ↑ATs: higher extraction from cell matrix favored by TT HPP: ↑F3OLs/PACs: release from the disrupted cell walls	[40]

Table 1. *Cont.*

Material	Treatment: Conditions	Storage	Impact of Processing Conditions on Polyphenols [a]	Impact of Storage Conditions on Polyphenols [b]	Mechanisms	Ref.
Strawberry puree	TT: 90 °C for 15 min HPP: 300 and 500 MPa/ 1, 5, 15 min at 0 °C	No	TT ATs: ↓44% cya-3-glu; ↓43% pel-3-glu; ↓49% pel-3-rut; ↓*44% T.ATs* EA: ↑30.5% EA HPP – 0 °C (all HPP conditions) ATs: ↓5% cya-3-glu; ↓7% pel-3-glu; ↓15% pel-3-rut; ↓*7% T. ATs* EA: ≈ EA (300 and 600 MPa) HPP + 50 °C ATs: ↓14% cya-3-glu; ↓13% pel-3-glu; ↓31% pel-3-rut; ↓*14% T.ATs* EA: ↑28.4% (300 MPa); ↑15.5% (600 MPa)	-	TT ↓ATs: cleavage of covalent bonds, polymerization, and derivatization ↑EA: release from the achenes favored by TT HPP – 0 °C ↓ATs: insufficient enzyme inactivation (PPO and POD) → oxidation HPP + 50 °C ↓ATs: formation of colorless chalcones ↑EA: release from ETs	[41]

Table 1. *Cont.*

Material	Treatment: Conditions	Storage	Impact of Processing Conditions on Polyphenols [a]	Impact of Storage Conditions on Polyphenols [b]	Mechanisms	Ref.
Strawberry Puree [c] (3 varieties)	TT: 88 °C/2 min HPP: 600 MPa/5 min at 20 °C	3 months 4 °C	TT Camarosa var. ATs: ↓16% cya-3-glu; ↓23% pel-3-glu; ↓26% pel-3-rut; ↓22% *T.ATs* Rubygem var. ATs: ↓42% cya-3-glu; ↓24% pel-3-glu; ↓29% pel-3-rut; ↓25% *T.ATs* Festival ATs: ↓27% cya-3-glu; ↓26% pel-3-glu; ↓26% pel-3-rut; ↓26% *T.ATs* HPP Camarosa var. ATs: ↓22% cya-3-glu; ↓26% pel-3-glu; ↓28% pel-3-rut; ↓26% *T.ATs* Rubygem var. ATs: ↓42% cya-3-glu; ↓27% pel-3-glu; ↓32% pel-3-rut; ↓28% *T.ATs* Festival ATs: ↓17% cya-3-glu; ↓20% pel-3-glu; ↓18% pel-3-rut; ↓20% *T.ATs*	TT Camarosa var. ATs: ↓66% cya-3-glu; ↓69% pel-3-glu; ↓60% pel-3-rut; ↓68% *T.ATs* Rubygem var. ATs: ↓52% cya-3-glu; ↓69% pel-3-glu; ↓59% pel-3-rut; ↓68% *T.ATs* Festival ATs: ↓59% cya-3-glu; ↓65% pel-3-glu; ↓59% pel-3-rut; ↓65% *T.ATs* HPP Camarosa var. ATs: ↓69% cya-3-glu; ↓72% pel-3-glu; ↓69% pel-3-rut; ↓72% *T.ATs* Rubygem var. ATs: ↓62% cya-3-glu; ↓75% pel-3-glu; ↓71% pel-3-rut; ↓75% *T.ATs* Festival ATs: ↓73% cya-3-glu; ↓77% pel-3-glu; ↓76% pel-3-rut; ↓76% *T.ATs*	↓ATs: partially due to variety effect PPO and POD; oxidation and co-oxidation; non-enzymatic reactions; cleavage of covalent bonds	[42]

Table 1. *Cont.*

Material	Treatment: Conditions	Storage	Impact of Processing Conditions on Polyphenols [a]	Impact of Storage Conditions on Polyphenols [b]	Mechanisms	Ref.
Strawberry pure	TT: 90 °C for 15 min HPP: 500 MPa/15 min at 50 °C	12 weeks 6 °C	-	TT ATs: ↓17% cya-3-glu; ↓19% pel-3-glu; ↓19% pel-3-rut; *↓19% T.ATs* EA: ↑ 56% EA until week 10 HPP+50 °C ATs: ↓72% cya-3-glu; ↓68% pel-3-glu; ↓72% pel-3-rut; *↓69% T.ATs* EA: ↑43% EA until the end of storage	TT-Storage ↑EA: release from the achenes; ETs hydrolysis; low pH increased ETs hydrolysis HPP Storage ↓ATs: not enough enzyme inactivation PPO and POD → oxidation	[43]
Strawberry pulp	TT: 70 °C/2 min HPP: 400, 500, 600 MPa/ 5, 10, 15, 20, 25 min at 25 °C	No	TT F3OLs/PACs: ↑42% CA ATs: ↓17% cy-3-glu, ↓23% pel-3-glu, ↓21% pel-3-rut; ↓22% *T.ATs* EA: ↑17% HPP F3OLs/PACs: ≈CA (500 MPa/20, 25 min; 600 MPa/5–25 min); ↓7–23% CA (400 MPa/5–25 min; 500 MPa/5–15 min); from ≈ *to ↓23% T.F3OLs/PACs* ATs: ≈cy-3-glu, ≈pel-3-glu, ≈pel-3-rut, *≈T.ATs (all HPP conditions)* EA: ↓2–37% (400MPa/20 min; 500 MPa/5, 20 min; 600 MPa/10, 20, 25 min)	-	TT ↑F3OLs/PACs: extraction from the achenes favored by TT ↓ATs: condensation reactions with other phenols → browning; PPO and POD oxidation ↑EA: release from the achenes; ETs hydrolysis HPP ↓F3OLs/PACs: no complete enzyme inactivation (PPO and POD) → oxidation	[44]

Table 1. *Cont.*

Material	Treatment: Conditions	Storage	Impact of Processing Conditions on Polyphenols [a]	Impact of Storage Conditions on Polyphenols [b]	Mechanisms	Ref.
Clear and cloud strawberry juices	TT (SB): 100 °C/1 min HPP: 600 MPa/4 min at 43 °C	6 months at 4 and 25 °C	-	Clear juice at 4 °C ATs: ↓10% cy-3-glu, ↓6% pel-3-glu, ↓9% pel-3-rut; *↓7% T.ATs* Cloudy juice at 4 °CATs: ↓26% cy-3-glu, ↓33% pel-3-glu, ↓21% pel-3-rut; *↓30% T.ATs* Clear and cloudy juice at 25 °C: T.ATs: ↓*> 95%*	↓ATs: PPO and POD oxidation; condensation with other phenols → colorless compounds; oxidative degradation of ascorbic acid (especially at higher storage temperature)	[45]

↑: increment vrs. control; ↓: diminution vrs. control; ≈: unchanged vrs. control; >: higher than; TT: thermal treatment; SB: steam blanching; P: Pasteurization;HPP: high-pressure processing; F3OLs/PACs: flavan-3-ols/proanthocyanidins; ATs: anthocyanins; EA: ellagic acid; T.: total; CA: catechin; EC: epicatechin: ECG: epicatechin gallate; EGC: epigallocatechin; EGCG: epigallocatechin gallate; pel: pelargonidin; cya: cyanidin; glu: glucoside; rut: rutinoside; mal-glu: malonylglucoside; ace-glu: acetylglucoside; pen: pentoside; ara: arabinoside; PPO: polyphenol oxidase; POD: peroxidase; [a] Total percentage of change (in italic) was calculated by adding the concentration of all polyphenols from the same family and comparing it before and after processing. These data are represented in Figures 3 and 4. Control samples were the same matrixes (juices, puree, etc.) just before TT or HPPP; [b] To study storage effects, samples after TT or HPPP at time 0 were used as control; [c] Control sample was the fresh fruit; [d] Studies without concentration data to calculate the percentage of change (kinetics or graphs).

Table 2. Effects of different TT and HPP treatments and storage on the main phenolic compounds in apple products.

Material	Treatment: Conditions	Storage	Impact of Processing Conditions on Polyphenols [a]	Impact of Storage Conditions on Polyphenols [b]	Mechanisms	Ref.
Apple pureec	TTwith O_2: 90 °C/30 min + O2 TT ∅ O_2: 90 °C/30 min ∅ O_2	No	TTwith O_2 F3OLs/PACs: ≈ EC; ≈procyanidin-dimer 1; ↓45% proanthocyanidin trimer; ↓62% CA; ↓30% T.F3OLs/PACs HCAs: ↓44% chlorogenic acid TT ∅ O_2 F3OLs/PACs: ≈EC, ↑35% procyanidin-dimer 1, ≈proanthocyanidin trimer; ≈CA; ↑7% T.F3OLs/PACs HCAs: ≈chlorogenic acid	-	↓F3OLs/PACs and HCAs: oxidation reactions during heating	[46]
Clear apple juice	TT: 25, 35, 45, 55, 65 and 75 °C/20 min	No	F3OLs/PACs: ↓EC and CA HCAs: ↓chlorogenic acid	-	↓F3OLs/PACs and HCAs: enzymatic oxidation	[47] [d]
Apple juice	TT1: 71.7 °C/0.4 min TT2: 90 °C/14.8 min	No	TT1 F3OLs/PACs: ↑71% *T.F3OLs/PACs* HCAs: ↑244% chlorogenic acid; ↑156% p-coumaoylquinic acid; ↑*205% T.HCAs* DHCs: ↑156% phloretin xyloglucoside; ↑192% phloridzin; ↑*165% T.DHCs* FOLs: ↑39% que-3-O-gal; ↑988% que-3-O-hex; ↑50% que-3-O-xyl; ↑7% que-3-O-rha; ↑33% que-3-O-pen; ↑49% *T.FOLs* TT2 F3OLs/PACs: ↑*1800% T.F3OLs/PACs* HCAs: ↑1352% chlorogenic acid; ↑389% p-coumaoylquinic acid; ↑*925% T.HCAs* DHCs: ↑752% phloretin xyloglucoside; ↑808% phloridzin; ↑*767% total DHCs* FOLs: ↑67% que-3-O-gal; ↑1113% que-3-O-hex; ↑92% que-3-O-xyl; ↑14% que-3-O-rha; ↑48% que-3-O-pen; ↑*69% T.FOLs*	-	↑F3OLs/PACs: cleavage → release of dimers and monomers ↑FOLs: release from cells walls; ↓ PPO activity ↑DHCs: enhanced release from peel and seeds ↑HCAs: release from cells walls favored by TT; enzyme inactivation	[48]

Table 2. *Cont.*

Material	Treatment: Conditions	Storage	Impact of Processing Conditions on Polyphenols [a]	Impact of Storage Conditions on Polyphenols [b]	Mechanisms	Ref.
Apple juice (2 years of harvest)	TT: 85 °C	360 days 4, 20, and 37 °C	-	4 and 20 °C: ≈ ∑ flavanols, DHCs, FOLs, and phenol carboxylic acids 37 °C: ↓ ∑ flavanols, DHCs, FOLs, and phenol carboxylic acids	↓ ∑ flavanols, DHCs, FOLs, and phenol carboxylic acids: Higly influenced by storage temperature	[49] [d]
Apple juice	TT1: 98 °C/30 sec TT2: 98 °C/30 sec	No	TT1 F3OLs/PACs: ↓32% CA; ↓31% EC; ↓33% EGC; ↓43% ECG; ↓18% procyanidin B2; *↓18% T.F3OLs/PACs* FOLs: ↓25% rutin; ↓50% hyperin; ↓27% que; *↓32% T.FOLs* DHCs: ↓ 18% phloridzin HCAs: ↓16% chlorogenic acid TT1+TT2 F3OLs/PACs: ↓58% CA; ↓56% EC; ↓59% EGC; ↓70% ECG; ↓37% procyanidin B2; *↓48% T.F3OLs/PACs* FOLs: ↓64% rutin; ↓86% hyperin; ↓55% que; *↓63% T.FOLs* DHCs: ↓ 48% phloridzin HCAs: ↓30% chlorogenic acid	-	↓F3OLs/PACs: TT conditions degraded heat-labile flavan-3-ols ↓FOLs: due to the discard of peel solids from the juice; the remaining enzyme activity ↓DHCs: due to thermal degradation	[50]
Apple sauce [c] (12 varieties)	TT1: 95 °C/2 min TT2: 95 °C/5 min	No	F3OLs/PACs: ↓20–85% procyanidins oligomers; ↓22–59% CA; ↓13–74% EC; *↓20–75% T.F3OLs/PACs* FOLs: from *↑6–63% to ↓4–57% T.FOLs;* DHCs: from ↓50% to ↑54% phloridzin; ↑1–1285% phloretin-2-xyloglucoside; from ↓8–14% to *↑%8–325% T.DHCs* HCAs: ↓1–47% 5′-caffeoylquinic acid (9 varieties); ↑4–30% 5′-caffeoylquinic acid (4 varieties); from *↓4–49% to ↑7–27% T. HCAs*		↑ and ↓ of phenols highly related to the apple variety ↓F3OLs/PACs: due to oxidation ↑ FOLs: diffusion of quercetin glycosides from the peel to the applesauce ↑DHCs: diffusion from the seeds to the applesauce ↑HCAs: release from cells walls favored by TT; Enzyme inactivation	[7]

Table 2. *Cont.*

Material	Treatment: Conditions	Storage	Impact of Processing Conditions on Polyphenols [a]	Impact of Storage Conditions on Polyphenols [b]	Mechanisms	Ref.
Cloudy apple juice [c]	TT: 80–145 °C/over 7200 sec	No	F3OLs/PACs: ↓7 procyanidin oligomers (mainly B type); ↑ CA; ↑EC; ↑dimeric compounds	-	↓F3OLs/PACs oligomers: cleavage → release of dimers and monomers ↓FOLs: glycosidic bond hydrolisis in an acidic matrix	[51] [d]
Apple juice	HPP: 300, 300 (x3), 450, 600 MPa/5 min at 20 °C	12 weeks 4 °C	F3OLs/PACs: ≈CA (all HPP); ↓EC (all HPP); ↑8% procyanidin B2 (300 and 400 MPa); ↑18% procyanidin B2 (300 x3 and 600 MPa); from ≈ *to ↑8% T. F3OLs/PACs* DHCs: ≈ phloridzin (300, 300x3, 450 MPa; 600 MPa) FOLs: ≈ que (300, 300x3, 450 MPa); ↑25% que (600 MPa); *↑1–25% T. FOLss* HCAs: ≈chlorogenic acid (300 and 300x3 MPa); ↑5% chlorogenic acid (450 and 600 MPa)	F3OLs/PACs: ∅ C and procyanidin B2 after 6 weeks (all HPP); ↓77% EC (except in 300 MPa) FOLs: ∅ quercetin after 2 weeks (all HPP) DHCs: ↓71–84% phloridzin (all HPP) HCAs: ↓66–77% chlorogenic acid (300x3, 450, 600 MPa)	HPP: higher pressurization → higher extraction from apple tissue ↑FOLs: release from the disrupted cell walls Storage ↓F3OLs/PACs and FOLs: oxidation reactions	[52]

Table 2. *Cont.*

Material	Treatment: Conditions	Storage	Impact of Processing Conditions on Polyphenols [a]	Impact of Storage Conditions on Polyphenols [b]	Mechanisms	Ref.
Apples (Spain and Italy)	HPP: 400, 500, 600 MPa/5 min at 35 °C	No	Spanish apples: *400 MPa (best treatment)* F3OLs/PACs: ≈CA; ≈EC; ≈dimers; ≈trimers; ↑4% procyanidin B2; ≈ *T.F3OLs/PACs* FOLs: ≈que-3-rut; ↑35% que-3-gal; ↑22% que-3-glu; ↑30% que-3-ara; ↑32% que-3-xyl; ↑33% que-3-rha; ↑*30% T.FOLs* DHCs: ≈phloridzin; ↓9% phloretin-2-xyloglucoside; ↓*2% T.DHCs* HCAs: ↓44% chlorogenic acid; ↓9% neochlorogenic acid; ↓10% cryptochlorogenic acid; ↓17% coumaroyl quinic acid; ↓*39% T. HCAs* *500–600 MPa* F3OLs/PACs: ≈CA; ↓8–13% procyanidin B2; ≈trimers; ≈dimers; ↓11–17% EC; ↓*11% T.F3OLs/PACs* FOLs:↓40–50% que-3-rut; ↓33–53% que-3-gal; ↓24–46% que-3-glu; ↓3–16% que-3-ara; ↓6–23% que-3-xyl; ↓15% que-3-rha; ↓*16–33% T.FOLs* DHCs: ↓16–20% phloridzin; ↓14–17% phloretin-2-xyloglucoside; ↓*15–19% T.DHCs* HCAs: ↓15–24% chlorogenic acid; ↓4% neochlorogenic acid; ↓12–14% cryptochlorogenic acid; ↓12–19% coumaroyl quinic acid; ↓*14–22% T. HCAs* Italian apples: *600 MPa (best treatment)*	-	Differences highly influenced by apple origin ↓Oligomers PACs → epimerization changes and depolymerization ↑FOLs: enhanced extraction by higher permeability or disruption of cell walls ↓FOLs: residual enzyme activity (PPO and POD) → oxidation	[12]

Table 2. *Cont.*

Material	Treatment: Conditions	Storage	Impact of Processing Conditions on Polyphenols [a]	Impact of Storage Conditions on Polyphenols [b]	Mechanisms	Ref.
Apples (Spain and Italy)	HPP: 400, 500, 600 MPa/5 min at 35 °C	No	F3OLs/PACs: ↑30% CA; ↑39% procyanidin B2; ↑70% trimers; ↑242% dimers; ↑45% EC; *↑58% T.F3OLs/PACs* FOLs: ↑ 88% que-3-rut; ↑107% que-3-gal; ↑78% que-3-glu; ↑59% que-3-ara; ↑68% que-3-xyl; ↑61% que-3-rha; *↑75% T.FOLs* DHCs: ↑ 67% phloridzin; ↑ 51% phloretin-2-xyloglucoside; *↑63% T.DHCs* HCAs: ↑31% chlorogenic acid; ↑4% neochlorogenic acid; ↑5% cryptochlorogenic acid; ↑51% coumaroyl quinic acid; *↑29% T. HCAs* *400–500 MPa* F3OLs/PACs: ↑2–13% CA; ↓10–19% procyanidin B2; ≈trimers; ↑131–161% dimers; ↓4–14% EC; *≈T.F3OLs/PACs* FOLs: ↑24% que-3-rut (400 MPa); ↑10–44% que-3-gal; ↓35% que-3-glu (400 MPa); ↓7–21% que-3-ara; ↓7–25% que-3-xyl; ↓7–23% que-3-rha; *↑5–29% T.FOLs* DHCs: ↓15% phloridzin; ↓1–14% phloretin-2-xyloglucoside; *↓11–↑8% T.DHCs* HCAs: ↓14–17% chlorogenic acid; ↓8–13% neochlorogenic acid; ↓8–16% cryptochlorogenic acid; ↓3–7% coumaroyl quinic acid; *↓13–16% T. HCAs*	-	Differences highly influenced by apple origin ↓Oligomers PACs → epimerization changes and depolymerization ↑FOLs: enhanced extraction by higher permeability or disruption of cell walls ↓FOLs: residual enzyme activity (PPO and POD) → oxidation	[12]

Table 2. *Cont.*

Material	Treatment: Conditions	Storage	Impact of Processing Conditions on Polyphenols [a]	Impact of Storage Conditions on Polyphenols [b]	Mechanisms	Ref.
Apples (Spain)	HPP: 400 MPa-/5 min at 35 °C	No	F3OLs/PACs: ≈procyanidin B1, EC trimers and tetramers; ↓4% EC; ↑10% CA, ↑4% procyanidin B2; ↑65% EC-dimer; ↑3% *T.F3OLs/PACs* FOLs: ↑9% que-3-rut; ↑35% que-3-gal; ↑22% que-3-glu; ↑30% que-3-ara; ↑32% que-3-xyl; ↑33% que-3-rha; ↑*30% T.FOLs* DHCs: ≈phloridzin; ↓9% phloretin-2-xyloglucoside; ≈*T.DHCs* HCAs: ↓13% chlorogenic acid; ↓9% neochlorogenic acid; ↓10% cryptochlorogenic acid; ↓17% coumaroyl quinic acid; ↓*12% T. HCAs*		↓Oligomers PACs → epimerization changes and depolymerization	[53]
Cloudy apple juice	HPP: 600 MPa/5 minat 25 °C	12 weeks 4 °C	F3OLs/PACs: ≈ CA; ↓13% EC; ↓45% procyanidin B1; ↓*15% T. F3OLs/PACs* DHCs: ↓18% phloridzin HCAs: ≈ chlorogenic acid	F3OLs/PACs: ↓92% procyanidin B1; ↓61%; ↓15% CA DHCs: ↓71% phloridzin HCAs: ↓53% chlorogenic acid	↓F3OLs/PACs: residual PPO and POD activity → oxidation ↓DHCs: residual enzyme activity → oxidation	[54]

↑: increment vrs. control; ↓: diminution vrs. control; ≈: unchanged vrs. control; ∅: absence; O_2: oxygen; ∑: sum total; Ctrl: control; TT: thermal treatment; HPP: high-pressure processing; F3OLs/PACs: proanthocyanidins; HCAs: hydroxycinnamics; DHCs: dihydrochalcones; FOLs: flavonols: T.: total; CA: catechin; EC: epicatechin: ECG: epicatechin gallate; EGC: epigallocatechin; que: quercetin; gal: galactoside; hex: hexoside; xyl; xyloside; rha: rhamnoside; pen: pentoside; ara: arabinoside; PPO: polyphenol oxidase; POD: peroxidase; [a] Total percentage of change (in italic) was calculated by adding the concentration of all polyphenols from the same family and comparing it before and after processing. These data are represented in Figures 3 and 4. Control samples were the same matrixes (juices, puree, etc.) just before TT and HPP; [b] To study storage effects, samples after TT or HPPP at time 0 were used as control; [c] Control sample was the fresh fruit; [d] Studies without concentration data to calculate the percentage of change (kinetics or graphs).

Table 3. Impact of TT and HPP treatments on different apple and strawberry phenolic compounds. (F3OLs/PAC) flavan-3-ols + proanthocyanidins; (ATs) anthocyanins; (EA) ellagic acid; (FOLs) flavonols; (DHCs) dihydrochalcones; (HCAs) hydroxycinnamic acids.

	TT		HPP	
	8 studies/30 trials		**6 studies/28 trials**	
	Positive (10)	**Negative (20)**	**Positive (13)**	**Negative (15)**
F3OLs/ PACs	• Apple puree/juice • Pasteurized strawberries • Strawberry pulp/puree	• Apple sauce/juice • Strawberry puree	• Pressurized apples • Cloudy apple juice • Strawberry puree	• Strawberry puree/pulp • Pressurized apples • Apple juice
	7 studies/23 trials		**6 studies/40 trials**	
	Positive (5)	**Negative (18)**	**Positive (16)**	**Negative (24)**
ATs	• Strawberry puree	• Pasteurized strawberries • Strawberry pulp/puree	• Strawberry pulp/puree	• Strawberry pulp/puree
	5 studies/10 trials		**3 studies/27 trials**	
	Positive (6)	**Negative (4)**	**Positive (13)**	**Negative (14)**
EA	• Strawberry puree/pulp • Pasteurized strawberries	• Strawberry puree	• Strawberry puree	• Strawberry pulp
	3 studies/17 trials		**3 studies/11 trials**	
	Positive (14)	**Negative (3)**	**Positive (9)**	**Negative (2)**
FOLs	• Applesauce • Apple juice	• Applesauce • Apple juice	• Pressurized apples • Apple juice	• Pressurized apples
	3 studies/16 trials		**6 studies/12 trials**	
	Positive (12)	**Negative (4)**	**Positive (6)**	**Negative (6)**
DHCs	• Applesauce • Apple juice	• Applesauce • Apple juice	• Pressurized apples • Apple puree	• Pressurized apples • Apple juice
	4 studies/19 trials		**4 studies/12 trials**	
	Positive (5)	**Negative (14)**	**Positive (5)**	**Negative (7)**
HCAs	• Apple puree/juice	• Applesauce/puree • Apple juice	• Apple juice	• Pressurized apples • Apple juice

F3OLs/PACs: proanthocyanidins; ATs: anthocyanins; EA: ellagic acid; FOLs: flavonols; DHCs: dihydrochalcones; HCAs: hydroxycinnamics.

For TT, a total of 8 studies with 30 trials including heat treatments from 70 to 98 °C during 0.4 to 15 min provided concentration data to calculate percentages of change of total F3OLs/PACs. Increments (10 trials) from 4% to 1800% were observed on total F3OLs/PACs in apple puree and juice, pasteurized strawberry and strawberry pulp, and puree [32,33,40,44,46,48]. Conversely, reductions (20 trials) from 16% to 75% were reported in strawberry puree, applesauce, and apple juice [7,33,46,50]. Whereas for HPP, a total of six studies with 28 trials under the following pressurization conditions, 300 to 600 MPa for 1 to 15 min at 22–35 °C were examined. Positive effects (13 trials) from no differences to 58% increments were observed in pressurized apples, apple juice, and strawberry puree [12,40,44,52,53]. However, decreases (15 trials) from 7% to 23% were reported in strawberry pulp, pressurized apples, and cloudy apple juice [12,44,54] (Table 3).

Overall, HPP maintained the content of F3OLs/PACs closer to the fresh control in both fruits. In contrast, the impact of TT on the degree of change of F3OLs/PACs was positive or negative depending on the study, influenced by the matrix, the TT conditions, and the response of the fruit cultivar to the processing technology [6,7] (Figures 3A and 4). In addition, it is important to keep in mind that the type of extraction highly influences the reported F3OLs/PACs concentrations and the analytical method used.

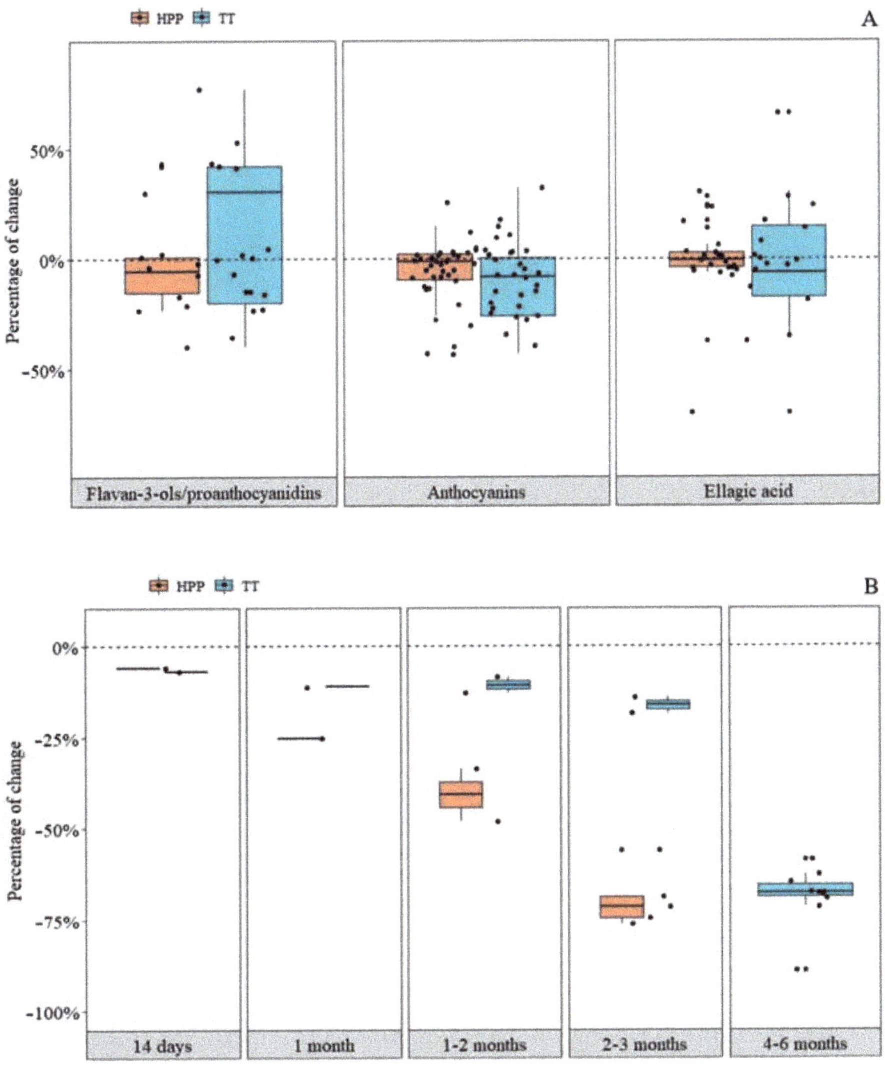

Figure 3. Impact of thermal treatment (TT) and high-pressure processing (HPP) on total flavan-3-ols/proanthocyanidins,

anthocyanins, ellagic acid after different processing conditions (**A**), and on anthocyanins storage at 4–6 °C (**B**) in strawberry products. Each point represents the results of a trial. The percentages of change with the processing of each polyphenol family were calculated by adding the concentration of all polyphenols from the same family and comparing it before and after processing.

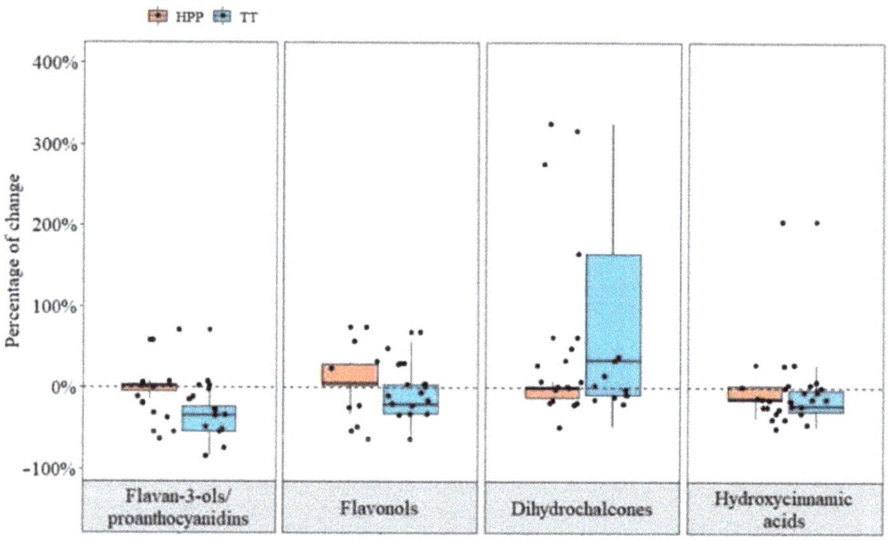

Figure 4. Impact of thermal treatment (TT) and high-pressure processing (HPP) on total flavan-3-ols/proanthocyanidins, flavonols, dihydrochalcones, and hydroxycinnamic acids after different processing conditions in apple products. Each point represents the results of a trial. The percentages of change with the processing of each polyphenol family were calculated by adding the concentration of all polyphenols from the same family and comparing it before and after processing. In TT, one trial with a percentage of change of F3OLs/PACs of 1800%, another with a percentage of change in DHCs of 768%, and another one with a percentage of change of HCAs of 925% were removed from the figures for scale reasons.

For TT, Stübler et al. [40] found a 122% increase in catechin and 33% in proanthocyanidin B1 levels after processing strawberry puree at 72 °C for 1 min [40]. An increase in catechin (42%) was also observed in strawberry pulp treated at 70 °C for 2 min [44]. In the same direction, Oliveira et al. [31,32] reported increments on (+)-catechin, (−)-epigallocatechin, (−)-epicatechin, and (−)-epigallocatechin gallate levels in entire strawberries pasteurized at 90 °C for 5 min. However, after 360 days of storage at −20 °C and 90 days at 23 °C, the concentration of the monomers decreased compared to the content evaluated immediately after processing, although they were still higher than the fresh control [31,32]. Conversely, a general increase in polymeric PACs was observed in thermally treated (90 °C/2 min) cloudy strawberry juice kept at 4 and 20 °C for six months [37]. In this study, the PACs increment during storage could be explained by the protective effect exerted by the food matrix since cloudy juice contains pectin, which formed colloidal suspensions that limit their degradation. Consistent with the positive impact of TT on strawberry products, Alongi et al. [48] recorded a 71% increase in total flavan-3-ols after mild pasteurization (71.7 °C/0.4 min) of apple juice and an 1800% increase after intense pasteurization (90 °C/14.8 min). This vast increase was not represented in Figure 4 for reasons of scale. An increase in flavan-3-ol monomers together with a decrease in seven PAC oligomers was also reported in cloudy apple juice treated at 80–145 °C for 120 min [51]. The positive effects observed in flavan-3-ols were attributed to the heat treatment, which could have promoted the increase in these flavan-3-ol monomers, aiding in the release from the cellular tissue or favoring the cleavage of complex PACs molecules into their structural monomers [32,40,48,51]. On the contrary, other studies have reported reductions in F3OLs/PACs, mainly monomers, in strawberry puree after treatment at 85 °C/3 min [33], in clear apple juice treated 20 min at different temperatures (25–75 °C) [47] and in apple juice after pre-pasteurization and pasteurization process (98 °C/3sec) [50]. A wide degradation range (20–75%) of total F3OLs/PACs was also observed in pasteurized applesauce from 12 varieties, showing the influence of the variety in the thermal treatment effects [7]. PACs degradation might be explained either by oxidation reactions due to insufficient inactivation of enzymes or because the monomers released into the food matrix are more prone to non-enzymatic reactions after processing and storage [32]. Kim et al. [46] proposed that not only heating (90 °C for 30 min) but also the presence of oxygen during processing affected the decreases observed for flavan-3-ols in apple puree.

As mentioned before, in general, HPP showed minor variation in F3OLs/PACs levels compared to control. In the case of strawberry puree pressurized at 600 MPa for 1 min, Stübler et al. [40] detected a rise of 19% on procyanidin B1 and 68% on catechin, which could be the result of the release of phenols to the food matrix from the disrupted cell walls. Similarly, Szczepańska et al. [52] registered about 8% increments in procyanidin B2 after 300 and 450 MPa, and 18% in apple juices subjected to 600 MPa and multi-pulsed pressurization (300 MPa × 3 pulses). An explanation for these results could be that higher pressurization induced a better extraction of procyanidin B2 from the tissue. However, catechin and epicatechin concentration decreased after all HPP treatments, and after 12 weeks of storage at 4 °C, procyanidin B2 and catechin were not detected in all the pressurized juices [52]. Similarly, Cao et al. [44] reported lower catechin concentrations (4–23%) after HPP (at 400, 500, and 600 MPa for 5, 10, 15, 20, and 25 min) than in the fresh strawberry pulp. Another study also found significant decreases of 13% and 45% on epicatechin and procyanidin B in cloudy apple juice after HPP (600 MPa for 5 min at 25 °C). However, no significant differences were reported on catechin [54]. Further degradations were reported for catechin, epicatechin, and procyanidin B1 after 12 weeks of storage at 4 °C. The highest degradation was recorded for procyanidin B1 (92%), which was not detected after week 4. In general, the degradation of flavan-3-ol monomers, oligomers, and polymers after pressurization and subsequent storage might be attributed to insufficient inactivation of PPO and POD, which led to oxidation reactions and brown polymer formation. As occurred with the TT, apple origin and variety showed a high impact on the processing effect by HPP. Fernández-Jalao et al. [12] reported a different behavior after pressurization in Spanish and Italian apples. In Spanish apples, 400 MPa showed increases of 4% in procyanidin B2, whereas 500 and 600 MPa resulted in degradation of all F3OLs/PACs. In apples from Italy, pressurization at 600 MPa, resulted in increments of 30% in catechin, 39% in procyanidin B2, 45% in epicatechin, 70% in trimers, and 240% in epicatechin dimers.

3.2. Effects on the Stability of Anthocyanins in Strawberry Products

Anthocyanins (ATs) are water-soluble compounds found in different tissues of the plant, such as leaves, roots, flowers, and fruits [55]. These play an essential role in the sensory attributes of food products, as they are responsible for the characteristic red, purple, and blue coloration in fruits [56]. Anthocyanins can be found as aglycones or glycosylated derivatives. The six most frequently found in foods are pelargonidin, cyanidin, malvidin, delphinidin, petunidin, and peonidin [57]. The highest AT levels are found in berries, currants, grapes, and tropical fruits [58]. In the case of fresh strawberries, AT concentrations from 3.7 to 64.9 mg/100 g FW have been reported [3,6]. Numerous in vitro and in vivo studies have recognized the potential effect on preventing neurodegenerative and cardiovascular diseases and antioxidant, anti-inflammatory, anti-obesity, anti-diabetic, and chemopreventive properties [59–61].

For TT, 7 studies with 23 trials comprising temperatures from 70 to 100 °C for 1 to 15 min provided concentration data to calculate percentages of change of total ATs. Reductions (18 trials) from 5% to 44% were observed in pasteurized strawberry, strawberry purees and pulp [32,33,36,41,42,44]. On the contrary, other authors reported increments (5 trials) from 2% to 32% in strawberry puree [33,40].

Regarding HPP, a total of 6 studies, including 40 trials with pressurization conditions from 100 to 600 MPa for 1 to 20 min at 0–50 °C, were analyzed. Degradations (24 trials) from 7% to 28% were reported in strawberry pulp and puree [38,39,41,44]. In contrast, positive effects (16 trials), from no differences to 15% increase, were recorded in the same products [38,40,44] (Table 3).

In general, just after pressurization, the levels of ATs were preserved close to those found in fresh strawberries. In contrast, a more significant impact was observed in the degree of change of ATs due to TT, with a downward trend in both cases (Figure 3A).

Since ATs are very chemically sensitive compounds, most of the studies registered the degradation of ATs after processing and storage. For instance, pasteurization of entire strawberries at 90 °C for 5 min resulted in a reduction of 5–35% of cyanidin-3-glucoside, pelargonidin-3-rutinoside and pelargonidin-3-glucoside [31,32]. Higher degradations of 87–92% of these ATs were observed during 90 days of storage at 23 °C [32]. Similar results with losses of around 90–93% were observed after storage for eight weeks at 25 °C of strawberry puree treated at 100 °C for 10 min [35]. In another study, the storage of strawberry puree (heated at 90 °C for 15 min) during 12 weeks at 6 °C led to a mean degradation of about 19% of these ATs [43]. Degradation of total ATs (22%) was also observed in the strawberry pulp after heating at 70 °C for 2 min [44] and in strawberry puree pasteurized 90 °C for 15 min (44% of reduction) [41]. In general, degradation of ATs can follow three possible processes; cleavage of covalent bonds, polymerization and derivatization [32,36,41,43]. Moreover, this decrease could be partially caused by condensation reactions of ATs with other phenolics to produce colored polymeric pigments, resulting in strawberry pulp browning.

Interestingly, many authors have reported different degradation percentages depending on the strawberry variety, concluding that the effect of TT and subsequent storage on ATs also depends on the response of the fruit variety to the treatment [36,37,42]. The stability of ATs is also highly influenced by the food matrix properties. High pH and ascorbic acid content in the food matrix accelerates the degradation rate of ATs [43,62–64].

Although overall TT led to degradation on ATs, some studies reported a positive impact after processing. Garzoli et al. [33] reported a slightly higher content (2–18%) on total ATs in pasteurized puree (85 °C/3 min) compared with fresh puree. In agreement, Stübler et al. [40] showed that heat-treated (72 °C for 1 min) strawberry puree incremented the individual ATs as follows: cyanidin-3-O-glucoside (40%), pelargonidin-3-O-glucoside (26%), pelargonidin-3-O-rutoside (22%), pelargonidin-3-O-malonyl-glucoside (34%), and pelargonidin-3-O-acetylglucoside (39%). These increments in ATs might be attributed to a higher extraction of ATs from the matrix due to the heat treatment [40]. As mentioned before, and can be observed in Figure 3A, the changes in ATs content after HPP were minor compared with TT. Some studies reported no significant changes after pressurization. For instance, Bodelón et al. [38] found no significant differences in ATs levels in strawberry puree after HPP at 100, 200, 300, and 400 MPa at 20 °C compared with the untreated puree. However, a slightly higher decrease in ATs was observed in the puree pressurized at 50 °C compared with the untreated control. In line with this, the levels of individual and total ATs in strawberry pulp showed no significant changes after HHP treatments, regardless of the applied pressures or treatment times [44]. These results further support the idea that ATs were stable after pressurization. Stübler et al. [40] reported minor non-significant increases (8–12%) on all the soluble individual ATs, in strawberry puree after pressurization at 600 MPa for 1 min at room temperature. One explanation for this apparent increase in ATs could be the release of these compounds from the intact cells to the surrounding matrix. In other works, slight but significant decreases of about 7% on total ATs were reported in strawberry puree after HPP at 300 and 500 MPa for 1, 5, and 15 min at 0 °C. Under the same pressurization conditions combined with 50 °C, there was a degradation of 14% on ATs [41]. In the samples pressurized without heat treatment, the oxidative enzyme activity

(PPO and POD) did not change significantly, leading to oxidation reactions and thus ATs degradation. However, the higher degradation of ATs in the pressurized puree combined with heat treatment might be due to the formation of colorless chalcones and reduction in flavylium cations and quinoid bases as a consequence of the thermal treatment. In another study, significant losses of 15 and 21% of total ATs were recorded in strawberry puree from two-year crops treated with HPP at 300 and 600 MPa for 15 min at 50 °C [39]. In the same line, Terefe et al. [42] also reported losses of ATs after HPP at 600 MPa for 5 min at 20 °C at different percentages depending on the variety: Camarosa (22–28%), Ruby Gem (27–42%), and Festival varieties (17–20%). Similar behavior for all varieties was observed after three months of refrigerated storage with losses between 62% and 77%. Losses up to 69% were also observed when pressurized strawberry puree (500 MPa for 15 min at 50 °C) was stored for 12 weeks at 6 °C showing that even HPP in combination with temperature was not enough to inactivate enzymes and led to a significant degradation during storage [43].

The effect of refrigerated storage (4–6 °C), from 14 to 180 days, on total ATs after different TT and HPP conditions is presented in Figure 3B. Storage at 4–6 °C after TT was examined in three studies [37,42,43]. Two studies were discussed to analyze the effect of storage after HPP, but only one used HPP without heat treatment [42], and the other combined HPP with heat treatment of 50 °C [43]. The losses of ATs during TT and HPP storage were progressive with the time, observing, in general, higher losses in the storage of HPP products. Storage from three to six months after HPP was only reported in a study with clear and cloudy strawberry juices that were not included in the figure because samples were also subjected to a previous blanching process at 100 °C/1 min in order to inactive enzymes [45].

3.3. Effects on the Stability of Ellagic Acid in Strawberry Products

Ellagitannins (ETs) constitute complex molecules with variable water solubility structured by one or more hexahydroxydiphenoyl (HHDP) moieties, which could be hydrolyzed to release ellagic acid (EA) [65]. ETs and EA are predominantly found in pomegranates, berry fruits, oak-aged red wine, tropical fruits, and nuts. ETs concentration in strawberries ranges from 7.18 to 28.85 mg/100 g FW and EA between 0.9 and 14.8 mg/100 g FW [3,5] with a more significant contribution from achenes than from flesh [5,66]. The potential health benefits of ETs and EA are associated with the metabolites (urolithins) produced by the human gut microbiota [67,68]. The biological functions attributed to urolithins comprise anti-oxidant, neuroprotective, anti-microbial, anti-inflammatory, and anticancer properties [67,69,70]. Notably, there were no differences in urolithins' production and urinary excretion between volunteers ingesting either fresh strawberries or thermally treated puree [71].

For TT, 5 studies with 10 trials including temperatures from 55 to 90 °C for 1 to 15 min provided concentration data to calculate percentages of change of EA. Increases (6 trials) from 8% to 66% were observed in pasteurized strawberries, strawberry pulp, and strawberry purees [32,40,41,44]. In contrast, only Álvarez-Fernández et al. [34] reported a slight EA degradation (4 trials) from 8% to 35% in strawberries. Whereas for HPP, 3 studies with 27 trials with pressurization conditions ranging from 100 to 600 MPa for 1 to 25 min at 0–50 °C were analyzed. Increments (13 trials) from non-differences to 28% were reported in strawberry puree [40,41], whereas 2 to 37% decreases (14 trials) were observed in strawberry pulp [44], showing that the increases observed were most probably due to the enhanced extraction from the achenes (Table 3).

In general, TT and HPP influenced the level of EA both positively and negatively depending on the conditions. Still, the most significant increases were observed as a result of heat treatment (Figure 3A).

For TT, Cao et al. [44] reported a 17% increment in EA content after heating strawberry pulp at 70 °C for 2 min. The rise of EA levels after processing could have resulted from the hydrolysis of ETs and the release from the cellular structures. In agreement, pasteurization of strawberry puree at 90 °C for 15 min led to an increase of 31% on EA [41], and only a limited increase of 8% was observed when a milder TT of 72 °C for 1 min was applied [40]. Likewise, pasteurized (90 °C/5 min) strawberries had 143% higher levels of EA when compared with fresh fruit. However, after 360 days of storage at −20 °C, EA decreased 65% compared to the control [31]. Another study showed a zero-order kinetic model for EA degradation, with a final reduction of 32% in pasteurized strawberries stored for 90 days at 23 °C. Degradation due to storage might be due to increased exposure of EA released from cell walls, leading to non-enzymatic oxidation reactions [32]. On the other hand, during storage of pasteurized strawberry pure (90 °C/15 min) at 6 °C for 12 weeks, EA levels raised until week 10 and then slightly decreased toward the end of storage [43]. An explanation for the increase in EA during storage might be attributed to progressive release from high molecular weight ETs present in the puree [43]. Although the general trend is to increase EA after TT, Álvarez-Fernández et al. [34] reported decreases in EA (8–35%) during the strawberry puree processing and in the final product in the 2011 and 2012 harvest.

For HPP, Cao et al. [44] reported significant losses from 2% to 37% on EA levels in strawberry puree after HPP at 400 MPa for 5, 10, 15, 20, and 25 min, and 500 MPa for 5, 10, and 15 min at room temperature. However, a minor and non-significant decrease of 3% EA was observed in strawberry puree pressurized at 600 MPa for 1 min at room temperature [40]. In agreement, Marszalek et al. [41] also reported that the levels of EA in fresh strawberry puree did not change significantly after HPP at 300 and 500 MPa for 1, 5, and 15 min at 0 °C. However, when combining the same pressurization conditions at 50 °C, a significant increase of 28% EA concentration was observed [41]. These increments could be due to a release of EA from ETs due to the combination of HPP with temperature [72]. An EA increase of 43% was reported when strawberry puree pressurized at 500 MPa for 15 min at 50 °C was stored for 12 weeks at 6 °C [43].

3.4. *Effects on the Stability of Flavonols in Apple Products*

Flavonols (FOLs) are plant secondary metabolites that could be found in foods as aglycones or much more frequently as glycosidic conjugates [73]. The most commonly found in foodstuff are quercetin, kaempferol, myricetin, and isorhamnetin [74]. Although these are present in various fruits and vegetables, the major sources of FOLs are capers, saffron, onion, and tea [74,75]. Among the FOLs, quercetin glycosides are the most frequently found in apples. These are located mainly in the peel in concentrations from 5.3 to 119.7 mg/100 g FW [4]. However, some studies also reported minor amounts in the flesh [9,13]. Numerous studies have reported the benefits of FOL consumption on preventing cardiovascular diseases, diabetes, inflammation, viral infections, neurodegeneration, and cancer [76–78].

For TT, 3 studies with 17 trials including temperatures from 71 to 98 °C during 0.4 to 15 min provided concentration data to calculate percentages of change of total FOLs. Increments ranging from 4% to 69% were observed in five trials with apple juice and applesauce [7,48]. In contrast, reductions (12 trials) from 32% to 63% were detected in other studies with apple juice and applesauce [7,50]. A total of 3 studies with 11 trials under the following pressurization conditions, 300 to 600 MPa for 5 to 15 min at 22–35 °C, were analyzed after HPP. Positive effects ranged between 1% and 75% in nine trials after pressurization in entire apples and apple juice [12,52,53]. Conversely, decreases (2 trials) from 16% to 33% were also observed in pressurized apples [12] (Table 3).

Overall, HPP had positive effects on FOLs concentration. With TT, the results depended on the conditions and the apple varieties with a general trend to decrease their content (Figure 4).

For instance, FOLs contents increased 49% and 69% after mild (71 °C/0.4 min) and intense (90 °C/14.8 min) pasteurization of apple juice, respectively, which indicates that thermal treatment favored the release of phenolic compounds and reduced PPO activity [48]. Additionally, the thermal treatment also increased the bioaccessibility of FOLs compared to the fresh sample [79]. On the contrary, pre-pasteurization (98 °C/30 s) and pasteurization

(98 °C/50 sec) of apple juice from Red Fuji variety reduced rutin (64%), hyperin (86%), and quercetin levels (55%). A possible explanation for this decline might be the incomplete inactivation of PPO and POD enzymes by milder heat treatments [50]. The occurrence of these enzymes should be determined after the TT conditions. In another study with applesauce (12 varieties), it was impossible to calculate exactly the change percentage of FOLs after TT. Concentration before processing was reported separately in fresh flesh and peel [7]. In this case, 95% of flesh and 5% of peel were considered to calculate the concentration in the fresh fruit (control sample). Like in apple peel, six quercetin glycosides were found in applesauce in this order of importance: quercetin-3-galactoside > quercetin-3-arabinopyranoside > quercetin-3-rhamnoside > quercetin-3-glucoside > quercetin-3-xyloside > quercetin-3-rutinoside. After crushing, cooking (95 °C/2 min), and pasteurization (90 °C/5 min), different behavior was observed depending on the variety: increments of FOLs from 4% to 57% were observed in three varieties, and decreases from 6% to 63% were observed in 9 varieties. FOLs in the applesauce could result from the diffusion of quercetin derivatives from the peel during crushing or from the presence of small particles of peel retained in the sauce. Differences in the percentages between varieties could be explained by differences in the skin of apples, different structures, mechanical resistance, and cuticle thickness.

About HPP, no significant differences were reported in quercetin levels after apple juice pressurization at 300 and 450 MPa for 5 min and in multi-pulsed HPP at 300 MPa × 3 pulses, each 5 min. However, after HPP treatment at 600 MPa for 5 min, a minor but significant increase from 0.73 to 0.91 mg/L (25%) was quantified for quercetin [52]. These results could be attributed to an enhanced extraction from the juice tissue due to the high pressurization. Quercetin was not detected after two weeks of storage at 4 °C in all the treatments, and this degradation might be attributed to residual enzyme activity, which led to oxidation reactions [52]. Increases (9–35%) in different glycoside conjugates of quercetin were also found between the untreated and pressurized apples at 400 MPa for 5 min [12]. As for other phenolic compounds, FOLs stability also depends on the fruit variety, degree of ripening, and food matrix characteristics (pH, sugar content, and presence/absence of oxygen), which influence the behavior of the enzymes [12,53]. In this sense, Fernández-Jalao et al. [12] concluded that individual and total FOLs levels after pressurization were affected by HPP conditions and the apple origin. In terms of total FOLs concentration, the best results were obtained after pressurization at 400 MPa in Spanish apples, increasing 30%. However in 500 and 600 MPa led to significant reductions of quercetin-3-rutinoside (40–50%), quercetin-3-galactoside (33–53%) and quercetin-3-glucoside (24–46%) [12]. The degradation of quercetin glycosides might result from oxidation reactions caused by the residual activity of PPO and POD. In Italian apples, all the pressurization treatments increased quercetin glycosides, but the highest increase of about 75% was reported for HPP at 600 MPa [12]. The increase in individual and total FOLs content after HPP might be attributed to a change in the cell walls permeability and/or by disruption of the cell membranes, promoting a better extractability from cellular tissues [53].

3.5. Effects on the Stability of Dihydrochalcones in Apple Products

The dihydrochalcones (DHCs), phloretin and phloridzin, are characteristic compounds of apple and apple products. Since these compounds are exclusive of apples, these have been used to detect adulterations [80]. DHCs levels in entire apples ranged from 0.011 to 0.043 mg/100 g FW [11,12]. Although the DHCs could be found in flesh and peel, the highest concentration is located in the seeds, ranging from 24.1 to 86.4 mg/100 g FW [8]. Numerous studies have shown that both phloridzin and phloretin exert antibacterial, anti-inflammatory, antihyperglycemic, anti-diabetic and anticancer activities, and cardioprotective, neuroprotective hepatoprotective, and immunomodulatory properties [81–83].

For TT, 3 studies with 16 trials comprising thermal treatments from 71 to 98 °C during 0.4 to 15 min provided concentration data to calculate percentages of change of total DHCs. Increments (12 trials) ranged from 8% to 767% in applesauce and apple juice [7,48]. Conversely, reductions (4 trials) from 8% to 48% in apple juice and applesauce were recorded [7,50]. For HPP, four studies with 12 trials at pressurization conditions from 300 to 600 MPa for 5 to 15 min at 22–35 °C were analyzed. Positive effects (6 trials) showing from no differences to 63% increase were noted in pressurized apples and apple juice [12,52]. Conversely, reductions (6 trials) from 2% to 19% were reported for pressurized apples and cloudy apple juice [12,53,54] (Table 3).

Overall, TT and HPP changed the content of DHCs of apple products (Figure 4) positively. These differences are mainly explained by the use of whole apples during crushing in industrial processing, which contributes to releasing these compounds from the peel and seeds [84].

For TT, Le Bourvellec et al. [7] reported high variability on change percentages of DHCs in applesauce depending on the variety. Increments up to 325% were observed in nine varieties after crushing, cooking (95 °C/2 min) and pasteurization (90 °C/5) min compared with fresh control (calculated as 95% of flesh + 5% of peel). An explanation for the higher DHCs levels after processing could be the little peel and seed particles in the applesauce [7,85]. One interesting finding in this study was identifying a colorless phloridzin oxidation product, which indicated limited enzymatic oxidation due to processing. In line with this, Alongi et al. [48] detected increases of 165% and 767% on total DHCs in apple juice after mild (71 °C/0.4 min) and intense (90 °C/14.8 min) pasteurizations, respectively. These results confirm the association that crushing and heat treatment increase the release of phenols from peel and seeds. In contrast, Tian et al. [50] reported 18% and 48% reduction in phloridzin levels after pre-pasteurization (98 °C/30 sec) and pasteurization (98 °C/30 s) respectively, probably due to a thermal degradation.

For HPP, Szczepańska et al. [52] reported no significant differences in phloridzin concentration in apple juice pressurized at 300, 450, and 600 MPa, and multi-pulsed 300 MPa × 3 pulses for 5 min. However, after 12 weeks of refrigerated storage, phloridzin concentration was reduced between 71% and 84% depending on the HPP conditions. In another study, after HPP at 600 MPa for 5 min at 25 °C, phloridzin levels changed from 48.8 to 40.2 mg/L, representing losses of 18%. After two weeks of storage at 4 °C, 51% decreases were observed, reaching up to 71% losses at the end of 12 weeks of storage [54]. As for FOLs, the degradation in DHCs was mainly due to HPP conditions that were insufficient to inactivate enzymes (PPO and POD). Therefore, the phenolic compounds underwent oxidation reactions. Fernández-Jalao et al. [12] indicated that the apple origin conditioned the effects of HPP on DHCs. In Spanish apples, 500 and 600 MPa treatments led to reductions in phloridzin (16–20%) and phloretin-2′-xylosylglucoside (14–17%) levels. Conversely, in Italian apples, HPP at 600 MPa increased 51% and 67% the phloretin-2′-xylosylglucoside and phloridzin concentrations, respectively.

3.6. Effects on the Stability of Hydroxycinnamic Acids in Apple Products

Hydroxycinnamic acids (HCAs) are the major subgroup of phenolic acids and may occur either in their free or conjugated forms, including amides, esters, and glycosides [86]. The main aglycones identified in foodstuff are *p*-coumaric, caffeic, ferulic, and sinapic acids [87]. More frequently found in coffee, various fruits, some vegetables, and whole grains [88]. In apples, chlorogenic acid (5-O-caffeoylquinic) is the predominant compound. The HCA levels vary depending on the parts of the fruit. In apples, the flesh is the structure characterized for the higher levels of HCAs [89]. Ranges in whole apples were from 2.5 to 23.1 mg/100 g FW [11,12]. Whereas in apple flesh, HCAs levels varied from 0.7 to 14.3 mg/100 g FW [9,10]. Some potential health benefits have been documented from HCAs intakes, such as anti-microbial, anti-diabetic, antioxidant activity, prevention of cardiovascular and neurodegenerative diseases, and some cancer conditions [87,90,91].

For TT, four studies with 19 trials comprising heat treatments from 71 to 98 °C for 0.4 to 30 min provided concentrations data to calculate percentages of change of total HCAs. Increments (5 trials) ranged from 8% to 925% in apple sauce and apple juice [7,48]. On the other hand, reductions (14 trials) fluctuated from 4% to 49% in apple puree, sauce, and juice [7,46,50]. For HPP, four studies with 12 trials with pressurization conditions from 300 to 600 MPa for 5 to 15 min at 22 to 35 °C were reviewed. Positive effects (5 trials) from no significant differences to increments up to 29% were recorded in entire apples and apple juice [12,52]. Conversely, losses (7 trials) varied from 12% to 39% in whole apples and cloudy apple juice [10,53,54] (Table 3).

In general, both TT and HPP caused degradation of HCAs after processing. Nevertheless, although a higher degree of degradation was observed after heat treatment, it also caused the most significant increase in two trials with apple juice (the highest one is not shown in the graph for scale reasons) (Figure 4).

For TT, Alongi et al. [48] identified the highest increases in HCAs after heat treatment, observed in both chlorogenic acid and *p*-coumaroylquinic acid. A total of 205% and 925% increments on total HCAs were reported after the mild (71 °C/0.4 min) and intense (90 °C/15 min) pasteurization of apple juice, respectively. The increase in HCAs could result from the favoring effect of thermal treatment in enzyme inactivation and the release of chlorogenic acid from the cell walls to the food matrix [7,48]. These results corroborate the finding of De Paepe et al. [51], who concluded that phloretin-2′-O-glucoside and 3-O-caffeoylquinic acid were thermal-resistant compounds in cloudy apple juice isothermally treated from 80 to 145 °C during 7200 s. In contrast, Tian et al. [50] reported a 15% and 30% degradation of chlorogenic acid levels in apple juice after pre-pasteurization (98 °C/30 s) and pasteurization (98 °C/50 sec) respectively. In agreement with these results, a significant decrease of 44% in chlorogenic acid was recorded in apple puree heated at 90 °C for 30 min in the presence of oxygen after storage. However, no effect was observed in the heat-treated puree in the absence of oxygen [46], suggesting that oxidation reactions during heating are the main reason for the HCAs degradation [46,50]. In addition, Le Bourvellec et al. [7] reported that the higher differences on 5′-caffeoylquinic acid and *p*-coumaroylquinic acid after TT were due to apple variety. After cooking (95 °C/2 min) and pasteurization (90 °C/5 min), nine varieties experienced reductions from 1% to 48% in 5′-caffeoylquinic acid, whereas increases of 8%, 4%, and 30% were observed in Golden Delicious, Freiberg, and Granny Smith varieties, respectively.

For HPP, Marszalek et al. [54] found no significant differences in chlorogenic acid concentration of cloudy apple juice due to pressurization at 600 MPa for 5 min at 25 °C. However, there was a progressive degradation up to 53% after 12 weeks of storage at 4 °C. Accordingly, Szczepańska et al. [52] reported no differences in chlorogenic acid levels after static HPP at 300 MPa for 5 min and multi-pulsed pressurization (300 MPa × 3 pulses). However, a slight increase of around 5% was observed on the pressurized juices at 450 and 600 MPa for 5 min. Refrigerated storage for 12 weeks produced substantial losses of chlorogenic acid (66–77%). Reductions of 12% for total HCAs were reported for apples pressurized at 400 MPa for 5 min [53]. In another study, Fernández-Jalao et al. [12] reported that the HPP effect on individual and total HCAs was associated with the apple origin. In Spanish apples, pressurization reduced all individual HCAs, neochlorogenic, cryptochlorogenic, coumaroylquinic, and chlorogenic acids. The latter is the one that suffered the most significant degradations, with losses of 44% at 400 MPa, 24% at 500 MPa, and 15% at 600 MPa. Whereas in Italian apples, HPP at 600 MPa increased all the individual HCAs, resulting in a total increment of 29%. In contrast, HPP at 400 and 500 MPa reduced total HCAs by 16% and 13%, respectively.

4. Conclusions

This review analyzed the effects of TT and HPP treatments on the phenolic compounds in strawberry (F3OLs/PACs, ATs, and EA) and apple (F3OLs/PACs, FOLs, DHCs, and HCAs) products. Our findings show that the effect on polyphenols content (positive or negative) was contingent upon the type of processing, type of fruit, polyphenol family, and the shelf-life conditions (time and temperature during storage) of the final product. The impact of TT relied mainly on the food matrix and the thermal stability of the different phenolic compounds.

TT had positive effects in strawberry products, as was observed for F3OLs/PACs and EA. However, it had negative effects on the ATs content. These were due to the thermal instability of ATs and the enhanced extraction of EA and condensed tannins from achenes with thermal treatments. It is well known that ATs are heat-labile compounds, susceptible to oxidation and condensation reactions, which was confirmed in most of the studies where degradation of ATs was observed after TT. On the contrary, TT increased EA content due to the hydrolysis from ETs and release from the lignified matrix of the achenes.

In apple products, TT had positive effects, significantly promoting the release of FOLs and DHCs from peel and seeds. However, most of the studies observed a negative effect of F·OL/PACs, principally due to the extraction processes that quantified flavan-3-ols oligomers rather than monomers. Nevertheless, the final concentrations were variable depending on the variety, and the TT conditions applied. HPP treatments maintained the concentrations of phenolic compounds closer to those of the fresh or unprocessed samples regardless of the food matrix.

The impact of storage after TT and HPP had only been described for ATs in strawberry products. In general, TT has positive effects preserving better the ATs during storage. However, in the only study that examined the effect of HPP on ATs during storage, a negative effect was described showing a higher degradation than TT. Negative effects were also observed as the fast and pronounced degradation of FOLs, DHCs, and HCAs in apple products after HPP and storage.

The phenolic compounds' degradation after storage of HPP products could be due to the limitations in oxidative enzyme inactivation when thermal treatments were insufficient. From the industrial perspective, manufacturers aiming to preserve the natural content of polyphenols need to find the sweet spot between polyphenol stability and product shelf life. Further studies are recommended to compare how both technologies influence the content of the different families of polyphenols and their bioavailability and bioactivity.

Author Contributions: Conceptualization, L.M.S.-S. and F.A.T.-B.; methodology, L.M.S.-S. and R.G.-V.; formal analysis, G.L.S.-O.; investigation, G.L.S.-O., R.G.-V., L.M.S.-S., and F.A.T.-B.; resources, L.M.S.-S. and F.A.T.-B.; data curation, G.L.S.-O. and R.G.-V., and writing—original draft preparation, G.L.S.-O.; writing—review and editing, R.G.-V., L.M.S.-S., and F.A.T.-B.; supervision, R.G.-V., L.M.S.-S., and F.A.T.-B.; funding acquisition, L.M.S.-S. and F.A.T.-B. All authors have read and agreed to the published version of the manuscript.

Funding: This research was funded by HERO Group under the FOODPRINT project: Understanding and reframing minimal processing.

Institutional Review Board Statement: Not applicable.

Informed Consent Statement: Not applicable.

Data Availability Statement: Data sharing is not applicable to this article.

Conflicts of Interest: L.M.S.S. is a member of the Research and Nutrition Department of Hero Group, a Swiss international food manufacturer.

References

1. Tomás-Barberán, F.A.; Espín, J.C. Effect of Food Structure and Processing on (Poly) phenol—Gut Microbiota Interactions and the Effects on Human Health. *Annu. Rev. Food Sci. Technol.* **2019**, *10*, 221–239. [CrossRef] [PubMed]
2. Tomás-Barberán, F.A.; Espín, J.C. Phenolic compounds and related enzymes as determinants of quality in fruits and vegetables. *J. Sci. Food Agric.* **2001**, *81*, 853–876. [CrossRef]
3. Nowicka, A.; Kucharska, A.Z.; Sokół-Łętowska, A.; Fecka, I. Comparison of polyphenol content and antioxidant capacity of strawberry fruit from 90 cultivars of Fragaria × ananassa Duch. *Food Chem.* **2019**, *270*, 32–46. [CrossRef] [PubMed]
4. Guo, S.; Guan, L.; Cao, Y.; Li, C.; Chen, J.; Li, J.; Liu, G.; Li, S.; Wu, B. Diversity of polyphenols in the peel of apple (Malus sp.) germplasm from different countries of origin. *Int. J. Food Sci. Technol.* **2016**, *51*, 222–230. [CrossRef]
5. Aaby, K.; Skrede, G.; Wrolstad, R.E. Phenolic Composition and Antioxidant Activities in Flesh and Achenes of Strawberries (Fragaria ananassa). *J. Agric. Food Chem.* **2005**, *53*, 4032–4040. [CrossRef]
6. Buendía, B.; Gil, M.I.; Tudela, J.A.; Gady, A.L.; Medina, J.J.; Soria, C.; López, J.M.; Tomás-Barberán, F.A. HPLC-MS analysis of proanthocyanidin oligomers and other phenolics in 15 strawberry cultivars. *J. Agric. Food Chem.* **2010**, *58*, 3916–3926. [CrossRef]

7. Le Bourvellec, C.; Bouzerzour, K.; Ginies, C.; Renard, C.M.G.C.; Regis, S.; Ple, Y. Phenolic and polysaccharidic composition of applesauce is close to that of apple flesh. *J. Food Compos. Anal.* **2011**, *24*, 537–547. [CrossRef]
8. Xu, Y.; Fan, M.; Ran, J.; Zhang, T.; Sun, H.; Dong, M.; Zhang, Z.; Zheng, H. Variation in phenolic compounds and antioxidant activity in apple seeds of seven cultivars. *Saudi J. Biol. Sci.* **2016**, *23*, 379–388. [CrossRef]
9. Jakobek, L.; García-Villalba, R.; Tomás-Barberán, F.A. Polyphenolic characterisation of old local apple varieties from Southeastern European region. *J. Food Compos. Anal.* **2013**, *31*, 199–211. [CrossRef]
10. Akagić, A.; Vranac, A.; Gaši, F.; Drkenda, P.; Spaho, N.; Žuljević, O.; Kurtović, M.; Musić, O.; Murtić, S.; Hudina, M. Sugars, acids and polyphenols profile of commercial and traditional apple cultivars for processing. *Acta Agric. Slov.* **2019**, *113*, 239–250. [CrossRef]
11. Łata, B.; Trampczynska, A.; Paczesna, J. Cultivar variation in apple peel and whole fruit phenolic composition. *Sci. Hortic. (Amsterdam)* **2009**, *121*, 176–181. [CrossRef]
12. Fernández-Jalao, I.; Sánchez-Moreno, C.; De Ancos, B. Effect of high-pressure processing on flavonoids, hydroxycinnamic acids, dihydrochalcones and antioxidant activity of apple 'Golden Delicious' from different geographical origin. *Innov. Food Sci. Emerg. Technol.* **2019**, *51*, 20–31. [CrossRef]
13. Masi, E.; Taiti, C.; Vignolini, P.; William, A.; Giordani, E.; Heimler, D.; Romani, A.; Mancuso, S. Polyphenols and aromatic volatile compounds in biodynamic and conventional 'Golden Delicious' apples (Malus domestica Bork). *Eur. Food Res. Technol.* **2017**, *243*, 1519–1531. [CrossRef]
14. Román, S.; Sánchez-Siles, L.M.; Siegrist, M. The importance of food naturalness for consumers: Results of a systematic review. *Trends Food Sci. Technol.* **2017**, *67*, 44–57. [CrossRef]
15. Sanchez-Siles, L.M.; Michel, F.; Román, S.; Bernal, M.J.; Philipsen, B.; Haro, J.F.; Bodenstab, S.; Siegrist, M. The Food Naturalness Index (FNI): An integrative tool to measure the degree of food naturalness. *Trends Food Sci. Technol.* **2019**, *91*, 681–690. [CrossRef]
16. Debelo, H.; Li, M.; Ferruzzi, M.G. Processing influences on food polyphenol profiles and biological activity. *Curr. Opin. Food Sci.* **2020**, *32*, 90–102. [CrossRef]
17. Pérez-Andrés, J.M.; Charoux, C.M.G.; Cullen, P.J.; Tiwari, B.K. Chemical Modifications of Lipids and Proteins by Nonthermal Food Processing Technologies. *J. Agric. Food Chem.* **2018**, *66*, 5041–5054. [CrossRef]
18. Misra, N.N.; Koubaa, M.; Roohinejad, S.; Juliano, P.; Alpas, H.; Inácio, R.S.; Saraiva, J.A.; Barba, F.J. Landmarks in the historical development of twenty fi rst century food processing technologies. *Food Res. Int.* **2017**, *97*, 318–339. [CrossRef]
19. Koszucka, A.; Nowak, A. Thermal processing food-related toxicants: A review. *Crit. Rev. Food Sci. Nutr.* **2019**, *59*, 3579–3596. [CrossRef]
20. Van Boekel, M.; Fogliano, V.; Pellegrini, N.; Stanton, C.; Scholz, G.; Lalljie, S.; Somoza, V.; Knorr, D.; Jasti, P.R.; Eisenbrand, G. A review on the beneficial aspects of food processing. *Mol. Nutr. Food Res.* **2010**, *54*, 1215–1247. [CrossRef]
21. Fam, S.N. High-Pressure Processing in Food. *Biointerface Res. Apllied Chem.* **2020**, *11*, 11553–11561.
22. Wael, E.; Johari, E.; Yus, Y.; Rosnah, S.; Anvarjon, A. High Pressure Processing Technology and Equipment Evolution: A Review. *J. Eng. Sci. Technol. Rev.* **2015**, *8*, 75–83.
23. Martinez-Monteagudo, S.I.; Saldan, M.D.A. Chemical Reactions in Food Systems at High Hydrostatic Pressure. *Food Eng. Rev.* **2014**, *6*, 105–127. [CrossRef]
24. Makita, T. Application of high pressure and thermophysical properties of water to biotechnology. *Fluid Phase Equilib.* **1992**, *76*, 87–95. [CrossRef]
25. Tinello, F.; Lante, A. Recent advances in controlling polyphenol oxidase activity of fruit and vegetable products. *Innov. Food Sci. Emerg. Technol.* **2018**, *50*, 73–83. [CrossRef]
26. Aaby, K.; Grimsbo, I.H.; Hovda, M.B.; Rode, T.M. Effect of high pressure and thermal processing on shelf life and quality of strawberry purée and juice. *Food Chem.* **2018**, *260*, 115–123. [CrossRef]
27. Benito-Román, O.; Sanz, M.T.; Illera, A.E.; Melgosa, R.; Beltrán, S. Polyphenol oxidase (PPO) and pectin methylesterase (PME) inactivation by high pressure carbon dioxide (HPCD) and its applicability to liquid and solid natural products. *Catal. Today.* **2019**, *346*, 112–120. [CrossRef]
28. Smeriglio, A.; Barreca, D.; Bellocco, E.; Trombetta, D. Proanthocyanidins and hydrolysable tannins: Occurrence, dietary intake and pharmacological effects. *Br. J. Pharmacol.* **2017**, *174*, 1244–1262. [CrossRef]
29. Vrhovsek, U.; Rigo, A.; Tonon, D.; Mattivi, F. Quantitation of polyphenols in different apple varieties. *J. Agric. Food Chem.* **2004**, *52*, 6532–6538. [CrossRef]
30. Rauf, A.; Imran, M.; Abu-Izneid, T.; Iahtisham-Ul-Haq; Patel, S.; Pan, X.; Naz, S.; Sanches Silva, A.; Saeed, F.; Rasul Suleria, H.A. Proanthocyanidins: A comprehensive review. *Biomed. Pharmacother.* **2019**, *116*. [CrossRef]
31. Oliveira, A.; Coelho, M.; Alexandre, E.M.C.; Almeida, D.P.F.; Pintado, M. Long-Term Frozen Storage and Pasteurization Effects on Strawberry Polyphenols Content. *Food Bioprocess Technol.* **2015**, *8*, 1838–1844. [CrossRef]
32. Oliveira, A.; Almeida, D.P.F.; Pintado, M. Changes in Phenolic Compounds During Storage of Pasteurized Strawberry. *Food Bioprocess Technol.* **2014**, *7*, 1840–1846. [CrossRef]
33. Garzoli, S.; Cairone, F.; Carradori, S.; Mocan, A.; Menghini, L.; Paolicelli, P.; Ak, G.; Zengin, G.; Cesa, S. Effects of Processing on Polyphenolic and Volatile Composition and Fruit Quality of Clery Strawberries. *Antioxidants* **2020**, *9*, 632. [CrossRef] [PubMed]

34. Álvarez-Fernández, M.A.; Hornedo-Ortega, R.; Cerezo, A.B.; Troncoso, A.M.; García-Parrilla, M.C. Effects of the strawberry (fragaria ananassa) purée elaboration process on non-anthocyanin phenolic composition and antioxidant activity. *Food Chem.* **2014**, *164*, 104–112. [CrossRef]
35. Howard, L.R.; Brownmiller, C.; Prior, R.L. Improved color and anthocyanin retention in strawberry puree by oxygen exclusion. *J. Berry Res.* **2014**, *4*, 107–116. [CrossRef]
36. Holzwarth, M.; Korhummel, S.; Kammerer, D.R.; Carle, R. Thermal inactivation of strawberry polyphenoloxidase and its impact on anthocyanin and color retention in strawberry (Fragaria x ananassa Duch) pure. *Eur. Food Res. Technol.* **2012**, *235*, 1171–1180. [CrossRef]
37. Teleszko, M.; Nowicka, P.; Wojdyło, A. Effect of cultivar and storage temperature on identification and stability of polyphenols in strawberry cloudy juices. *J. Food Compos. Anal.* **2016**, *54*, 10–19. [CrossRef]
38. Bodelón, O.G.; Avizcuri, J.; Dizy, M. Pressurization and cold storage of strawberry purée: Colour, anthocyanins, ascorbic acid and pectin methylesterase. *LWT Food Sci. Technol.* **2013**, *52*, 123–130. [CrossRef]
39. Marszałek, K.; Woźniak, L.; Skąpska, S. The application of high pressure–mild temperature processing for prolonging the shelf-life of strawberry purée. *High Press. Res.* **2016**, *36*, 220–234. [CrossRef]
40. Stübler, A.S.; Lesmes, U.; Juadjur, A.; Heinz, V.; Rauh, C.; Shpigelman, A.; Aganovic, K. Impact of pilot-scale processing (thermal, PEF, HPP) on the stability and bioaccessibility of polyphenols and proteins in mixed protein- and polyphenol-rich juice systems. *Innov. Food Sci. Emerg. Technol.* **2020**, *64*, 102426. [CrossRef]
41. Marszałek, K.; Mitek, M.; Skąpska, S. The effect of thermal pasteurization and high pressure processing at cold and mild temperatures on the chemical composition, microbial and enzyme activity in strawberry purée. *Innov. Food Sci. Emerg. Technol.* **2015**, *27*, 48–56. [CrossRef]
42. Terefe, N.S.; Kleintschek, T.; Gamage, T.; Fanning, K.J.; Netzel, G.; Versteeg, C.; Netzel, M. Comparative effects of thermal and high pressure processing on phenolic phytochemicals in different strawberry cultivars. *Innov. Food Sci. Emerg. Technol.* **2013**, *19*, 57–65. [CrossRef]
43. Marszałek, K.; Woźniak, Ł.; Skąpska, S.; Mitek, M. High pressure processing and thermal pasteurization of strawberry purée: Quality parameters and shelf life evaluation during cold storage. *J. Food Sci. Technol.* **2017**, *54*, 832–841. [CrossRef]
44. Cao, X.; Zhang, Y.; Zhang, F.; Wang, Y.; Yi, J.; Liao, X. Effects of high hydrostatic pressure on enzymes, phenolic compounds, anthocyanins, polymeric color and color of strawberry pulps. *J. Sci. Food Agric.* **2011**, *91*, 877–885. [CrossRef]
45. Cao, X.; Bi, X.; Huang, W.; Wu, J.; Hu, X.; Liao, X. Changes of quality of high hydrostatic pressure processed cloudy and clear strawberry juices during storage. *Innov. Food Sci. Emerg. Technol.* **2012**, *16*, 181–190. [CrossRef]
46. Kim, A.N.; Lee, K.Y.; Rahman, M.S.; Kim, H.J.; Kerr, W.L.; Choi, S.G. Thermal treatment of apple puree under oxygen-free condition: Effect on phenolic compounds, ascorbic acid, antioxidant activities, color, and enzyme activities. *Food Biosci.* **2021**, *39*, 100802. [CrossRef]
47. Murtaza, A.; Iqbal, A.; Marszałek, K.; Iqbal, M.A.; Ali, S.W.; Xu, X.; Pan, S.; Hu, W. Enzymatic, phyto-, and physicochemical evaluation of apple juice under high-pressure carbon dioxide and thermal processing. *Foods* **2020**, *9*, 243. [CrossRef]
48. Alongi, M.; Verardo, G.; Gorassini, A.; Anese, M. Effect of pasteurization on in vitro α -glucosidase inhibitory activity of apple juice. *LWT Food Sci. Technol.* **2018**, *98*, 366–371. [CrossRef]
49. Knebel, T.; Braun, P.; Dietrich, H. Degradation kinetics of anthocyanins and polyphenols during storage of red apple juice produced from red-fleshed apples. *Eur. Food Res. Technol.* **2018**, *244*, 1741–1750. [CrossRef]
50. Tian, Y.; Sun, L.; Yang, Y.; Gou, X.; Niu, P.; Guo, Y. Changes in the physicochemical properties, aromas and polyphenols of not from concentrate (NFC) apple juice during production. *CyTA J. Food* **2018**, *16*, 755–764. [CrossRef]
51. De Paepe, D.; Valkenborg, D.; Coudijzer, K.; Noten, B.; Servaes, K.; De Loose, M.; Voorspoels, S.; Diels, L.; Van Droogenbroeck, B. Thermal degradation of cloudy apple juice phenolic constituents. *Food Chem.* **2014**, *162*, 176–185. [CrossRef]
52. Szczepańska, J.; Pinto, C.A.; Skąpska, S.; Saraiva, J.A.; Marszałek, K. Effect of static and multi-pulsed high pressure processing on the rheological properties, microbial and physicochemical quality, and antioxidant potential of apple juice during refrigerated storage. *LWT Food Sci. Technol.* **2021**, *150*, 112038. [CrossRef]
53. Fernández-Jalao, I.; Balderas, C.; Sánchez-Moreno, C.; De Ancos, B. Impact of an in vitro dynamic gastrointestinal digestion on phenolic compounds and antioxidant capacity of apple treated by high-pressure processing. *Innov. Food Sci. Emerg. Technol.* **2020**, *66*, 102486. [CrossRef]
54. Marszalek, K.; Szczepańska, J.; Starzonek, S.; Wozniak, L.; Trych, U.; Skapska, S.; Rzoska, S.; Saraiva, J.A.; Lorenzo, J.M.; Barba, F.J. Enzyme inactivation and evaluation of physicochemical properties, sugar and phenolic profile changes in cloudy apple juices after high pressure processing, and subsequent refrigerated storage. *Food Process. Eng.* **2019**, *42*, e13034. [CrossRef]
55. Castañeda-Ovando, A.; Pacheco-Hernández, M.d.L.; Páez-Hernández, M.E.; Rodríguez, J.A.; Galán-Vidal, C.A. Chemical studies of anthocyanins: A review. *Food Chem.* **2009**, *113*, 859–871. [CrossRef]
56. Krga, I.; Milenkovic, D. Anthocyanins: From Sources and Bioavailability to Cardiovascular-Health Benefits and Molecular Mechanisms of Action. *J. Agric. Food Chem.* **2019**, *67*, 1771–1783. [CrossRef]
57. Bueno, J.M.; Sáez-Plaza, P.; Ramos-Escudero, F.; Jiménez, A.M.; Fett, R.; Asuero, A.G. Analysis and Antioxidant Capacity of Anthocyanin Pigments. Part II: Chemical Structure, Color, and Intake of Anthocyanins. *Crit. Rev. Anal. Chem.* **2012**, *42*, 126–151. [CrossRef]

58. Tena, N.; Martín, J.; Asuero, A.G. State of the art of anthocyanins: Antioxidant activity, sources, bioavailability, and therapeutic effect in human health. *Antioxidants* **2020**, *9*, 451. [CrossRef]
59. Mattioli, R.; Francioso, A.; Mosca, L.; Silva, P. Anthocyanins: A Comprehensive Review of Their Chemical Properties and Health Effects on Cardiovascular and Neurodegenerative Diseases. *Molecules* **2020**, *25*, 3809. [CrossRef]
60. Motilva, M.; Bars-cortina, D.; Sakhawat, A.; Pi, C. Chemopreventive effects of anthocyanins on colorectal and breast cancer: A review. *Semin. Cancer Biol.* **2021**, *6*, 1203. [CrossRef]
61. Li, D.; Wang, P.; Luo, Y.; Zhao, M.; Chen, F. Health benefits of anthocyanins and molecular mechanisms: Update from recent decade. *Crit. Rev. Food Sci. Nutr.* **2017**, *57*, 1729–1741. [CrossRef] [PubMed]
62. Skrede, G.; Wrolstad, R.E.; Lea, P.; Enersen, G. Color Stability of Strawberry and Blackcurrant Syrups. *J. Food Sci.* **1992**, *57*. [CrossRef]
63. Nikkhah, E.; Khaiamy, M.; Heidary, R.; Azar, A.S. The effect of ascorbic acid and H2O2 treatment on the stability of anthocyanin pigments in berries. *Turk J. Biol.* **2010**, *34*, 47–53. [CrossRef]
64. Pacheco-Palencia, L.A.; Hawken, P.; Talcott, S.T. Phytochemical, antioxidant and pigment stability of ac (Euterpe oleracea Mart) as affected by clarification, ascorbic acid fortification and storage. *Food Res. Int.* **2007**, *40*, 620–628. [CrossRef]
65. Bakkalbasi, E.; Mentes, O.; Artik, N. Food ellagitannins-occurrence, effects of processing and storage. *Crit. Rev. Food Sci. Nutr.* **2009**, *49*, 283–298. [CrossRef]
66. Williner, M.R.; Pirovani, M.E.; Güemes, D.R. Ellagic acid content in strawberries of different cultivars and ripening stages. *J. Sci. Food Agric.* **2003**, *83*, 842–845. [CrossRef]
67. Tomás-Barberán, F.A.; González-Sarrías, A.; García-Villalba, R.; Núñez-Sánchez, M.A.; Selma, M.V.; García-Conesa, M.T.; Espín, J.C. Urolithins, the rescue of "old" metabolites to understand a "new" concept: Metabotypes as a nexus among phenolic metabolism, microbiota dysbiosis, and host health status. *Mol. Nutr. Food Res.* **2017**, *61*. [CrossRef]
68. Cerdá, B.; Periago, P.; Espín, J.C.; Tomás-Barberán, F.A. Identification of urolithin A as a metabolite produced by human colon microflora from ellagic acid and related compounds. *J. Agric. Food Chem.* **2005**, *53*, 5571–5576. [CrossRef]
69. Al-Harbi, S.A.; Abdulrahman, A.O.; Zamzami, M.A.; Khan, M.I. Urolithins: The Gut Based Polyphenol Metabolites of Ellagitannins in Cancer Prevention, a Review. *Front. Nutr.* **2021**, *8*, 1–15. [CrossRef]
70. Okuda, T.; Yoshida, T.; Hatano, T. Ellagitannins as active constituents of medicinal plants. *Planta Med.* **1989**, *55*, 117–122. [CrossRef]
71. Truchado, P.; Larrosa, M.; García-Conesa, M.T.; Cerdá, B.; Vidal-Guevara, M.L.; Tomás-Barberán, F.A.; Espín, J.C. Strawberry processing does not affect the production and urinary excretion of urolithins, ellagic acid metabolites, in humans. *J. Agric. Food Chem.* **2012**, *60*, 5749–5754. [CrossRef]
72. Landete, J.M. Ellagitannins, ellagic acid and their derived metabolites: A review about source, metabolism, functions and health. *Food Res. Int.* **2011**, *44*, 1150–1160. [CrossRef]
73. Materska, M. Quercetin and its derivatives: Chemical structure and bioactivity—A review. *Polish J. Food Nutr. Sci.* **2008**, *58*, 407–413.
74. Kamiloglu, S. Effect of different freezing methods on the bioaccessibility of strawberry polyphenols. *Int. J. Food Sci. Technol.* **2019**, *54*, 2652–2660. [CrossRef]
75. Herrmann, K. Flavonols and flavones in food plants: A review. *Int. J. Food Sci. Technol.* **1976**, *11*, 433–448. [CrossRef]
76. Lesjak, M.; Beara, I.; Pinta, D.; Majki, T.; Bekvalac, K.; Or, D.; Mimica-duki, N. Antioxidant and anti-in fl ammatory activities of quercetin and its derivatives. *J. Funct. Foods* **2018**, *40*, 68–75. [CrossRef]
77. Panthi, V.K.; Kaushal, S.; Adhikari, B.; Basnet, N.; Chaudary, D.; Parajuli, R.; Pokhrel, P. A Review of Quercetin: Anticancer Activity. *Int. J. Innov. Res. Rev.* **2020**, *4*, 1–7.
78. Islam, S.; Quispe, C.; Hossain, R.; Islam, M.T. Neuropharmacological Effects of Quercetin: A Literature-Based Review. *Front. Pharmacol.* **2021**, *12*, 1–16. [CrossRef]
79. Alongi, M.; Verardo, G.; Gorassini, A.; Lemos, M.A.; Hungerford, G.; Cortella, G.; Anese, M. Phenolic content and potential bioactivity of apple juice as affected by thermal and ultrasound pasteurization. *Food Funct.* **2019**, *10*, 7366–7377. [CrossRef]
80. Tomás-Barberán, F.; Garcia-viguera, C.; Nieto, J.L.; Ferreres, F. Dihydrochalcones from apple juices and jams. *Food Chem.* **1993**, *46*, 33–36. [CrossRef]
81. Tian, L.; Cao, J.; Zhao, T.; Liu, Y.; Khan, A.; Cheng, G. The bioavailability, extraction, biosynthesis and distribution of natural dihydrochalcone: Phloridzin. *Int. J. Mol. Sci.* **2021**, *22*, 962. [CrossRef]
82. Viet, M.H.; Chen, C.Y.; Hu, C.K.; Chen, Y.R.; Li, M.S. Discovery of dihydrochalcone as potential lead for Alzheimer's disease: In Silico and in Vitro study. *PLoS ONE* **2013**, *8*, e79151. [CrossRef]
83. Raphaelli, C.d.O.; Azevedo, J.G.; Pereira, E.d.S.; Vinholes, J.R.; Camargo, T.M.; Hoffmann, J.F.; Ribeiro, J.A.; Vizzotto, M.; Rombaldi, C.V.; Wink, M.R.; et al. Phenolic-rich apple extracts have photoprotective and anti-cancer effect in dermal cells. *Phytomedicine Plus* **2021**, *1*, 100112. [CrossRef]
84. Tomás-Barberán, F.; Clifford, M. Flavanones, chalcones and dihydrochalcones—Nature, occurrence and dietary burden. *J. Sci. Food Agric.* **2000**, *80*, 1073–1080. [CrossRef]
85. Oszmianski, J.; Wolniak, M.; Wojdylo, A.; Wawer, I. Influence of apple puree preparation and storage on polyphenol contents and antioxidant activity. *Food Chem.* **2008**, *107*, 1473–1484. [CrossRef]

86. Teixeira, J.; Gaspar, A.; Garrido, E.M.; Garrido, J.; Borges, F. Hydroxycinnamic Acid Antioxidants: An Electrochemical Overview. *Biomed. Res. Int.* **2013**, *2013*, 1–11. [CrossRef]
87. Sova, M. Activities and Health Benefits of Hydroxycinnamic Acids and Their Metabolites. *Nutrients* **2020**, *12*, 2190. [CrossRef]
88. Vinholes, J.; Silva, B.M.; Silva, L.R. Hydroxycinnamic acids (HCAS): Structure, Biological properties and health effects. *Adv. Med. Biol.* **2019**, *88*, 1–33.
89. Tsao, R.; Yang, R.; Young, J.C.; Zhu, H. Polyphenolic profiles in eight apple cultivars using high-performance liquid chromatography (HPLC). *J. Agric. Food Chem.* **2003**, *51*, 6347–6353. [CrossRef]
90. Coman, V.; Vodnar, D.C. Hydroxycinnamic acids and human health: Recent advances. *Soc. Chem. Ind.* **2019**. [CrossRef]
91. Rocha, L.D.; Monteiro, M.C.; Teodoro, A.J. Anticancer Properties of Hydroxycinnamic Acids -A Review. *Cancer Clin. Oncol.* **2012**, *1*, 109–121. [CrossRef]

foods

MDPI

Article

High-Pressure Processing on Whole and Peeled Potatoes: Influence on Polyphenol Oxidase, Antioxidants, and Glycaemic Indices

Konstantina Tsikrika [1], Aine Muldoon [2], Nora M. O'Brien [2] and Dilip K. Rai [1,*]

1 Teagasc Food Research Centre Ashtown, Department of Food BioSciences, D15 DY05 Dublin, Ireland; kontsikrika@gmail.com
2 School of Food and Nutritional Sciences, University College Cork, T12 YN60 Cork, Ireland; 118222208@umail.ucc.ie (A.M.); nob@ucc.ie (N.M.O.)
* Correspondence: dilip.rai@teagasc.ie; Tel.: +353-(0)1805-9569

Abstract: Polyphenol oxidase (PPO) inactivation in five whole and peeled Irish potato cultivars was investigated using high-pressure processing (HPP) at 400 MPa and 600 MPa for 3 min. PPO activity was significantly lower in most of the HPP-treated samples, while the highest PPO inactivation was observed after HPP at 600 MPa. No significant ($p > 0.05$) changes were observed on the total phenolic content and antioxidant activity of all the HPP-treated potatoes. Regarding individual phenolic acids, chlorogenic acid was decreased significantly ($p < 0.05$) in all studied varieties with a concomitant increase ($p < 0.05$) in caffeic and quinic acid. Similarly, ferulic acid was also increased ($p < 0.05$) in all studied varieties after the HPP treatment, while there was a variation in rutin and 4-coumaric acid levels depending on the cultivar and the sample type. Anthocyanins in the coloured whole potato varieties (i.e., Kerr's Pink and Rooster), tentatively identified as pelargonidin-*O*-ferulorylrutinoside-*O*-hexoside and pelargonidin-*O*-rutinoside-*O*-hexoside, also exhibited significantly ($p < 0.05$) higher levels in the HPP-treated samples as opposed to those untreated. Glycaemic indices of the potatoes treated with HPP did not differ with the corresponding untreated cultivars.

Keywords: high-pressure processing; potatoes; polyphenol oxidase; polyphenols; antioxidant activity

Citation: Tsikrika, K.; Muldoon, A.; O'Brien, N.M.; Rai, D.K. High-Pressure Processing on Whole and Peeled Potatoes: Influence on Polyphenol Oxidase, Antioxidants, and Glycaemic Indices. *Foods* **2021**, *10*, 2425. https://doi.org/10.3390/foods10102425

Academic Editors: Milan Houška and Francesco Donsì

Received: 22 July 2021
Accepted: 9 October 2021
Published: 13 October 2021

1. Introduction

Potato consumption has been associated with beneficial properties such as anti-cancer, hypocholesterolemic, anti-inflammatory, anti-obesity, and anti-diabetic properties in human cell culture, experimental animals, and human clinical studies [1]. Although potatoes have high carbohydrate levels that can result in high glycaemic load (GL), recent reviews present inconsistent effects regarding potato consumption on the risk of type 2 diabetes, obesity, and cardio-metabolic health [2]. In fact, studies have shown that glycaemic index (GI) values decrease in cooled potatoes after cooking [3,4], and are greatly influenced by the cultivar [5]. Furthermore, whole potato consumption can significantly improve the cardioprotective fibre intake [6].

The processing of potatoes prior to their consumption is necessary mostly due to the presence of indigestible ungelatinised starch [7] and the fact that the potato industry has seen promising growth in the demand of minimally processed potatoes, i.e., pre-peeled, fresh cut, or sliced [8]. However, peeling and cutting tubers lead to changes in colour, which could also impact on their sensory properties such as flavour [9]. Polyphenol oxidase (PPO) is the main enzyme causing browning in cut potatoes. PPO catalyses the oxidation of polyphenols that are present in potatoes to quinones, which then undergo further polymerisation into melanin pigments resulting in the aforementioned undesirable changes [10]. Therefore, the inactivation of PPO is essential in order to maintain the quality of potatoes. Although thermal blanching is commonly used to inactivate PPO, its usage has shown a detrimental impact on the nutritional value, physicochemical, and organoleptic characteristics of food [11].

Consumers' demand for fresh-like products in tandem with enhanced legislation restrictions have led to advancements in food processing technologies that enable a wholesome, fresh product with extended shelf-life [7,12]. In this context, non-thermal processing such as high-pressure processing (HPP) offers a feasible choice against conventional thermal processes [13]. HPP involves applying pressures ranging from 100 to 1000 MPa, commonly at room temperatures and for a short time to packaged foods using water as a medium to transmit pressure [14]. HPP has been applied in foods in order to inhibit the growth of food-borne microorganisms [15–17] and enzymes responsible for browning and anaerobic metabolism [18–20], while a decreased respiratory activity has also been observed upon the HPP treatment [21].

Previous work of this group has shown that HPP at 600 MPa on whole potatoes exhibited promising results with regard to PPO inactivation while having insignificant impacts on the antioxidant activity, proximate composition, and phytochemical constituents [22,23]. This study further explores the application of two different high-pressure treatments in whole and peeled potatoes of baby potatoes (Maris Piper), two coloured (Kerr's Pink, Rooster) and two white (Cultra, Maris Piper) varieties that are popular to consumers and widely used by the industry in the island of Ireland. Specifically, the impact of HPP at 400 MPa and 600 MPa was investigated on PPO inactivation, antioxidant activity, polyphenols (including anthocyanins), and their glycaemic indices were determined following the HPP treatments.

2. Materials and Methods

2.1. Samples

Potatoes of five different cultivars (Kerr's Pink, Rooster, Maris Piper, Cultra, and Maris Piper baby) and at their commercial maturity were bought from a local supermarket in Dublin, Ireland and were stored in the dark at 4 °C prior to processing.

2.2. Chemicals

Hydrochloric acid (HCl), sodium phosphate monobasic, sodium phosphate dibasic, catechol, Folin–Ciocalteu's phenol reagent, gallic acid, methanol, ethanol, sodium carbonate, formic acid, 6-hydroxy-2,5,7,8-tetramethylchroman-2-carboxylic acid (Trolox), 2,4,6-tripyridyl-striazine, Iron(III) chloride hexahydrate, sodium acetate anhydrous, acetic acid, sodium hydroxide, potassium hydroxide (KOH), guar gum, methanol, 2,2-diphenyl-1-picrylhydrazyl (DPPH), phytochemical standards (quinic acid, chlorogenic acid, caffeic acid, rutin, 4-coumaric acid, ferulic acid), and the enzymes (pepsin, invertase, amyloglucosidase, and pancreatin) were purchased from Merck Ltd. (formerly Sigma Aldrich Ltd., Wicklow, Ireland).

2.3. HPP Treatment

Potatoes free of defects were selected, washed, dried, and half of them were peeled. Whole (not-peeled) and peeled potatoes (~1 kg) were packaged separately in polyethylene/polyamide pouches and vacuum-sealed. HPP treatment was performed at 400 MPa and 600 MPa for 3 min with initial temperature at 7.0 °C and reached 10.6 °C (max. temperature reached) (Supplementary Materials Figure S1). Then, 8 kg each of vacuum-packed whole and peeled potatoes of each cultivar were loaded in the 420 Litre Hiperbaric vessel that operated in a horizontal mode on a commercial-scale high pressure process (Hiperbaric 55HT, Miami, FL, USA) that was used located at HPP Tolling (St. Margaret's, Co., Dublin, Ireland).

2.4. Potato PPO Extraction and Activity Assay

The extraction and activity assay of PPO were carried out as previously mentioned [17,23]. Briefly, potato (20 g) was blended with cold phosphate buffer (40 mL; 0.1 M; pH 6.5; 4 °C) for 3 min, and the mixture was centrifuged at 10,000× *g* at 4 °C for 25 min. The supernatant was used to monitor PPO activity. PPO activity assay mixture consisted of phosphate buffer (1.5 mL; 0.1 M; pH 5.5), catechol (1 mL; 0.2 M), and crude PPO extract (0.5 mL). The absorbance was measured at 410 nm for 2 min by a spectrophotometer (UV-1700, Shimadzu, Nanjing, China). The residual activity (RA) of PPO was calculated, using the following Equation (1):

$$RA\ (\%) = (A_t/A_0) \times 100 \quad (1)$$

where A_t and A_0 are PPO activity after and before the treatment, respectively.

2.5. Extraction of Phytochemicals

The extraction of phytochemicals was conducted following previously published literature [17,23]. Potato samples were cut into cubes and frozen overnight on silver-foil trays. Freeze drying of the frozen cubes were performed in a Cuddon freeze-drier, model FD80 (Cuddon Freeze Dry, Blenheim, New Zealand) at a temperature of −54 °C and a pressure of 0.064 mbar for 72 h followed by grinding into powders. The extraction of phytochemicals from the lyophilised whole and peeled potatoes was carried out overnight in 80% methanol containing 0.01% formic acid (1:5 *w*/*v*) at 4 °C. This was followed by sonication at 30 °C for 30 min and then centrifuging at 10,000× *g* for 30 min. The extraction process was repeated on the residue. The supernatants were pooled and filtered through 0.45 µm syringe filters.

2.6. Total Phenolic Content

The total phenolic content (TPC) of whole and peeled potatoes was assessed using the Folin–Ciocalteu reagent (FCR) method [24] adapted to a microplate reader (FLUOstar Omega Microplate Reader, BMG Labtech GmbH, Offenburg, Germany) as previously described in [17,23]. Potato extract (100 µL) was added to a solution containing FCR (100 µL), sodium carbonate (100 µL; 20% *w*/*v*), and methanol (100 µL). The sample–reagent mixture was allowed to react in the dark for 20 min, and then, it was centrifuged at 13,000 rpm for 3 min. The absorbance was monitored at 735 using gallic acid as a standard, while methanol was used as a blank. The results were calculated as µg of gallic acid equivalent per g of the extract's dry weight (µg GAE/g dw).

2.7. Antioxidant Activity (AOA)

2.7.1. Ferric-Reducing Antioxidant Power (FRAP)

The FRAP assay was carried out following an amended method of Benzie and Strain (1996) [25] by Stratil et al. (2006) [26] and Ou et al. (2002) [27]. The FRAP solution contained acetate buffer (100 mL; 0.3 M; pH 3.6), ferrous chloride hexahydrate (10 mL; 0.01 M) and TPTZ (10 mL; 0.01 M in 0.04 M HCl). Potato extract (20 µL) was added to FRAP solution (180 µL) in the microplate well, and the mixture was kept at 37 °C for 40 min before reading the absorbance at 593 nm. Trolox and methanol were used as standard and blank, respectively. The results were reported as µg of Trolox equivalent per g of the extract's dry weight (µg TE/g dw).

2.7.2. DPPH Radical Scavenging Capacity

The DPPH radical scavenging capacity of HPP treated and untreated peeled and whole potatoes was assayed as described in Goupy et al. (1999) [28]. A stock solution of DPPH (11.9 mg) and methanol (50 mL) was prepared followed by further a 1:5 dilution of the stock solution for the assay. The diluted DPPH solution (100 µL) was mixed with potato extract (100 µL) and kept in the dark for 30 min. Then, the absorbance was measured at 515 nm in the microplate reader, while Trolox and methanol were used as standard and blank, respectively. The results were reported as µg of Trolox equivalent per g of the extract's dry weight (µg TE/g dw).

2.8. Liquid Chromatography-Mass Spectrometry Analysis

The identification of polyphenols was performed by HPLC-QToF mass spectrometry [23,29] and their quantification by ultra-high performance liquid chromatography coupled to a tandem quadrupole mass spectrometer (UPLC-TQD, Waters Corp., Milford, MA, USA). In addition, two anthocyanins, namely pelargonidin-*O*-ferulorylrutinoside-*O*-hexoside and pelargonidin-*O*-rutinoside-*O*-hexoside, were tentatively identified based on their molecular mass and fragmentation pattern [30]. The separation of the natural compounds in the potato extract was performed on a Waters Acquity UPLC HSS T3 column (100 × 2.1 mm, 1.8 µm). The binary solvents constituted water containing 0.1% formic acid and acetonitrile containing 0.1% formic acid at 0.5 mL/min [31,32]. A multiple reaction monitoring (MRM) mode analysing at least two transitions per compound was employed to detect and quantify the phenolic compounds in the UPLC-TQD. IntelliStart^{TM} software (Masslynx 4.1, Waters Corp., Milford, MA, USA) was used in order to optimise the cone voltages and collision energies for each MRM transition. As for the two tentative anthocyanins, MRM transitions were obtained from Kim et al. 2018 [30]. Quadruplicate extracts were used for the analyses, while the target compounds were quantified using standard calibration curves of concentrations ranging from 10–25 µg/mL. The two anthocyanins were quantified as rutin equivalents. The results were reported as µg compound per g of extract in dry weight (µg/g dw).

2.9. In Vitro Determination of Glycaemic Index (GI)

2.9.1. In Vitro Digestion

The enzyme mixture (0.5 g pepsin, 0.5 g guar gum, 100 mL 0.05 mol/L HCl) was prepared, and the reaction mixture was maintained at room temperature. Then, a potato sample (1 g) was added to pepsin–guar gum mixture (10 mL) followed by incubation in a shaking water bath at 37 °C for 30 min. Five glass balls (0.5 cm diameter) were mixed with sodium acetate (10 mL; 0.25 mol/L), along with the second enzyme mixture (5 mL), which contained invertase (2000 units/mL), amyloglucosidase, and pancreatin. Then, samples were incubated in a 37 °C water bath, and after 20 min and 120 min, an aliquot of 0.2 mL was taken out and mixed with 20 mL of 66% ethanol for glucose analysis (these were the samples, which were used to calculate the rapidly available glucose (RAG) and slowly available glucose (SAG). The remaining mixture was vortexed and then incubated for 30 min at 100 °C. The mixture was cooled for 20 min in ice bath; then, we added KOH (10 mL; 7 mol/L) followed by an incubation at 0 °C in a shaking ice bath for 30 min. The mixture was vortexed, and 1 mL of solution mixture was transferred into a heat-resistant beaker, which contained acetic acid (10 mL; 0.5 mol/L) and amyloglucosidase (0.2 mL). Then, this mixture was incubated for 30 min at 70 °C in a shaking water bath. The mixture was taken out from the water bath, after which 40 mL of distilled water was added, and the mixture was vortexed. This was the sample used to determine total glucose (TG), which was performed immediately [33].

2.9.2. Glucose Analysis

Glucose was determined using the Megazyme D-glucose assay kit (GOPOD-format). In glass heat-resistant tubes, 3 mL of GOPOD reagent was mixed with 0.1 mL of sample solution (RAG, SAG, TG, and FG (free glucose)). Then, the sample-reagent mixture was incubated for 20 min at 50 °C, and the absorbance was measured at 510 nm on the spectrophotometer.

Three independent experiments were carried out in all studies. Each sample was tested in duplicate (n = 6).

Equations (2)–(6) as described by Englyst et al. (2000) [33] were used to calculate the GI, GL, total starch (TS), RAG, and SAG:

$$RAG_{rel} = \frac{RAG \times 100}{TG} \quad (2)$$

$$SAG_{rel} = \frac{SAG \times 100}{TG} \quad (3)$$

$$TS = 0.9 \times (TG - FG) \quad (4)$$

$$GI = 17.7 + 77.9\frac{RAG}{TS + 2FG} \quad (5)$$

$$GL = GI\frac{TS + 2FG}{100}. \quad (6)$$

2.10. Statistical Analysis

Experiments were performed in duplicates, and all analyses (except for GI) were repeated four times (n = 8). Results are reported as means ± standard deviation (SD). One-way ANOVA followed by a Tukey's post hoc test was performed for the statistical analysis of PPO RA before/after treatment, while a Games–Howell post hoc test was used for the rest of the data. SPSS Statistics 26 (IBM-Armonk, New York, NY, USA) was used for all the statistical analyses. The significance level was set at $p < 0.05$.

3. Results

3.1. PPO Inactivation

The effect of HPP treatments on the activity of PPO expressed as residual activity (RA) is shown in Figure 1. Although HPP at 400 MPa and 600 MPa for 3 min each decreased the PPO activity significantly ($p < 0.05$) in both the peeled and whole potatoes, the HPP at 600 MPa was more effective against PPO, where the RA ranged from approximately 21% to 51% in peeled and 23% to 53% in whole potatoes. As with the HPP at 400 MPa, the RA of the PPO varied from approximately 83% to 93% in peeled and 86% to 97% in whole samples of the studied cultivars, while no significant changes ($p > 0.05$) were observed in the peeled baby Maris Piper and whole Rooster samples as compared to the controls. In general, a higher PPO activity was found in the whole than in the peeled potatoes irrespective of different pressure (400 MPa and 600 MPa) treatments, which was also noted by Thygesen et al. (1995) [34]. The PPO RA values of whole potatoes are similar to our previous study where it was found to vary from 31% to 51% after HPP at 600 MPa [23]. Similarly, a 30% inactivation of PPO by HPP at 600 MPa has been reported in carrot juice [19], while HPP at 400 MPa and 600 MPa decreased PPO by 11% and 30%, respectively in strawberry puree [35]. A significant ($p < 0.05$) reduction in PPO RA was also observed in a commercial PPO isolated from mushroom under HPP at 600 MPa for 5 min [36], and in Peruvian carrot and cocoyam puree, cubes, and extract after HPP treatment at 600 MPa for 5 or 30 min. On the contrary, PPO activity in HPP-treated sweet potato was much higher, reaching even four times higher RA than the controls, while a higher PPO activity was also observed in HPP-treated avocado slices than in those untreated [21]. These findings indicate the importance and relevance of the food matrix in PPO inactivation by HPP.

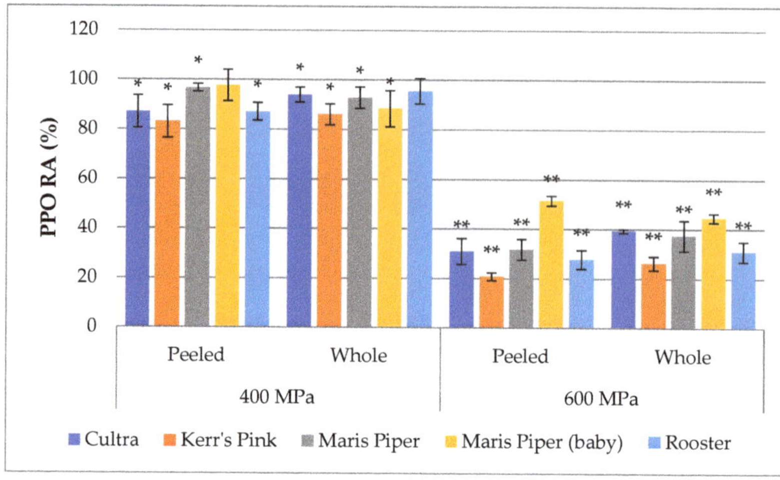

Figure 1. Residual activity (%) of PPO from whole and peeled potatoes upon HPP treatment at 400 MPa and 600 MPa. '*' indicates statistically significant ($p < 0.05$) difference from the control. '**' indicates statistically significant difference ($p < 0.05$) within the same cultivar and sample type (peeled/whole).

HPP-induced enzyme inactivation is a multifaceted phenomenon that has not been fully determined yet. Potential mechanisms involve the formation and/or disruption of various intramolecular interactions, hydration of charged groups, interference of bound water, the stabilisation of hydrogen bonds, as well as alterations in the native structure of enzymes by folding and/or unfolding [36]. HPP may lead to reversible or irreversible and partial or complete unfolding of the native structure of the enzyme, resulting in a change in enzyme activity, as its specificity is very much associated to its active sites in the protein structure [37]. From a thermodynamic perspective, HPP may influence enzyme-catalysed reactions by modifying the equilibrium and the rate constants [38]. One or a combination of these factors might have contributed to the observed loss of PPO activity in the current study.

3.2. Total Phenolic Content

Figure 2 shows the total phenolic content (TPC) of peeled and whole potatoes of the studied cultivars before and after HPP treatment. TPC in untreated potatoes was found to range from 568 to 853 μg GAE/g dw in peeled, while it was higher in whole samples (686 to 938 μg GAE/g dw), which is consistent with the literature [39]. As shown in Figure 2, the impact of HPP on TPC of peeled and whole potatoes exhibits a variation, mainly depending on the cultivar. Nevertheless, the changes are not significant ($p > 0.05$). These results contradict previous findings where TPC levels were significantly ($p < 0.05$) higher in whole potatoes after HPP at 600 MPa than in those untreated [23]. This difference may be due to the fact that the levels of phenolic compounds in the potatoes as well as their stability and consequently the effect of a processing method on them are highly dependent on several factors such as genetic (cultivar), environmental, agronomic, the stage of ripeness, and the post-harvest handling and storage [40,41]. No significant changes in TPC have also been reported in HPP treated mango nectars [42], litchi juice [43], and acai juice [44] with respect to the controls. On the contrary, a significant increase was observed in TPC in pumpkin slices after HPP at 450 MPa for 15 min and 550 MPa for 10 min [16] and in HPP-treated pomegranate juice [33] than in the corresponding untreated samples, which was associated with a higher cell permeability, as a consequence of the disruption of the cell walls and the cell membrane hydrophobic bonds, and thereby leading to mass release of matrix-bound phenolic compounds.

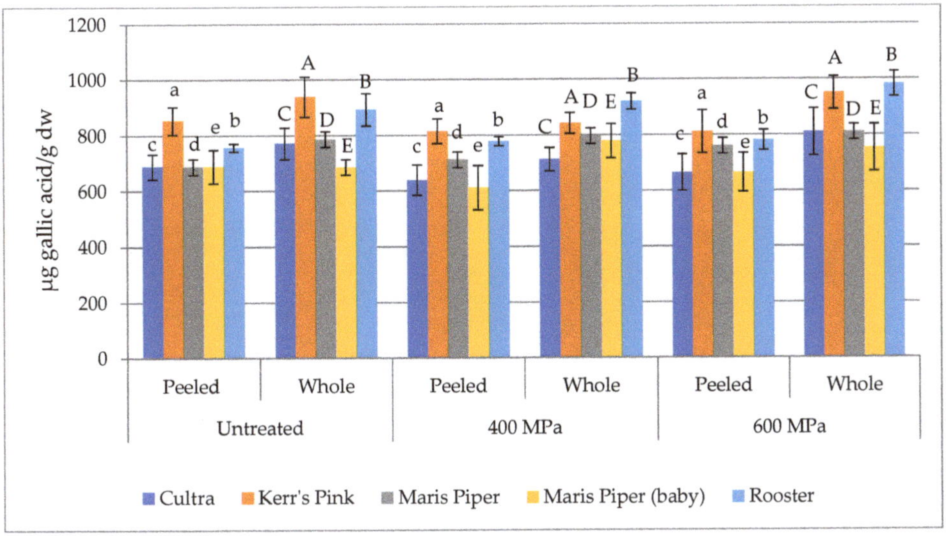

Figure 2. Total phenolic content (TPC) as expressed as μg gallic acid equivalent (GAE)/g dry weight in untreated and HPP treated peeled and whole potatoes. Different letters above bars indicate a statistically significant ($p < 0.05$) difference within the same cultivar and sample type (peeled and whole).

3.3. Antioxidant Activity

The antioxidant activity (AOA) of untreated and HPP treated, peeled and whole potato samples as examined by FRAP assay and DPPH radical scavenging capacity is shown in Figure 3a,b, respectively. Similar to the TPC results, AOA assayed by either DPPH or FRAP showed a variation after the HPP treatments, depending on the treatment pressure, cultivar, and the type (peeled or whole) of the sample. Nevertheless, the changes in AOA were insignificant ($p > 0.05$), which was also observed in our previous work following 600 MPa treatment of whole potatoes [22]. Studies on other food products, namely aronia berry puree [45], mango nectars [42], and purple sweet potato nectars [46], have also shown no significant ($p > 0.05$) changes in their AOA following HPP treatments ranging from 400 to 600 MPa and for 1–10 min. The authors of these studies linked their findings to unaltered levels of the TPC, which can also explain the results of the current study. Conversely, AOA was slightly but significantly ($p < 0.05$) increased in smoothies [14] and in pumpkin [16] upon HPP, whereas a significant ($p < 0.05$) decrease was found in *Aloe Vera* gel after HPP at 150, 250, 350, 450, and 550 MPa for 5 min [47]. These findings indicate that the AOA in HPP-treated products is dependent on the state of the substrate studied, where the AOAs of those that are in liquid or semi-liquid forms can be altered by the HPP treatment.

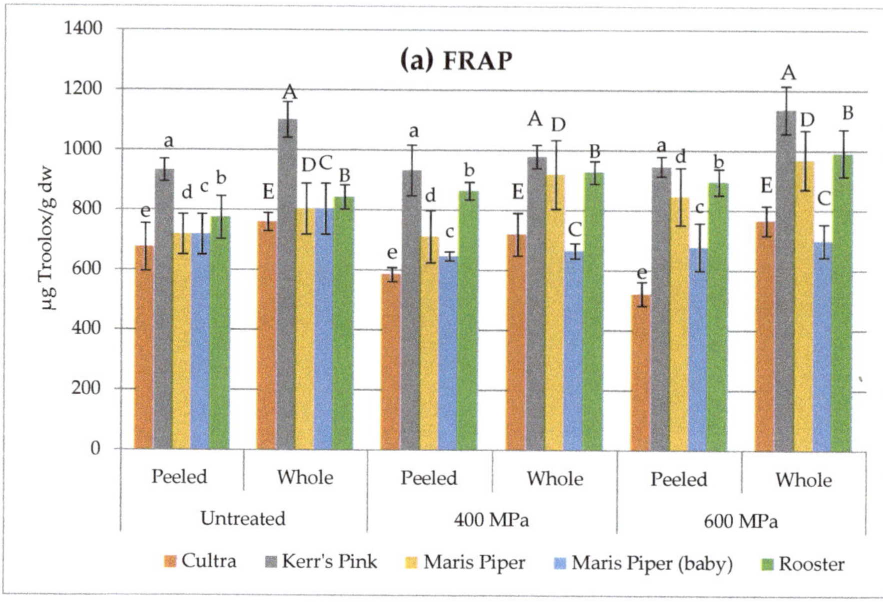

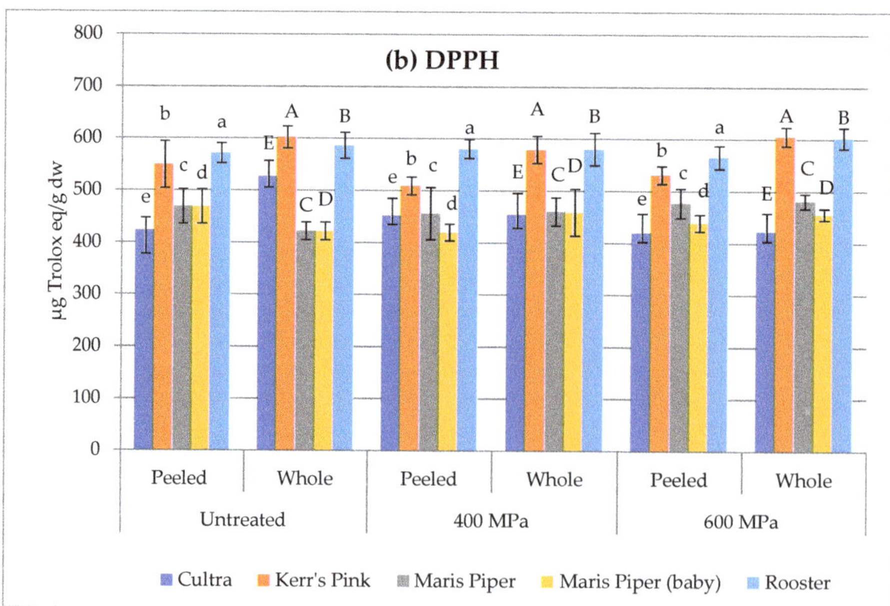

Figure 3. DPPH radical scavenging capacity (**a**) and FRAP (**b**) as expressed as µg Trolox equivalent (TE)/g dry weight in peeled and whole, untreated, and HPP-treated potatoes. Different letters above bars indicate a statistically significant ($p < 0.05$) difference within the same cultivar and sample type (peeled and whole).

3.4. Effect of HPP on Polyphenols

Tables 1 and 2 show the effect of HPP on individual polyphenols in whole and peeled potatoes. Chlorogenic acid was the most abundant polyphenol in all of the studied samples, and it varied from approximately 36 to 97 µg/g dw and 35 to 51 µg/g dw in whole and peeled samples, respectively as observed by other authors [48]. A significant ($p < 0.05$) decrease in chlorogenic acid both in peeled and whole potatoes was noted after the HPP treatments. A simultaneous increase ($p < 0.05$) of chlorogenic acid's constituent phenolic acids, i.e., caffeic acid and quinic acid, was also observed, which suggests that HPP induces degradation. Similarly, levels of rutin, which is a rutinoside of quercetin, were slightly reduced in the HPP-treated peeled samples as opposed to the untreated peeled samples, although the changes were not significant ($p > 0.05$). A significant ($p < 0.05$) increase was found in the ferulic acid content in most of the HPP-treated samples, and in particular, Maris Piper cultivars showed almost six times higher levels as compared to those untreated, suggesting that ferulic acid occurs possibly also in bound forms in these varieties. It is known that hydroxycinnamic acids (i.e., ferulic acid, 4-coumaric acid, chlorogenic acid, isochlorogenic acid, caffeic acid, and sinapic acid) and some flavonoids ((-)epicatechin, quercetin, and rutin) occur in free and bound forms, and that HPP can alter their concentrations [49].

Table 1. Means ± SD (n = 8) of the phytochemical content (μg/g dry weight) in untreated and HPP-treated whole potatoes.

Cultivar	Treatment	Polyphenols						Anthocyanins (Rutin Equivalent)	
		Ferulic Acid	Chlorogenic Acid	Caffeic Acid	Quinic Acid	4-Coumaric Acid	Rutin	Pelargonidin-*O*-Feruloylrutinoside-*O*-Hexoside	Pelargonidin-*O*-Rutinoside-*O*-Hexoside
Cultra	Untreated	0.4 ± 0.02 ^{b}	35.9 ± 1.1 ^{ab}	18.0 ± 1.6 ^{b}	71.3 ± 2.9 ^{ab}	3.6 ± 0.5 ^{b}	0.66 ± 0.1 ^{b}	ND	ND
	400 MPa	3.1 ± 0.7 ^{a}	37.5 ± 3.1 ^{ab}	23.5 ± 4.5 ^{a}	99.0 ± 1.8 ^{a}	10.1 ± 1.5 ^{a}	0.85 ± 0.2 ^{b}	ND	ND
	600 MPa	2.7 ± 0.3 ^{a}	34.8 ± 2.1 ^{ab}	19.7 ± 4.2 ^{a}	86.2 ± 2.7 ^{b}	9.0 ± 0.5 ^{a}	1.85 ± 0.2 ^{a}	ND	ND
Kerr's Pink	Untreated	0.5 ± 0.2 ^{b}	54.0 ± 1.3 ^{a}	18.3 ± 2.5 ^{ab}	32.5 ± 3.7 ^{b}	2.4 ± 0.8 ^{ab}	0.5 ± 0.1 ^{a}	1.8 ± 0.3 ^{b}	2.0 ± 0.2 ^{a}
	400 MPa	0.6 ± 0.2 ^{b}	50.4 ± 1.8 ^{b}	48.3 ± 1.3 ^{b}	75.2 ± 3.5 ^{a}	23.6 ± 1.1 ^{a}	0.5 ± 0.1 ^{a}	3.5 ± 1.0 ^{a}	2.7 ± 0.6 ^{a}
	600 MPa	0.9 ± 0.3 ^{b}	47.5 ± 3.3 ^{b}	60.4 ± 1.9 ^{a}	72.9 ± 3.2 ^{a}	20.5 ± 0.1 ^{b}	0.6 ± 0.1 ^{a}	3.0 ± 0.3 ^{a}	2.5 ± 0.5 ^{a}
Maris Piper	Untreated	0.4 ± 0.1 ^{b}	55.3 ± 4.1 ^{a}	28.3 ± 3.6 ^{ab}	48.8 ± 1.7 ^{ab}	10.4 ± 0.8 ^{b}	5.1 ± 0.3 ^{a}	ND	ND
	400 MPa	3.4 ± 0.2 ^{a}	38.3 ± 9.80 ^{ab}	56.8 ± 7.9 ^{b}	57.0 ± 5.4 ^{b}	15.2 ± 0.6 ^{a}	5.6 ± 0.2 ^{a}	ND	ND
	600 MPa	3.2 ± 0.1 ^{a}	38.9 ± 4.8 ^{ab}	68.2 ± 8.9 ^{a}	88.8 ± 4.6 ^{a}	15.5 ± 0.6 ^{a}	5.4 ± 0.2 ^{a}	ND	ND
Maris Piper (baby)	Untreated	0.4 ± 0.1 ^{b}	55.3 ± 4.1 ^{a}	28.3 ± 3.6 ^{ab}	48.8 ± 1.7 ^{ab}	10.4 ± 0.8 ^{ab}	5.1 ± 0.3 ^{a}	ND	ND
	400 MPa	3.3 ± 0.1 ^{a}	43.9 ± 3.8 ^{b}	80.1 ± 3.8 ^{a}	77.9 ± 2.2 ^{b}	17.4 ± 0.8 ^{b}	4.9 ± 0.4 ^{a}	ND	ND
	600 MPa	3.6 ± 0.3 ^{a}	44.6 ± 3.6 ^{b}	77.3 ± 4.0 ^{b}	107 ± 12 ^{a}	19.6 ± 1.2 ^{a}	4.4 ± 0.7 ^{a}	ND	ND
Rooster	Untreated	0.5 ± 0.01 ^{b}	60.8 ± 2.0 ^{a}	16.3 ± 2.0 ^{ab}	50.1 ± 2.9 ^{ab}	9.4 ± 1.1 ^{ab}	1.8 ± 0.4 ^{a}	9.9 ± 0.5 ^{b}	6.5 ± 1.7 ^{b}
	400 MPa	0.4 ± 0.03 ^{b}	36.6 ± 5.1 ^{ab}	21.3 ± 4.3 ^{b}	86.8 ± 7.0 ^{a}	24.5 ± 3.6 ^{a}	2.3 ± 0.4 ^{a}	14.7 ± 3.2 ^{a}	11.1 ± 5.4 ^{a}
	600 MPa	0.4 ± 0.03 ^{b}	30.7 ± 2.6 ^{ab}	24.5 ± 0.7 ^{a}	76.5 ± 8.2 ^{b}	16.1 ± 2.3 ^{b}	2.1 ± 0.2 ^{a}	18.2 ± 3.5 ^{a}	11.8 ± 3.0 ^{a}

Values with different superscript letters within the same column and cultivar are significantly different ($p < 0.05$).

Two anthocyanins, namely pelargonidin-*O*-feruloylrutinoside-*O*-hexoside and pelarg onidin-*O*-coumarylrutinoside-*O*-hexoside, have been tentatively identified in whole Kerr's Pink and Rooster samples based on mass spectral data and literature [50], and their levels were measured. Both HPP treatments led to a significant ($p < 0.05$) increase of these two anthocyanins in these coloured cultivars. PPO inactivation has been associated with the stability of the anthocyanins in HPP-treated (800 MPa, 18–22 °C for 15 min) red raspberry and strawberry fruits [51], which is consistent with the findings of this study where PPO RA was very low in Kerr's Pink and Rooster potatoes after the HPP at 600 MPa. As discussed earlier in Sections 3.2 and 3.3, HPP might cause cell-wall damage resulting in the release of phytochemicals including anthocyanins into the extracellular environment.

Table 2. Means ± SD (n = 8) of the phytochemical content (μg/g dry weight) untreated and HPP-treated peeled potatoes.

Cultivar	Treatment	Polyphenols					
		Ferulic Acid	Chlorogenic Acid	Caffeic Acid	Quinic Acid	4-Coumaric Acid	Rutin
Cultra	Untreated	ND	35.2 ± 2.5 [a]	5.7 ± 1.1 [a]	32.9 ± 2.6 [ab]	2.3 ± 0.4 [a]	0.5 ± 0.1 [a]
	400 MPa	ND	19.5 ± 1.0 [b]	5.9 ± 1.1 [a]	58.3 ± 4.5 [b]	1.9 ± 0.4 [a]	0.3 ± 0.1 [b]
	600 MPa	ND	2.5 ± 0.5 [ab]	0.4 ± 0.1 [b]	111.2 ± 6.9 [a]	ND	0.3 ± 0.1 [b]
Kerr's Pink	Untreated	0.6 ± 0.1 [a]	47.5 ± 8.3 [a]	6.3 ± 1.0 [a]	36.1 ± 3.1 [ab]	1.9 ± 0.3 [a]	0.6 ± 0.1 [a]
	400 MPa	0.5 ± 0.05 [a]	26.4 ± 4.4 [b]	6.9 ± 1.7 [a]	48.7 ± 2.3 [b]	1.3 ± 0.2 [b]	0.2 ± 0.1 [b]
	600 MPa	0.5 ± 0.05 [a]	1.1 ± 0.2 [ab]	0.2 ± 0.05 [b]	57.9 ± 2.5 [a]	1.5 ± 0.1 [b]	0.2 ± 0.1 [b]
Maris Piper	Untreated	0.6 ± 0.1 [b]	38.1 ± 4.9 [a]	34.1 ± 2.7 [ab]	27.1 ± 3.2 [ab]	1.4 ± 0.1 [a]	0.7 ± 0.1 [a]
	400 MPa	1.6 ± 0.5 [a]	4.1 ± 0.5 [b]	39.5 ± 1.3 [b]	85.2 ± 1.5 [b]	1.2 ± 0.1 [b]	0.3 ± 0.2 [b]
	600 MPa	2.3 ± 0.4 [a]	4.2 ± 0.2 [b]	43.5 ± 1.5 [a]	138 ± 7.9 [a]	1.0 ± 0.2 [b]	0.3 ± 0.1 [b]
Maris Piper (baby)	Untreated	0.5 ± 0.1 [b]	38.5 ± 3.7 [a]	31.4 ± 1.6 [a]	21.5 ± 6.3 [ab]	1.4 ± 0.3 [a]	0.8 ± 0.2 [a]
	400 MPa	2.3 ± 1.0 [a]	3.2 ± 0.6 [b]	26.7 ± 0.7 [b]	57.2 ± 1.8 [b]	1.1 ± 0.2 [b]	0.6 ± 0.1 [a]
	600 MPa	2.4 ± 0.8 [a]	2.5 ± 0.2 [b]	28.9 ± 1.4 [a]	116 ± 11 [a]	1.2 ± 0.2 [b]	0.7 ± 0.1 [a]
Rooster	Untreated	0.3 ± 0.05 [b]	50.9 ± 4.7 [a]	14.1 ± 0.8 [a]	36.1 ± 2.9 [ab]	2.5 ± 0.5 [a]	0.8 ± 0.2 [a]
	400 MPa	2.0 ± 0.1 [a]	10.6 ± 0.5 [b]	5.6 ± 0.4 [b]	63.2 ± 4.3 [b]	2.2 ± 0.3 [a]	0.4 ± 0.1 [b]
	600 MPa	2.5 ± 0.1 [a]	5.5 ± 0.7 [ab]	3.0 ± 0.2 [ab]	77.1 ± 1.7 [a]	2.1 ± 0.4 [a]	0.4 ± 0.2 [b]

Values with different superscript letters within the same column and cultivar are significantly different ($p < 0.05$).

3.5. Glycaemic Index (GI) of Potato Cultivars

The amount of total glucose that is released rapidly (RAG) is one of the key determinants of the GI of the food. All the untreated potato cultivars had a high GI value apart from Maris Piper, which had a medium value. According to the literature, the GI of British potatoes varies from 56 to 94 [5], whilst a variance of 53 to 103 was seen between Australian potato varieties [52]. It could be hypothesized that the variation in the GI of potato cultivars is due to structural differences of the starch in the tubers. For instance, the in vitro and in vivo studies investigating the GI of rice had shown that the cultivars with high levels of resistant starch and amylose decreased the GI [53,54].

In the present study, statistically significant decreases were observed in the GI values of both coloured potato cultivars (Roosters and Kerr's Pink) and Maris Piper (baby) following 600 MPa treatment. For the rest of the samples, there were no statistically significant ($p > 0.05$) differences in the GI values irrespective of HPP treatment or not (Table 3).

Previously published literature suggested that the HPP treatments might result in a decrease in GI [55,56]. However, in both these studies, they used in vivo methodologies as well as ready-to-eat foods with no further processing after the HPP, whilst in the present study, potatoes were cooked following the HPP before the in vitro digestion was carried out. Nasehi et al. (2012) [57] suggested that owing to its high water content, potato starch is more resistant to pressure than the starch present in rice, corn, or tapioca, which may explain the unaltered GI values in some of the potato cultivars studied here. The application of cycles of HPP and longer periods of HPP, or a long continuous HPP treatment instead of the single treatment for three minutes may cause the desired changes in the starch structure of the potato. This can subsequently impede the release of glucose as well as the breakdown of the starch and thereby reduce the GI and GL of the potato [58].

Table 3. Carbohydrate parameters, glycaemic index and glycaemic load of untreated and HPP-treated potatoes.

Cultivars	Treatment	Rapidly Available Glucose (g/100 g)	Total Glucose (g/100 g)	Glycaemic Index	Glycaemic Load
Cultra	Untreated	14.38 ± 1.02 [a]	19.31 ± 0.77 [a]	79.10 ± 6.79 [a]	14.09 ± 1.07 [a]
	400 MPa	11.73 ± 0.39 [b]	18.92 ± 2.27 [a]	68.36 ± 10.52 [a]	11.73 ± 0.72 [b]
	600 MPa	14.93 ± 0.75 [a]	20.18 ± 3.43 [a]	80.97 ± 7.72 [a]	14.92 ± 1.07 [a]
Kerr's Pink	Untreated	10.77 ± 0.18 [a]	11.39 ± 0.51 [a]	97.79 ± 3.42 [a]	10.24 ± 0.17 [a]
	400 MPa	9.23 ± 1.29 [a]	12.06 ± 4.15 [a]	85.00 ± 12.05 [a]	9.16 ± 1.69 [a]
	600 MPa	9.37 ± 1.95 [a]	17.33 ± 2.10 [b]	62.14 ± 10.65 [b]	9.86 ± 1.77 [a]
Maris Piper	Untreated	10.97 ± 0.66 [a]	17.87 ± 2.99 [a]	66.71 ± 13.05 [a]	11.06 ± 1.24 [a]
	400 MPa	11.39 ± 0.03 [a]	17.69 ± 3.04 [a]	73.20 ± 9.51 [a]	11.76 ± 0.45 [a]
	600 MPa	12.65 ± 3.36 [a]	18.21 ± 2.56 [a]	67.74 ± 3.02 [a]	11.61 ± 1.25 [a]
Maris Piper (baby)	400 MPa	12.18 ± 5.73 [a]	19.41 ± 2.98 [a]	117.70 ± 5.80 [a]	10.43 ± 3.91 [a]
	600 MPa	11.40 ± 1.18 [a]	17.78 ± 1.73 [a]	65.52 ± 0.02 [b]	10.77 ± 1.09 [a]
Rooster	Untreated	12.38 ± 2.76 [a]	16.38 ± 3.17 [a]	81.60 ± 5.36 [a]	12.31 ± 2.65 [a]
	400 MPa	12.23 ± 1.60 [a]	17.30 ± 1.77 [a]	78.74 ± 11.50 [a]	12.32 ± 1.18 [a]
	600 MPa	12.04 ± 0.84 [a]	20.97 ± 3.36 [a]	62.03 ± 8.96 [b]	11.75 ± 0.75 [a]

All data are the means ± SD (n = 6). Values with different letters within the same column and variety are statistical different ($p < 0.05$).

4. Conclusions

HPP at 600 MPa for 3 min is efficient in inhibiting PPO activity by as much as 79% in peeled potatoes. The HPP treatments do not alter total phenolic content and antioxidant activities of whole and peeled potatoes irrespective of the strength of the applied pressures. However, the HPP treatments can induce the degradation of (poly)phenolic acid conjugates such as chlorogenic acid, which is an ester of caffeic acid and quinic acid. Chlorogenic acid is the most abundant polyphenol in potatoes, which is significantly decreased upon HPP treatments with concurrent increases in caffeic acid and quinic acid. On the other hand, HPP led to an increase in ferulic acid, which is most commonly found bound to cell walls in whole and peeled samples. Similarly, the anthocyanins pelargonidin-*O*-ferulorylrutinoside-*O*-hexoside and pelargonidin-*O*-rutinoside-*O*-hexoside in the HPP-treated whole samples of the coloured potato varieties increased significantly, implying that the high hydrostatic pressure disrupts the cell walls, enabling the release of bound (poly)phenols. Despite the alterations of individual (poly)phenols by the impact of HPP, the overall antioxidant capacity was not affected. No significant effect was observed on the carbohydrate parameters, i.e., glycaemic index and glycaemic load of the HPP-treated potato cultivars. HPP is a promising technology in potato processing that can be applied to inhibit browning while maintaining nutritional and functional qualities. Further studies on the combination of HPP with other novel processing technology(ies) or natural antioxidant(s) are necessary to establish the validity as alternate to thermal and/or chemical processing of minimally processed potatoes, which is beyond the scope of the current study.

Supplementary Materials: The following is available online at https://www.mdpi.com/article/10.3390/foods10102425/s1. Figure S1: The temperature and pressure profile at 600 MPa.

Author Contributions: D.K.R. and N.M.O. conceptualise the research project and funding acquisition; K.T. and A.M. performed the experiments, analysed the data and drafted the manuscript; D.K.R. and N.M.O. curated the data and edited the manuscript. All authors have read and agreed to the published version of the manuscript.

Funding: This research was funded by Food Institutional Research Measure under the Irish Department of Agriculture, Food and Marine (DAFM), FIRM 17/F/299.

Institutional Review Board Statement: Not applicable.

Informed Consent Statement: Not applicable.

Acknowledgments: The authors would like to thank the Department of Agriculture, Food and Marine for funding the 'ProcessPotato' project (FIRM 17/F/299).

Conflicts of Interest: The authors declare no conflict of interest.

References

1. Burgos, G.; Zum Felde, T.; Andre, C.; Kubow, S. The Potato and Its Contribution to the Human Diet and Health. In *The Potato Crop: Its Agricultural, Nutritional and Social Contribution to Humankind*; Campos, H., Ortiz, O., Eds.; Springer International Publishing: Cham, Switzerland, 2020; pp. 37–74. ISBN 978-3-030-28683-5.
2. McGill, C.R.; Kurilich, A.C.; Davignon, J. The role of potatoes and potato components in cardiometabolic health: A review. *Ann. Med.* **2013**, *45*, 467–473. [CrossRef] [PubMed]
3. Nayak, B.; Berrios, J.D.J.; Tang, J. Impact of food processing on the glycemic index (GI) of potato products. *Food Res. Int.* **2014**, *56*, 35–46. [CrossRef]
4. Haase, N.U. The In Vitro Digestibility of Carbohydrates in Boiled and Processed Potatoes. *Potato Res.* **2015**, *58*, 91–102. [CrossRef]
5. Henry, C.J.K.; Lightowler, H.J.; Strik, C.M.; Storey, M. Glycaemic index values for commercially available potatoes in Great Britain. *Br. J. Nutr.* **2005**, *94*, 917–921. [CrossRef]
6. Lockyer, S.; Spiro, A.; Stanner, S. Dietary fibre and the prevention of chronic disease—Should health professionals be doing more to raise awareness? *Nutr. Bull.* **2016**, *41*, 214–231. [CrossRef]
7. Dourado, C.; Pinto, C.; Barba, F.J.; Lorenzo, J.M.; Delgadillo, I.; Saraiva, J.A. Innovative non-thermal technologies affecting potato tuber and fried potato quality. *Trends Food Sci. Technol.* **2019**, *88*, 274–289. [CrossRef]
8. Rocculi, P.; Romani, S.; Gómez Galindo, F.; Dalla Rosa, M. Effect of Minimal Processing on Physiology and Quality of Fresh-Cut Potatoes: A Review. *Food* **2009**, *3*, 18–30.
9. Taranto, F.; Pasqualone, A.; Mangini, G.; Tripodi, P.; Miazzi, M.M.; Pavan, S.; Montemurro, C. Polyphenol oxidases in crops: Biochemical, physiological and genetic aspects. *Int. J. Mol. Sci.* **2017**, *18*, 377. [CrossRef]
10. Yoruk, R.; Marshall, M.R. Physiochemical Properties and Function of Plant Polyphenol Oxidase. *J. Food Biochem.* **2003**, *27*, 361–422. [CrossRef]
11. Marques Silva, F.V.; Sulaiman, A. *Polyphenoloxidase in Fruit and Vegetables: Inactivation by Thermal and Non-Thermal Processes*; Elsevier: Amsterdam, The Netherlands, 2018; ISBN 9780128140451.
12. Alamar, M.C.; Tosetti, R.; Landahl, S.; Bermejo, A.; Terry, L.A. Assuring Potato Tuber Quality during Storage: A Future Perspective. *Front. Plant Sci.* **2017**, *8*, 2034. [CrossRef]
13. San Martín, M.F.; Barbosa-Canovas, G.V.; Swanson, B.G. Food processing by high pressure. *Crit. Rev. Food Sci. Nutr.* **2002**, *42*, 627–645. [CrossRef]
14. Andrés, V.; Villanueva, M.J.; Tenorio, M.D. The effect of high-pressure processing on colour, bioactive compounds, and antioxidant activity in smoothies during refrigerated storage. *Food Chem.* **2016**, *192*, 328–335. [CrossRef]
15. Stratakos, A.C.; Inguglia, E.S.; Linton, M.; Tollerton, J.; Murphy, L.; Corcionivoschi, N.; Koidis, A.; Tiwari, B.K. Effect of high pressure processing on the safety, shelf life and quality of raw milk. *Innov. Food Sci. Emerg. Technol.* **2019**, *52*, 325–333. [CrossRef]
16. Zhou, C.-L.; Liu, W.; Zhao, J.; Yuan, C.; Song, Y.; Chen, D.; Ni, Y.-Y.; Li, Q.-H. The effect of high hydrostatic pressure on the microbiological quality and physical–chemical characteristics of Pumpkin (*Cucurbita maxima* Duch.) during refrigerated storage. *Innov. Food Sci. Emerg. Technol.* **2014**, *21*, 24–34. [CrossRef]
17. Tsikrika, K.; Walsh, D.; Joseph, A.; Burgess, C.M.; Rai, D.K. High-Pressure Processing and Ultrasonication of Minimally Processed Potatoes: Effect on the Colour, Microbial Counts, and Bioactive Compounds. *Molecules* **2021**, *26*, 2614. [CrossRef] [PubMed]
18. Denoya, G.I.; Polenta, G.A.; Apóstolo, N.M.; Budde, C.O.; Sancho, A.M.; Vaudagna, S.R. Optimization of high hydrostatic pressure processing for the preservation of minimally processed peach pieces. *Innov. Food Sci. Emerg. Technol.* **2016**, *33*, 84–93. [CrossRef]
19. Szczepańska, J.; Barba, F.J.; Skąpska, S.; Marszałek, K. High pressure processing of carrot juice: Effect of static and multi-pulsed pressure on the polyphenolic profile, oxidoreductases activity and colour. *Food Chem.* **2020**, *307*, 125549. [CrossRef]
20. Denoya, G.I.; Nanni, M.S.; Apóstolo, N.M.; Vaudagna, S.R.; Polenta, G.A. Biochemical and microstructural assessment of minimally processed peaches subjected to high-pressure processing: Implications on the freshness condition. *Innov. Food Sci. Emerg. Technol.* **2016**, *36*, 212–220. [CrossRef]
21. Woolf, A.B.; Wibisono, R.; Farr, J.; Hallett, I.; Richter, L.; Oey, I.; Wohlers, M.; Zhou, J.; Fletcher, G.C.; Requejo-Jackman, C. Effect of high pressure processing on avocado slices. *Innov. Food Sci. Emerg. Technol.* **2013**, *18*, 65–73. [CrossRef]
22. Tsikrika, K.; Rai, D.K. The Effect of High Pressure Processing on Antioxidant Activity of Irish Potato Cultivars. *Proceedings* **2019**, *11*, 9. [CrossRef]
23. Tsikrika, K.; O'Brien, N.; Rai, D.K. The Effect of High Pressure Processing on Polyphenol Oxidase Activity, Phytochemicals and Proximate Composition of Irish Potato Cultivars. *Foods* **2019**, *8*, 517. [CrossRef] [PubMed]
24. Singleton, V.L.; Rossi, J.A.J. Colorimetry of total phenolics with phosphomolybdic-phosphotungstic acid reagents. *Am. J. Enol. Vitic.* **1965**, *16*, 144–158.
25. Benzie, I.; Strain, J. The ferric reducing ability of plasma as a measure of antioxodant. *Anal. Biochem.* **1996**, *239*, 70–76. [CrossRef]
26. Stratil, P.; Klejdus, B.; Kubáň, V. Determination of total content of phenolic compounds and their antioxidant activity in vegetables—Evaluation of spectrophotometric methods. *J. Agric. Food Chem.* **2006**, *54*, 607–616. [CrossRef] [PubMed]

27. Ou, B.; Huang, D.; Hampsch-Woodill, M.; Flanagan, J.A.; Deemer, E.K. Analysis of antioxidant activities of common vegetables employing oxygen radical absorbance capacity (ORAC) and ferric reducing antioxidant power (FRAP) assays: A comparative study. *J. Agric. Food Chem.* **2002**, *50*, 3122–3128. [CrossRef]
28. Goupy, P.; Hugues, M.; Boivin, P.; Amiot, M.J. Antioxidant composition and activity of barley (Hordeum vulgare) and malt extracts and of isolated phenolic compounds. *J. Sci. Food Agric.* **1999**, *79*, 1625–1634. [CrossRef]
29. Hossain, M.B.; Rai, D.K.; Brunton, N.P.; Martin-Diana, A.B.; Barry-Ryan, A.C. Characterization of phenolic composition in lamiaceae spices by LC-ESI-MS/MS. *J. Agric. Food Chem.* **2010**, *58*, 10576–10581. [CrossRef]
30. Kim, H.W.; Kim, S.R.; Lee, Y.M.; Jang, H.H.; Kim, J.B. Analysis of Variation in Anthocyanin Composition in Korean Coloured Potato Cultivars by LC-DAD-ESI-MS and PLS-DA. *Potato Res.* **2018**, *61*, 1–17. [CrossRef]
31. Hossain, M.B.; Tiwari, B.K.; Gangopadhyay, N.; O'Donnell, C.P.; Brunton, N.P.; Rai, D.K. Ultrasonic extraction of steroidal alkaloids from potato peel waste. *Ultrason. Sonochem.* **2014**, *21*, 1470–1476. [CrossRef]
32. Hossain, M.B.; Rai, D.K.; Brunton, N.P. Optimisation and validation of ultra-high performance liquid chromatographic-tandem mass spectrometry method for qualitative and quantitative analysis of potato steroidal alkaloids. *J. Chromatogr. B* **2015**, *997*, 110–115. [CrossRef]
33. Englyst, O.N.; Hudson, G.J.; Englyst, H.N. Starch analysis in food. In *Encyclopedia of Analytical Chemistry*, 1st ed.; John Wiley & Sons: Chichester, UK, 2000; pp. 4246–4262.
34. Thygesen, P.W.; Dry, I.B.; Robinson, S.P. Polyphenol oxidase in potato: A multigene family that exhibits differential expression patterns. *Plant Physiol.* **1995**, *109*, 525–531. [CrossRef]
35. Aaby, K.; Grimsbo, I.H.; Hovda, M.B.; Rode, T.M. Effect of high pressure and thermal processing on shelf life and quality of strawberry purée and juice. *Food Chem.* **2018**, *260*, 115–123. [CrossRef]
36. Marszałek, K.; Doesburg, P.; Starzonek, S.; Szczepańska, J.; Woźniak, Ł.; Lorenzo, J.M.; Skaopska, S.; Rzoska, S.; Barba, F.J. Comparative effect of supercritical carbon dioxide and high pressure processing on structural changes and activity loss of oxidoreductive enzymes. *J. CO2 Util.* **2019**, *29*, 46–56. [CrossRef]
37. Ludikhuyze, L.; Van Loey, A.; Denys, I.S.; Hendrickx, M.E.G. Effects of High Pressure on Enzymes Related to Food Quality. In *Ultra High Pressure Treatments of Foods*; Springer: Boston, MA, USA, 2001; pp. 115–166.
38. Chakraborty, S.; Kaushik, N.; Rao, P.S.; Mishra, H.N. High-pressure inactivation of enzymes: A review on its recent applications on fruit purees and juices. *Compr. Rev. Food Sci. Food Saf.* **2014**, *13*, 578–596. [CrossRef]
39. Valcarcel, J.; Reilly, K.; Gaffney, M.; O'Brien, N.M. Antioxidant Activity, Total Phenolic and Total Flavonoid Content in Sixty Varieties of Potato (*Solanum tuberosum* L.) Grown in Ireland. *Potato Res.* **2015**, *58*, 221–244. [CrossRef]
40. Akyol, H.; Riciputi, Y.; Capanoglu, E.; Caboni, M.F.; Verardo, V. Phenolic compounds in the potato and its byproducts: An overview. *Int. J. Mol. Sci.* **2016**, *17*, 835. [CrossRef] [PubMed]
41. Cilla, A.; Bosch, L.; Barberá, R.; Alegría, A. Effect of processing on the bioaccessibility of bioactive compounds—A review focusing on carotenoids, minerals, ascorbic acid, tocopherols and polyphenols. *J. Food Compos. Anal.* **2018**, *68*, 3–15. [CrossRef]
42. Liu, F.; Wang, Y.; Li, R.; Bi, X.; Liao, X. Effects of high hydrostatic pressure and high temperature short time on antioxidant activity, antioxidant compounds and color of mango nectars. *Innov. Food Sci. Emerg. Technol.* **2014**, *21*, 35–43. [CrossRef]
43. Zheng, X.; Yu, Y.; Xiao, G.; Xu, Y.; Wu, J.; Tang, D.; Zhang, Y. Comparing product stability of probiotic beverages using litchi juice treated by high hydrostatic pressure and heat as substrates. *Innov. Food Sci. Emerg. Technol.* **2014**, *23*, 61–67. [CrossRef]
44. Da Silveira, T.F.F.; Cristianini, M.; Kuhnle, G.G.; Ribeiro, A.B.; Filho, J.T.; Godoy, H.T. Anthocyanins, non-anthocyanin phenolics, tocopherols and antioxidant capacity of açaí juice (*Euterpe oleracea*) as affected by high pressure processing and thermal pasteurization. *Innov. Food Sci. Emerg. Technol.* **2019**, *55*, 88–96. [CrossRef]
45. Yuan, B.; Danao, M.G.C.; Stratton, J.E.; Weier, S.A.; Weller, C.L.; Lu, M. High pressure processing (HPP) of aronia berry purée: Effects on physicochemical properties, microbial counts, bioactive compounds, and antioxidant capacities. *Innov. Food Sci. Emerg. Technol.* **2018**, *47*, 249–255. [CrossRef]
46. Wang, Y.; Liu, F.; Cao, X.; Chen, F.; Hu, X.; Liao, X. Comparison of high hydrostatic pressure and high temperature short time processing on quality of purple sweet potato nectar. *Innov. Food Sci. Emerg. Technol.* **2012**, *16*, 326–334. [CrossRef]
47. Di Scala, K.; Vega-Gálvez, A.; Ah-Hen, K.; Nuñez-Mancilla, Y.; Tabilo-Munizaga, G.; Pérez-Won, M.; Giovagnoli, C. Chemical and physical properties of aloe vera (Aloe barbadensis Miller) gel stored after high hydrostatic pressure processing | Propriedades químicas e físicas do gel de aloe vera (Aloe barbadensis Miller) armazenado após processamento sob alta pressão hi. *Cienc. Tecnol. Aliment.* **2013**, *33*, 52–59. [CrossRef]
48. Friedman, M.; Levin, C.E. Analysis and Biological Activities of Potato Glycoalkaloids, Calystegine Alkaloids, Phenolic Compounds, and Anthocyanins. *Adv. Potato Chem. Technol.* **2009**, 127–161. [CrossRef]
49. Jeż, M.; Wiczkowski, W.; Zielińska, D.; Białobrzewski, I.; Błaszczak, W. The impact of high pressure processing on the phenolic profile, hydrophilic antioxidant and reducing capacity of purée obtained from commercial tomato varieties. *Food Chem.* **2018**, *261*, 201–209. [CrossRef] [PubMed]
50. Zhang, Y.; Cheng, S.; De Jong, D.; Griffiths, H.; Halitschke, R.; De Jong, W. The potato R locus codes for dihydroflavonol 4-reductase. *Theor. Appl. Genet.* **2009**, *119*, 931–937. [CrossRef] [PubMed]
51. Garcia-Palazon, A.; Suthanthangjai, W.; Kajda, P.; Zabetakis, I. The effects of high hydrostatic pressure on β-glucosidase, peroxidase and polyphenoloxidase in red raspberry (*Rubus idaeus*) and strawberry (Fragaria x ananassa). *Food Chem.* **2004**, *88*, 7–10. [CrossRef]

52. Ek, K.L.; Wang, S.; Copeland, L.; Brand-Miller, J.C. Discovery of a low-glycaemic index potato and relationship with starch digestion in vitro. *Br. J. Nutr.* **2014**, *111*, 699–705. [CrossRef]
53. Kumar, A.; Sahoo, U.; Baisakha, B.; Okpani, O.A.; Ngangkham, U.; Parameswaran, C.; Ngangkham, U.; Basak, N.; Kumar, G.; Sharma, S.G. Resistant starch could be decisive in determining the glycemic index of rice cultivars. *J. Cereal Sci.* **2018**, *79*, 348–353. [CrossRef]
54. Panlasigui, L.N.; Thompson, L.U.; Juliano, B.O.; Perez, C.M.; Yiu, S.H.; Greenberg, G.R. Rice varieties with similar amylose content differ in starch digestibility and glycemic response in humans. *Am. J. Clin. Nutr.* **1991**, *54*, 871–877. [CrossRef]
55. Elizondo-Montemayor, L.; Hernández-Brenes, C.; Ramos-Parra, P.A.; Moreno-Sánchez, D.; Nieblas, B.; Rosas-Pérez, A.M.; Lamadrid-Zertuche, A.C. High hydrostatic pressure processing reduces the glycemic index of fresh mango puree in healthy subjects. *Food Funct.* **2015**, *6*, 1352–1360. [CrossRef]
56. Huang, H.-W.; Hsu, C.-P.; Wang, C.-Y. Healthy expectations of high hydrostatic pressure treatment in food processing industry. *J. Food Drug Anal.* **2019**, *28*. [CrossRef]
57. Nasehi, B.; Javaheri, S. Application of high hydrostatic pressure in modifying functional properties of starches: A Review. *Middle-East J. Sci. Res.* **2012**, *11*, 856–861.
58. Colussi, R.; Kaur, L.; da Rosa Zavareze, E.; Dias, A.R.G.; Stewart, R.B.; Singh, J. High pressure processing and retrogradation of potato starch: Influence on functional properties and gastro-small intestinal digestion in vitro. *Food Hydrocoll.* **2018**, *75*, 131–137. [CrossRef]

foods

MDPI

Article

Thermosonication Combined with Natural Antimicrobial Nisin: A Potential Technique Ensuring Microbiological Safety and Improving the Quality Parameters of Orange Juice

Qinyu Zhao, Quyu Yuan, Chenxu Gao, Xiaoyang Wang, Bihe Zhu, Jiaqi Wang, Xiangyu Sun and Tingting Ma *

College of Food Science and Engineering, College of Enology, Northwest A&F University, Xianyang 712100, China; zqy1004@nwafu.edu.cn (Q.Z.); y23933@nwafu.edu.cn (Q.Y.); 2018013406@nwafu.edu.cn (C.G.); 18821712077@163.com (X.W.); zbhxinong@nwafu.edu.cn (B.Z.); hello-wjq@nwafu.edu.cn (J.W.); sunxiangyu@nwafu.edu.cn (X.S.)

* Correspondence: matingting@nwafu.edu.cn; Tel./Fax: +86-29-8709-2486

Abstract: Currently, thermal pasteurisation (TP) remains the most widely applied technique for commercial orange juice preservation; however, a high temperature causes adverse effects on the quality attributes of orange juice. In order to explore a novel non-thermal sterilization method for orange juice, the impacts of thermosonication combined with nisin (TSN) and TP treatments on the quality attributes including microbial and enzyme inactivation and the physicochemical, nutritional, functional, and sensory qualities of orange juice were studied. Both TP and TSN treatments achieved desirable bactericidal and enzyme inactivation effects, and nisin had a significant synergistic lethal effect on aerobic bacteria in orange juice ($p < 0.05$). Additionally, TSN treatment significantly improved the color attributes of orange juice and well maintained its physicochemical properties and sensory quality. More importantly, TSN treatment significantly increased the total polyphenols content (TPC) and total carotenoids (TC) by 10.03% and 20.10%, increased the ORAC and DPPH by 51.10% and 10.58%, and the contents of total flavonoids and ascorbic acid were largely retained. Correlation analysis of antioxidant activity showed that the ORAC and scavenging ability of DPPH radicals of orange juice are mainly attributed to TC and TPC. These findings indicate that TSN shows great potential application value, which could guarantee the microbiological safety and improve the quality attributes of orange juice.

Keywords: orange juice; nisin-assisted thermosonication; microbial and enzyme inactivation; sensory quality; bioactive properties

Citation: Zhao, Q.; Yuan, Q.; Gao, C.; Wang, X.; Zhu, B.; Wang, J.; Sun, X.; Ma, T. Thermosonication Combined with Natural Antimicrobial Nisin: A Potential Technique Ensuring Microbiological Safety and Improving the Quality Parameters of Orange Juice. *Foods* **2021**, *10*, 1851. https://doi.org/10.3390/foods10081851

Academic Editors: Marina Cano Lamadrid and Francisco Artés-Hernández

Received: 6 July 2021
Accepted: 7 August 2021
Published: 11 August 2021

1. Introduction

Orange juice has gained worldwide popularity owing to its high nutritional value, attractive color, and distinctive sweet and sour flavor [1]. It is rich in a variety of nutritional and bioactive phytochemicals, including a high concentration of vitamin C, folate, polyphenols, flavonoids, carotenoid, and limonoid [2–4]. Epidemiological studies have shown that the consumption of orange juice is beneficial to human health, e.g., by enhancing the antioxidant activity of blood serum [5] and lowering lipid peroxidation [6], improving risk factors associated with cardiovascular disease [7], and improving the cognitive function of the elderly [8]. According to the United States Department of Agriculture in 2016, the annual consumption of orange juice is about 1.8 billion liters globally [9].

Sterilization is an important unit operation in the production of orange juice. Currently, conventional thermal pasteurisation (TP) remains the most widely applied technique for commercial orange juice preservation and is considered the safest and most cost-effective way to inactivate microorganisms and enzymes [10]. However, a high temperature may cause the deterioration of the color and flavor of orange juice [11–13] and the loss of heat-sensitive nutrients and functional substances [14,15], which might impair the functional properties and sensory attributes of the final product, thus making it unattractive to the consumer. Thus, it is necessary to explore a new non-thermal sterilization method which can not only effectively inactivate microorganisms and enzymes, but also maintain the optimal sensory attributes and the nutritional and functional characteristics of orange juice.

Ultrasound (US) processing is one of the emerging non-thermal sterilization technologies recently being applied in liquid food. Its mechanism of inactivated microorganisms and enzymes is generally attributed to the combination of mechanical effects (e.g., cavitation, bubble rupture, and mechanical shear force) and chemical effects (e.g., the formation of free radicals and the decomposition of water vapor in collapsing bubbles) [13,16]. Recently, US has been intensively investigated as a means to preserve fruit and vegetable juice; however, some researchers have shown that using US alone in some cases may not be very efficient for the inactivation of some types of microorganisms and enzymes in juices [16–18]; meanwhile, US is combined with other sterilization techniques, for example, mild heat treatment [i.e., thermosonication (TS)] [16,19], ultraviolet treatment [20], a pulsed electric field [21], gassing [22], and TS with antibacterial agents [23,24]. These showed high application potential in the preservation of various juices. Various US-combined sterilization techniques not only showed a synergistic effect in the inactivation of microorganisms and enzymes, but also maintained the sensory, physicochemical, and nutritional attributes of juices. Among them, due to the low costs and a simple operation, combine use of TS and TS treatment with an antibacterial agent (for example, nisin) have been a wide concern [16,24,25]. Mild heat treatment, which uses a lower heating temperature during thermal processing (usually lower than 60 °C), could bring down, even remove the loss of heat-sensitive nutrient substances and bioactive phytochemicals, thus prevented the deterioration caused by high thermal temperatures to juice quality [19,24]. Nisin is a heat-stable antimicrobial peptide produced by strains of Lactococcus lactis subsp. lactis. It is hydrolyzed into amino acids by a protease in the digestive tract after consumption and does not produce resistance or allergic reactions [26]. It is a safe antimicrobial peptide for food preservation, which was recognized by the WHO (World Health Organization). The bactericidal mechanisms of nisin include inhibiting the cell wall biosynthesis, leading to ATP hydrolysis and ion leakage, membrane pore formation and disruption of the pH equilibrium and the proton motive force, and eventually to cell death [24,26].

Orange juice is very popular in the word, and significantly contributes to the daily fruit consumption in most countries; thus, the commercial importance of orange juice is self-evident. In recent years, in order to improve the overall quality of commercial orange juice, many non-thermal processing technologies for the sterilization process of orange juice have been widely reported, including high-pressure processing [27], pulsed electric fields [2], US processing [13], TS processing [28], cold plasma [29], ozone processing [29], dimethyl dicarbonate [30], and the ultrasound-assisted supercritical CO_2 system [22]. However, the research on the combination of TS and nisin is rarely reported. Therefore, the purpose of this study was to (i) compare the germicidal efficacy of TS, TS combined with nisin (TSN), and TP treatments on microorganisms in orange juice, (ii) compare the impacts of TSN and TP treatments on overall juice quality (enzyme inactivation, physicochemical, nutritional and functional and sensory qualities), and (iii) analyze the correlation between the antioxidant capacity, color characteristics, and functional substances of orange juice. The research aims to provide a theoretical basis and technical support for the high-quality orange juice produce.

2. Materials and Methods

2.1. Orange Juice Preparation

Fresh navel oranges grown in Jiangxi province (China) were used in the study. The oranges were washed and peeled, cut into four pieces, and pressed mechanically by a juice extractor. The pressed orange juice was filtered through a sterilized double-layered muslin cloth, homogenized at 4 °C and 150 bar, transferred into a sterile food-grade container, and stored at 4 °C until further treatment.

2.2. Chemicals and Reagents

All the standards, including 1,1-diphenyl-2-picrylhydrazyl (DPPH), gallic acid, 6-hydroxy-2,5,7,8-tetramethylchromane-2-carboxylic acid (Trolox), catechol, and Folin-Ciocalteu reagent were purchased from Sigma-Aldrich (St. Louis, MO, USA). All media, including the lauryl sulfate peptone broth, Bengal red medium, plate counting medium, and bright green lactose bile broth, were purchased from Hope Bio-Technology Co., Ltd. (Qingdao, China).

2.3. Preparation of Nisin Solution

Referring to the method of Ma et al. (2020) [24] with minor modifications, 0.5 g of commercial nisin Z (900 IU/mg) was dissolved in 20 mL of sterile water, then filtered through a 0.22-μm inorganic membrane to remove the microorganisms, and kept at 4 °C for the further experiments.

2.4. TS, TSN, and CTS Treatment

Fresh orange juices were treated with TS, TSN, and TP respectively, with fresh orange juice (no sterilization treatment) as the control. TS treatment [24,25] was conducted by a ATPIO-1000D built-in probe ultrasound device (Xianou Corporation of Nanjing, Jiangsu, China), operating with a frequency of 20–25 kHz, a maximum electrical power input of 1000 W, and a horn microtip diameter of 6 mm. Juice samples of 45 mL were subjected to a double-wall cylindrical vessel, in which water was circulated with a digital thermostatic bath (XODC-0515-II, Nanjing Xianou, Nanjing, China) to control the TS temperature at a constant 50 °C. The power was adjusted to 70% of the maximum power, and the processing time was 10 min, with the pulse duration time set as 2 s on and 3 s off. For the TSN treatment, adding 360 μL of nisin solution to samples to reach a final concentration of 200 ppm, then it was placed in a double-walled cylindrical vessel for the same operation as the TS treatment. For the TP treatment, orange juice samples of 45 mL were pasteurized at 80 °C for 10 min using an electro-thermostatic water bath [13]. Each treatment was carried out at least in triplicate (Table S1).

2.5. Microbiological Assay

The viable cells of natural microorganisms in the orange juice samples were measured based on the National Food Hygiene Standard of China. The total bacterial count (TBC), Escherichia coli, and mold were detected according to GB 4789.2-2016, GB 4789.3-2016, and GB 4789.3-2016, respectively. The specific experimental operation refers to the method of Ma et al. (2020) [31]. Results are expressed as the log CFU/mL and log MPN/mL.

2.6. Determination of the Activities of the Polyphenol Oxidase (PPO), Peroxidase (POD), and Pectin Methylesterase (PME)

PPO and POD was extracted by dissolving 4% (*w*/*v*) polyvinyl pyrrolidone (PVPP), 1% (*v*/*v*) Triton x-100, and 1 M NaCl in a 0.2 M phosphate buffer (pH = 6.5). Orange juice and enzyme extracts (1:1, *w*/*w*) were mixed, extracted at 4 °C for 2 h, and centrifuged at 10,000 g/min for 20 min at 4 °C. The supernatant was taken to determine the PPO and POD activities. The enzyme activities were measured based on the method of Ma et al. (2020) [24].

PME was extracted by mixing juice samples and NaCl (8.8% *w*/*v*) at a ratio of 4.5:15 (*w*/*v*) and centrifuged at 15,000× *g* for 20 min at 4 °C. The supernatant was collected to determine the PME. PME activities were measured exactly according to the method of Muthukumarappan et al. (2009) [32].

2.7. Physicochemical Indexes

The total soluble solids (TSS) of the orange juice were measured by a PAL-1 digital Abbe Refractometer (ATAGO Co., Tokyo, Japan). The preparation of the titratable acid (TA) extract and the measurement of TA using the acid–base titration method were based on

China National Standard GB/T12456-2008 [33]. The pH value was evaluated by a PHS-3E pH meter (Shanghai Leici Co. Ltd., Shanghai, China). The juice viscosity was determined using an NDJ-5S rotary viscometer (Jinan Jingtian Co. Ltd., Jinan, China).

2.8. Functional Indices and Antioxidant Activity

The ascorbic acid (AA) of juice samples was determined by the 2,6-dichlorindophenol method, the specific experimental operation according to the literature of Ma et al. (2020) [31]. The total polyphenol content (TPC) and total flavonoid content (TFC) were measured by the Folin–Ciocalteu colorimetric method (Ma et al., 2019) [34] and the aluminum chloride colorimetric assay (Aadil et al., 2013) [35], and results are expressed as milligrams of gallic acid equivalents (GAE) per liter (mg GAE/L) and milligrams of catechol equivalents per liter (mg CTE/L), respectively. The content of total carotenoids (TC) was measured based on the method of Abid et al. (2020) [36], and results are expressed as milligrams of β-carotene equivalents per liter.

The antioxidant capacity of the orange juice with different treatments was measured by the DPPH free radical scavenging ability, the ferric ion reducing antioxidant power (FRAP), and the oxygen radical antioxidant capacity (ORAC) method based on previous reports [34,37]. Results are expressed as millimoles of trolox equivalents per liter (mM TE/L).

2.9. Sensory Evaluation

2.9.1. Electronic Nose (E-Nose) Assay

The aroma profiles of the orange juice samples were obtained with a portable PEN 3 E-nose (Airsense Analytics, Schwerin, Germany) containing 10 metal oxide semiconductors. Each sensor has a certain degree of affinity for specific volatile compounds. Specific descriptions of the sensor and the specific operation steps are found in our previous work [24,31]. Each sample was measured at least 10 times.

2.9.2. Color Determination

The color characteristics of the orange juice samples were measured by an X-Rite ci7600 colorimeter in reflection mode. In the CIELab scale, L* represents lightness, a* represents greenness to redness, b* represents yellowness to blueness, and the total color difference (ΔE*), hue (H°), chroma (C*), L*, a*, and b* were automatically determined by the colorimeter or calculated by its own software.

2.9.3. Artificial Sensory Evaluation

Previously trained panelists Potential consumers (18 men and 18 women, from 20 to 45 years old) participated in the sensory evaluation. These panelists were previously trained with the basic knowledge of sensory evaluation, so that they could do a better sensory assessment. The overall quality of the juice samples was evaluated by 5 different attributes (appearance, color, smell odor, sweet and sour suitability, and overall acceptance) using a 9-point hedonic scale a quantitative descriptive analysis with 100 score (Tables S2 and S3). The specific experimental operation is in accordance with our previous work [31].

2.10. Statistical Analysis

Data analysis was performed using Excel 2016, SPSS 23, RStudio-1.1.463, and Origin 9.1. The experimental results are expressed as the means ± standard deviations (SD) of three replicates for each treatment.

3. Results and Discussion

3.1. Microbial Inactivation of Different Treatments

As shown in Figure 1, the TBC, Escherichia coli, and mold in the control group were 4.04 log CFU/mL, 2.07 log MPN/mL, and 1.4 log cfu/mL, respectively. After TP, TS, and TSN treatment, the TBC decreased to 0, 2.26, and 1.18 log CFU/mL, respectively, and no

Escherichia coli or mold were detected in the orange juice samples, which indicated that all three treatments showed a strong inactivation effect on Escherichia coli and mold in the orange juice. Undoubtedly, TP was still the most efficacious and thorough sterilization method. Meanwhile, the effect of TS on TBC was limited, while TSN significantly enhanced the bactericidal effect ($p < 0.05$), indicating that nisin and TS had synergistic inactivation effects on aerobic bacteria, which is consistent with Ma et al.'s (2020) and Liao et al.'s (2018) results. The effect of nisin on the goal bacteria occurs by inhibiting the cell wall biosynthesis and membrane pore formation and disrupting the pH equilibrium and the proton motive force, leading to ATP hydrolysis, ion leakage, and finally, to cell death [24,26]. Compared with another new non thermal processing pulsed electric field (PEF) treatment, under 24.8 kV/cm, 60 pulses, 169 μs treatment time, 53.8 °C PEF treatment, apple juice cannot be stored at room temperature, because the PEF treatment cannot provide enough energy to inactivate microorganisms under this condition. TSN can make orange juice meet the microbiological safety standards and it is a good and gentle sterilization method [38].

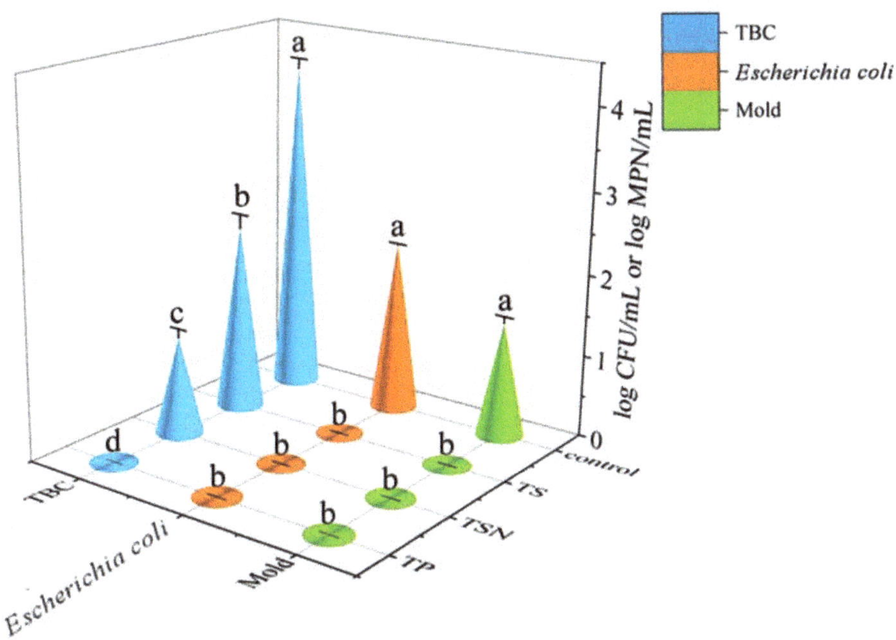

Figure 1. Lethal effects of different sterilization treatments on the microorganisms in orange juice. The *Escherichia coli* results are expressed as log MPN/mL, and the other results are expressed as log CFU/mL.

Based on the Chinese national standard GB7101-2015, when the TBC is less than 2 log CFU/mL, the number of Escherichia coli is less than 1 MPN/mL, and the juice is considered safe to drink. Therefore, in this study, in addition to TS treatment, both TSN and CTS treatments could ensure the microbiological safety of the orange juice (Figure 1). TSN and CTS treatments were selected for further research.

3.2. Residual Enzyme Activities of Different Treatments

The residual activities of PPO, POD, and PME with different sterilization treatments are shown in Figure 2. After TP treatment, the remainder activities of PPO, POD, and PME were 19.41%, 15.83%, and 24.75%, respectively. After TSN treatment, the residual activities of these three enzymes were 34.19%, 39.30%, and 45.60%, respectively. Obviously, the enzyme inactivation effect of TSN treatment was weaker than that of the TP treatment ($p <$

0.05), but compared with the control group, TSN still showed a desired inactivation effect, reducing the PPO, POD, and PME activities by 65.81%, 60.70%, and 54.40%, respectively, significantly reducing the activity of three endogenous enzymes in orange juice ($p < 0.05$).Therefore, the two treatments could effectively control the enzymatic browning of the orange juice and maintain its cloud stability. Compared with high isostatic pressure (HIP) treatment, Patrícia Martins de Oliveira and others have shown that the use of HIP to process mango–carrot juice mixed juice can effectively inactivate polyphenol oxidase (PPO), but it will increase peroxidase (POD) activity. It is not conducive to the long-term preservation of juice [39].

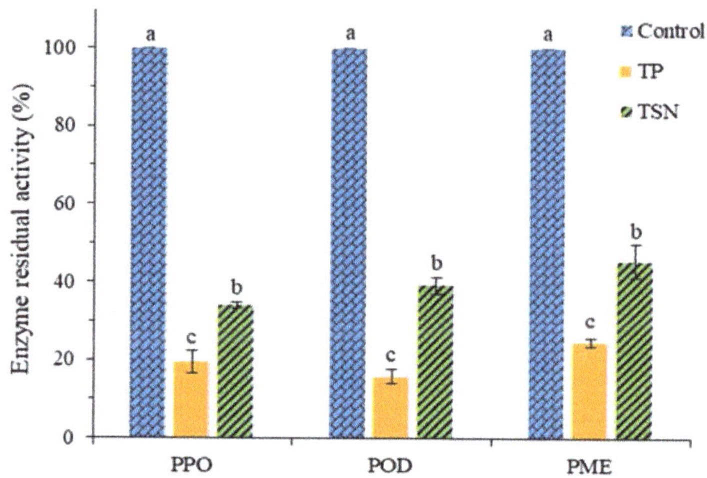

Figure 2. Passivation effects of different sterilization treatments on the enzyme activity of orange juice.

3.3. Physicochemical Properties of Different Treatments

The effects of TP and TSN on pH, TSS, TA, and viscosity are shown in Figure 3A–D. Compared with the control group, no significant change was observed in the pH, TSS, and TA of the TSN group ($p > 0.05$), which indicated that TSN did not alter pH, TSS, or TA of the orange juice, which is consistent with the findings of Liao et al. (2018) [25] and Walkling-Ribeiro et al. (2009) [40]. Previous studies have shown that TSN had no influence on the viscosity of grape juice [24] and green juice (celery stalk, apple, cucumber, parsley) [41]; however, in this study, TSN treatment significantly reduced the viscosity of the orange juice, and this was mainly due to the difference in operating parameters and juice matrix. In addition, TP treatment significantly reduced the TSS of the orange juice ($p < 0.05$). Ma et al. also obtained similar results with grape juice [24]. In the research on grape juice, it was also found that PEF treatment will not affect its pH, TSS, and acidity [42].

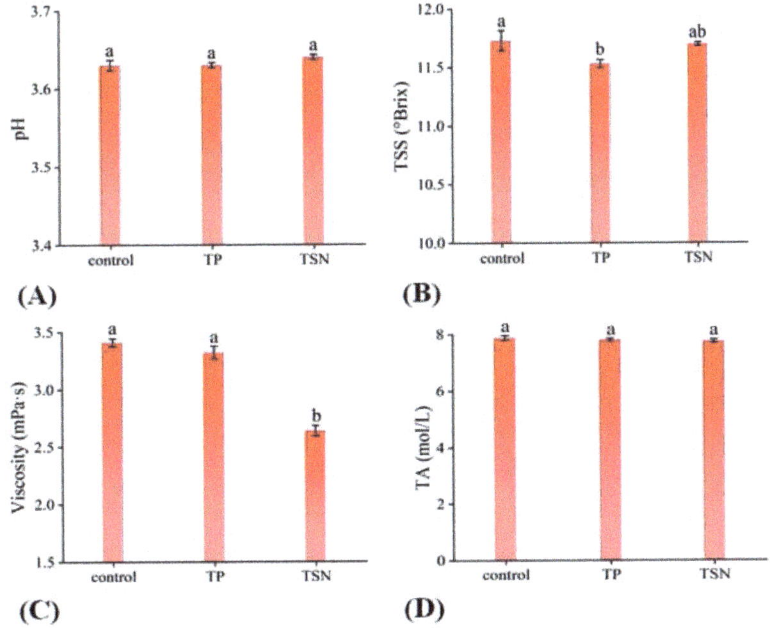

Figure 3. Effects of different sterilization treatments on the physicochemical properties of orange juice. (**A**) TSS; (**B**) pH; (**C**) viscosity; and (**D**) TA.

3.4. Functional Indices of Different Treatments

3.4.1. AA and TFC

The influence of TP and TSN on the nutritional and functional indicators is shown in Figure 4A–G. Orange juice is well known as an appreciable source of AA; unfortunately, AA in juice is unstable and is easily affected by environmental factors such as temperature, pH, dissolved oxygen, and metal ions [43,44]; thus, AA is easily lost during juice processing. As shown in Figure 4A, compared with the control group, the AA content decreased by 23.10% and 15.83% after TP and TSN treatment, respectively. Thus, the different sterilization treatments all caused a certain loss of AA, but compared with TP, TSN could significantly slow down the loss of AA ($p < 0.05$) and retain the AA to the greatest extent. Similar results were also reported by Liao et al. (2018) [25], who found an 87.42% retention of AA exposed to TSN treatment at 52 °C for 30 min. This was mainly due to the mild temperature (<60 °C) applied during the TSN processing. Ultrasonic cavitation reduced the level of dissolved oxygen in the orange juice, thus inhibiting the degradation and loss of AA [19].

As shown in Figure 4C, the effect of TP and TSN treatments on the TFC of orange juice was similar to that of AA (Figure 4A). TFC suffered 41.45% and 34.55% of losses after TP and TSN treatment, respectively. Nevertheless, the TFC retention rate of TSN-treated orange juice was much higher than that of TP-treated samples ($p < 0.05$). A recent study on the effects of different sterilization treatments on the quality attributes of grape juice also reached the same conclusion [24].

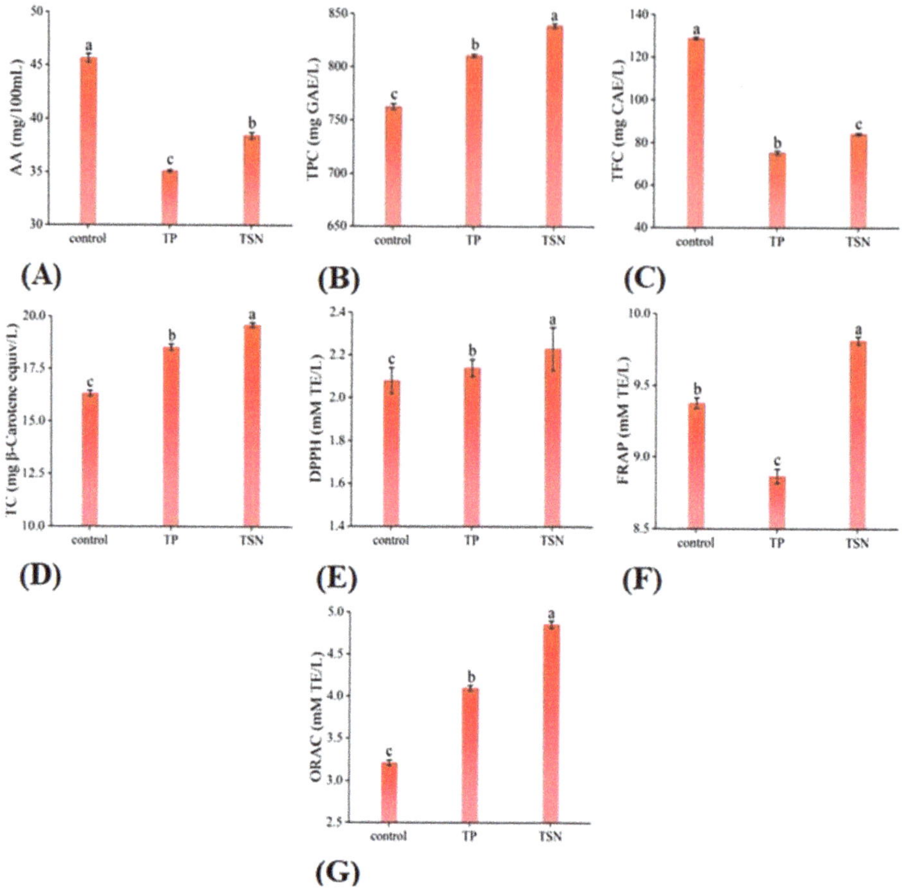

Figure 4. Effects of different sterilization treatments on the functional indicators of orange juice. (**A**) AA; (**B**) TPC; (**C**) TFC; (**D**) TC; (**E**) FRAP assays; (**F**) DPPH scavenging activity; (**G**) ORAC assays.

3.4.2. TPC and TC

Polyphenols are important antioxidant components in plants, and their content is particularly abundant in citrus fruits. As shown in Figure 4B, both TP and TSN treatments increased the TPC in orange juice, while TSN showed the highest effect, as it added TPC value from 762.41 mg GAE/L to 838.89 mg GAE/L, with an increase of 10.03% ($p < 0.05$). Similar results were also reported in TSN-treated grape juice [24], TS-treated hog plum juice [45], and TS-treated star fruit juice [46].

Similarly, both TP and TSN treatments significantly increased the TC in orange juice ($p < 0.05$); it increased by 13.48% and 20.10%, respectively, after TP and TSN treatment (Figure 4D). Due to the lipophilic nature of carotenoids and their specific localization in plant tissues, appropriate processing can promote the release and dissolution of carotenoids [47].

Figure 4A–D shows that TSN treatment significantly increased the functional components in orange juice, as TPC increased by 10.03%, and TC increased by 20.10%. This is principally because the cavitation effect during TSN treatment increases the mechanical disruption of the plant cell wall, so as to promote the dissolution and extraction efficiency of the nutrients and functional components of juice [45]. Meanwhile, TSN treatment significantly reduced the related oxidase activities and dissolved oxygen levels in juice; thus, it can be employed as a preservation technique for orange juice processing with a high retention of AA and TFC.

3.4.3. Antioxidant Activity

FRAP, ORAC, and DPPH are considered to be the most widely used in antioxidant activity measuring, as they can reflect the antioxidant capacity from different aspects. As shown in Figure 4E, the FRAP values of the control group, the TP group, and the TSN group were 9.37, 8.86, and 9.81 mM TE/L, respectively. TP treatment significantly reduced the FRAP of orange juice, whereas TSN treatment greatly enhanced the FRAP value. Additionally, both the TP and TSN treatments significantly increased the ORAC and DPPH values of orange juice compared with the control group (Figure 4F,G), but the TSN-treated juice showed the highest ORAC (4.85 mM TE/L) and DPPH (2.23 mM TE/L) values, which were 51.10% and 10.58% higher than in the control group and were 15.46% and 4.70% higher than in the TP group. A large number of published studies have also confirmed that US-related treatments, such as US, TS, and TSN, could significantly enhance the antioxidant activity of juice [17,24,45,46]. This is mainly because the ultrasonic process promotes the extraction and release of antioxidants, such as polyphenols and carotenoids [24,45].

Figure 4A–G shows that TSN treatment significantly increased or highly retained the contents of nutrients and functional substances in orange juice and significantly enhanced its antioxidant activity in three different systems. The enhancement of antioxidant content and antioxidant activity are important manifestations of the nutritional value of orange juice. Therefore, TSN treatment, as a new non-thermal sterilization method, can significantly improve the nutritional quality of orange juice.

3.5. Correlation Analysis of Antioxidant Activity

Previous studies have proven that several antioxidants such as ascorbic acid (AA), polyphenols, flavonoids, and anthocyanins might contribute to the antioxidant activity of various juice, and the color properties might also be related to the antioxidant activity [24,48,49]. In order to clarify the potential phytochemicals that contribute to the antioxidant capacity of orange juice, the correlations between the antioxidant capacity, color characteristics, and functional substances of orange juice were also analyzed, and the results are shown in Figure 5.

As shown in Figure 5, the closer the absolute value of the correlation coefficient is to 1, the closer the shape of the graph is to an ellipse. The closer the correlation coefficient is to 1, the redder the color is. On the contrary, the closer the correlation coefficient is to −1, the bluer the color. Thus, the indicators that are highly correlated with ORAC and DPPH are TC (RORAC = 0.99, RDPPH = 0.95), TPC (RORAC = 0.95, RDPPH = 0.88), L* (RORAC = 0.98, RDPPH = 1.00), b* (RORAC = 0.90, RDPPH = 0.96), and ΔE* (RORAC = 0.95, RDPPH = 0.99), which indicates that the antioxidant substances TPC and TC in orange juice and the color attributes L*, b*, and ΔE* make a greater contribution to the DPPH free radical scavenging ability and ORAC. Similar results were also observed by Zhao et al. (2018) [50], who found significant positive correlations between antioxidant activity and both TC and TFC in Lycium barbarum juice. However, in previous studies, Wang et al. (2019) [48] found that the antioxidant activity of strawberry juice was highly correlated with TPC, AA, and TFC; Ma et al. (2020) [24] reported that TPC plays a prominent role in the antioxidant activity of grape juice. Due to the complexity of the antioxidant activity of liquid foods, it includes different lipids and water-soluble compounds. Therefore, the correlation between the different methods used to determine antioxidant capacity mainly depends on the food matrix [51]. Furthermore, no significant correlation was observed between FRAP and the indicators measured in this study, which may be because the FRAP of orange juice was associated with multiple indicators, none of which was a major contributor.

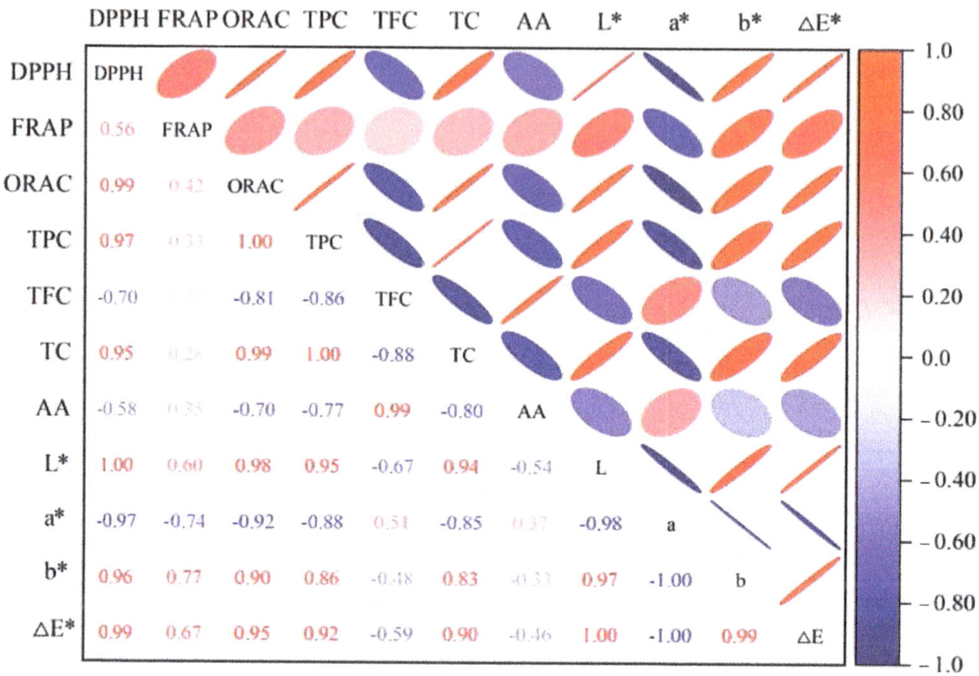

Figure 5. Correlation between the color attributes, antioxidants and antioxidant capacity of orange juice.

3.6. Effects of Different Sterilization Treatments on the Sensory Quality

3.6.1. Color Analysis

Juice color might affect consumers' buying decisions and affect the flavor sensory characteristics of juice [52]. Hence, the color of orange juice after different sterilization treatments is also an important factor in measuring its quality. Table 1 displays the color parameters changes of orange juice after different sterilization treatments.

Table 1. The color parameters of control, TSN, and TP treatment juices.

Processing Method	Color					
	L*	a*	b*	ΔE	H°	C*
Control	51.60 ± 0.03 [c]	-2.02 ± 0.01 [a]	29.96 ± 0.04 [b]	0.00 ± 0.00 [c]	93.86 ± 0.03 [b]	30.03 ± 0.04 [b]
TP	52.85 ± 0.45 [b]	-2.18 ± 0.17 [a]	30.44 ± 0.77 [b]	1.46 ± 0.55 [b]	94.11 ± 0.43 [b]	30.52 ± 0.75 [b]
TSN	55.07 ± 0.16 [a]	-2.92 ± 0.03 [b]	33.39 ± 0.12 [a]	4.96 ± 0.18 [a]	95.00 ± 0.06 [a]	33.51 ± 0.12 [a]

Note: Different letters represent significant difference ($p < 0.05$) from each other in the same column.

As Table 1 shows, compared with the control group, the a*, b*, H°, and C* values of orange juice with TP treatment did not change significantly ($p > 0.05$), and the ΔE* value is less than 2 CIELAB units. Fernández-Vázquez et al. (2013) [53] proposed that only when the ΔE* value between two orange juice samples is greater than 2.8 CIELAB units can consumers distinguish color. Therefore, juice color is basically unchanged after TP treatment. However, TSN treatment significantly increased the L*, b*, H°, and C* values of orange juice ($p < 0.05$), which indicated that the lightness, yellowness, and color saturation of orange juice increased significantly after TSN treatment, and this change caters to consumers' psychological expectations concerning the ideal color of orange juice [53]. Studies have shown that the color characteristics of most commercial orange juice are as follows: a L* value from 61 to 66, an H° value from 79 to 93, and a C* value from 42 to 60 [52,53]. This color range can be accepted by most consumers and is considered an ideal orange juice color. From Table 1, it is obvious that the L* and C* values of TSN-treated orange juice are closer to the ideal color. In general, TSN treatment significantly improved the color attributes and had a positive impact on the quality of the orange juice.

3.6.2. Artificial Sensory Evaluation

Sensory properties are very important for the consumers' acceptance or rejection of food. The sensory evaluation including color, odor, appearance, sweet and sour suitability, and overall acceptability of orange juice is illustrated in Figure 6. Among the three groups, the overall acceptability of samples in control was highest, mainly because of its optimal sweet and sour suitability and odor attribute. Conversely, the scores of the various sensory attributes of orange juice in the TP group were relatively low, thus its overall acceptability is the lowest. During the sensory experiment, it was found that the TP-treated orange juice had an obvious sour taste. Our previous research also proved that TP treatment could increase the acidity of grape juice [24]. The TSN-treated orange juice had the highest score of all color attributes, which is consistent with the result of the color analysis in Section 3.6.1. In addition, the odor attribute of the orange juice altered significantly after different sterilization treatments. The TP-treated orange juice had the lowest odor attribute score, and this is mainly because the high temperature during the TP process negatively affects its aroma profile [10,27]. The odor score of the TSN-treated juice was significantly higher than that treated by TP.

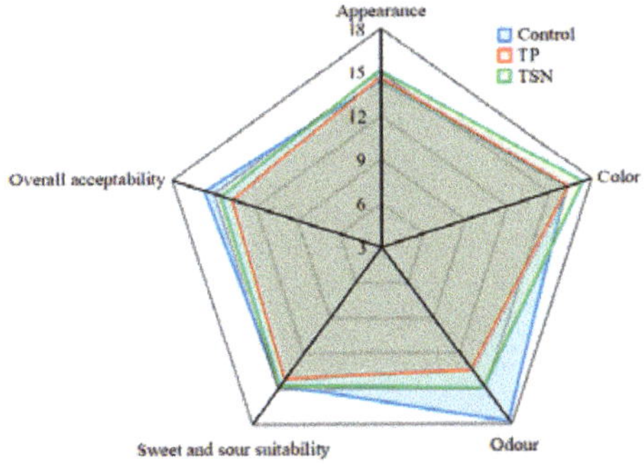

Figure 6. Spider plot of artificial sensory evaluation of orange juice under different sterilization treatments.

In general, TSN treatment significantly improved the color attributes of the orange juice. There was no significant difference between TSN and the control group in terms of the appearance and sweet and sour suitability, and the odor score of the TSN group was relatively high. Thus, the overall acceptability of the TSN group was significantly higher than that of the TP group. TSN treatment can well maintain the sensory quality of orange juice.

3.6.3. E-Nose Analysis

Figure 7C–E shows the records of the control, TP, and TSN groups. The LDA linear discriminant method was used to analyze the average stable signal of the ten sensor of E-nose at the last 5 s [24], and the results are shown in Figure 7A. It can be seen in Figure 7A that the two discriminant functions could explain 87.7% of the whole variance, of which LD1 explained 75.1%, and LD2 explained 12.6%. E-nose using the LDA model could clearly distinguish the different samples after different treatments.

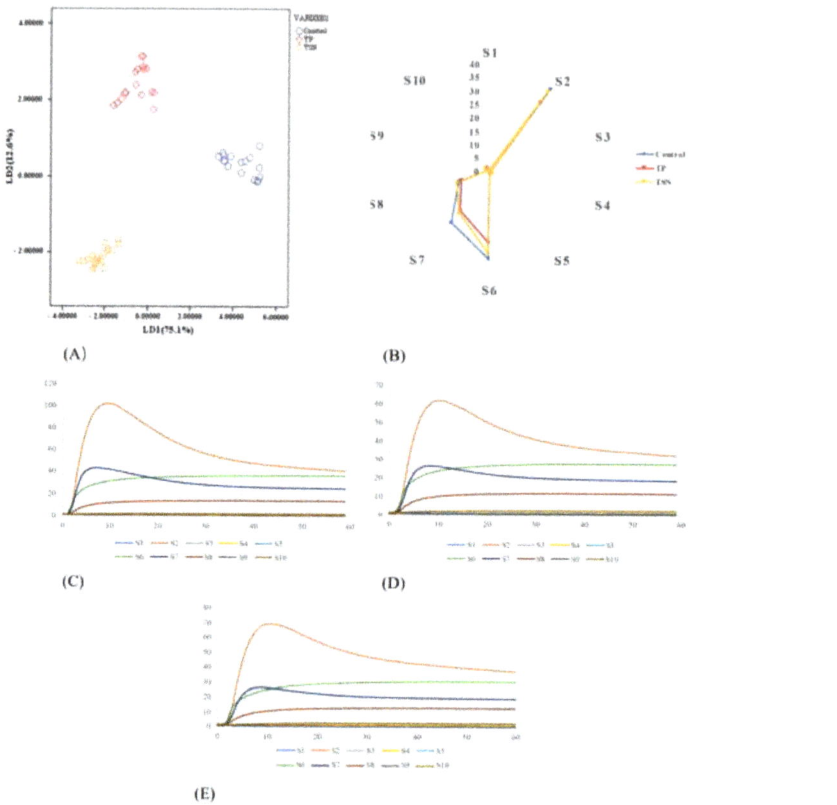

Figure 7. E-nose assay of orange juice after different sterilization treatments. (**A**) LDA results of orange juice under different sterilization treatments; (**B**) radar chart of E-nose response data of orange juice after different sterilization treatments; sensor responses recorded using ten sensors in the E-nose for (**C**) control group; (**D**) CTS treatment; and (**E**) TSN treatment.

In addition, a radar chart was drawn based on the response data of different sensors for the orange juice's odor profile. Figure 7B shows that the orange juice samples of the three groups only responded to Sensors S2 (sensitive to nitrogen oxides), S6 (sensitive to methane, in a broad range), and S7 (sensitive to many sulfur organic compounds and terpenes. The response values of the control group were all the highest, while the response values of each sensor reduced significantly after TP treatment. After TSN treatment, the response value of Sensor S7 decreased significantly, while the response values of Sensors S2 and S6 were not significant different from the control group. These indicated that both TSN and TP treatments caused varying degrees of odor substance loss in the orange juice. In general, TSN showed a close odor characteristics compared to control, which was in keeping with the artificial sensory evaluation.

4. Conclusions

The effects of TSN and TP on the overall quality of orange juice were systematically studied in this paper. The results indicate that both TSN and TP could guarantee the microbial safety firstly. Although the enzyme inactivation effect of TSN was weaker than that of TP, it also exhibited a desirable effect. In terms of the nutritional and functional characteristics, TSN treatment significantly increased the content of the functional components in orange juice. It increased the TPC by 10.03% and the TC by 20.10%, and the antioxidant capacity, e.g., the ORAC and DPPH values, increased by 50% and 10%, respectively. Meanwhile, the TAC and TFC were highly retained. In terms of sensory quality, TSN treatment significantly improved the color attributes of the orange juice, and the overall acceptability of the TSN-treated juice was significantly higher than that of the TP-treated juice. In addition, TSN treatment basically did not change the physicochemical indicators of the orange juice, while TP treatment significantly reduced the TSS.

In summary, based on ensuring the microbial safety firstly, TSN can not only well maintain the physicochemical indictors and sensory quality of orange juice, but TSN could also greatly improve the functional and nutritional characteristics of orange juice. Thus, TSN can be considered a novel non-thermal technique with the potential to assist producers to produce high-quality juice; thus, it8 has high commercial application prospects. In addition, in future studies, in order to explore the wider application of this method in other fruit and vegetable juice matrix, it is necessary to add an amount of specific microorganisms to measure its effectiveness.

Supplementary Materials: The following are available online at https://www.mdpi.com/article/10.3390/foods10081851/s1, Table S1: The treatment conditions of different treatments, Table S2: The standard table for sensory rating of orange juice, Table S3: The table of sample scoring.

Author Contributions: Conceptualization, T.M. and X.S.; methodology, Q.Z., Q.Y., and C.G.; investigation, X.W., and B.Z.; data curation, J.W.; writing original draft preparation, Q.Z.; writing review and editing, Q.Z., and T.M. All authors have read and agreed to the published version of the manuscript.

Funding: This study was supported by the National Nature Science Foundation Project (31801560), the National key research and development program (2019YFD1002502), the Natural Science Foundation of Ningxia (2021AAC02023, 2021BSB03003), the Science and Technology Innovation Project of Ningxia Agriculture and Forestry Academy (NKYJ-20-01, NGSB-2021-5-04), and the Innovation Capacity Support Plan of Shaanxi Province (2019TD-006, 2020TD-047).

Institutional Review Board Statement: Not applicable.

Informed Consent Statement: Not applicable.

Data Availability Statement: The datasets generated for this study are available on request to the corresponding author.

Conflicts of Interest: The authors declare no conflict of interest.

References

1. Wibowo, S.; Grauwet, T.; Santiago, J.S.; Tomic, J.; Vervoort, L.; Hendrickx, M.; Van Loey, A. Quality changes of pasteurised orange juice during storage: A kinetic study of specific parameters and their relation to colour instability. *Food Chem.* **2015**, *187*, 140–151. [CrossRef] [PubMed]
2. Agcam, E.; Akyıldız, A.; Evrendilek, G.A. Comparison of phenolic compounds of orange juice processed by pulsed electric fields (PEF) and conventional thermal pasteurisation. *Food Chem.* **2014**, *143*, 354–361. [CrossRef] [PubMed]
3. Ohrvik, V.; Witthoft, C. Orange juice is a good folate source in respect to folate content and stability during storage and simulated digestion. *Eur. J. Nutr.* **2008**, *47*, 92–98. [CrossRef]
4. Rampersaud, G.C.; Valim, M.F. 100% citrus juice: Nutritional contribution, dietary benefits, and association with anthropometric measures. *Crit. Rev. Food Sci. Nutr.* **2017**, *57*, 129–140. [CrossRef] [PubMed]
5. De Paiva, A.; Goncalves, D.; Ferreira, P.; Baldwin, E.; Cesar, T. Postprandial effect of fresh and processed orange juice on the glucose metabolism, antioxidant activity and prospective food intake. *J. Funct. Foods* **2019**, *52*, 302–309. [CrossRef]
6. Foroudi, S.; Potter, A.S.; Stamatikos, A.; Patil, B.S.; Deyhim, F. Drinking Orange juice increases total antioxidant status and decreases lipid peroxidation in adults. *J. Funct. Foods* **2014**, *17*, 612–617. [CrossRef]
7. Alhabeeb, H.; Sohouli, M.H.; Lari, A.; Fatahi, S.; Shidfar, F.; Alomar, O.; Salem, H.; Al-Badawi, I.A.; Abu-Zaid, A. Impact of orange juice consumption on cardiovascular disease risk factors: A systematic review and meta-analysis of randomized-controlled trials. *Crit. Rev. Food Sci. Nutr.* **2020**. [CrossRef]
8. Kean, R.J.; Lamport, D.J.; Dodd, G.F.; Freeman, J.E.; Williams, C.M.; Ellis, J.A.; Butler, L.T.; Spencer, J.P. Chronic consumption of flavanone-rich orange juice is associated with cognitive benefits: An 8-wk, randomized, double-blind, placebo-controlled trial in healthy older adults. *Am. J. Clin. Nutr.* **2015**, *101*, 506–514. [CrossRef] [PubMed]
9. United States Department of Agriculture. Citrus: World Market and Trade. 2021. Available online: http://apps.fas.usda.gov/psdonline/circulars/citrus.pdf (accessed on 1 July 2021).

10. Kim, M.K.; Jang, H.W.; Lee, K.G. Sensory and instrumental volatile flavor analysis of commercial orange juices prepared by different processing methods. *Food Chem.* **2018**, *30*, 217–222. [CrossRef]
11. Mastello, R.B.; Capobiango, M.; Chin, S.; Monteiro, M.; Marriott, P.J. Identification of odour-active compounds of pasteurised orange juice using multidimensional gas chromatography techniques. *Food Res. Int.* **2015**, *75*, 281–288. [CrossRef]
12. Perez-Cacho, P.R.; Rouseff, R. Processing and storage effects on orange juice aroma: A Review. *J. Agric. Food Chem.* **2008**, *56*, 9785–9796. [CrossRef]
13. Khandpur, P.; Gogate, P.R. Understanding the effect of novel approaches based on ultrasound on sensory profile of orange juice. *Ultrason. Sonochem.* **2015**, *27*, 87–95. [CrossRef]
14. Hocine, R.; Christian, M.; Amine, B.; Nawel, A.; Manuel, D.; Khodir, M. Degradation kinetic modelling of ascorbic acid and colour intensity in pasteurised blood orange juice during storage. *Food Chem.* **2015**, *173*, 665–673. [CrossRef]
15. Khandpur, P.; Gogate, P.R. Effect of novel ultrasound based processing on the nutrition quality of different fruit and vegetable juices. *Ultrason. Sonochem.* **2015**, *27*, 125–136. [CrossRef] [PubMed]
16. Anaya-Esparza, L.M.; Velazquez-Estrada, R.M.; Roig, A.X.; Garcia-Galindo, H.S.; Sayago-Ayerdi, S.G.; Montalvo-Gonzalez, E. Thermosonication: An alternative processing for fruit and vegetable juices. *Trends Food Sci. Technol.* **2017**, *61*, 26–37. [CrossRef]
17. Santhirasegaram, V.; Razali, Z.; Somasundram, C. Effects of thermal treatment and sonication on quality attributes of Chokanan mango (*Mangifera indica* L.) juice. *Ultrason. Sonochem.* **2013**, *20*, 1276–1282. [CrossRef]
18. Dias, D.D.R.C.; Barros, Z.M.P.; Carvalho, C.B.O.D.; Honorato, F.A.; Guerra, N.B.; Azoubel, P.M. Effect of sonication on soursop juice quality. *LWT Food Sci. Technol.* **2015**, *62*, 883–889. [CrossRef]
19. Abid, M.; Jabba, S.; Hu, B.; Hashim, M.M.; Wu, T.; Lei, S.C.; Khan, M.A.; Zeng, X.X. Thermosonication as a potential quality enhancement technique of apple juice. *Ultrason. Sonochem.* **2014**, *21*, 984–990. [CrossRef]
20. Wang, J.Y.; Liu, Q.D.; Xie, B.J.; Sun, Z.D. Effect of ultrasound combined with ultraviolet treatment on microbial inactivation and quality properties of mango juice. *Ultrason. Sonochem.* **2020**, *64*, 105000. [CrossRef] [PubMed]
21. Aadil, R.M.; Zeng, X.A.; Sun, D.W.; Wang, M.S.; Liu, Z.; Zhang, Z.H.; Liu, Z.W. Combined effects of sonication and pulsed electric field on selected quality parameters of grapefruit juice. *LWT Food Sci. Technol.* **2015**, *62*, 890–893. [CrossRef]
22. Paniagua-Martinez, I.; Mulet, A.; Garcia-Alvarado, M.A.; Benedito, J. Orange juice processing using a continuous flow ultrasound-assisted supercritical CO_2 system: Microbiota inactivation and product quality. *Innov. Food Sci. Emerg.* **2018**, *47*, 362–370. [CrossRef]
23. Park, J.S.; Ha, J.W. Ultrasound treatment combined with fumaric acid for inactivating food-borne pathogens in apple juice and its mechanisms. *Food Microbiol.* **2019**, *84*, 103277.1–103277.9. [CrossRef]
24. Ma, T.T.; Wang, J.Q.; Wang, L.K.; Yang, Y.H.; Yang, W.Y.; Wang, H.L.; Lan, T.; Zhang, Q.W.; Sun, X.Y. Ultrasound-combined sterilization technology: An effective sterilization technique ensuring the microbial safety of grape juice and significantly improving its quality. *Foods* **2020**, *9*, 1512. [CrossRef] [PubMed]
25. Liao, H.M.; Jiang, L.F.; Cheng, Y.L.; Liao, X.J.; Zhang, R.R. Application of nisin-assisted thermosonication processing for preservation and quality retention of fresh apple juice. *Ultrason. Sonochem.* **2018**, *42*, 244–249. [CrossRef] [PubMed]
26. Mok, J.H.; Pyatkovskyy, T.; Yousef, A.; Sastry, S.K. Synergistic effects of shear stress, moderate electric field, and nisin for the inactivation of *Escherichia coli* K12 and *Listeria innocua* in clear apple juice. *Food Control* **2020**, *113*, 107209. [CrossRef]
27. Mastello, R.B.; Janzantti, N.S.; Bisconsin, A.; Monteiro, M. Impact of HHP processing on volatile profile and sensory acceptance of Pera-Rio orange juice. *Innov. Food Sci. Emerg.* **2018**, *45*, 106–114. [CrossRef]
28. Guerrouja, K.; Sánchez-Rubiob, M.; Taboada-Rodríguezc, A.; Cava-Rodac, R.M.; Marín-Iniestab, F. Sonication at mild temperatures enhances bioactive compounds and microbiological quality of orange juice. *Food Bioprod. Process.* **2016**, *99*, 20–28. [CrossRef]
29. Alves, E.G.; Rodrigues, T.H.S.; Fernandes, F.A.N.; de Brito, E.S.; Cullen, P.J.; Frias, J.M.; Bourk, P.; Cavalcante, R.S.; Almeida, F.D.L.; Rodrigues, S. An untargeted chemometric evaluation of plasma and ozone processing effect on volatile compounds in orange juice. *Innov. Food Sci. Emerg.* **2019**, *53*, 63–69. [CrossRef]
30. Cheng, R.M.; Churey, J.J.; Worobo, R.W. Inactivation of Salmonella enterica and spoilage microorganisms in orange juice treated with dimethyl dicarbonate (DMDC). *Int. J. Food Microbiol.* **2018**, *285*, 152–157. [CrossRef]
31. Ma, T.T.; Wang, J.Q.; Wang, H.L.; Lan, T.; Liu, R.H.; Gao, T.; Yang, W.Y.; Yuan, Z.; Ge, Q.; Fang, Y.L.; et al. Is Overnight Fresh Juice Drinkable? The Shelf Life Prediction of Non-Industrial Fresh Watermelon Juice Based on the Nutritional Quality, Microbial Safety Quality, and Sensory Quality. *Food Nutr. Res.* **2020**, *64*, 4237. [CrossRef]
32. Muthukumarappan, K.; O'Donnell, C.P.; Cullen, P.J.; Tiwari, B.K. Inactivation kinetics of pectin methylesterase and cloud retention in sonicated orange juice. *Innov. Food Sci. Emerg.* **2009**, *10*, 166–171. [CrossRef]
33. The National Standard of China. *GB/T 12456-2008. Determination of Total Acid in Food*; CN-GB: Beijing, China, 2008.
34. Ma, T.T.; Lan, T.; Ju, Y.L.; Cheng, G.; Que, Z.L.; Geng, T.H.; Fang, Y.L.; Sun, X.Y. Comparison on the nutritional properties and biological activities of kiwifruit (*Actinidia*) and their different forms products: Towards making kiwifruit more nutritious and functional. *Food Funct.* **2019**, *10*, 1317–1329. [CrossRef]
35. Aadil, R.M.; Zeng, X.A.; Han, Z.; Sun, D.W. Effects of ultrasound treatments on quality of grapefruit juice. *Food Chem.* **2013**, *141*, 3201–3206. [CrossRef]
36. Abid, M.; Jabbar, S.; Wu, T.; Hashim, M.M.; Hu, B.; Lei, S.C.; Zeng, X.X. Sonication enhances polyphenolic compounds, sugars, carotenoids and mineral elements of apple juice. *Ultrason. Sonochem.* **2014**, *21*, 93–97. [CrossRef] [PubMed]

37. Zulueta, A.; Esteve, M.J.; Frigola, A. ORAC and TEAC assays comparison to measure the antioxidant capacity of food products. *Food Chem.* **2009**, *114*, 310–316. [CrossRef]
38. Sulaiman, A.; Farid, M.; Silva, F.V. Quality stability and sensory attributes of apple juice processed by thermosonication, pulsed electric field and thermal processing. *Int. J. Food Sci. Technol.* **2017**, *23*, 265–276. [CrossRef] [PubMed]
39. Oliveira, P.M.; Castro-Leite-Júnior, B.R.; Martins, E.M.F.; Cristianini, M.; Martins, M.L.; Vieira, É.N.R.; Binoti, M.L.; Almeida-Paula, D.; Ramos, A.M. Impact of high pressure and thermal processing on probiotic mixed mango and carrot juices. *J. Food Process. Pres.* **2020**, *44*, 1–11. [CrossRef]
40. Walkling-Ribeiro, M.; Noci, F.; Cronin, D.A.; Lyng, J.G.; Morgan, D.J. Shelf life and sensory evaluation of orange juice after exposure to thermosonication and pulsed electric fields. *Food Bioprod. Process.* **2009**, *87*, 102–107. [CrossRef]
41. Oner, M.E. The effect of high-pressure processing or thermosonication in combination with nisin on microbial inactivation and quality of green juice. *J. Food Process. Pres.* **2020**, *44*, 1–12. [CrossRef]
42. Marsellés-Fontanet, R.; Puig-Pujol, A.; Olmos, P.; Mínguez-Sanz, S.; Martín-Belloso, O. A Comparison of the Effects of Pulsed Electric Field and Thermal Treatments on Grape Juice. *Food Bioprocess Technol.* **2013**, *6*, 978–987. [CrossRef]
43. Gabriel, A.A.; Cayabyab, J.E.C.; Tan, A.L.K.L.; Corook, M.L.F.; Ables, E.J.O.; Tiangson-Bayaga, C.L.P. Development and validation of a predictive model for the influences of selected product and process variables on ascorbic acid degradation in simulated fruit juice. *Food Chem.* **2015**, *177*, 295–303. [CrossRef] [PubMed]
44. Lu, Q.; Peng, Y.; Zhu, C.H.; Pan, S.Y. Effect of thermal treatment on carotenoids, flavonoids and ascorbic acid in juice of orange cv. Cara Cara. *Food Chem.* **2018**, *265*, 39–48. [CrossRef] [PubMed]
45. Adebola, O.O.; Folasade, O.A.; Titilola, A.O.; Olaide, R.A. Effect of thermosonication on quality attributes of hog plum (*Spondias mombin* L.) juice. *Ultrason. Sonochem.* **2021**, *70*, 105316. [CrossRef]
46. Nayak, P.K.; Chandrasekar, C.M.; Kesavan, R.K. Effect of thermosonication on the quality attributes of star fruit juice. *J. Food Process Eng.* **2018**, *41*, e12857. [CrossRef]
47. Lemmens, L.; Colle, I.; Buggenhout, S.V.; Palmero, P.; Loey, A.V.; Hendrickx, M. Carotenoid bioaccessibility in fruit- and vegetable-based food products as affected by product (micro)structural characteristics and the presence of lipids: A review. *Trends Food Sci. Technol.* **2014**, *38*, 125–135. [CrossRef]
48. Wang, J.; Wang, J.; Ye, J.H.; Vanga, S.K.; Raghavan, V. Influence of high-intensity ultrasound on bioactive compounds of strawberry juice: Profiles of ascorbic acid, phenolics, antioxidant activity and microstructure. *Food Control* **2019**, *96*, 128–136. [CrossRef]
49. Kwaw, E.; Ma, Y.k.; Tchabo, W.; Apaliya, M.T.; Wu, M.; Sackey, A.S.; Xiao, L.L.; Tahir, H.E. Effect of lactobacillus strains on phenolic profile, color attributes and antioxidant activities of lactic-acid-fermented mulberry juice. *Food Chem.* **2018**, *250*, 148–154. [CrossRef]
50. Zhao, F.; Mei, X.; Zhang, Y.; Liao, X.; Wang, Y. Effect of High Hydrostatic Pressure and Heat Sterilization on the Quality of Lycium barbarum Juice. *J. Chin. Inst. Food Sci. Technol.* **2018**, *18*, 169–178. [CrossRef]
51. Barba, F.J.; Esteve, M.J.; Tedeschi, P.; Brandolini, V.; Frígola, A. A Comparative Study of the Analysis of Antioxidant Activities of Liquid Foods Employing Spectrophotometric, Fluorometric, and Chemiluminescent Methods. *Food Anal. Method* **2013**, *6*, 317–327. [CrossRef]
52. Wei, S.T.; Ou, L.C.; Luo, M.R.; Hutchings, J.B. Optimisation of food expectations using product colour and appearance. *Food Qual. Prefer.* **2012**, *23*, 49–62. [CrossRef]
53. Fernández-Vázquez, R.; Stinco, C.M.; Hernanz, D.; Heredia, F.J.; Vicario, I.M. Colour training and colour differences thresholds in orange juice. *Food Qual. Prefer.* **2013**, *30*, 320–327. [CrossRef]

MDPI
St. Alban-Anlage 66
4052 Basel
Switzerland
Tel. +41 61 683 77 34
Fax +41 61 302 89 18
www.mdpi.com

Foods Editorial Office
E-mail: foods@mdpi.com
www.mdpi.com/journal/foods

www.ingramcontent.com/pod-product-compliance
Lightning Source LLC
Chambersburg PA
CBHW061808210926
51850CB00006B/642
9783036568300